Advancement in Dietary Assessment and Self-Monitoring Using Technology

Advancement in Dietary Assessment and Self-Monitoring Using Technology

Special Issue Editors

Tracy Burrows
Megan Rollo

MDPI • Basel • Beijing • Wuhan • Barcelona • Belgrade

Special Issue Editors

Tracy Burrows	Megan Rollo
Priority Research Centre in	Priority Research Centre in
Physical Activity and Nutrition,	Physical Activity and Nutrition,
School of Health Sciences,	School of Health Sciences,
Faculty of Health,	Faculty of Health and Medicine,
University of Newcastle	University of Newcastle
Australia	Australia

Editorial Office
MDPI
St. Alban-Anlage 66
4052 Basel, Switzerland

This is a reprint of articles from the Special Issue published online in the open access journal *Nutrients* (ISSN 2072-6643) from 2018 to 2019 (available at: https://www.mdpi.com/journal/nutrients/special_issues/advancement-dietary-assessment-selfmonitoring).

For citation purposes, cite each article independently as indicated on the article page online and as indicated below:

LastName, A.A.; LastName, B.B.; LastName, C.C. Article Title. *Journal Name* **Year**, *Article Number*, Page Range.

ISBN 978-3-03928-058-2 (Pbk)
ISBN 978-3-03928-059-9 (PDF)

Contents

About the Special Issue Editors

Tracy Burrows is Associate Professor at the School of Health Sciences (Nutrition and Dietetics), University of Newcastle. Tracy is an Advanced Accredited Practising Dietitian and was awarded her PhD in 2008. Tracy has expertise in the areas of the assessment of dietary intake and conducting high-quality research trials. She has more recently initiated investigations into the area of food addition. Tracy has >60 peer-reviewed journal publications, supervises 6 PhD, 1 master's, and 3 honours students and is involved in research studies regarding dietary validation and obesity treatment and an expertise in working with paediatric populations.

Megan Rollo is a Postdoctoral Research Fellow and Senior Lecturer at the School of Health Sciences and Priority Research Centre for Physical Activity and Nutrition, University of Newcastle. She holds a Bachelor of Applied Science (2001), Bachelor of Health Science (Nutrition and Dietetics) (2008), and Doctor of Philosophy (Nutrition and Dietetics) (2012), all from the Queensland University of Technology. Dr. Rollo is an Accredited Practising Dietitian with research interests in technology-assisted dietary assessment and personalised behavioural nutrition interventions. To date, she has published 75 published/accepted publications, been awarded >\$3.3 million in research funding, and currently supervises 6 PhD candidates.

Editorial

Advancement in Dietary Assessment and Self-Monitoring Using Technology

Tracy L. Burrows [1,2,*] and Megan E. Rollo [1,2]

[1] School Health Science, Faculty of Health and Medicine, University of Newcastle, Callaghan, NSW 2308, Australia
[2] Priority Research Centre for Physical Activity and Nutrition, University of Newcastle, Callaghan, NSW 2308, Australia
* Correspondence: tracy.burrows@newcastle.edu.au; Tel.: +61-2-4921-5514

Received: 17 July 2019; Accepted: 18 July 2019; Published: 19 July 2019

1. Introduction

On the surface, some methods to assess and self-monitor dietary intake may be considered similar; however, the intended function of each is quite distinct. Methods used in the assessment of dietary intake aim to measure food and nutrient intake and/or derive dietary patterns for determining diet-disease relationships, conduct population surveillance, or determine the effectiveness of interventions [1]. In comparison, dietary self-monitoring primarily aims to create awareness and reinforcement of individual eating behaviours, in addition to tracking foods consumed, and has been particularly useful in the context of weight management [2]. Advancements in the capabilities of technologies, such as smartphones and wearable devices, have enhanced the proficiencies of collection, analysis, and interpretation of dietary intake data in both contexts across the spectrum of users, including consumers, clinicians and researchers.

In this issue, a range of new articles are presented, and we are fortunate to have a collection of reviews and empirical studies to assist in the development of understanding and attainment of new knowledge to assist in progressing this area of research on the use of technology in dietary assessment methods.

This special issue includes five review papers. Two articles reviewed mobile/smartphone applications [3,4], including the potential of mHealth apps to increase fruit and vegetable intake [3]. This specific review included eight studies, six of which were effective in increasing fruit and/or vegetable intake [3]. Additionally, a second paper included a review of recipe functions in 12 popular dietary smartphone apps and found a large variation in their energy and macronutrient calculations [4]. The main variation between apps occurred at the analysis phase due to the type of food composition table used to generate nutrient values [4].

A narrative review of new methods for assessing food and energy intake [5] is presented along with a review on the evaluation of new technology-based tools for dietary intake assessment [6]. This review of technology-based diet assessment tools, which included tools categorised for both research and consumer use, showed that the majority (79%) relied on self-reported dietary intakes. Most (91%) used text entry, 33% used image-based methods, 65% had integrated databases to estimate energy or nutrients, and less than 50% had customisation features [6]. Technology-based dietary assessment offers many advantages for research, and is often preferable to consumers over more traditional methods.

In addition, a narrative review in this special issue presents a synthesis of data on the dietary assessment of shared plate eating, which is reported as a missing link within a large proportion of methods that collect or focus on individual intake only [7]. Shared plate eating is reported as a particular issue for low-and lower-middle income countries where this type of eating behavior is common. Most studies used 24-h recalls—many used tools to assist in quantifying food intake, including food photographs and images of portion sizes [7]. The gap in this area of research was identified, as well as practical set of recommendations provided to move the field forward.

Finally, a systematic review on upper limb sensors for the assessment of eating behaviour summarises the findings from 69 studies [8]. To date, the majority of studies in the area have been conducted in controlled environments, among young, healthy individuals, and using accelerometers in combination with gyroscopes to detect eating activity. Heydarian and colleagues suggest the development of large datasets are paramount to advancing the field, particularly with regard to the use of deep learning for the detection of different eating gestures.

The empirical studies in this special issue can be classified into a number of key areas and are summarised below:

2. Web-Based Dietary Assessment Methods

Three articles are presented around the use of web-based methods, which in this special issue were applied in two of three studies to the 24-h recall method [9–11]. Interestingly, two studies explored the useability of technology-based methods in populations where use and acceptability might be questioned [9,10]. In one study by Polfuss et al., the usability of a technology-based 24-h recall was explored in individuals with and without disabilities, showing the methods were acceptable [10]. Another study in low income adults identified a range of useability issues with the automated self-administered dietary assessment tool (ASA24), including the misunderstanding of questions and uncertainties concerning how to proceed to the next step [9]. These papers provide very practical suggestions when applying technology to dietary assessment methods in these populations.

3. Image-Based or Image-Assisted Methods

Image-based or image-assisted methods were used to capture dietary intake and are were reported in this special issue in three studies [12–14], two of which were carried out in pediatric populations [13,14]. A validation study in young infants investigated the accuracy of image-assisted food records versus regular food records compared to the objective marker of doubly labelled water (DLW) method [13]. Another study in children of primary school age (9–12 years) investigated the accuracy of an electronic image-based food diary compared with a paper-based food diary over a four-day collection period [14]. The image-based food diary used a combination of photographs and written descriptions of foods consumed. Similar results were found for macro-and-micro nutrients for both methods. However, the image-based food diary was less burdensome for researchers and participants—it was also preferred by the children, and they required less help completing it [14].

4. Mobile/Smartphone Applications for Capturing Intake or Self-Monitoring

An interesting collection of articles are presented which range from quantitative, qualitative, and mixed method evaluations of the use of applications for dietary self-monitoring. For quantitative evaluation, the relative validity of the eat and track (EaT) smartphone app for the collection of dietary intake data was explored in young adults aged 18 to 30 years [15]. This population group is often difficult to engage in dietary and lifestyle interventions despite their known weight gain trajectory to be higher than any other population group. In this group the app was compared with dietitian-administered 24-h recalls. Significant differences in dietary energy were found but an agreement for most nutrient densities were reported at the group level. In another study, the effectiveness of the nutritional app "MyNutriCart" was reported and compared to a traditional face-to-face counselling session in order to determine the differences in food choices related to purchase and dietary behavior [16]. While in this pilot study there were no differences between groups, "MyNutriCart" did lead to significant improvements in household purchasing behaviours and individual intakes compared to baseline [16].

The Bridge2U mobile app food log was compared to control meal and dietary recall methods in another study [17]. While carried out in small population group ($n = 14$), the Bridge2U was reported as a good dietary assessment method for the assessment of intake at the group level, but data was reported to be highly variable for individual assessment [17]. Qualitative data provides very useful insights that often cannot be obtained through quantitative measurements only. An interesting study in this special issue reports on a qualitative evaluation of the eaTracker® Mobile App [18]. Structured interviews of 26 participants were analysed to evaluate ways to improve the eaTracker and provide information for those looking to develop apps to facilitate positive behaviour change. A number of positive aspects, challenges and suggestions for improvement of the app were collected and reported [18]. An evaluation of mixed methods is reported in a study by den Braber et al. to comprehensively determine the requirements of "The Diameter": an app to monitor diet, physical activity, and glucose values in patients with Type 2 Diabetes [19]. The study provides useful insight for this population group.

Mobile apps can be used to collect images, but they can also be used to collect voice recordings. Voice recordings can be used to add details to images which may often not be apparent and/or be used to collect information where an image of a food or drink maybe missed. An interesting study describes a voice operated app to determine the accuracy of automatic carbohydrate, protein, fat, and calorie counting based on the voice descriptions of meals in people with Type 1 Diabetes [20]. In 30 patients, insulin doses were estimated by a physician using dietary data obtained from VoiceDiab ($n = 16$) and this was compared to dietary data provided by a dietitian ($n = 14$). No significant differences in insulin doses or glycaemic control were reported using either system [20].

Wearable cameras are considered a passive technology as opposed to active capture whereby an individual still needs to be actively involved in the process. Passive measures can reduce the burden for participants in collecting dietary intake data, however, researcher burden still exists in other stages of dietary assessment, such as image processing and quantification. In one study of this special issue, a wearable system called the automatic ingestion monitor (AIM) was used to detect and monitor participant food intake ($n = 40$) for three days [21]. This was validated by a comparison with video observation that was annotated by three researchers to report activities, resting, walking, chewing, and biting during each eating and drinking episode [21].

If we look at the technology being applied to the analysis part of dietary assessment rather than the collection phase, when many other papers report on this in this special issue, one study compared the nutrient estimates based on food volume versus weight [22]. The weights of 35 individual food volumes were measured (control) and compared to the USDA-SR weights. Significant differences were found for 80% of foods which suggests that USDA-SR may not provide accurate estimates of dietary intake when assessed using food volumes [22].

This special issue presents a great set of articles regarding technology-based issues in the collection, analysis, and interpretation of dietary data.

Author Contributions: T.L.B. and M.E.R. conceptualized and co-wrote this Article.

Funding: This research received no external funding. Burrows is supported by a University of Newcastle Brawn research fellowship.

Conflicts of Interest: The authors declare no conflict of interest.

References

1. Thompson, F.; Subar, A. Chapter 1—Dietary assessment methodology. In *Nutrition in the Prevention and Treatment of Disease*; Coulston, A., Boushey, C., Ferruzzi, M., Delahanty, L., Eds.; Academic Press: Cambridge, MA, USA, 2017; pp. 5–48.
2. Burke, L.; Wang, J.; Sevick, M. Self-monitoring in weight loss: A systematic review of the literature. *J. Am. Diet. Assoc.* **2011**, *111*, 92–102. [CrossRef] [PubMed]
3. Mandracchia, F.; Llauradó, E.; Tarro, L.; del Bas, J.; Valls, R.; Pedret, A.; Radeva, P.; Arola, L.; Solà, R.; Boqué, N. Potential use of mobile phone applications for self-monitoring and increasing daily fruit and vegetable consumption: A systematized review. *Nutrients* **2019**, *11*, 686. [CrossRef] [PubMed]
4. Zhang, L.; Nawijn, E.; Boshuizen, H.; Ocké, M. Evaluation of the recipe function in popular dietary smartphone applications, with emphasize on features relevant for nutrition assessment in large-scale studies. *Nutrients* **2019**, *11*, 200. [CrossRef]

5.   Herrera, C.; Chan, C. Narrative review of new methods for assessing food and energy intake. *Nutrients* **2018**, *10*, 1064. [CrossRef]

6.   Eldridge, A.; Piernas, C.; Illner, A.-K.; Gibney, M.; Gurinović, M.; De Vries, J.; Cade, J. Evaluation of new technology-based tools for dietary intake assessment—An ilsi europe dietary intake and exposure task force evaluation. *Nutrients* **2019**, *11*, 55. [CrossRef] [PubMed]

7.   Burrows, T.; Collins, C.; Adam, M.; Duncanson, K.; Rollo, M. Dietary assessment of shared plate eating: A missing link. *Nutrients* **2019**, *11*, 789. [CrossRef]

8.   Heydarian, H.; Adam, M.; Burrows, T.; Collins, C.; Rollo, M.E. Assessing eating behaviour using upper limb mounted motion sensors: A systematic review. *Nutrients* **2019**, *11*, 1168. [CrossRef]

9.   Kupis, J.; Johnson, S.; Hallihan, G.; Olstad, D. Assessing the usability of the automated self-administered dietary assessment tool (asa24) among low-income adults. *Nutrients* **2019**, *11*, 132. [CrossRef]

10. Polfuss, M.; Moosreiner, A.; Boushey, C.; Delp, E.; Zhu, F. Technology-based dietary assessment in youth with and without developmental disabilities. *Nutrients* **2018**, *10*, 1482. [CrossRef]

11. Brassard, D.; Lemieux, S.; Charest, A.; Lapointe, A.; Couture, P.; Labonté, M.-E.; Lamarche, B. Comparing interviewer-administered and web-based food frequency questionnaires to predict energy requirements in adults. *Nutrients* **2018**, *10*, 1292. [CrossRef]

12. Fang, S.; Shao, Z.; Kerr, D.; Boushey, C.; Zhu, F. An end-to-end image-based automatic food energy estimation technique based on learned energy distribution images: Protocol and methodology. *Nutrients* **2019**, *11*, 877. [CrossRef] [PubMed]

13. Johansson, U.; Venables, M.; Öhlund, I.; Lind, T. Active image-assisted food records in comparison to regular food records: A validation study against doubly labeled water in 12-month-old infants. *Nutrients* **2018**, *10*, 1904. [CrossRef] [PubMed]

14. Davison, B.; Quigg, R.; Skidmore, P. Pilot testing a photo-based food diary in nine- to twelve- year old-children from dunedin, new zealand. *Nutrients* **2018**, *10*, 240. [CrossRef] [PubMed]

15. Wellard-Cole, L.; Chen, J.; Davies, A.; Wong, A.; Huynh, S.; Rangan, A.; Allman-Farinelli, M. Relative validity of the eat and track (eat) smartphone app for collection of dietary intake data in 18-to-30-year olds. *Nutrients* **2019**, *11*, 621. [CrossRef]

16. Palacios, C.; Torres, M.; López, D.; Trak-Fellermeier, M.; Coccia, C.; Pérez, C. Effectiveness of the nutritional app "mynutricart" on food choices related to purchase and dietary behavior: A pilot randomized controlled trial. *Nutrients* **2018**, *10*, 1967. [CrossRef] [PubMed]

17. Lemacks, J.; Adams, K.; Lovetere, A. Dietary intake reporting accuracy of the bridge2u mobile application food log compared to control meal and dietary recall methods. *Nutrients* **2019**, *11*, 199. [CrossRef]

18. Lieffers, J.; Valaitis, R.; George, T.; Wilson, M.; Macdonald, J.; Hanning, R. A qualitative evaluation of the eatracker® mobile app. *Nutrients* **2018**, *10*, 1462. [CrossRef]

19. Den Braber, N.; Vollenbroek-Hutten, M.; Oosterwijk, M.; Gant, C.; Hagedoorn, I.; van Beijnum, B.-J.; Hermens, H.; Laverman, G. Requirements of an application to monitor diet, physical activity and glucose values in patients with type 2 diabetes: The diameter. *Nutrients* **2019**, *11*, 409. [CrossRef]

20. Ladyzynski, P.; Krzymien, J.; Foltynski, P.; Rachuta, M.; Bonalska, B. Accuracy of automatic carbohydrate, protein, fat and calorie counting based on voice descriptions of meals in people with type 1 diabetes. *Nutrients* **2018**, *10*, 518. [CrossRef]

21. Farooq, M.; Doulah, A.; Parton, J.; McCrory, M.; Higgins, J.A.; Sazonov, E. Validation of sensor-based food intake detection by multicamera video observation in an unconstrained environment. *Nutrients* **2019**, *11*, 609. [CrossRef]

22. Partridge, E.; Neuhouser, M.; Breymeyer, K.; Schenk, J. Comparison of nutrient estimates based on food volume versus weight: Implications for dietary assessment methods. *Nutrients* **2018**, *10*, 973. [CrossRef]

Article

Mobile Phone Text Message Intervention on Diabetes Self-Care Activities, Cardiovascular Disease Risk Awareness, and Food Choices among Type 2 Diabetes Patients

Martha J. Nepper [1], Jennifer R. McAtee [2], Lorey Wheeler [3]  and Weiwen Chai [2,*

[1]   Nebraska Methodist Health System, 8111 Dodge Street, Omaha, NE 68114, USA; martha.nepper@nmhs.org

[2]   Department of Nutrition and Health Sciences, University of Nebraska-Lincoln, 1700 N 35th Street, Lincoln, NE 68583, USA; jennifer.mcatee@huskers.unl.edu

[3]   Nebraska Academy for Methodology, Analytics & Psychometrics, Nebraska Center for Research on Children, Youth, Families, and Schools, University of Nebraska-Lincoln, Lincoln, NE 68583, USA; lorey@unl.edu

*   Correspondence: wchai2@unl.edu; Tel.: +1-402-472-7822

Received: 27 February 2019; Accepted: 8 May 2019; Published: 11 June 2019

Abstract: This study examines the effects of educational text messages on diabetes self-care activities, cardiovascular disease (CVD) risk awareness, and home food availabilities related to food choices among patients with type 2 diabetes. Quasi-experimental design was used with 40 patients (58.0 ± 10.6 years) in the intervention group and 39 (55.7 ± 12.2 years) in the control group. In addition to the usual care provided for all participants, the intervention group received three educational text messages weekly for 12 weeks. Pre- and post-intervention measures were collected for both groups. Ninety-four percent of the participants receiving text messages indicated the usefulness of this program. The intervention group either maintained the same level or demonstrated small improvements in diabetes self-care activities after the intervention. Significant increases in scores of CVD risk awareness (57% increase; $p = 0.04$) and availabilities of fresh fruits (320% increase; $p = 0.01$) and fresh vegetables (250% increase; $p = 0.02$) in the home and weekly total (16% increase; $p = 0.02$) and moderate/vigorous (80% increase; $p = 0.006$) physical activity levels were observed for the intervention group relative to the control group. The pilot results suggest the feasibility and usefulness of the text message program for diabetes education. The study is registered with Clinical Trials.gov (NCT03039569).

Keywords: text messages; type 2 diabetes; diabetes self-care activities; cardiovascular disease risk awareness; food availability; food choices

---

1. Introduction

Type 2 diabetes is a complex and chronic illness affecting approximately 30.3 million people in the United States [1]. Adults with type 2 diabetes have a two to four-fold increase in the risk of developing cardiovascular disease (CVD), a leading cause of morbidity and mortality in this population [2]. Despite the strong link between type 2 diabetes and CVD, studies have found that patients with type 2 diabetes are unaware of their risk for developing CVD [3,4]. Nutrition and physical activity remain critical in the management of type 2 diabetes and are considered key in achieving optimal glycemic control and reducing major consequences such as CVD, foot damage, and kidney failure [5]. Since poor dietary practice may lead to insulin resistance, which further elevates blood glucose and lipid levels [6], self-inventory of household foods may offer a practical method for diabetes patients to monitor their food choices and dietary intake [7], thereby helping patients successfully manage the disease [6]. In addition to nutrition and physical activity, diabetes self-care activities include healthy coping skills, medication adherence, testing and managing blood glucose, problem solving, and strategies of reducing risk for health complications such as CVD [8]. These self-care activities have been used as a framework for patient-centered diabetes education.

Successfully managing type 2 diabetes requires life-long behavioral changes that can be challenging for many diabetes patients. Increasing evidence suggests that patients who take a more active role in their care achieve better health outcomes [9]. Conventional diabetes education, such as clinic visits with a health care provider, is commonly used for patients with type 2 diabetes to manage blood glucose and improve self-care skills. However, patients may have infrequent contact with health care providers because of their lack of time and transportation, expensive office visits and/or extended time between appointments. The use of text messages via cellular phones to convey health information and education represents a novel opportunity and low-cost method for improving diabetes self-care skills and increasing contact with health care providers. Previous work has demonstrated that text message interventions improve eating patterns, physical activity, and blood glucose management among patients with type 2 diabetes [10–13]. In addition, it has been shown that patients who receive short text messages find this type of intervention feasible and useful for managing the disease [12,13]. Evidence further suggests that cellular phone-based or web-based tools are particularly useful for patients living in rural regions with few specialized hospitals and limited access to health care clinics [13–15]. However, when addressing the effectiveness of using text messages for patients with type 2 diabetes, there is little research on whether a text message intervention would also increase the awareness of consequences of type 2 diabetes, for example, the risk of developing CVD. In addition, the impact of using text messages on patients' home food environment such as the availability of healthy or unhealthy foods in the home has not yet been examined. The home food environment is particularly important to investigate for diabetes patients since it is likely to reflect patients' food choices and purchase habits. In the current study, diabetes self-care education was performed using unidirectional text messages. Thus, the objective of the study was to examine the effect of educational text messages on diabetes self-care activities (general diet for healthy eating, specific diet for healthy eating, exercise, medication adherence, blood glucose testing, and foot care) and awareness of CVD risk among patients with type 2 diabetes. Additionally, the study also sought to assess whether the utilization of educational text messages had an influence on the availability of participants' food choices in the home that were relevant to type 2 diabetes.

2. Materials and Methods

2.1. Participants

Study participants were recruited from the Methodist Health System Center for Diabetes and Nutritional Health in Omaha, Nebraska, from February to December 2017. The center is an ambulatory outpatient clinic for treatment of patients with diabetes. Inclusion criteria were English-speaking adults with type 2 diabetes aged 30 years or older, self-reported hemoglobin A1C (HbA1C) greater than 6.5%, and having a cellular phone with the ability to receive text messages. Eligible participants who came into the clinic for outpatient diabetes care were identified and asked to participate in the study. All participants who elected to participate in the study were required to provide written informed consent. Seventy-nine patients (40 in the intervention group and 39 in the control group) with type 2 diabetes were enrolled in the study. Thirty-five participants in the intervention group and 35 participants in the control group completed the post-intervention surveys. Participants who did not complete the post-intervention surveys ($N = 9$) were either lost to follow-up, were no longer interested in participating, or did not respond to follow-up contact attempts (Figure 1). There were no differences in demographics and other factors relevant to type 2 diabetes between participants who completed the post-intervention surveys ($N = 70$) and those who did not ($N = 9$).

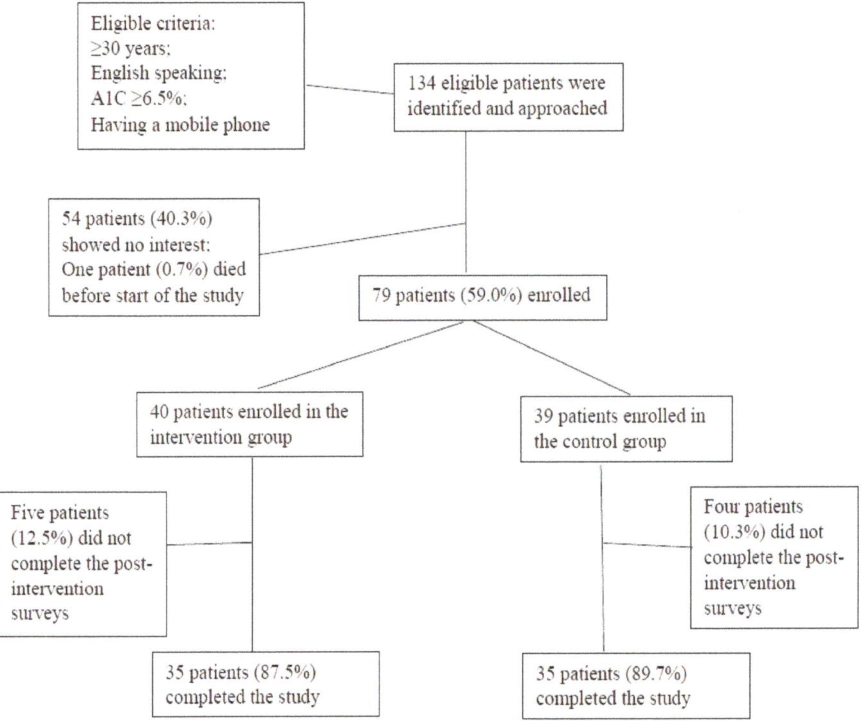

Figure 1. Flowchart of study participants.

2.2. Study Design

A quasi-experimental design was used in this initial pilot study due to timeframe restraints. The intervention group (N = 40 patients) started approximately two months earlier than the control group (N = 39). The first 40 participants recruited were assigned to the intervention group. A survey regarding participants' demographics and relevant risk factors for type 2 diabetes was completed by all the participants at baseline. Participants in both groups received the usual care for type 2 diabetes including an initial visit and follow-up visits from either a registered dietitian or a certified diabetes educator. The intervention group received three different educational text messages weekly (on Monday, Wednesday, and Friday) for 12 weeks (36 text messages in total). The text messages were sent in the late morning or early afternoon (between 11:00 am and 2:00 pm). The message topics were different each week during the first six weeks (Weeks 1 to 6; 18 text messages total) and repeated for the remaining weeks (Weeks 7 to 12; 18 text messages total). The messages consisted of strategies for healthy eating, being physically active, improving diabetes self-care skills including testing and managing blood glucose, taking medication, and increasing awareness of the risk of diabetes complications such as CVD. The text messages developed by the primary investigators were derived from the wording, topics, and guidelines provided by the American Association of Diabetes Educators (AADE7™) [16]. Each text message was comprised of a short message and a link (which was a novel approach) to a specific AADE7™ handout that allowed participants to open and retrieve the specific AADE7™ information (Table 1). Text message contents were not piloted before the start of the study. However, the AADE7™ handouts are available to the patients of the Methodist Health System Center for Diabetes and Nutritional Health as part of their usual care and diabetes education. Unidirectional text messages were sent by the project investigators to the participants in the intervention group via a computer-based text message program through a password protected computer which was only accessed by the investigators. Participants' phone numbers used in the intervention were kept confidential and participants were advised not to reply to the text messages. If participants had a medical concern, they were advised to contact their physician or call 911. The control group (usual care group) did not receive text messages. Participants in both the intervention group (usual care and receiving text messages) and the control group (usual care only) completed surveys regarding their diabetes self-care activities, dietary intake, physical activity, awareness of CVD risk, and self-inventory of household foods at baseline and at the 12-week follow-up (conclusion of the intervention). The participants in the intervention group completed an additional survey to evaluate their satisfaction with receiving educational text messages for managing diabetes after the intervention was concluded. A $25 gift card was offered to all study participants. The research project was approved by the Institutional Review Boards of the University of Nebraska-Lincoln and Nebraska Methodist Health System. The study is registered with Clinical Trials.gov (NCT03039569).

Table 1. Contents of educational text messages for type 2 diabetes patients.

AADE7™ Handout Titles/Topics	Contents of Text Messages [a]
Healthy Eating (Weeks 1 and 7)	1) Eat breakfast every day! 2) There are only three main types of nutrients in foods: carbohydrates, proteins, and fats. A healthy meal will include all three of these. 3) Do not skip meals! Remember to eat regular meals and snacks every day.
	For more info: *https://www.diabeteseducator.org/patient-resources/aade7-self-care-behaviors/healthy-eating*
Healthy Coping (Weeks 2 and 8)	1) Think positive! Feeling down? Remember your successes and feel good about your progress with diabetes. 2) Build healthy relationships. You are not alone when you have diabetes. 3) If you are sad, anxious or stressed, go for a walk or stand up and stretch.
	For more info: *https://www.diabeteseducator.org/patient-resources/aade7-self-care-behaviors/healthy-coping*
Monitoring (Weeks 3 and 9)	1) Checking your blood sugars gives you vital information about your diabetes control. 2) Monitoring your blood sugars helps you know when they are on target. 3) Call your doctor or diabetes educator if you are concerned about your blood sugars.
	For more info: https://www.diabeteseducator.org/patient-resources/aade7-self-care-behaviors/ aade7-self-care-behaviors-monitoring
Being Active (Weeks 4 and 10)	1) Being active has many health benefits, like improving blood pressure and blood sugars. 2) If you haven't exercised for a while, start with a five minute walk and increase gradually. 3) Break activity into three ten minute sessions.
	For more info: *https://www.diabeteseducator.org/patient-resources/aade7-self-care-behaviors/being-active*
Taking Medication and Problem Solving (Weeks 5 and 11)	1) Take notes when you visit with your doctor about your medication. 2) Learn what causes your blood sugar to go above or below target. 3) Talk to your doctor about how to improve your blood sugar.
	For more info: https://www.diabeteseducator.org/patient-resources/aade7-self-care-behaviors/ taking-medication *https://www.diabeteseducator.org/patient-resources/aade7-self-care-behaviors/problem-solving*
Reducing Risks (Weeks 6 and 12)	1) See your eye doctor at least once a year 2) Keep a wallet card that lists all of the tests you should be regularly getting and the targets for each. 3) Lowering your cholesterol can decrease your risk for a stroke. Talk to your doctor about what you can do.
	For more info: *https://www.diabeteseducator.org/patient-resources/aade7-self-care-behaviors/reducing-risks*

[a] Each text message included a link to a specific American Association of Diabetes Educator (AADE7™) handout (36 text messages total).

2.3. Outcome Measures

2.3.1. Diabetes Self-Care Activities

Diabetes self-care activities were measured using a previously validated Summary of Diabetes Self-Care Activities including categories of general diet for healthy eating (following a healthful eating plan; following one's eating plan), specific diet for healthy eating (eating ≥5 servings of fruits and vegetables; avoiding consuming high-fat foods), exercise (participating in at least 30 minutes of physical activity per day; participating in a specific exercise session), testing blood glucose (testing blood glucose; testing blood glucose the number of times recommended by one's health care provider per day), medication adherence (taking one's recommended diabetes medication); and foot care (checking one's feet; inspecting the inside of one's shoes; washing one's feet; avoiding soaking one's feet; drying between one's toes after washing). Respondents reported on the frequency with which they performed various self-care activities over the past seven days (how many days per week) [17]. The responses were scored from 0–7 accordingly; reverse scoring was used for a negative item. Individual items in each of the diabetes self-care activity categories (general diet for healthy eating, specific diet for healthy eating, exercise, blood glucose testing, medication adherence, and foot care) were combined to create an average score for the respective category.

2.3.2. Dietary Intake and Physical Activity

In addition to diet and exercise items included in the aforementioned diabetes self-care activities, we additionally measured individual's dietary intake and physical activity. Participants' dietary intake was measured using a previously validated Block Fat-Sugar-Fruit-Vegetable Screener [18]. This screener contained 55 questions about frequency of food eaten (none or less than one day, one day, two days, three to four days, five to six days, or every day/per week) and portion sizes of 32 food items during the past month. Daily nutrient intakes including total calories were determined based on the data from the screener. Weekly physical activity levels were measured using the Block Physical Activity Screener [19]. This brief screening tool contained 11 items including job-related as well as daily life and leisure activities based upon National Human Activities Patterns Survey data. Total metabolic equivalent of task (MET) minutes per week for all the activities as well as for moderate/vigorous physical activities were calculated using the Ainsworth Compendium [20].

2.3.3. CVD Risk Awareness

The CVD risk awareness questions were derived from a questionnaire used in a previous study [21]. Three questions were asked about how seriously a participant was concerned about having a CVD event in the next five years and in their lifetime (level of concern of CVD risk). The responses to the questions were scored from 0–3, indicating "no concern", "low-level of concern", "somewhat concerned", and "highly concerned", respectively. In this study, we summed the response scores for these three questions and calculated the mean score for the category. There was an additional question about how often a participant had a concern about having a CVD event with responses including "never" (zero times per week), "rarely" (one to two times per week), "sometimes" (three to four times per week), and "always" (five to seven times per week).

2.3.4. Home Food Self-Inventory

A previously validated home food self-inventory checklist was used to assess the presence and absence of foods relevant to obesity and type 2 diabetes [22]. The checklist contained a total of 65 healthy and unhealthy food and beverage items including sweet and savory snacks, beverages, breakfast cereal/oatmeal, breads/pastas, dairy foods, and individual fruits and vegetables. Of these 65 food and beverage items/categories, there were 19 fruit and 16 vegetable items. Each fruit or vegetable item includes its fresh, canned/jarred/dried, and frozen forms. A "yes/no" format was used to indicate the availability of the food in the home with "1" indicating "yes" and "0" indicating "no".

The classification of "healthy" and "unhealthy" foods and beverages were derived from previous home food inventory tools [23,24] and followed the "We Can: Go, Slow, Whoa" food system, in which "Go" foods were considered healthy and "Whoa" foods were unhealthy [25].

Food items on the checklist were grouped into the following categories: all fruits, fresh fruits, canned/jarred/dried fruits, frozen fruits, all vegetables, fresh vegetables, canned/jarred/dried vegetables, frozen vegetables, all healthy foods (including fruits and vegetables), and all unhealthy foods. In addition, we also categorized foods on the checklist into high, medium, and low glycemic index (GI) foods according to the American Diabetes Association guidelines: low GI foods, GI $\leq$55; medium GI foods, GI ranging from 56 to 69; high GI foods, GI $\geq$70 [26]. GI measures how a food that contains carbohydrate influences blood glucose in comparison to a reference food (e.g., glucose or white bread) [26]. Similarly, the availability scores for food items in each of the above food categories were summed and the average score was calculated for the category.

2.4. Data Analysis

To make our quasi-experiment more rigorous, we followed recommendations by Shadish and colleagues [27], such as testing for baseline differences between groups and including the pre-test measure of outcomes to address selection bias resulting from not randomizing participants into groups. Preliminary analyses included comparing baseline characteristics between groups using *t*-tests for continuous variables and chi-square analyses for categorical variables. Our primary analyses included multivariate analysis of covariance (MANCOVA) to assess intervention effects by examining the differences between the intervention and the control groups at the 12 week follow-up. Any baseline differences between groups were controlled for in the test of effects at the 12 week follow-up by including the baseline measure of the outcome in the analyses. A proc GLM (generalized linear model) procedure was used to estimate MANCOVA. To control for experiment-wise error, the outcome/dependent variables were clustered into the following groups: diabetes self-care activities, fruit and vegetable availabilities, all healthy and unhealthy food availabilities, and GI-based fruit and vegetable availabilities. With MANCOVA, individual variables in each group mentioned above were analyzed together as group-based outcome/dependent variables (multivariate analysis). For outcome/dependent variables that were not categorized into a group (CVD risk awareness, MET minutes for total or moderate/vigorous physical activities, and intakes of dietary nutrients), analysis of covariance (ANCOVA) estimated by the Proc Glimmix procedure was used to assess the effects of intervention on these variables. We also used absolute change (time and treatment interaction), a more stringent test, to estimate intervention effects. Absolute change was determined as follows: absolute change = [(intervention group follow-up) – (intervention group baseline)] – [(control group follow-up) – (control group baseline)]. Since this is a pilot study, we used the results from both tests (MANCOVA/ANCOVA and absolute change) as supporting preliminary evidence for the intervention effects. Further, to provide perspective on the magnitude of the intervention effects, relative change, defined as (absolute change/intervention group baseline) x100%, was calculated. The covariates included in the models were age (continuous), sex, race/ethnicity (white, black, Hispanic, Asian, or other), education (college graduates or non-college graduates), baseline self-report HbA1C values, and the length of time of having had type 2 diabetes (<1 year, 1–5 years, or $\geq$5 years). For daily nutrient intake (carbohydrate, sugar, added sugar, total fat, saturated fat, and protein), we repeated the analyses with additional adjustment for total calorie intake and the results did not change substantially. An a priori power estimate suggests that our sample size ($N = 35$ in each group) was adequate for finding large effects ($d = 0.8$) and had 70% power for detecting medium effects ($d = 0.5$), assuming $\alpha = 0.05$ (two-tailed) based on Cohen's recommendations [28]. SAS software version 9.4 (SAS Institute, Cary, NC, USA) was used for all analyses. We conservatively used two-tailed tests and $p < 0.05$ was considered statistically significant.

3. Results

3.1. Characteristics of Study Participants and Usefulness of Educational Text Messages

Overall, the mean ages of patients in the intervention and the control groups were 58.0 ± 10.6 and 55.7 ± 12.2 years, respectively. The majority of the participants were female (65% for the intervention group; 67% for the control group). The intervention and control groups had similar characteristics at baseline except for racial/ethnic distribution; a larger proportion of Hispanic participants was observed in the intervention group than the control group (p = 0.01). In addition, the intervention group had higher percentages of college graduates (49% vs. 34%) and those who had had type 2 diabetes for at least five years (73% versus 58%) compared to the control group (Table 2).

Table 2. Baseline characteristics of study participants with type 2 diabetes [a]

Characteristics	Intervention	Control	p Value [b]
N	40	39	
Age (year)	58.0 ± 10.6	55.7 ± 12.2	0.21
Sex (%)			0.82
Male	34.7	32.9	
Female	65.3	67.1	
Race/ethnicity (%)			0.01
White	84.0	93.2	
Black	2.7	5.4	
Hispanic	8.0	0	
Asian	2.7	0	
Other	2.7	1.4	
BMI (kg/m^2)	34.6 ± 9.05	35.9 ± 6.1	0.30
Hemoglobin A1C (%) [c]	7.8 ± 1.4	8.2 ± 1.9	0.09
College graduate (%)	49.3	33.8	0.05
Having diabetes <1 year (%)	12.0	20.3	0.17
Having diabetes ≥5 years (%)	73.3	58.1	0.05
Taking diabetes medication (%)	92.0	91.9	0.98
Current smoker (%)	4.0	8.0	0.30

[a] Data are given as mean ± standard deviation unless otherwise specified. [b] p value for difference between the intervention and control groups by t test for continuous variables and chi-square test for categorical variables. [c] Hemoglobin A1C values were based on self-report values by study participants.

Participants in the intervention group (N = 35) completed a satisfaction survey regarding the feasibility and usefulness of the educational text messages in helping them with diabetes self-care management. The majority of the participants (94%) reported the text message intervention program was useful and stated that they would highly recommend this program to others with type 2 diabetes.

3.2. Diabetes Self-Care Activities, Dietary Intake, Physical Activity, and Awareness of CVD Risk

Overall, there were no statistically significant differences in changes of scores on diabetes self-care activities after the 12-week text message intervention. However, the intervention group in general maintained the same level or showed small improvements at the 12-week follow-up compared to the control group. In addition, weekly MET minutes for both in total (5548 versus 2877; 16% increase; p = 0.02) and moderate/vigorous physical activity (3163 versus 405; 80% increase; p = 0.006) were significantly higher for the intervention group than the control group at the 12-week follow-up after taking into account the baseline values for these variables. There were no significant changes in the intakes of relevant nutrients after intervention (Table 3).

Table 3. Diabetes self-care activities, cardiovascular disease (CVD) risk awareness, physical activity (PA) and dietary intake at baseline and 12-week follow-up.

Variable	Baseline Mean (SE)[c]	p [d]	12-Week Follow-Up Mean (SE) [c]	p [d]	Absolute Change [a] Mean (SE)[c]	p [e]	Relative Change [b]
Diabetes Self-Care Activities (day/week)							
General diet for healthy eating							
Intervention	4.66 (0.43)	0.14	5.18 (0.46)	0.61	0.76 (0.52)	0.15	16%
Control	5.22 (0.46)		4.98 (0.48)				
Specific diet for healthy eating							
Intervention	4.20 (0.40)	0.93	4.03 (0.43)	0.62	−0.21 (0.48)	0.67	−5%
Control	4.17 (0.43)		4.21 (0.45)				
Exercise							
Intervention	5.14 (0.61)	0.67	4.45 (0.65)	0.77	−0.07 (0.74)	0.93	−1%
Control	4.91 (0.65)		4.29 (0.68)				
Blood glucose testing							
Intervention	5.06 (0.62)	0.08	5.28 (0.66)	0.49	0.57 (0.75)	0.45	11%
Control	6.01 (0.67)		5.66 (0.69)				
Medication adherence							
Intervention	6.05 (0.35)	0.51	6.35 (0.38)	0.09	0.33 (0.42)	0.44	5%
Control	5.85 (0.38)		5.82 (0.39)				
Foot care							
Intervention	4.23 (0.41)	0.79	4.86 (0.43)	0.08	0.53 (0.49)	0.28	13%
Control	4.13 (0.43)		4.23 (0.45)				
CVD Risk Awareness							
Intervention	1.01 (0.23)	0.21	1.26 (0.25)	0.13	0.58 (0.29)	0.04	57%
Control	1.27 (0.25)		0.94 (0.26)				
PA (MET [f] minute/week)							
Total PA							
Intervention	4806 (1451)	0.08	5548 (1645)	0.02	768 (1533)	0.62	16%
Control	2904 (1541)		2877 (1602)				
Moderate/vigorous PA							
Intervention	2112 (1435)	0.25	3163 (1555)	0.006	1688 (1300)	0.20	80%
Control	1042 (1430)		405 (1532)				
Dietary Intake							
Total calories (kcal/day)							
Intervention	1598 (152)	0.43	1499 (161)	0.79	−134 (176)	0.45	−8%
Control	1417 (161)		1452 (9168)				
Carbohydrates (g/day)							
Intervention	162.3 (17.3)	0.54	153.4 (18.2)	0.81	−12.5 (20.0)	0.53	−8%
Control	140.2 (18.3)		143.9 (19.1)				
Total sugar (g/day)							
Intervention	58.9 (8.4)	0.52	52.4 (8.9)	0.50	0.4 (9.7)	0.96	1%
Control	49.6 (8.9)		44.7 (9.3)				
Added sugar (g/day)							
Intervention	34.8 (7.3)	0.39	29.6 (7.7)	0.25	2.1 (8.4)	0.80	6%
Control	30.1 (7.7)		22.9 (8.0)				
Total fat (g/day)							
Intervention	67.3 (7.0)	0.28	61.1 (7.4)	0.89	−7.0 (8.1)	0.39	−10%
Control	60.1 (7.4)		60.9 (7.7)				
Saturated fat (g/day)							
Intervention	23.1 (2.5)	0.30	21.0 (2.6)	0.98	−2.1 (2.9)	0.47	−9%
Control	21.3 (2.7)		21.3 (2.8)				
Protein (g/day)							
Intervention	70.4 (6.9)	0.71	68.3 (7.2)	0.62	−5.1 (7.9)	0.53	−7%
Control	62.4 (7.3)		65.3 (7.6)				

[a] Absolute change = [(intervention group follow-up) − (intervention group baseline)] − [(control group follow-up) − (control group baseline)]. [b] Relative change = (absolute change / intervention group baseline) × 100%. [c] Adjusted mean is presented. [d] p value for difference between the intervention and the control groups by MANCOVA or ANCOVA adjusting for age, sex, race/ethnicity, education, self-report hemoglobin A1C, and length of time having had type 2 diabetes at baseline. [e] p value for absolute change adjusting for age, sex, race/ethnicity, education, self-report hemoglobin A1C, and length of time having had type 2 diabetes at baseline. [f] MET = metabolic equivalent of task.

With respect to CVD risk awareness, there was a statistically significant improvement in the score regarding how seriously a participant was concerned about having a CVD event (the level of concern) in the intervention group compared to the control group (57% increase; $p = 0.04$). However, the average score was nevertheless at the lower end of the scale (1.26), being between "low-level of concern" and "somewhat concerned" for the intervention group (Table 3). Similarly, at the baseline, a majority of the study participants in both groups reported that they were never or rarely concerned about a CVD event (70% for the intervention group; 68% for the control group). After 12 weeks of intervention, "never" or "rarely concerned" about CVD risk was reported by 68% of the participants in the intervention group and 82% of those in the control group.

3.3. Home Food Availabilities Related to Food Choices

The intervention group had significant increases in availability scores for fresh fruits (320% increase; $p = 0.01$) and fresh vegetables (250% increase; $p = 0.02$) in the home after the intervention compared to the control group. When the food availabilities were assessed based on GI values, there was a significant increase in the score for high GI fruit availability (431% increase; $p = 0.001$) and a decrease in the score for medium GI vegetable availability (40% decrease; $p = 0.03$) for the intervention group relative to the control group at the 12-week follow-up. It appeared that high GI vegetables were more likely to be available in the home among participants in both groups (Table 4).

Table 4. Availability of fruits and vegetables in the home at baseline and 12-week follow-up.

Food Availability	Baseline Mean (SE) [c]	p [d]	12-Week Follow-Up Mean (SE) [c]	p [d]	Absolute Change [a] Mean (SE) [c]	p [e]	Relative Change [b]
Fruits							
Total							
Intervention	0.06 (0.04)	0.29	0.08 (0.04)	0.001	0.04 (0.03)	0.15	67%
Control	0.03 (0.04)		0.01 (0.04)				
Fresh							
Intervention	0.05 (0.09)	0.94	0.17 (0.08)	0.0005	0.16 (0.06)	0.01	320%
Control	0.05 (0.09)		0.01 (0.09)				
Canned/jarred/dried							
Intervention	0.06 (0.06)	0.15	0.04 (0.05)	0.35	−0.02 (0.04)	0.63	−33%
Control	0.01 (0.06)		0.02 (0.06)				
Frozen							
Intervention	0.05 (0.04)	0.18	0.02 (0.04)	0.27	−0.008 (0.03)	0.78	−16%
Control	0.02 (0.04)		−0.005 (0.04)				
Vegetables							
Total							
Intervention	0.13 (0.05)	0.79	0.11 (0.05)	0.98	−0.009 (0.04)	0.82	−7%
Control	0.12 (0.05)		0.11 (0.05)				
Fresh							
Intervention	0.06 (0.09)	0.94	0.18 (0.09)	0.0008	0.15 (0.07)	0.02	250%
Control	0.06 (0.09)		0.02 (0.09)				
Canned/jarred/dried							
Intervention	0.06 (0.06)	0.12	0.05 (0.05)	0.31	−0.02 (0.04)	0.62	−33%
Control	0.01 (0.05)		0.02 (0.06)				
Frozen							
Intervention	0.04 (0.03)	0.21	0.02 (0.03)	0.11	0.003 (0.02)	0.90	8%
Control	0.02 (0.03)		−0.005 (0.03)				
All healthy foods [f]							
Intervention	1.00 (0.06)	0.45	1.08 (0.06)	0.68	0.01 (0.06)	0.82	1%
Control	1.04 (0.06)		1.10 (0.06)				
All unhealthy foods							
Intervention	1.41 (0.09)	0.83	1.50 (0.09)	0.75	0.04 (0.10)	0.70	3%
Control	1.42 (0.09)		1.47 (0.10)				

Table 4. *Cont.*

Food Availability	Baseline Mean (SE) [c]	p [d]	12-Week Follow-Up Mean (SE) [c]	p [d]	Absolute Change [a] Mean (SE) [c]	p [e]	Relative Change [b]
Based on Glycemic Index (GI) [g]							
Fruits							
Low GI							
Intervention	11.03 (2.15)	0.25	10.68 (2.17)	0.05	1.43 (2.35)	0.54	13%
Control	9.06 (2.23)		7.28 (2.34)				
Medium GI							
Intervention	14.75 (3.15)	0.43	17.51 (3.17)	0.004	5.27 (3.43)	0.13	36%
Control	12.78 (3.27)		10.27 (3.42)				
High GI							
Intervention	8.07 (9.68)	0.02	40.02 (9.75)	0.03	34.79 (10.54)	0.001	431%
Control	25.96 (10.04)		23.13 (10.50)				
Vegetables							
Low GI							
Intervention	8.90 (1.34)	0.77	9.09 (1.35)	0.69	0.74 (1.46)	0.61	8%
Control	9.31 (1.39)		8.67 (1.45)				
Medium GI							
Intervention	47.83 (7.85)	0.67	32.24 (7.91)	0.009	−19.33 (0.03)	0.03	−40%
Control	45.13 (8.15)		48.87 (8.52)				
High GI							
Intervention	91.58 (15.60)	0.90	65.08 (15.72)	0.08	−23.68(16.99)	0.17	−26%
Control	90.01 (16.20)		87.20 (16.93)				

[a] Absolute change = [(intervention group follow-up) − (intervention group baseline)] − [(control group follow-up) − (control group baseline)]. [b] Relative change = (absolute change / intervention group baseline) x 100%. [c] Adjusted mean is presented. [d] p value for difference between the intervention group and the control group by MANCOVA adjusting for age, sex, race/ethnicity, education, self-report hemoglobin A1C, and length of time having had type 2 diabetes at baseline. [e] p value for absolute change adjusting for age, sex, race/ethnicity, education, self-report hemoglobin A1C, and length of time having had type 2 diabetes at baseline. [f] Including fruits and vegetables. [g] Low GI foods: GI ≤55; medium GI foods: GI between 56–69; high GI foods: GI ≥70.

4. Discussion

Using text messages via a cellular phone device is a low-cost and simple method of delivering health information and education. In this pilot study, the intervention group either maintained the same level or showed small improvements in diabetes self-care activities after 12 weeks of the text message intervention. Improvements in adherence to following a specific diet plan for diabetes [29], eating habits [30], physical activity [13,30], and self-care management skills [29] among type 2 diabetes patients using text messages have been documented previously. In the current study, each patient in the intervention group received a short text message three days (one message per day) per week. Each text message also had a link that directed patients to the AADE7[TM] handout to provide patients additional information and strategies of diabetes self-care skills, which was a novel approach to diabetes education. Based on the feedback from study participants, the current educational text message program was perceived as useful and beneficial (94% responded that yes it was) for helping type 2 diabetes patients with self-care management, suggesting the feasibility and usefulness of the program.

There are possible explanations for the non-statistically significant improvements in diabetes self-care activities. First, patients in both intervention and control groups had been receiving the usual care for type 2 diabetes (clinic visits with registered dietitians or certified diabetes educators) and therefore might have already been working on their self-care management skills before the intervention. This was suggested by the baseline data showing patients in both groups having an average of five days or more per week of engaging in self-care activities such as following eating plans, checking blood glucose levels, and taking medications. Thus, the positive changes due to the text message intervention would be more substantial for diabetes patients who do not receive routine care for the disease, such as those living in rural areas with limited access to health care or having other conditions resulting in infrequent contacts with health care providers or diabetes educators. Second, the participants in the control group were not prohibited from seeking diabetes self-care and other health information online or through other resources, thereby potentially mitigating differences from the text message intervention. Third, the relatively short intervention period (12 weeks) may have also contributed to non-statistically significant results. Promisingly, our results (from the ANCOVA tests) suggest significant improvements in weekly MET minutes for both total and moderate/vigorous physical activity in the intervention group compared to the control group at the 12-week follow-up. Although the results from the more stringent test for absolute change were not statistically significant, given the pilot study nature, the significant findings based on ANCOVA nevertheless provide preliminary evidence for the positive effect of the current intervention program on patients' physical activity levels as each participant in the program received six messages total on physical activity.

It is established that adults with type 2 diabetes have a two- to four-fold increase in the risk of developing CVD; however, the majority of study participants in the intervention and the control groups did not have a high awareness of CVD risk at baseline. This finding was consistent with previous studies [3,4] that reported adults with type 2 diabetes were unaware of their risk for developing CVD. The intervention group indeed had a significant improvement in the score regarding the level of a participant's concern over having a CVD event after the intervention. However, despite the increase, the average score was still at the lower end of the scale between "low-level of concern" and "somewhat concerned" for the intervention group after receiving text messages for 12 weeks. In addition, being "never" or "rarely concerned" about CVD risk was reported by most of the participants after the intervention (68% for the intervention group and 82% for the control group). In the current study, the intervention group received six messages total on the topic of reducing the risk of complications associated with diabetes. Although each message included a link to an AADE7^{TM} handout addressing the direct relationship between type 2 diabetes and CVD risk, it is possible that participants did not click on the added link in the text messages to learn more about this information. Therefore, future interventions using educational text messages should focus more on increasing participants' awareness of CVD risk. For example, when creating text messages, one may consider phrasing the messages with extra emphasis on the strong link between type 2 diabetes and CVD. Furthermore, extending the intervention period from 12 weeks to six months and including more text messages on the topic may enhance the impact of intervention on CVD risk awareness. Nevertheless, the significant improvement, although small as observed in this study, suggests that the current text message program to some extent made participants aware of CVD risk, a first step towards achieving the ultimate goal that is to reduce the risk of developing CVD and other diabetes-related complications.

The results from the current study suggest that the 12-week text message intervention had promising effects on the participants' food choices that were reflected by the presence or absence of foods relevant to type 2 diabetes in the home. The study observed significant increases in the availability scores for fresh fruits and fresh vegetables in the intervention group after receiving educational text messages for 12 weeks. Although the availability of high GI fruits also increased after the intervention and the participants in both groups were more likely to store high GI vegetables in the home, we should not make dietary recommendations for healthy eating solely based on GI values since GI itself does not reflect the likely quantity an individual would eat and high GI fruits and vegetables contain other beneficial compounds such as fiber, vitamins, minerals, and polyphenols. Future diabetes educational programs using text messages should educate patients on the health benefits of increasing fruit and vegetable intake (e.g., fiber, vitamins, minerals, and polyphenol content). In addition, educational messages should address the influence of fruit and vegetable intake on blood glucose levels when eaten in the appropriate portion sizes to help patients make wise food choices, since GI does not address portion sizes which are relevant for managing blood glucose levels.

Our study had limitations. The non-randomized, quasi-experimental study design may have increased baseline differences between the intervention and the control groups due to selection bias. However, there were no major differences in the relevant characteristics at baseline between the two groups and the current analyses were adjusted for the relevant covariates to address potential selection bias. Participants who agreed to enroll in the study may have been more interested in improving their diabetes self-care skills and healthy habits relative to those who did not. There is a possibility that participants in the intervention group deleted the text messages without reading the message or clicking on the AADE7™ handout link or did not receive some text messages, thus negating any significant health behavior changes. However, all possible attempts were made to ensure that the participants were receiving and reading the messages. For example, investigators called participants several times during the study and visited with them when they came into the clinic on whether they were receiving the text messages or had any problems with opening the AADE7™ handout link. During the study, no problems of undeliverable messages were encountered. The results might be underestimated or overestimated due to loss at follow-up that occurred in the study. However, there were no differences in demographics and relevant factors at baseline between participants who completed the post-intervention surveys and those who did not. In addition, participants might know when to "expect" the messages, which may have some effects on the effectiveness of the intervention. Lastly, although validated self-report measures were used, objective indicators may be more accurate for assessing the intervention effects.

5. Conclusions

The results from this pilot study suggest the feasibility and usefulness of using educational text messages for patients with type 2 diabetes to maintain or improve their diabetes self-care skills. Further, the current text message program can benefit patients living in rural areas with limited access to health care or having other conditions resulting in infrequent contacts with health care providers.

The pilot results also demonstrate a small but statistically significant increase in CVD risk awareness as well as significant increases in physical activity and the availabilities of fresh fruits and vegetables in the home among participants receiving text messages. Although these results need to be confirmed by randomized experimental trials in the future, our findings, especially the ones related to CVD risk awareness and home food self-inventory, add to the growing body of literature on using text messages to deliver health information to patients with health concerns, including type 2 diabetes. For future interventions, approaches such as extending the length of the intervention, increasing the frequency of delivering such messages to participants or combining with other strategies such as a telephone-based coaching approach may enhance the impact of the program. When revising the content of the educational messages, one may need to increase the focus on reducing CVD risk by highlighting the direct relationship between type 2 diabetes and CVD. Also, text messages should address the importance of including fruits and vegetables in a patient's daily food intake for health benefits, and how portion sizes influence blood glucose levels, which may help patients make healthy food choices.

Author Contributions: All of the authors made substantial contributions to the study concept and design or analysis and interpretation of the data. Specifically, M.J.N. designed the study, recruited participants, collected data, and was the primary author of every section of the text. W.C. was instrumental in the design of the study and helped to draft the manuscript. J.R.M. analyzed data, helped with data collection and co-wrote the manuscript. L.W. helped to design the study's analytic strategy and commented on the manuscript. All authors read, contributed to, and approved the final manuscript.

Funding: This investigation was supported by the University of Nebraska-Lincoln Layman Faculty Seed Grant.

Acknowledgments: The authors thank the patients who participated in the study and research assistant Christina Gregory for her contribution to the study.

Conflicts of Interest: The authors declare no conflict of interest.

References

1.   Centers for Disease Control and Prevention. National Diabetes Statistics Report, 2017. Atlanta, GA: Centers for Disease Control and Prevention, U.S. Department of Health and Human Services. 2017. Available online: https://www.cdc.gov/diabetes/pdfs/data/statistics/nationaldiabetes-statistics-report.pdf (accessed on 21 March 2018).

2. Nesto, R. CHD: A major burden in type 2 diabetes. *Acta Diabetol.* **2001**, *38*, S3–S8. [CrossRef]

3. Choi, S.; Rankin, S.; Steward, A.; Oka, R. Perceptions of Coronary Heart Disease risk in Korean immigrants with type 2 diabetes. *Diabetes Educ.* **2008**, *34*, 484–492. [CrossRef]

4. Carroll, C.; Naylor, E.; Marsden, P.; Dornan, T. How do people with type 2 diabetes perceive and respond to cardiovascular risk? *Diabet. Med.* **2003**, *20*, 355–360. [CrossRef]

5. American Diabetes Association. Prevention of delay of type 2 diabetes: Standards of medical care in diabetes—2018. *Diabetes Care* **2018**, *41* (Suppl. 1), S51–S54. [CrossRef]

6. American Diabetes Association. Evidence-based nutrition principles and recommendations for the treatment and prevention of diabetes and related complications. *Diabetes Care* **2002**, *25*, S50–S60. [CrossRef]

7. Miller, C.; Edwards, L. Development and validation of a shelf inventory to evaluate household food purchases among older adults with diabetes mellitus. *J. Nutr. Educ. Behav.* **2002**, *34*, 261–267. [CrossRef]

8. Tomky, D.; Cypress, M.; Dang, D.; Maryniuk, M.; Peyrot, M.; Mensing, C. AADE Position Statement. AADE7 Self-Care Behaviors. *Diabetes Educ.* **2008**, *34*, 445–449. [CrossRef]

9. Ancker, J.S.; Senathirajah, Y.; Kukafka, R.; Starren, J.B. Design features of graphs in health risk communication: A systematic review. *J. Am. Med. Inform Assoc.* **2006**, *13*, 608–618. [CrossRef]

10. Fjeldsoe, B.S.; Marshall, A.L.; Miller, V.D. Behavior change interventions delivered by mobile telephone short-message service. *Am. J. Prev. Med.* **2009**, *35*, 165–173. [CrossRef]

11. Kim, H.S.; Hwang, Y.; Lee, J.H.; Oh, H.Y.; Kim, Y.J.; Kwon, H.Y.; Kang, H.; Kim, H.; Park, R.W.; Kim, J.H. Future prospects of health management systems using cellular phones. *Telemed. J. E Health* **2014**, *20*, 544–551. [CrossRef]

12. Fortmann, A.L.; Gallo, L.C.; Garcia, M.I.; Taleb, M.; Euyoque, J.A.; Clark, T.; Skidmore, J.; Ruiz, M.; Dharkar-Surber, S.; Schultz, J.; et al. Dulce Digital: An mHealth SMS-Based Intervention Improves Glycemic Control in Hispanics With Type 2 Diabetes. *Diabetes Care* **2017**, *40*, 1349–1355. [CrossRef]

13. Ramirez, M.; Wu, S. Phone messaging to prompt physical activity and social support among low-income Latino patients with Type 2 diabetes: A randomized pilot study. *J. Med. Internet Res.* **2017**, *2*, e8. [CrossRef]

14. Sawesi, S.; Rashrash, M.; Phalakornkule, K.; Carpenter, J.S.; Jones, J.F. The impact of information technology on patient engagement and health behavior change: A systematic review of the literature. *J. Med. Internet Res.* **2016**, *4*, e1. [CrossRef]

15. Arora, S.; Peters, A.L.; Burner, E.; Lam, C.N.; Menchine, M. Trial to examine text message-based mHealth in emergency department patients with diabetes (TExT-MED): A randomized controlled trial. *Ann. Emerg. Med.* **2014**, *63*, 745–754. [CrossRef]

16. American Association of Diabetes Educators (AADE) AADE7. Self-Care Behaviors. Available online: https://www.diabeteseducator.org/patient-resources/aade7-self-care-behaviors (accessed on 21 March 2018).

17. Toobert, D.J.; Hampson, S.E.; Glasgow, R.E. The summary of diabetes self-care activities measure. *Diabetes Care* **2000**, *23*, 943–950. [CrossRef]

18. NutritionQuest. Assessment and Analysis Services. Block Food Screeners; Fat/Sugar/Fruit/Vegetable Screener. Available online: http://www.nutritionquest.com/assessment/ (accessed on 21 March 2018).

19. NutritionQuest. Assessment and Analysis Services. Block Food Screeners; Physical Activity Screener. Available online: http://www.nutritionquest.com/assessment/ (accessed on 21 March 2018).

20. Ainsworth, B.E.; Haskell, W.L.; Herrmann, S.D.; Meckes, N.; Bassett, D.R., Jr.; Tudor-Locke, C.; Greer, J.L.; Vezina, J.; Whitt-Glover, M.C.; Leon, A.S. 2011 Compendium of Physical Activities: A Second Update of Codes and MET Values. *Med. Sci. Sports Exerc.* **2011**, *43*, 1575–1581. [CrossRef]

21. Becker, D.M.; Levine, D.M. Risk perception, knowledge, and lifestyles in siblings of people with premature coronary disease. *Am. J. Prev. Med.* **1987**, *3*, 45–50. [CrossRef]

22. Nepper, M.J.; Chai, W. Validation of a home food checklist to assess the home food environment of school-age children. *Health Behav. Policy Rev.* **2016**, *3*, 348–360. [CrossRef]

23. Boles, R.E.; Scharf, C.; Filigno, S.S.; Saelens, B.E.; Stark, L.J. Differences in home food and activity environments between obese and healthy weight families of preschool children. *J. Nutr. Educ. Behav.* **2013**, *45*, 222–231. [CrossRef]

24. Fulkerson, J.A.; Nelson, M.C.; Lytle, L.; Moe, S.; Heitzler, C.; Pasch, K.E. The validation of a home food inventory. *Int. J. Behav. Nutr. Phys. Act* **2008**, *5*. [CrossRef]

25. U.S. Department of Health & Human Services, National Heart, Lung, and Blood Institute. Choosing foods for your family: Go, Slow, Whoa Foods. Adapted from Catch: Coordinated Approach to School Heart. Available online: https://www.nhlbi.nih.gov/health/educational/wecan/eat-right/choosing-foods.htm (accessed on 19 April 2016).
26. American Diabetes Association. Glycemic Index and Diabetes. Available online: http://www.diabetes.org/food-and-fitness/food/what-can-i-eat/understanding-carbohydrates/glycemic-index-and-diabetes.html (accessed on 29 January 2019).
27. Shadish, W.; Cook, T.D.; Campbell, D.T. *Experimental and Quasi-Experimental Designs for Generalized Causal Inference*; Houghton, Mifflin and Company: Boston, MA, USA, 2002.
28. Cohen, J. A power primer. *Psychol. Bull.* **1992**, *112*, 115–159. [CrossRef]
29. Nundy, S.; Mishra, A.; Hogan, P.; Lee, S.M.; Solomon, M.C.; Peek, M.E. How do mobile phone diabetes programs drive behavior change? *Diabetes Educ.* **2014**, *40*, 806–819. [CrossRef]
30. Tamban, C.; Isip-Tan, I.T.; Jimeno, C. Use of short message services (SMS) for the management of Type 2 diabetes: A randomized controlled trial. *JASEAN Fed. Endocr. Soc.* **2013**, *28*, 143–148. [CrossRef]

Article

An End-to-End Image-Based Automatic Food Energy Estimation Technique Based on Learned Energy Distribution Images: Protocol and Methodology

Shaobo Fang [1], Zeman Shao [1], Deborah A. Kerr [2,3], Carol J. Boushey [4,5] and Fengqing Zhu [1,*]

[1] School of Electrical and Computer Engineering, Purdue University, West Lafayette, IN 47907, USA; fang29@purdue.edu (S.F.); shao112@purdue.edu (Z.S.)
[2] School of Public Health, Curtin University, Perth, WA 6845, Australia; d.kerr@curtin.edu.au
[3] Curtin Institute of Computation, Curtin University, Perth, WA 6845, Australia
[4] Cancer Epidemiology Program, University of Hawaii Cancer Center, Honolulu, HI 96813, USA; cjboushey@cc.hawaii.edu
[5] Department of Nutrition, Purdue University, West Lafayette, IN 47907, USA
*   Correspondence: zhu0@ecn.purdue.edu; Tel.: +1-7654960407

Received: 8 March 2019; Accepted: 12 April 2019; Published: 18 April 2019

Abstract: Obtaining accurate food portion estimation automatically is challenging since the processes of food preparation and consumption impose large variations on food shapes and appearances. The aim of this paper was to estimate the food energy numeric value from eating occasion images captured using the mobile food record. To model the characteristics of food energy distribution in an eating scene, a new concept of "food energy distribution" was introduced. The mapping of a food image to its energy distribution was learned using Generative Adversarial Network (GAN) architecture. Food energy was estimated from the image based on the energy distribution image predicted by GAN. The proposed method was validated on a set of food images collected from a 7-day dietary study among 45 community-dwelling men and women between 21–65 years. The ground truth food energy was obtained from pre-weighed foods provided to the participants. The predicted food energy values using our end-to-end energy estimation system was compared to the ground truth food energy values. The average error in the estimated energy was 209 kcal per eating occasion. These results show promise for improving accuracy of image-based dietary assessment.

Keywords: dietary assessment; food energy estimation; generative models; generative adversarial networks; image-to-energy mapping; neural networks; regressions

---

1. Introduction

Leading causes of death in the United States, including cancer, diabetes, and heart disease, can be linked to diet [1,2]. Measuring accurate dietary intake is considered to be an open research problem, and developing accurate methods for dietary assessment and evaluation continues to be a challenge. Underreporting is well documented amongst dietary assessment methods. Compared to traditional dietary assessment methods that often involve detailed handwritten reports, technology-assisted dietary assessment approaches reduce the burden of keeping such a detailed report and are preferred over traditional written dietary record for monitoring everyday activity [3].

In recent years, mobile telephones have emerged and provide unique mechanisms to monitor personal health and to collect dietary information [4]. Image-based approaches integrating application technology for mobile devices have been developed which aim at capturing all eating occasions by images as the primary record of dietary intake [3]. To date, these image-based approaches have primarily relied on trained analysts to estimate energy intake from the food images. Validation studies of the trained analyst have shown limited accuracy within and between the trained analysts [5,6]. Although automated methods are not sufficiently advanced to entirely replace the trained analyst, these methods hold promise to ultimately improve accuracy and reduce participant and researcher burden. Several mobile dietary assessment systems have been developed, such as the Technology Assisted Dietary Assessment (TADA^{TM}) system [7,8], FoodLog [9], FoodCam [10], DietCam [11], and Im2Calories [12], to address some of the challenges of automatically-determined food types and energy consumed based on image processing and analysis methods. However, developing automatic dietary assessment techniques remains an open research problem.

Estimating food energy from a single-view food image is an ill-posed problem, as most 3D information has been lost when the eating scene is projected from 3D world coordinates onto 2D image coordinates. Several methods have been proposed to estimate food portions from a single-view image. In Chen et al. [13], 3D models were manually fitted onto a 2D food image in order to estimate the food portion sizes. However, manual fitting does not scale with larger data sets. Another method used was participants placing their thumbs in their images as a size reference to estimate the food area and then the portion size of the food [14]. The inconsistency in the sizes of thumbs is an obvious issue. The model proposed by Zhang et al. [15] counts the pixels of each food segmentation in the image to estimate food portion. No 3D information is incorporated into the model. In the approach used by Aizawa et al. [16], the food image is divided into sub-regions and then food portions are estimated based on predetermined serving size classifications. Food portion estimation, in this case, is a task of selecting from limited discrete portion size choices.

We previously developed a 3D geometric-model based method for food portion estimation [17]. Our technique did not require manual tuning of model parameters, and we were able to obtain accurate food portion estimates [17]. Later, we showed that accurate food portions could be estimated using geometric models for food objects with well-defined 3D shapes [18]. To further improve the accuracy of food portion estimation, we incorporated the contextual dietary information of food portion co-occurrence patterns [19]. However, geometric-model-based techniques estimate food volumes rather than food energy. With food volumes estimated, food density is still required to compute the food weights which can then be mapped to food energy using a food composition resource, such as, the United States Department of Agriculture (USDA) Food and Nutrient Database for Dietary Studies (FNDDS) [20]. In addition, geometric-model-based techniques require food labels and food segmentation masks (i.e., location of foods in the image). Errors from automatic food classification and image segmentation can propagate into the final portion estimation. Therefore, new approaches that can directly link food images to food energy in the image would be desirable.

Recently, deep learning [21] techniques, especially techniques based on Convolutional Neural Networks (CNN) [22] have shown substantial success in many computer vision techniques, such as object detection [23–25], object segmentation [26], and image to image transfer [27–29]. Meyers et al. [12] proposed a food portion estimation method based on the predicted depth maps [30] of the eating scene. We have shown there is a tendency of over-estimation using depth image-based techniques, and an accurate estimation is not always guaranteed, even when depth information is available [18]. Ege et al. [31] used a multi-task CNN [32] architecture for identification of food, ingredients, and cooking directions. Food energy estimation is treated as a regression task [31], and only one unit in the last fully-connected layer in the VGG-16 architecture [23] is used for energy estimation. Further analysis of where the error may come from for energy estimation becomes difficult. Techniques based on CNN rely highly on well-constructed training data sets with sufficient samples and properly designed neural network architecture. In this paper, we focused on automatic dietary assessment of food energy estimation. We used single-view food images captured by users before and after eating their meals.

We proposed the concept of an "energy distribution image", which was one approach to establish the relationships between the food image and how food energy was distributed in the food image [33]. Each pixel in the energy distribution image represented the relative food energy weights at the corresponding pixel location. The use of an "energy distribution image" enabled us to first visualize how food energy estimation was spatially distributed across the eating scene.

Generative models learn from real data distribution and can generate samples that are similar to those in the real data distribution by taking random noises (for example, generate fake faces that look realistic [34]). In addition, generative models can also take prior information when generating new samples [27]. Therefore, they are suitable for tasks of image-to-image translation. We used generative models to predict energy distribution image based on eating occasion image, as generative models are a natural fit for solving image-to-image translations by its proven capability of learning the correspondences from one data distribution to another [27]. The aim of this paper was to develop a novel dietary assessment method to estimate the food energy numeric value from eating occasion images.

2. Methods

To estimate food portions (in energy), the energy distribution image is a new approach to visualize where foods are in the image and how much relative energy is presented at different food regions. We used Generative Adversarial Network (GAN) architecture to train the generative model that predicts the food energy distribution images based on eating occasion images. We built a food image data set with paired images for the training of the GAN [33]. To complete the end-to-end task of estimating food energy value based on a single-view eating occasion image, we used a CNN based regression model to estimate the numeric food energy value using the learned energy distribution images.

2.1. Image-to-Energy Data Set

Food images were collected using the mobile food record (mFR^{TM}) as part of the Food in Focus study, which was a community dwelling study of 45 adults (15 men and 30 women) between 21 and 65 years of age in a 7-day study period [35]. Pre-weighed food pack-outs were distributed to the participants and uneaten foods were returned and weighted. Briefly, participants captured images of each eating occasion over the entire period using the mFR^{TM}. Providing known foods and amounts supported the objective of being able to identify the foods consumed and their amounts, which were used as ground truth for evaluating the proposed method. The food categories provided for breakfast, lunch, and dinner are listed in Table 1.

Since there is no public data set available for training our generative model, the data set of image pairs, consisting of eating occasion images and corresponding energy distribution images, were constructed using the Food in Focus study. The purpose of this data set was to learn the mappings from food images to the food energy distribution images [33]. This data set was based on the ground truth food labels, segmentation masks, and energy information from the study where known foods and amounts were provided [35]. To build this data set, ground truth food labels, segmentation masks, food energy information, and the presence of the known size fiducial marker were required. To the best of our knowledge, we are the only group that has collected such a food image data set with all required information listed above. We used GAN [34] architecture to train the generative model for the task predicting the food energy distribution image, as GAN has shown impressive success in training generative models [27–29,36,37]. In addition, GAN is able to effectively reduce the adversarial space during training [34] compared to other generative models, such as Variational Autoencoders (VAEs) [38]. Our image-to-energy data set described in Section 2.1 could not cover all food types, eating scenes, and all possible food combinations. Therefore, GAN's characteristic reducing adversarial space was important for our task, as it reduced the chance of the generative model overfitting on training image pairs. The energy value of the meal image is estimated based on the learned food energy distribution image by training a CNN. Figure 1 shows the design of the proposed end-to-end food energy estimation based on a single-view eating occasion image.

Table 1. Type of food items in eating occasion images separated by breakfast, lunch, and dinner.

Breakfast	Lunch	Dinner
Bagel	Apple	Apple
Banana	Bagel	Banana
English muffin	Carrot	Broccoli
Grape	Celery	Celery
Margarine	Cherry	Cherry
Mayonnaise	Chicken wrap	Doritos
Milk	Chocolate chip	Fruit cocktail
Orange	Ding Dong	Garlic bread
Orange juice	Doritos	Garlic toast
Pancake	Grape	Grape
Peanut butter	Ham sandwich	Lasagna
Ranch dressing	Mashed potato	Margarine
Saltines	Mayonnaise	Mashed potato
Sausage	Milk	Mayonnaise
Strawberry	Mustard	Milk
Syrup	No fat dressing	Muffin
Water	Noodle soup	Orange
Wheaties	Peas	Peas
Yogurt	Pizza	Ranch dressing
	Potato	Rice crispy bar
	Potato chip	Salad mix
	Ranch dressing	Strawberry
	Salad mix	String cheese
	Saltines	Tomato
	Snicker doodle	Water
	Strawberry	Watermelon
	String cheese	Wheat bread
	Tea	Yogurt
	Tomato	
	Water	
	Watermelon	
	Yogurt	

To train the GAN for the task of mapping eating occasion images to energy distribution images, eating occasion image and energy distribution image pairs were required. There is no device that can be used to directly capture the "energy distribution image". We constructed the image-to-energy distribution data set using food images collected from the Food in Focus study [35]. Each food item and each eating occasion image were manually labeled and segmented in the data set. The ground truth energy information of each weighed food item in each eating occasion image was estimated using the energy values in the USDA Food and Nutrient Database for Dietary Studies.

In order to construct the energy distribution image, we first detected the location of the fiducial marker [39]. A fiducial marker is a colored checkerboard, as shown in Figure 2a, which is included in each eating occasion scene image. The marker is used to correct the color of the acquired images to match the reference colors during food identification and for camera calibration in portion size estimation [40,41]. The image-to-energy distribution data set could not be constructed if any of the above components (ground truth food labels, segmentation masks, food energy information, and the presence of the known size fiducial marker) were missing.

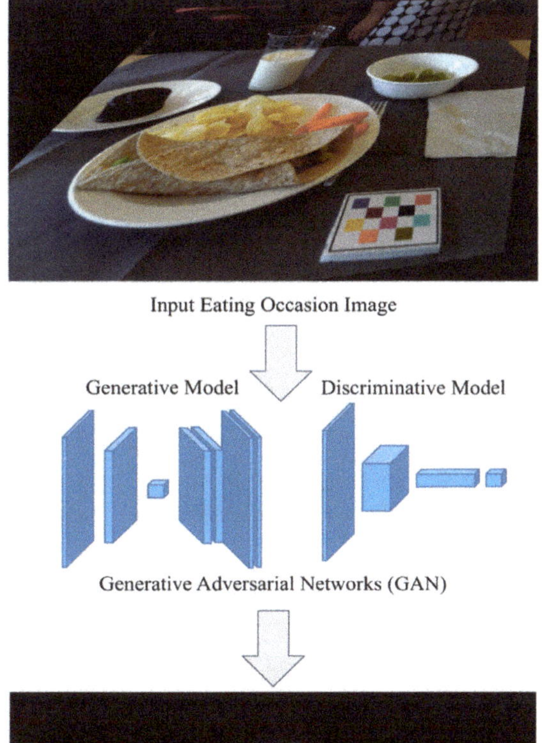

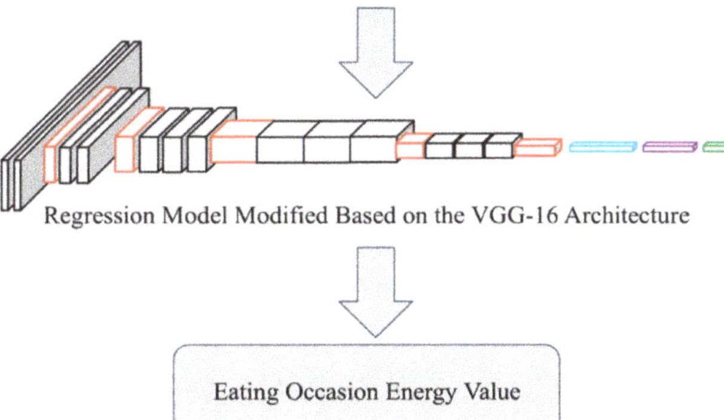

Figure 1. End-to-end system design of food energy estimation based on a single-view RGB eating occasion image.

With the reference of the known size fiducial marker, we removed the projective distortion in the original image using Direct Linear Transform (DLT) [42] based on the estimated homography matrix H to create a rectified image. Suppose I is the original eating occasion image; we denote $\hat{I}$ as the rectified image that is obtained: $\hat{I} = H^{-1}I$. Following the same rule of notation, for each food k and its associated segmentation mask S_k, the rectified segmentation can be expressed as: $\hat{S}_k = H^{-1}S_k$. For each pixel location $(\hat{i}, \hat{j}) \in \hat{S}_k$, a scale factor $\hat{w}_{i,j}$ is assigned to reflect the distance between the pixel location $(\hat{i}, \hat{j})$ to the centroid of the segmentation mask $\hat{S}_k$. Based on the scale factor $\hat{w}_{i,j}$ assigned to each pixel location in $\hat{S}_k$, the weighted segmentation masks $\hat{S}_k$ can be projected back to the original pixel coordinates denoted as $\bar{S}_k$, where: $\bar{S}_k = H\hat{S}_k$, and learn the parameter P_k such that:

$$c_k = P_k \sum\nolimits_{\forall(\bar{i},\bar{j})\in\bar{S}_k} \bar{w}_{\bar{i},\bar{j}}, \tag{1}$$

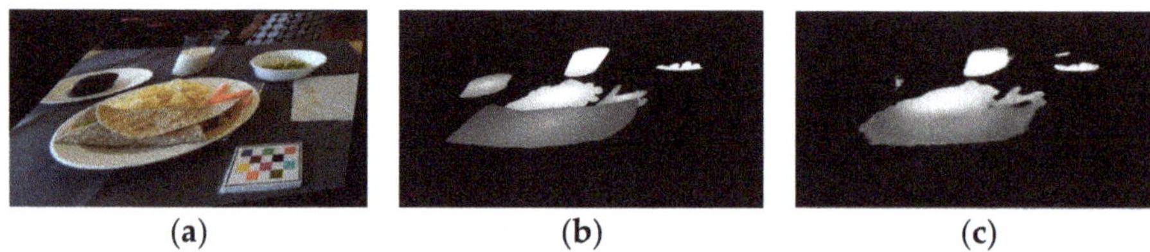

(a)	**(b)**	**(c)**

Figure 2. Learning image-to-energy translation using generative models. (a) Eating occasion image $\bar{I}$. (b) Ground truth energy distribution image $\overline{W}$. (c) Estimated energy distribution image $\tilde{W}$.

where c_k is the ground truth energy associated with food k, P_k is the energy mapping coefficient for $\bar{S}_k$, and $\bar{w}_{\bar{i},\bar{j}}$ is the energy weight factor at each pixel that makes up the ground truth energy distribution image. We can then update the energy weight factors $\bar{w}_{\bar{i},\bar{j}}$ in $\bar{S}_k$ as:

$$\bar{w}_{\bar{i},\bar{j}} = P_k \cdot \bar{w}_{\bar{i},\bar{j}}, \forall(\bar{i}, \bar{j}) \in \bar{S}_k. \tag{2}$$

Repeat the above process for all $k \in \{1, \ldots, M\}$, where M is total number of food items in the eating occasion image, and then overlay all segments $\bar{S}_k$ onto the ground truth energy distribution image $\overline{W}$, whose size is the same as image $\bar{I} = H\hat{I}$. Here, we show a pair of image $\bar{I}$ and the energy distribution image $\overline{W}$, as shown in Figure 2a,b, accordingly. The estimated energy distribution image shown in Figure 2c is denoted as $\tilde{W}$, which is learned from training on pairs of images $\bar{I}$ and the ground truth energy distribution image $\overline{W}$.

2.2. Generative Adversarial Networks (GAN)

GAN architecture has shown impressive success in training generative models [27–29,36,37]. In GAN, two models are trained simultaneously: a generative model G that captures the data distribution, and a discriminative model D that determines the probability that a sample came from the training data rather than G [34]. The common analogy for the GAN architecture is a game between producing counterfeits (generative models) and detecting counterfeits (discriminative model) [34]. To formulate the GAN, we specified the cost functions. We use $\theta^{(G)}$ to denote the parameters of generative model G and $\theta^{(D)}$ to denote the parameters of discriminative model D. The generative model G attempts to minimize the cost function:

$$J^{(G)}\left(\theta^{(D)}, \theta^{(G)}\right) \tag{3}$$

where the discriminative model D attempts to minimize the cost function:

$$J^{(D)}\left(\theta^{(D)}, \theta^{(G)}\right) \tag{4}$$

In a zero-sum game, we have:

$$J^{(G)}\left(\theta^{(D)}, \theta^{(G)}\right) = -J^{(D)}\left(\theta^{(D)}, \theta^{(G)}\right) \tag{5}$$

Therefore, the overall cost can be formulated as:

$$J^{(D)}\left(\theta^{(D)}, \theta^{(G)}\right) = -\frac{1}{2}E_{x\sim p_{data}}(x)[\log D(x)] - \frac{1}{2}E_{z\sim p_z(z)}[\log D(1 - (G(z)))] \tag{6}$$

where x is sampled from the true data p_{data} and z is random noise generated by distribution p_z. The generative model takes z and generates fake sample $G(z)$. The goal of the minimax game would then be:

$$\min_{\theta(G)} \max_{\theta(D)} - J^{(D)}\left(\theta^{(D)}, \theta^{(G)}\right) \tag{7}$$

Adversarial samples are those data which can easily lead neural networks to make mistakes. The GAN takes adversarial training samples by its nature, therefore, it could significantly reduce the adversarial space for the generative models to make mistakes. As a result, the use of GAN architecture can greatly reduce the training samples needed to model the statistical insights of the true data. During each update of the generative model G, the generated fake sample $G(z)$ will become more like the true sample x. Therefore, after sufficient epochs of training, the discriminator D is unable to differentiate between the two distributions x and $G(z)$ [34].

2.3. The Use of Conditional GAN (cGAN) for Image Mappings

We used conditional GAN (cGAN) [27] to estimate the energy distribution image [33], as cGAN is a natural fit for predicting an image output based on an input image. A cGAN attempts to learn the mapping from a random noise vector z to a target image y conditioned on the observed image x: $G(x, z) \rightarrow y$. The objective of a cGAN can be expressed as:

$$\mathcal{L}_{cGAN}(G, D) = \mathbb{E}_{x,y\sim p_{data}(x,y)}[\log D(x, y)] + \mathbb{E}_{x\sim p_{data}(x),z\sim p_z(z)}[\log(1 - D(x, G(x, z)))] \tag{8}$$

Otherwise, an additional conditional loss $\mathcal{L}_{conditional}(G)$ [27] is added to further improve $G(x, z) \rightarrow y$:

$$\mathcal{L}_{conditional}(G) = \mathbb{E}_{x,y\sim p_{data}(x,y),z\sim p_z(z)}[D(y, G(x,z))], \tag{9}$$

Common criteria used in $D(y, G(x, z))$ to measure the distance between y and $G(x, z)$ are the L_2 distance [43]:

$$D(y, G(x, z)) = \frac{1}{n}\sum_{i=1}^{n}(y_i - G(x_i, z_i))^2 \tag{10}$$

the L_1 distance [27]:

$$D(y, G(x, z)) = \frac{1}{n}\sum_{i=1}^{n}|(y_i - G(x_i, z_i))| \tag{11}$$

and a smooth version of the L1 distance:

$$D(y, G(x, z)) = \begin{cases} \frac{(y_i - G(x_i, z_i))^2}{2} & if|y_i - G(x_i, z_i)| < 1 \\ |y_i - G(x_i, z_i)| & otherwise. \end{cases} \tag{12}$$

So, the final objective [27,34] is:

$$G^* = arg\min_{G}\max_{D} \mathcal{L}_{cGAN}(G, D) + \lambda\mathcal{L}_{conditional}(G) \tag{13}$$

where the generative model G^* is used to estimate the energy distribution image $\tilde{W}$ based on the input eating occasion image $\bar{I}$.

2.4. Food Energy Estimation Based on Energy Distribution Images

We were able to obtain the energy distribution image [33] for each RGB eating occasion image using generative model G trained by GAN. An example of an original food image and an estimated energy distribution image is shown in Figure 2a,c. Energy distribution images represent how food energy is distributed in the eating scene. Our goal was to estimate food energy (a numerical value) based on the estimated energy distribution image. This is essentially a regression task as shown in Figure 3. We used a CNN-based regression model to conduct the task of estimating energy from energy distribution images. For the regression model, we used a VGG-16-based [23] architecture, as shown in Figure 4. As VGG-16 has shown impressive results on object detection tasks, VGG-16 is sufficient for learning complex image features. We modified the original VGG-16 architecture and added an additional linear layer, as shown in Figure 4, so that the CNN-based architecture was suitable for the energy value regression task. Instead of using random initialization for VGG-16 and training from scratch, we used pre-trained weights of VGG-16 architecture on ImageNet [44]. The pre-trained weights are indicated in the dash bounding box in Figure 4. We used random initialization for the linear layer. We then fine-tuned the pre-trained weights of the VGG-16 network for energy value prediction task based on the building blocks of complex features originally learned from ImageNet [44]. With the regression model, we can predict the energy of the foods in a single-view eating occasion image.

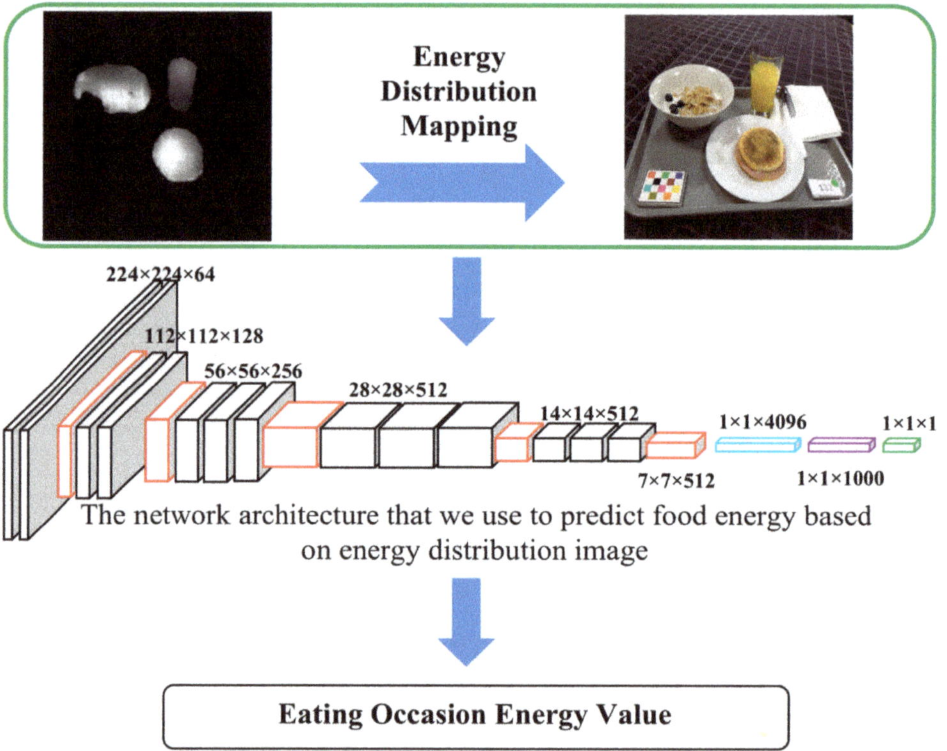

Figure 3. Estimating food energy of a meal based on predicted energy distribution image.

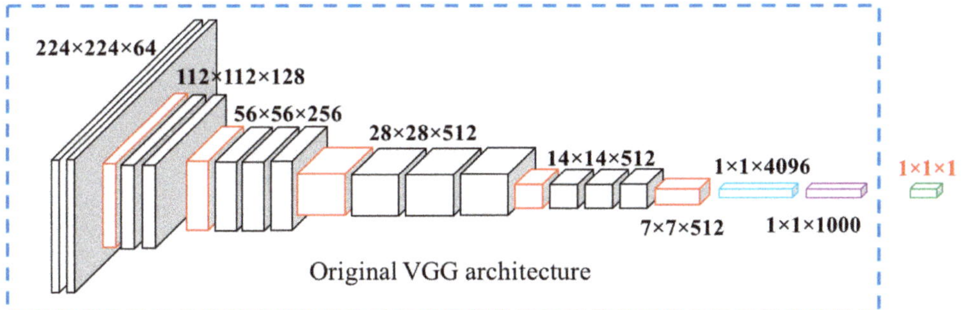

Figure 4. The network architecture used to predict food energy based on energy distribution image.

3. Experimental Results

3.1. Learning Image-to-Energy Mappings

We used 202 food images [35] that were manually annotated with ground truth segmentation masks and labels which we used for training. Data augmentation techniques, such as rotating, cropping, and flipping, were used to expand the database. In total, there were 1875 paired images (an image pair contains one eating occasion image and its corresponding energy distribution image) used to train the cGAN and 220 paired images for testing.

Once the cGAN estimated the energy distribution image $\widetilde{W}$, we could then determine the energy for a food image (portion size estimation) as:

$$EstimatedEnergy = \sum_{\forall (i,j) \in I} \left(\widetilde{W}_{i,j} \right) \tag{14}$$

To compare the estimated energy image $\widetilde{W}$ (Figure 2c) with the ground truth energy image $\overline{W}$ (Figure 2b), we defined the error between $\overline{W}$ and $\widetilde{W}$ as:

$$Energy\ Estimation\ Error\ Rate = \frac{\sum_{\forall (i,j) \in \overline{I}} \left(\widetilde{W}_{i,j} - \overline{W}_{i,j} \right)}{\sum_{\forall (i,j) \in \overline{I}} \left(\overline{W}_{i,j} \right)} \tag{15}$$

We compared the energy estimation error rates at different epochs for the two different cGAN models we used, the encoder-decoder architecture (Figure 5) and the U-Net architecture (Figure 6). Compared to the encoder-decoder architecture (Figure 5), the U-Net architecture (Figure 6) was more accurate and stable. The reason is that information from the "encoder" can be directly copied to the "decoder" layers in the U-Net architecture to provide precise locations [45], which is an idea similar to ResNet [25].

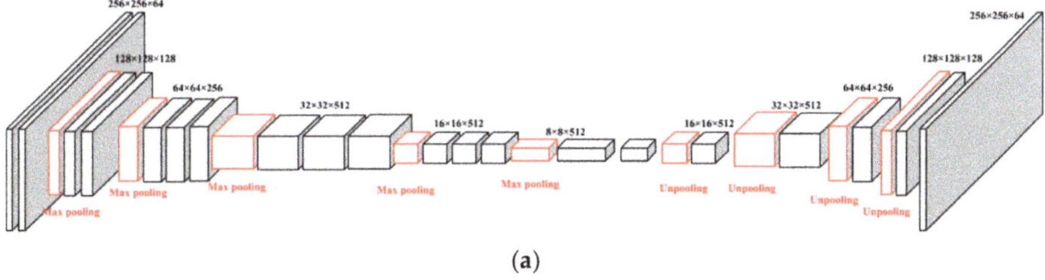

(a)

Figure 5. *Cont.*

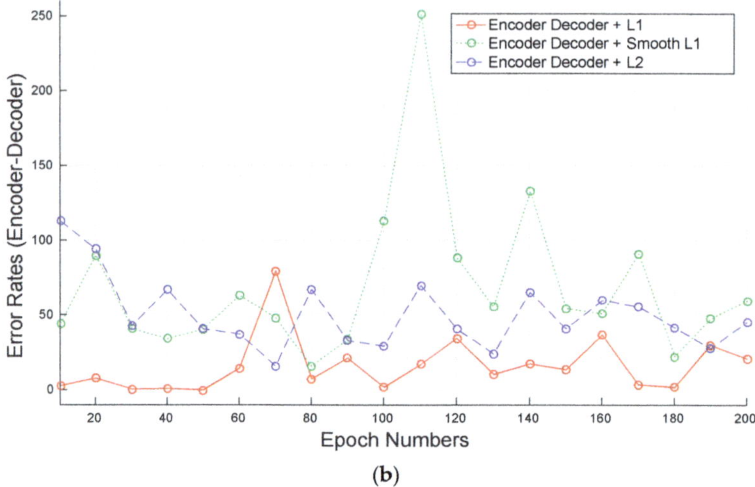

(b)

Figure 5. Generative model: encoder-decoder. (**a**) Architecture of encoder-decoder. (**b**) Error rate of encoder-decoder.

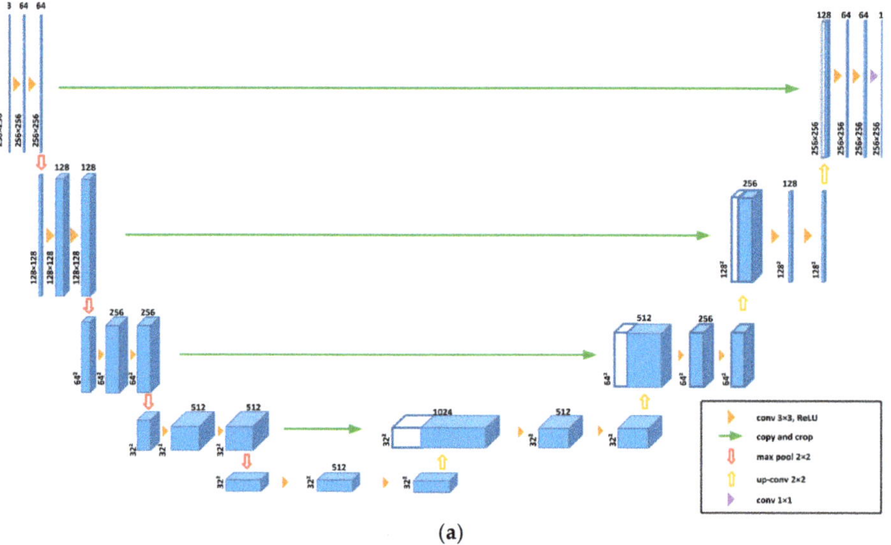

(a)

Figure 6. *Cont.*

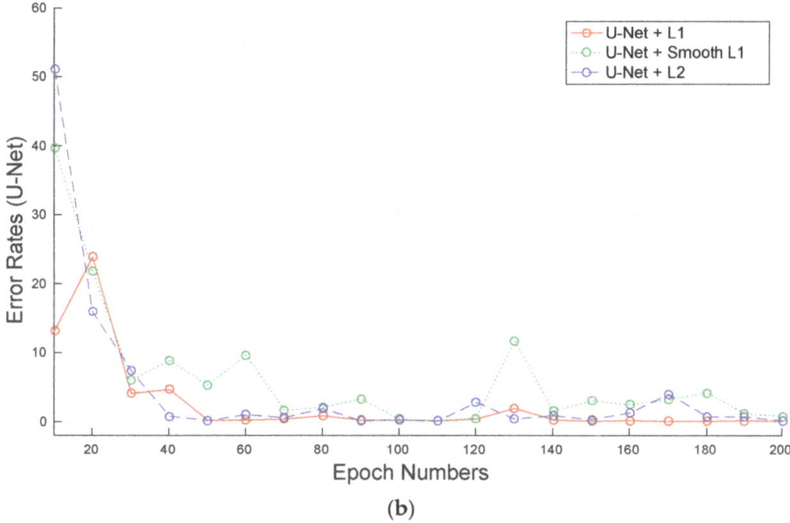

Figure 6. Generative model: U-Net. (**a**) Architecture of U-Net. (**b**) Error rate of U-Net.

We also compared the energy estimation error rates under different conditional loss settings, $\mathcal{L}_{conditional}(G)$, using U-Net. We used the batch size of 16 with $\lambda = 100$ in Equation (13), the Adam [46] solver with initial learning rate $\alpha = 0.0002$, and momentum parameters $\beta_1 = 0.5$, $\beta_2 = 0.999$ [27]. We observed that distance measure $D(y, G(x, z))$ as defined in Equations (10)–(12) using the L_1 or L_2 norms is better than using smoothed L_1 norm. At epoch 200, the energy estimation error rates are 10.89% (using L_1 criterion) and 12.67% (using L_2 criterion), respectively. In the experiments, we included food types whose shapes are difficult to define (for example, fries). Predicting the energy for these food types is very challenging using a geometric-model-based approach [17].

3.2. Food Energy Estimation Based on Energy Distribution Images

We predicted the food energy of each eating occasion image based on its energy distribution generated by generative model. The dimension for the predicted energy distribution image was 256 by 256. We resized the predicted energy distribution image from 256 by 256 to 224 by 224 to fit the input image size of VGG-16 architecture. To resize the output from generative model, we used OpenCV implementation of image resize, which is based on linear interpolation. The food energy estimation was then compared to the ground truth food energy from the Food in Focus study. We used 1390 eating occasion images also collected from the Food in Focus study [35], with ground truth food energy (kilocalories) for each food item in the eating occasion image. A total of 1043 of these eating occasion images were used for training and 347 of them for testing. The images selected for training and testing were selected by random sampling. All of the eating occasion images were captured by the users sitting naturally at a table. There were no extreme changes in the viewing angle. The errors for predicted food energy in Figure 7 are defined as:

$$\text{Error} = \text{Estimated Food Energy} - \text{Ground Truth Food Energy} \tag{16}$$

Figure 8 shows the relationship between the ground truth food energy and the food energy estimation of the eating occasion images in the testing data set. The dash line in Figure 8 indicates the ground truth and estimated energy are the same, i.e., estimation error is equal to zero. Therefore, the points above this line are overestimated, and the points below this line are underestimated. Figures 9 and 10 show examples of food energies the have been over- and underestimated, and we use

"+" and "−" to indicate over- and underestimation, respectively. The average ground truth of an eating occasion image in the testing data set was 538 kilocalories. We observed that the estimation was more accurate for the eating occasion image with ground truth energy around average, when compared to those with extremely high or low ground truth energy, such as zero kilocalories. This is due to the fact that there were not sufficient eating occasion images in our data set with very high or low ground truth energy provided to the neural networks for training.

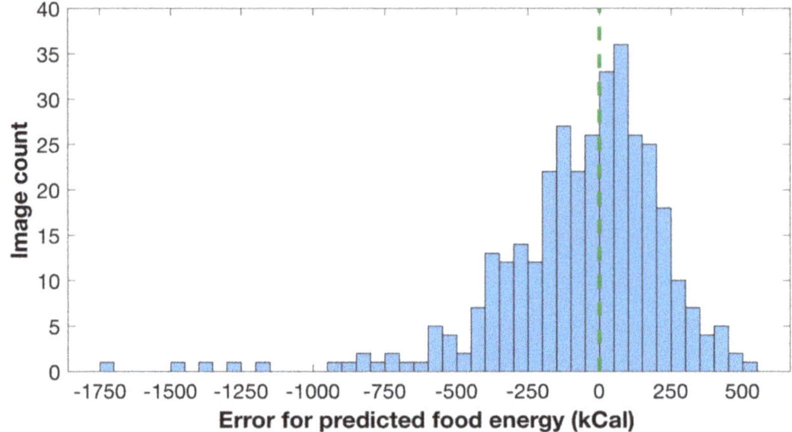

Figure 7. Error distribution of predicted food energy for all eating occasion images.

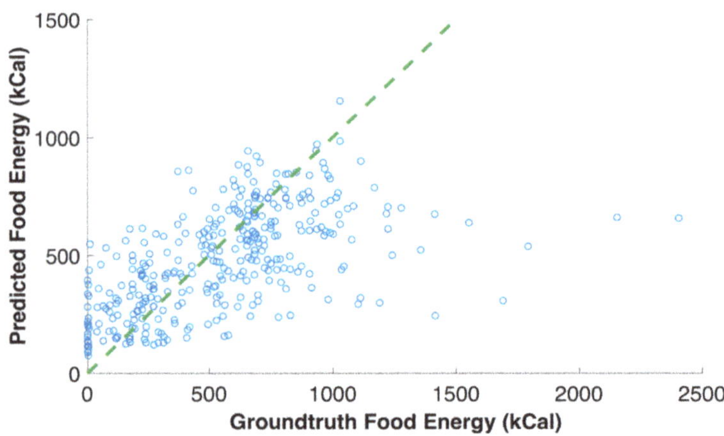

Figure 8. Relationship between the ground truth food energy and the food energy predicted for each eating occasion.

The error distribution of predicted food energies for 347 eating occasion images is shown in Figure 7. We found that the average energy estimation error was 209 kilocalories. An overestimation is displayed as a positive number. The average ground truth for all eating occasion images was 546 kilocalories, and the average ground truth for breakfast, lunch, and dinner eating occasion images was 531 kilocalories, 603 kilocalories, and 506 kilocalories, respectively. The average energy estimation error we obtained was 209 kilocalories, and the average energy estimation error for breakfast, lunch, and dinner eating occasion images was 204 kilocalories, 211 kilocalories, and 210 kilocalories, respectively. Several sample eating occasion images for overestimated food energy are shown in Figure 9, and eating occasion images for underestimated food energy are shown in Figure 10 accordingly.

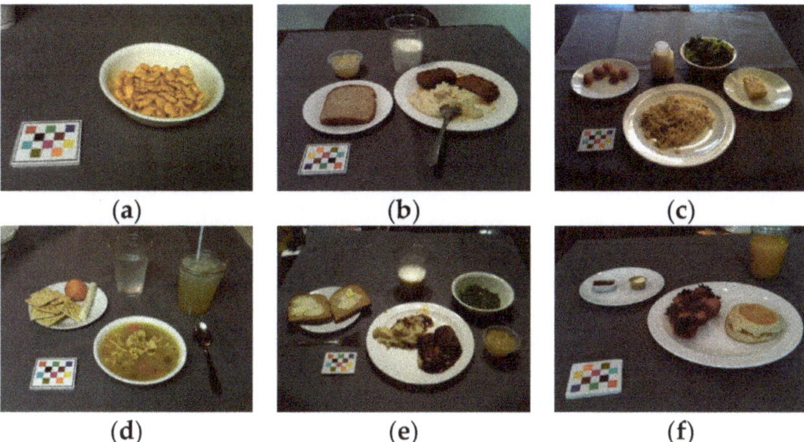

Figure 9. Examples of over-estimated food energy. (**a**) Ground truth energy: 287 kCal Predicted energy: 314 kCal Energy error: +27 kCal. (**b**) Ground truth energy: 520 kCal Predicted energy: 621 kCal Energy error: +101 kCal. (**c**) Ground truth energy: 653 kCal Predicted energy: 875 kCal Energy error: +222 kCal. (**d**) Ground truth energy: 498 kCal Predicted energy: 579 kCal Energy error: +81 kCal. (**e**) Ground truth energy: 705 kCal Predicted energy: 893 kCal Energy error: +188 kCal. (**f**) Ground truth energy: 354 kCal Predicted energy: 425 kCal Energy error: +71 kCal.

Figure 10. Examples of under-estimated food energy. (**a**) Ground truth energy: 542 kCal Predicted energy: 472 kCal Energy error: −70 kCal. (**b**) Ground truth energy: 990 kCal Predicted energy: 732 kCal Energy error: −258 kCal. (**c**) Ground truth energy: 508 kCal Predicted energy: 504 kCal Energy error: −4 kCal. (**d**) Ground truth energy: 508 kCal Predicted energy: 474 kCal Energy error: −34 kCal. (**e**) Ground truth energy: 749 kCal Predicted energy: 629 kCal Energy error: −120 kCal. (**f**) Ground truth energy: 1084 kCal Predicted energy: 708 kCal Energy error: −376 kCal.

4. Discussion

We have advanced the field of research for automatic food portion estimation by developing a novel food image based end-to-end system to estimate food energy using learned energy distribution images. The contributions of this work can be summarized as the following: We introduced a method for modeling the characteristics of energy distribution in an eating scene using generative models. Based on the predicted food energy distribution image, we designed a CNN-based regression model to estimate the energy value based on the learned energy distribution images. We designed and implemented a novel end-to-end system to estimate food energy based on a single-view RGB eating occasion image. The results were validated using data generated from the Food in Focus study using data from the 45 community-dwelling men and women between 21–65 years old consuming known foods and amounts over 7 days [35].

The advantage of our technique compared to a geometric model-based technique is that the system is training based. The pre-defined geometric models were limited to cover only certain types of food with known shapes, which is no longer an issue for training-based methods. In addition, the "energy distribution image" we introduced enabled us to first visualize how food energy estimation is spatially distributed across the eating scene (for example, regions of the image containing apple should have smaller weights due to lower energy (in kcal) compared to regions in the image containing cheese). Therefore, not only the final estimated numeric energy values could be used to analyze where the error may have come from, but also the intermediate results of the "energy distribution image" could be used.

As our end-to-end food portion estimation is a training based system, the limitation of the system is mainly determined by the training data. Expanding the training data set with a larger sample size, capturing images over a longer period of time, and more food types could improve the accuracy of automatic food portion estimation. For wider application, future studies need to include diverse eating styles and patterns, thus broadening the application of these methods to diverse population groups. These results point to the importance of controlled feeding studies using known foods and amounts. The results of such studies, on a wider scale, would contribute to wider application of these automated image-based methods with the benefit of improving accuracy of results. The use of an image-based system, such as TADA^{TM}, which uses the mFR^{TM}, is necessary for the automatic food portion estimation.

There are several reasons that may have led to the food energy estimation errors observed. Firstly, although we used 1875 paired food images to train the generative model using GAN architecture [33], the amount of food images did not cover all different eating occasions. Similarly, to train the regression model for numeric energy value prediction, 1043 eating occasion images were used where using more eating occasion images and food types could improve the accuracy of the end-to-end system. Secondly, when building the image-to-energy data set [33], the energy distribution images were synthetic images defined by handcrafted energy spread functions, rather than incorporating real 3D structures or depth information. Neither depth nor real 3D structure information was available when the study was conducted to capture eating occasion images [3]. To further improve the accuracy and address this challenge, we are currently investigating techniques to incorporate depth information into the end-to-end system where the 3D structures features of the foods in the images can also be learned by the neural networks.

5. Conclusions

In this work, we proposed a novel end-to-end system to directly estimate food energy using automatic food portion estimation from eating occasion images captured with an image-based system. Our system first estimated the image to energy mappings using a GAN structure. Based on the predicted food energy distribution image, we designed a CNN-based regression model to further estimate the energy value based the learned energy distribution images. To our knowledge, this method represents a paradigm shift in dietary assessment. The proposed method was validated using data collected by 45 men and women between 21–65 years old. We were able to obtain accurate food energy estimation with an average error of 209 kilocalories for eating occasion images collected from the Food in Focus study using the mFR^{TM}. The training-based technique for end-to-end food energy estimation no longer requires fitting geometric models onto the food objects that may have issues scaling up, as we need a large amounts of geometric models to fit different food types in many food images. In the future, combining automatically detected food labels, segmentation masks, and contextual dietary information has the potential to further improve the accuracy of such end-to-end food portion estimation system.

Author Contributions: The manuscript represents the collaborative work of all the authors. The work was conceptualized by S.F. and F.Z., S.F. and Z.S. developed the methodology and performed the analysis with supervision from F.Z. The Food in Focus study was designed and conducted by C.J.B and D.A.K. The original draft was prepared by S.F. and Z.S., and all authors reviewed and edited the manuscript. All authors read and approved the final manuscript.

Funding: This work was partially sponsored by the National Science Foundation under grant 1657262, NIH, NCI (1U01CA130784-01); NIH, NIDDK (1R01-DK073711-01A1) for the mobile food record and by the endowment of the Charles William Harrison Distinguished Professorship at Purdue University. Address all correspondence to Fengqing Zhu, zhu0@ecn.purdue.edu or see www.tadaproject.org.

Conflicts of Interest: The authors declare no conflicts of interest. The funders had no role in the design of the study; in the collection, analyses, or interpretation of data; in the writing of the manuscript, or in the decision to publish the results.

References

1. Liese, A.D.; Krebs-Smith, S.M.; Subar, A.F.; George, S.M.; Harmon, B.E.; Neuhouser, M.L.; Boushey, C.J.; Schap, T.E.; Reedy, J. The Dietary Patterns Methods Project: Synthesis of Findings across Cohorts and Relevance to Dietary Guidance. *J. Nutr.* **2015**, *145*, 393–402. [CrossRef] [PubMed]

2. Harmon, B.E.; Boushey, C.J.; Shvetsov, Y.B.; Reynolette Ettienne, J.R.; Wilkens, L.R.; Marchand, L.L.; Henderson, B.E.; Kolonel, L.N. Associations of key diet-quality indexes with mortality in the Multiethnic Cohort: The Dietary Patterns Methods Project. *Am. J. Clin. Nutr.* **2015**, 587–597. [CrossRef] [PubMed]

3. Boushey, C.J.; Spoden, M.; Zhu, F.M.; Delp, E.J.; Kerr, D.A. New mobile methods for dietary assessment: Review of image-assisted and image-based dietary assessment methods. *Proc. Nutr. Soc.* **2017**, *76*, 283–294. [CrossRef] [PubMed]

4. Six, B.; Schap, T.; Zhu, F.; Mariappan, A.; Bosch, M.; Delp, E.; Ebert, D.; Kerr, D.; Boushey, C. Evidence-based development of a mobile telephone food record. *J. Am. Diet. Assoc.* **2010**, *110*, 74–79. [CrossRef] [PubMed]

5. Howes, E.; Boushey, C.J.; Kerr, D.A.; Tomayko, E.J.; Cluskey, M. Image-based dietary assessment ability of dietetics students and interns. *Nutrients* **2017**, *9*, 114. [CrossRef]

6. Williamson, D.A.; Allen, R.; Martin, P.D.; Alfonso, A.J.; Gerald, B.; Hunt, A. Comparison of digital photography to weighed and visual estimation of portion sizes. *J. Am. Diet. Assoc.* **2003**, *103*, 1139–1145. [CrossRef]

7. Zhu, F.; Bosch, M.; Woo, I.; Kim, S.; Boushey, C.; Ebert, D.; Delp, E.J. The Use of Mobile Devices in Aiding Dietary Assessment and Evaluation. *IEEE J. Sel. Top. Signal Process.* **2010**, *4*, 756–766. [CrossRef]

8. Zhu, F.; Bosch, M.; Khanna, N.; Boushey, C.; Delp, E. Multiple Hypotheses Image Segmentation and Classification with Application to Dietary Assessment. *IEEE J. Biomed. Health Inform.* **2015**, *19*, 377–388. [CrossRef]

9. Kitamura, K.; Yamasaki, T.; Aizawa, K. FoodLog: Capture, Analysis and Retrieval of Personal Food Images via Web. In Proceedings of the ACM Multimedia Workshop on Multimedia for Cooking and Eating Activities, Beijing, China, 23 October 2009; pp. 23–30.

10. Joutou, T.; Yanai, K. A Food Image Recognition System with Multiple Kernel Learning. In Proceedings of the IEEE International Conference on Image Processing, Cairo, Egypt, 7–10 November 2009; pp. 285–288.

11. Kong, F.; Tan, J. DietCam: Automatic dietary assessment with mobile camera phones. *Pervasive Mob. Comput.* **2012**, *8*, 147–163. [CrossRef]

12. Meyers, A.; Johnston, N.; Rathod, V.; Korattikara, A.; Gorban, A.; Silberman, N.; Guadarrama, S.; Papandreou, G.; Huang, J.; Murphy, K.P. Im2Calories: Towards an Automated Mobile Vision Food Diary. In Proceedings of the IEEE International Conference on Computer Vision, Santiago, Chile, 7–13 December 2015; pp. 1233–1241.

13. Chen, H.; Jia, W.; Li, Z.; Sun, Y.; Sun, M. 3D/2D model-to-image registration for quantitative dietary assessment. In Proceedings of the IEEE Annual Northeast Bioengineering Conference, Philadelphia, PA, USA, 16–18 March 2012; pp. 95–96.

14. Pouladzadeh, P.; Shirmohammadi, S.; Almaghrabi, R. Measuring Calorie and Nutrition from Food Image. *IEEE Trans. Instrum. Meas.* **2014**, *63*, 1947–1956. [CrossRef]

15. Zhang, W.; Yu, Q.; Siddiquie, B.; Divakaran, A.; Sawhney, H. Snap-n-Eat Food Recognition and Nutrition Estimation on a Smartphone. *J. Diabetes Sci. Technol.* **2015**, *9*, 525–533. [CrossRef]

16. Aizawa, K.; Maruyama, Y.; Li, H.; Morikawa, C. Food Balance Estimation by Using Personal Dietary Tendencies in a Multimedia Food Log. *IEEE Trans. Multimed.* **2013**, *15*, 2176–2185. [CrossRef]

17. Fang, S.; Liu, C.; Zhu, F.; Delp, E.; Boushey, C. Single-View Food Portion Estimation Based on Geometric Models. In Proceedings of the IEEE International Symposium on Multimedia, Miami, FL, USA, 14–16 December 2015; pp. 385–390.

18. Fang, S.; Zhu, F.; Jiang, C.; Zhang, S.; Boushey, C.; Delp, E. A Comparison of Food Portion Size Estimation Using Geometric Models and Depth Images. In Proceedings of the IEEE International Conference on Image Processing, Phoenix, AZ, USA, 25–28 September 2016; pp. 26–30.

19. Fang, S.; Zhu, F.; Boushey, C.; Delp, E. The use of co-occurrence patterns in single image based food portion estimation. In Proceedings of the IEEE Global Conference on Signal and Information Processing, Montreal, QC, Canada, 14–16 November 2017; pp. 462–466.

20. *USDA Food and Nutrient Database for Dietary Studies, 1.0*; Agricultural Research Service, Food Surveys Research Group: Beltsville, MD, USA, 2004.

21. LeCun, Y.; Bengio, Y.; Hinton, G. Deep Learning. *Nature* **2015**, *521*, 436–444. [CrossRef]

22. LeCun, Y.; Bottou, L.; Bengio, Y.; Haffner, P. Gradient-based learning applied to document recognition. *Proc. IEEE* **1998**, *86*, 2278–2324. [CrossRef]

23. Simonyan, K.; Zisserman, A. Very Deep Convolutional Networks for Large-Scale Image Recognition. *arXiv* **2014**, arXiv:1409.1556.

24. Ren, S.; He, K.; Girshick, R.; Sun, J. Faster R-CNN: Towards real-time object detection with region proposal networks. In Proceedings of the Advances in Neural Information Processing Systems, Montreal, QC, Canada, 7–12 December 2015; pp. 91–99.

25. He, K.; Zhang, X.; Ren, S.; Sun, J. Deep Residual Learning for Image Recognition. In Proceedings of the IEEE Conference on Computer Vision and Pattern Recognition, Las Vegas, NV, USA, 27–30 June 2016; pp. 770–778.

26. He, K.; Gkioxari, G.; Dollar, P.; Girshick, R. Mask R-CNN. In Proceedings of the IEEE International Conference on Computer Vision, Venice, Italy, 22–29 October 2017; pp. 2980–2988.

27. Isola, P.; Zhu, J.Y.; Zhou, T.; Efros, A.A. Image-to-Image Translation with Conditional Adversarial Networks. In Proceedings of the IEEE Conference on Computer Vision and Pattern Recognition, Honolulu, HI, USA, 21–26 July 2017; pp. 5967–5976.

28. Wang, T.; Liu, M.; Zhu, J.; Tao, A.; Kautz, J.; Catanzaro, B. High-Resolution Image Synthesis and Semantic Manipulation with Conditional GANs. *arXiv* **2017**, arXiv:1711.11585.

29. Zhu, J.; Park, T.; Isola, P.; Efros, A.A. Unpaired Image-to-Image Translation using Cycle-Consistent Adversarial Networks. In Proceedings of the International Conference on Computer Vision, Venice, Italy, 22–29 October 2017; pp. 2223–2232.

30. Silberman, N.; Kohli, P.; Hoiem, D.; Fergus, R. Indoor Segmentation and Support Inference from RGBD Images. In Proceedings of the European Conference on Computer Vision, Florence, Italy, 7–13 October 2012; pp. 746–760.

31. Ege, T.; Yanai, K. Image-Based Food Calorie Estimation Using Knowledge on Food Categories, Ingredients and Cooking Directions. In Proceedings of the Workshops of ACM Multimedia on Thematic, Mountain View, CA, USA, 23–27 October 2017; pp. 367–375.

32. Abdulnabi, A.H.; Wang, G.; Lu, J.; Jia, K. Multi-Task CNN Model for Attribute Prediction. *IEEE Trans. Multimed.* **2015**, *17*, 1949–1959. [CrossRef]

33. Fang, S.; Shao, Z.; Mao, R.; Fu, C.; Delp, E.J.; Zhu, F.; Kerr, D.A.; Boushey, C.J. Single-view food portion estimation: Learning image-to-energy mappings using generative adversarial networks. In Proceedings of the IEEE International Conference on Image Processing, Athens, Greece, 7–10 October 2018; pp. 251–255.

34. Goodfellow, I.; Pouget-Abadie, J.; Mirza, M.; Xu, B.; Warde-Farley, D.; Ozair, S.; Courville, A.; Bengio, Y. Generative Adversarial Nets. In Proceedings of the Advances in Neural Information Processing Systems 27, Montreal, QC, Canada, 8–13 December 2014; pp. 2672–2680.

35. Boushey, C.J.; Spoden, M.; Delp, E.J.; Zhu, F.; Bosch, M.; Ahmad, Z.; Shvetsov, Y.B.; DeLany, J.P.; Kerr, D.A. Reported energy intake accuracy compared to doubly labeled water and usability of the mobile food record among community dwelling adults. *Nutrients* **2017**, *9*, 312. [CrossRef]

36. Radford, A.; Metz, L.; Chintala, S. Unsupervised Representation Learning with Deep Convolutional Generative Adversarial Networks. *arXiv* **2015**, arXiv:1511.06434.

37. Liu, M.; Breuel, T.; Kautz, J. Unsupervised Image-to-Image Translation Networks. In Proceedings of the Advances in Neural Information Processing Systems, Long Beach, CA, USA, 4–9 December 2017; pp. 700–708.

38. Kingma, D.P.; Welling, M. Auto-Encoding Variational Bayes. In Proceedings of the International Conference on Learning Representations, Banff, AB, Canada, 14–16 April 2014.

39. Zhang, Z. A flexible new technique for camera calibration. *IEEE Trans. Pattern Anal. Mach. Intell.* **2000**, *22*, 1330–1334. [CrossRef]

40. Xu, C.; He, Y.; Khanna, N.; Boushey, C.J.; Delp, E.J. Model-based food volume estimation using 3D pose. In Proceedings of the IEEE International Conference on Image Processing, Melbourne, Australia, 15–18 September 2013; pp. 2534–2538.

41. Xu, C.; Zhu, F.; Khanna, N.; Boushey, C.J.; Delp, E.J. Image enhancement and quality measures for dietary assessment using mobile devices. In Proceedings of the SPIE 8296, Computational Imaging X, Burlingame, CA, USA, 22–26 January 2012.

42. Hartley, R.I.; Zisserman, A. *Multiple View Geometry in Computer Vision*, 2nd ed.; Cambridge University Press: Cambridge, UK, 2004.

43. Pathak, D.; Krahenbuhl, P.; Donahue, J.; Darrell, T.; Efros, A. Context Encoders: Feature Learning by Inpainting. In Proceedings of the IEEE Conference on Computer Vision and Pattern Recognition, Las Vegas, NV, USA, 27–30 June 2016; pp. 2536–2544.

44. Russakovsky, O.; Deng, J.; Su, H.; Krause, J.; Satheesh, S.; Ma, S.; Huang, Z.; Karpathy, A.; Khosla, A.; Bernstein, M.; et al. ImageNet Large Scale Visual Recognition Challenge. *Int. J. Comput. Vis.* **2015**, *115*, 211–252. [CrossRef]

45. Ronneberger, O.; Fischer, P.; Brox, T. U-Net: Convolutional Networks for Biomedical Image Segmentation. In Proceedings of the Medical Image Computing and Computer-Assisted Intervention, Munich, Germany, 5–9 October 2015; pp. 231–241.

46. Kingma, D.P.; Ba, J. Adam: A Method for Stochastic Optimization. *arXiv* **2014**, arXiv:1412.6980.

Article

Relative Validity of the Eat and Track (EaT) Smartphone App for Collection of Dietary Intake Data in 18-to-30-Year Olds

Lyndal Wellard-Cole *, Juliana Chen †, Alyse Davies †, Adele Wong, Sharon Huynh, Anna Rangan and Margaret Allman-Farinelli

Nutrition and Dietetics Group, School of Life and Environmental Science, Charles Perkins Centre, The University of Sydney, NSW 2006, Australia; jche6526@uni.sydney.edu.au (J.C.); adav5418@uni.sydney.edu.au (A.D.); adele.wlp@gmail.com (A.W.); shuy9672@uni.sydney.edu.au (S.H.); anna.rangan@sydney.edu.au (A.R.); margaret.allman-farinelli@sydney.edu.au (M.A.-F.)
* Correspondence: lwel3754@uni.sydney.edu.au; Tel.: +61-2-8627-4854
† These authors contributed equally to this work.

Received: 25 January 2019; Accepted: 8 March 2019; Published: 14 March 2019

Abstract: (1) Background: Smartphone dietary assessment apps can be acceptable and valid data collection methods but have predominantly been validated in highly educated women, and none specifically measured eating-out habits in young adults. (2) Methods: Participants recorded their food and beverage consumption for three days using the Eat and Track (EaT) app, and intakes were compared with three dietitian-administered 24-h recall interviews matched to the same days as the reference method. Wilcoxon signed-rank or *t*-tests, correlation coefficients and Bland–Altman plots assessed agreement between the two methods for energy and percentage energy from nutrients (%E). (3) Results: One hundred and eighty nine of 216 participants (54% females, 60% resided in higher socioeconomic areas, 49% university-educated) completed the study. There were significant differences in median energy intake between methods ($p < 0.001$), but the EaT app had acceptable agreement for most nutrient densities at the group level. Correlation coefficients ranged from r = 0.56 (%E fat) to 0.82 (%E sugars), and between 85% and 94% of participants were cross-classified into the same or adjacent quartiles. Bland–Altman plots showed wide limits of agreement but no obvious biases for nutrient densities except carbohydrate in males. (4) Conclusions: The EaT app can be used to assess group nutrient densities in a general population of 18-to-30-year olds.

Keywords: diet assessment; relative validity; smartphone; young adults; apps

---

1. Introduction

Young adults (aged 18 to 30 years) have experienced the fastest rate of weight gain of any birth cohort in Australia [1]. One factor that appears to influence the diets of people in this age group is the amount of foods eaten prepared away from home, such as fast foods. More frequent consumption of fast foods has been associated with less healthy eating habits [2]. Young adult Australians consume fast foods more frequently than other age groups [3], and spend the highest proportion of their household income on eating out [4].

There have been no recent surveys on the amount and types of foods prepared and eaten away from home by young adults in Australia. The Measuring Young adults' Meals (MYMeals) study aims to fill this research gap [5]. Central to determining what young people are eating are valid and feasible dietary intake data collection methods.

Smartphone dietary intake methods can be acceptable and valid ways of collecting dietary data [6]. Considering that 95% of 18-to-34-year olds in Australia own a smartphone [7], it is an accessible way to collect dietary data from this age group, including those living in rural and remote locations [6,8].

Smartphone applications (apps) can be used as alternatives to traditional pen and paper or telephone food records or recalls [6]. Advantages of using electronic methods to collect dietary intake data are that entries can be completed more quickly than traditional methods [8], nutritional analysis can be conducted in real time, and researcher burden can be significantly reduced [9].

A number of commercial diet collection apps are available, however, validation studies show that these do not have good agreement with established dietary methods for some nutrients, including energy, protein, total fat, sugars, fibre and sodium [10,11]. Such differences may be due to the underlying nutrition composition databases, inadequacy of food listings available and no accounting for food preparation methods. In addition, many of the commercial apps were developed with American audiences in mind and therefore contain foods that are different to those available in Australia [10]. This makes it difficult for Australians to find and log the correct foods, which may reduce the accuracy of the nutritional data captured [12]. Further, these apps are mostly designed for weight management, and provide continuous feedback on the amounts of energy and/or nutrients consumed that may change behaviour, reducing their validity in research settings [6].

There have been three smartphone dietary recording apps (My Meal Mate, electronic Dietary Intake Assessment (eDIA) and Easy Diet Diary) that have been validated for the research setting using 24-h recalls as the reference method [13–16]. An additional app, electronic Carnet Alimentaire (e-CA, or "food record" in French), has also been evaluated favourably in a small study of 50 participants [17]. Two of these studies were conducted mostly in women in older age groups [16,17]. One was conducted in young adults, but the participants were almost exclusively university educated and of high socioeconomic status [14].

The Eat and Track (EaT) smartphone application is a new app for collection of dietary intake data, purpose-designed by the research team [18]. The aim of this study was to assess the relative validity of the EaT app with dietitian-administered 24-h recalls, examining energy and nutrient densities in a sample more inclusive of the Australian young adult population with respect to education and socioeconomic status.

2. Materials and Methods

2.1. Sample

Potential participants completed a screening and demographics questionnaire with questions on age group, educational attainment and residential postcode, to allow the socioeconomic status to be determined using Socio-Economic Indexes for Areas [19]. Participants were recruited from the overall MYMeals study population [5]. To satisfy ethics, participants had to give separate consent to opt into the validation study. This subgroup of participants was randomly allocated to complete the validation until 20% of the entire MYMeals sample was included. Potential participants were recruited across New South Wales (NSW), Australia's most populous state. They were eligible to participate in both the MYMeals study and the present validation study if they were aged 18 to 30 years, owned a working smartphone, were English-speaking, and consumed at least one meal, snack or drink purchased outside the home per week. Participants were excluded if they did not meet the aforementioned criteria, had ever been diagnosed with an eating disorder, were not able to complete the three days required for the study or were pregnant and/or breastfeeding. Potential participants completed a screening questionnaire through the online research management platform, REDCap [20], and provided consent. The study was approved by The University of Sydney Human Research Ethics Committee (project 2016/546).

2.2. Eat and Track Smartphone Application (EaT App)

The EaT app was developed by nutrition and information technology experts at the University of Sydney for the purposes of data collection for the MYMeals study specifically, and is based on the e-DIA app the researchers developed previously that was validated for nutrients and food groups [14,15]. Key usability modifications to the e-DIA app were the addition of a large, branded fast food database and improved usability functions (for more information on the development of the EaT app, see [18]). The nutrition database underpinning the EaT app included 4046 foods and beverages from the Australian Bureau of Statistics' AUSNUT 2011–2013 database [21], and 2229 food items from the largest chain outlets in Australia [18,22]. The fast food items are categorised by outlet name and the range of portion sizes of foods and beverages available at the outlet, for example, small, medium or large fries, to enhance recording and overcome previously reported difficulties in portion size estimation [14,15,17]. The fast food nutritional composition data is currently restricted to energy, protein, total and saturated fats, carbohydrates, sugars, and sodium, and does not contain micronutrients [23].

Participants were provided with written and video instructions on how to use the EaT app prior to starting the study, and could access these resources throughout the study period from the MYMeals study website [5]. Participants using the EaT app selected an eating occasion (Breakfast; Lunch; Dinner; or Snacks and Drinks) from the landing screen of the EaT app [18]. A free-text box appeared, and participants typed in the food they had eaten. Shortlists of foods appeared, and participants could scroll through the provided options, or use keyword prompts to find the food they consumed. Once the food was selected, participants chose the amount and unit of food (e.g., gram, millilitre, slice, cup, etc.), and where the food was sourced. Participants also received a portion measures booklet [24] to assist with estimating portion sizes during recording [18]. If a participant was unable to find a particular food they consumed, they could enter it manually. When entering a food manually, the app prompted participants to enter the food or individual ingredients, amounts and units consumed.

2.3. Procedures

After obtaining consent, participants were emailed links to download the EaT app from either the Apple App Store or Google Play, and the instructional videos on how to use the app to log their dietary intake. Participants were required to record all foods and beverages they consumed for three consecutive days. The researchers instructed participants on the days they must record their intakes. The starting days were staggered across the population to facilitate an even spread of days over the week. Participants received daily email and/or SMS prompts to remind them to log their intakes during the study period.

The participants also completed three 24-h recall telephone interviews with research dietitians. To allow all foods to be captured by both methods, the 24-h recalls were conducted the following day, but captured data for the same days that the app was used. The automated, online ASA-24 Australia [9,25] was used to conduct the recalls so that the interview process was standardised. This computerised method involved the dietitian recording all foods and drinks consumed throughout the day into the ASA-24 Australia as they interviewed the participants. Multiple passes prompt for additional information on food form, preparation methods, portion size and omitted items. The three 24-h recalls were conducted on the days following each of the data collection days. The ASA-24 Australia uses the AUSNUT 2011–2013 database [9,26], but differs from the EaT app with respect to the number of fast food items available [18].

At the conclusion of the three 24-h recalls, participants completed an online demographics questionnaire that included questions on self-reported height and weight data [5], to enable body mass index (BMI) to be calculated. Participants received a $100AUD gift voucher as compensation for their time after they had completed all study requirements.

2.4. Data Cleaning

All EaT app entries were checked by research dietitians in the week following the data collection days, and participants were contacted to clarify the additional manually entered food items, any obvious errors such as incorrect unit sizes, and skipped meals. However, to give a true indication of the relative validity of the EaT app, minimal changes to the data were made. Manually entered foods ($n = 33$) were matched to the nearest entry from the EaT app by one research dietitian, then checked by two others. If the participant stated brand names for entered items that were not in the original database, the Nutrition Information Panel data for that item was added to the database by the research dietitian (2% of total entries). Two Accredited Practising Dietitians each checked all the data independently for any discrepancies. These were identified and clarified until agreement was reached. All entries for the ASA-24 Australia recall were downloaded and checked.

2.5. Data Analysis

Daily totals for energy and each nutrient were summed, then means were calculated for each participant for the three study days. Group means and medians for energy and nutrient densities (percentage energy (%E) from protein, total and saturated fat, total carbohydrate and sugars, and sodium per 1000 kJ) were determined for both the EaT app and 24-h recall data [27]. Paired *t*-tests were conducted on normally distributed data and the Wilcoxon signed-rank test was conducted on non-parametric data to compare the three days of data from each method.

Correlations between the EaT app and 24-h recalls were assessed using Pearson product-moment correlation or Spearman rank correlation coefficients for skewed data. Quartiles of intake from each method were calculated for energy and each nutrient density. Cross-classification was calculated by the proportion of participants classified into the same, adjacent or extreme quartiles of energy or nutrient density intake by both methods.

Bland–Altman plots [28] were constructed to assess the agreement between the EaT app and 24-h recalls for the mean energy and nutrient densities.

Participants' basal metabolic rate (BMR) was calculated using the Schofield equation [29], based on the participants' self-reported weight, age and gender from the demographics questionnaire. Under- and over-reporters were identified using Goldberg's cut-offs [30]. Any participants who reported consuming an average energy intake over the three days of less than $1.0\times$ BMR were considered as under-reporters, and if they reported more than $2.4\times$ BMR they were deemed over-reporters [31]. Twenty-eight participants (14.8%) were classified as under-reporters and six participants were classified as over-reporters (3.2%) by the reference 24-h recall data. The full sample was used for analysis, as removing mis-reporters did not significantly change results.

IBM SPSS Statistics, version 24 was used to conduct all statistical analyses, and *p*-values < 0.05 were considered statistically significant.

3. Results

In total, 216 participants were recruited into the validation study. Of these, five withdrew from the study for personal or employment reasons and 20 did not complete all three days of data collection, while two were deemed to fail selection criteria, leaving a final sample size of 189 participants. The mean BMI of participants was 24.9 (SD 5.0). The characteristics of the included participants are shown in Table 1. It should be noted there are slightly fewer males than the Australian population proportion of 49%, and the proportion with post-school qualifications (65%) is more than the 56% reported by the Australian Bureau of Statistics [32].

3.1. Comparing Intakes between 24-h Recalls and EaT App

Significantly more energy was recorded using the 24-h recalls than the EaT app for the total sample ($p < 0.001$), females ($p < 0.01$) and males ($p < 0.001$) (Table 2). However, there were no significant differences in %E from protein, total or saturated fat, or sodium densities. The EaT app recorded significantly more %E carbohydrate than the recall for the total sample (p = 0.03) and for males (p = 0.01).

Table 1. Sample characteristics.

Participant Characteristics		N (%) [a]
Gender	Female	102 (54)
	Male	87 (46)
Age bracket	18–24 years	105 (56)
	25–30 years	84 (44)
Body mass index	Underweight ($\leq$18.49 kg/m^2)	4 (2)
	Healthy weight (18.5–24.9 kg/m^2)	116 (61)
	Overweight (25–29.9 kg/m^2)	47 (25)
	Obese ($\geq$30 kg/m^2)	22 (12)
Highest education attained	Primary school or less	2 (1)
	Secondary school	64 (34)
	Trade or diploma qualification	31 (16)
	University degree	92 (49)
Socioeconomic status [a]	Higher	114 (60)
	Lower	75 (40)

[a] From Socio-Economic Indexes for Areas [19] based on residential postcode, lowest five deciles = lower, highest five deciles = higher.

Table 2. Differences in energy and nutrient density intakes recorded by the 24-h recalls and Eat and Track (EaT) app.

Energy and Nutrient Densities	Median 24-h Recall (IQR) [c]	Median EaT App (IQR)	p [d]
Entire Sample n = 189			
Total energy, kJ [a]	9611 (7947–11,764)	8813 (7051–10,828)	<0.001 *
Protein, % energy [b]	18.3 (15.2–21.6)	18.0 (15.1–21.7)	0.14
Total fat, % energy [a]	35.8 (32.0–40.5)	35.6 (31.4–40.5)	0.47
Saturated fat, % energy [a]	12.8 (10.6–15.5)	12.3 (10.5–15.1)	0.21
Carbohydrate, % energy [a]	40.4 (35.3–45.7)	41.8 (35.0–47.4)	0.03 *
Sugars, % energy [b]	15.4 (11.7–21.4)	16.4 (11.8–19.2)	0.81
Sodium, mg/1000 kJ [b]	294.3 (239.5–349.3)	294.5 (237.2–362.3)	0.89
Females n = 102			
Total energy, kJ [a]	9001 (7752–11,122)	8209 (6818–10,399)	<0.01 *
Protein, % energy [a]	17.5 (14.9–20.3)	17.6 (14.8–21.0)	0.14
Total fat, % energy [a]	36.2 (32.0–41.1)	36.6 (32.0–40.8)	0.97
Saturated fat, % energy [a]	12.9 (10.6–16.0)	12.4 (10.6–15.6)	0.39
Carbohydrate, % energy [a]	41.3 (35.6–47.1)	42.2 (34.6–47.6)	0.57
Sugars, % energy [a]	18.1 (12.9–22.4)	17.2 (12.2–21.0)	0.14
Sodium, mg/1000 kJ [a]	282.8 (229.0–354.8)	282.0 (225.3–363.9)	0.56
Males n = 87			
Total energy, kJ [a]	10479 (8424–12985)	9140 (7359–11740)	<0.001 *
Protein, % energy [a]	19.0 (15.5–22.7)	19.2 (15.4–21.9)	0.92
Total fat, % energy [a]	34.9 (32.0–40.0)	34.9 (30.6–39.8)	0.29
Saturated fat, % energy [a]	12.7 (10.5–14.7)	12.3 (9.8–14.6)	0.36
Carbohydrate, % energy [a]	40.1 (35.1–43.7)	40.6 (35.9–47.2)	0.01 *
Sugars, % energy [b]	14.2 (11.0–18.1)	15.0 (11.6–18.5)	0.13
Sodium, mg/1000 kJ [b]	297.1 (249.1–349.0)	301.1 (245.0–362.1)	0.91

[a] *t*-tests for normally distributed data. [b] Wilcoxon signed-rank test for non-parametric data. [c] IQR = interquartile range. [d] $p \leq 0.05$ considered significant, * denotes significant results.

3.2. Correlation Coefficients and Cross-Classification

Table 3 shows the correlation coefficients between the 24-h recalls and EaT app. All correlation coefficients were positive and statistically significant ($p < 0.001$). Correlations ranged from 0.56 (%E total fat) to 0.82 (%E sugars) for the total sample. Quartile cross-classification of energy and nutrient densities with 24-h recalls and the EaT app placed 84% (%E fat) to 96% (%E sugars) of participants into the same or adjacent quartile. The proportion of participants classified into the extreme quartile ranged from 0% for %E carbohydrate to 4% for %E fat.

Table 3. Correlation coefficients and cross-classification of energy and nutrient densities between the 24-h recall and Eat and Track (EaT) app.

Energy and Nutrient Densities	Correlation Coefficients [c]	Cross-Classification into Quartiles (%)		
		Same	Same or Adjacent	Extreme
Entire Sample $n = 189$				
Total energy, kJ [a]	0.67	50.3	90.5	2.1
Protein, % energy [b]	0.73	53.4	93.7	2.1
Total fat, % energy [a]	0.56	46.0	84.1	4.2
Saturated fat, % energy [a]	0.59	49.2	84.7	3.7
Carbohydrate, % energy [a]	0.79	52.4	95.2	0
Sugars, % energy [b]	0.82	59.8	95.8	1.1
Sodium, mg/1000 kJ [b]	0.56	43.3	84.7	3.2
Females $n = 102$				
Total energy, kJ [a]	0.69	46.1	90.2	2.0
Protein, % energy [a]	0.71	52.9	93.1	1.0
Total fat, % energy [a]	0.61	48.0	86.3	2.9
Saturated fat, % energy [a]	0.62	56.9	86.3	2.9
Carbohydrate, % energy [a]	0.83	55.9	95.1	0
Sugars, % energy [a]	0.82	53.9	88.2	0
Sodium, mg/1000 kJ [a]	0.51	42.2	84.3	2.9
Males $n = 87$				
Total energy, kJ [a]	0.64	54.0	85.1	2.3
Protein, % energy [a]	0.72	56.3	90.8	2.3
Total fat, % energy [a]	0.50	36.8	80.5	4.6
Saturated fat, % energy [a]	0.53	43.7	85.1	4.6
Carbohydrate, % energy [a]	0.75	50.6	93.1	1.1
Sugars, % energy [b]	0.74	58.6	90.8	2.3
Sodium, mg/1000 kJ [b]	0.56	40.2	85.1	4.6

[a] Pearson's correlation coefficients. [b] Spearman's rank correlation. [c] All correlations were significant ($p < 0.01$).

3.3. Bland–Altman Plots for 24-h Recalls and EaT App

Bland–Altman plots showing the agreement between EaT app and 24-h recalls for energy for the total sample, males and females are presented in Figure 1. Males had a higher mean difference than females. Agreement between 24-h recalls and the EaT app for the nutrient density for carbohydrate are shown in Figure 2 because these were the nutrient densities for which a difference was found between medians. For males, carbohydrate showed underestimation at lower intakes and overestimation at higher intakes with the app compared with 24-h recalls. There were no biases detected for the other nutrient densities (plots not shown). The mean difference and 95% limits of agreement between the EaT app and 24-h recalls for energy and all nutrient densities can be seen in Table 4.

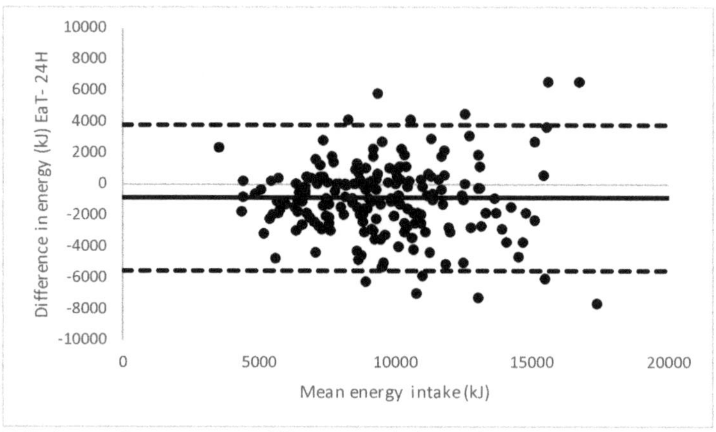

(a) Entire sample energy, mean difference −878 kJ, limits of agreement −5510 to 3755 kJ.

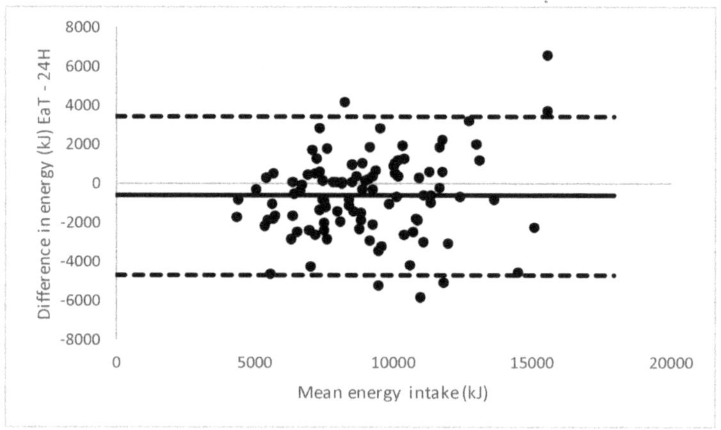

(b) Females for energy, mean difference −607 kJ, limits of agreement −4705 to 3492 kJ.

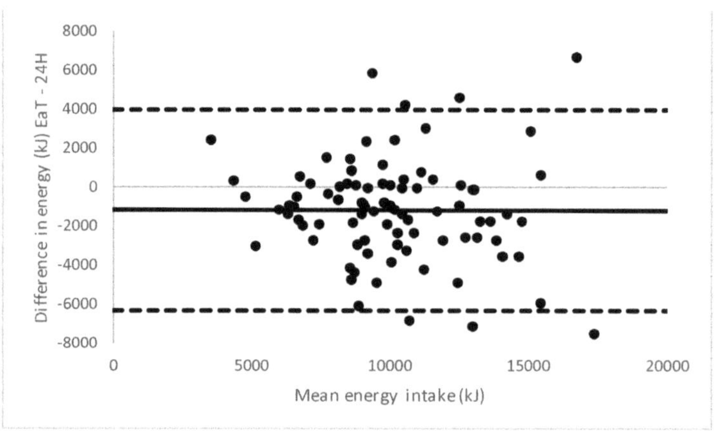

(c) Males for energy, mean difference 1195 kJ, limits of agreement −6339 to 3948 kJ.

Figure 1. Bland–Altman plot of 24-h recalls (24H) and Eat and Track (EaT) app for energy intake. (**a**) Entire sample, (**b**) females and (**c**) males.

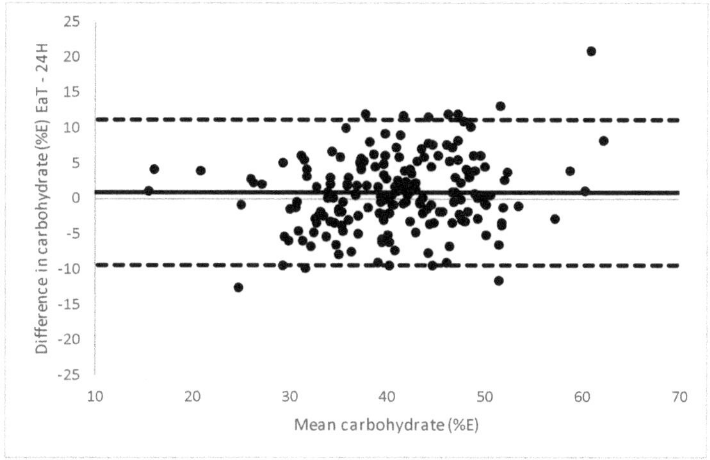

(**a**) Entire sample %E carbohydrate, mean difference 0.9. Limits of agreement −9.5 to 11.2.

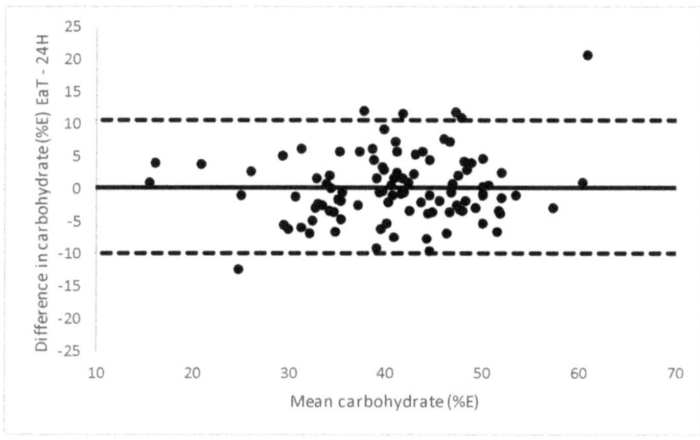

(**b**) Females %E carbohydrate, mean difference 0.3, limits of agreement −9.9 to 10.5.

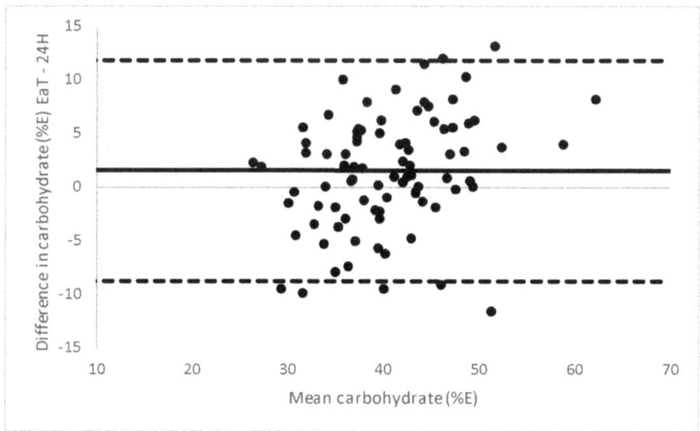

(**c**) Males %E carbohydrate, mean difference 1.5, limits of agreement −8.8 to 11.9.

Figure 2. Bland–Altman plot of 24-h recalls (24H) and Eat and Track (EaT) app for %E carbohydrate. (**a**) Entire sample, (**b**) females and (**c**) males.

Table 4. Agreement between the means of the three days of recording with Eat and Track (EaT) app with the 24-h recalls.

Nutrient	EaT Mean (SD)	24-h Recall Mean (SD)	Mean Difference (SD)	95% LOA [a]
Total energy, kJ	9071 (2908)	9949 (2916)	−878 (2363)	(−5510, 3755)
Protein, % energy	18.8 (5.0)	18.5 (4.5)	0.3 (3.6)	(−6.8, 7.4)
Total fat, % energy	36.0 (7.0)	36.3 (6.8)	−0.3 (6.5)	(−13.0, 12.3)
Saturated fat, % energy	12.7 (3.4)	13.0 (3.4)	−0.3 (3.1)	(−6.3, 5.7)
Carbohydrate, % energy	41.3 (8.6)	40.5 (7.6)	0.9 (5.3)	(−9.5, 11.2)
Sugars, % energy	16.5 (6.5)	16.7 (6.4)	−0.2 (4.1)	(−8.2, 7.9)
Sodium, mg/1000 kJ	299.9 (89.4)	303.3 (102.5)	−3.4 (97.5)	(−194.5, 187.7)

[a] LOA, limits of agreement.

4. Discussion

The present study showed generally good agreement between the EaT app and 24-h recalls for nutrient densities. This finding is based on nonsignificant differences in group intakes with the exception of carbohydrates, acceptable correlation coefficients and cross-classification results. Further, the lack of bias in the Bland–Altman plots, except for carbohydrate in males, suggests that the EaT app is suitable for measuring intakes at the group level. Though there was poor agreement for energy intake, it is well established that self-reported energy intake is not a good measure of true energy intake [33]. However, energy adjustment can be used to improve estimation of nutrients [33], as has been applied in our study.

Similar to the EaT app, the apps that have been the subject of validation studies have shown good correlation with 24-h recalls, though with wide limits of agreement on Bland–Altman tests [13,14,16]. Another study in young adults validated a smartphone app that included text description, and spoken and photographic descriptions of the foods eaten using the objective measure of energy expenditure using the Sensewear armband [34]. The study reported high correlations between the methods [34]. However, it needs to be noted that of 90 participants, 13 either failed to record food intakes or wear the armband for a sufficient period of time and 21 (27%) participants were removed from the analysis because of energy misreporting [34]. As in other validation studies, the sample was mostly young educated women [34].

Due to the issues with reporting of energy intake [33], 24-h recalls are not a true 'gold standard' reference method of dietary intake collection. This study found low levels of underreporting via the dietitian-administered 24-h recall. In the latest Australian Health Survey, the rate of underreporting was 19% of males and 23% of females [35], higher than the rate found in our study (14.8% overall). To better assess the true validity of the EaT app, future studies using biomarkers or doubly labelled water should be conducted [33].

A strength of our study is that our sample included higher proportions of males than previous studies, various education levels and participants from both higher and lower socioeconomic status areas. This 'real-world' approach shows that the EaT app is likely to be useful in a diversity of population groups and may also be developed further into an app for members of the public to record and monitor their intakes.

An advantage of the EaT app in measuring diet with a focus on eating out is that participants are able to choose from a greater number of fast food options than was possible with the 24-h recall, thus increasing their likelihood of selecting the correct item. Using actual portion sizes from the fast food chains should enable better recording of these foods. For other foods, participants received a portions booklet used in national nutrition surveys to estimate serving sizes, but moving forward, inclusion of images within the app may be advantageous. There are some inherent limitations. Due to the ever-changing food supply, databases may only be accurate at one time point and quickly become outdated [36]. Fast food chains frequently offer new menu items to encourage customers into their outlets [37]. In addition, some of the differences between the EaT app and 24-h recalls may be explained by the differences in the databases used for the 24-h recalls with the ASA-24 Australia containing mostly generic fast food options [18]. Not only were there many more fast foods to choose from in the EaT app, but the fast foods in the EaT app had greater nutrient ranges and higher maximum values [18,22].

Prospective dietary data collection methods, such as the EaT app, do not rely on participants' memories, which may be advantageous [6]. However, the selection of this method also introduces a limitation [14]. Requiring participants to record their intake in the EaT app in real time may improve the accuracy of the following 24-h recalls. However, the EaT app clears its history at 3:00 am each day [18], so participants were not able to access their data from the day before.

Overall, EaT is a promising method of collecting dietary intake data of young adults, with a particular focus on eating out. The EaT app could be used to collect data investigating the types and contributions of nutrients from different types of food outlets, and investigate effects of environmental interventions in fast food chain outlets.

5. Conclusions

The Eat and Track smartphone application is a valid way of collecting group nutrient density intake data in 18-to-30-year olds, with a specific focus on the nutrients of interest when frequently eating out, that is, sugars, saturated fat and sodium. To further assess the validity of the app, additional methods that do not rely on food and beverage capture and nutrient databases, such as biomarker or doubly labelled water studies, should be conducted.

Author Contributions: M.A.-F. and L.W.-C. conceived the study. L.W.-C., A.D., A.W. and S.H. collected the data. L.W.-C., A.D. and J.C. cleaned the data. A.W. and S.H. conducted preliminary analysis. L.W.-C., J.C., A.R. and M.A.-F. finalised the analysis approach, and L.W.-C., A.D., A.R. and J.C. conducted the final analysis. L.W.-C. drafted the manuscript. J.C., A.R. and M.A.-F. provide significant input into the manuscript. All authors critically reviewed and approved the manuscript.

Funding: This research was funded by and Australian Research Council Linkage Grant, grant number LP150100831 and Cancer Council NSW.

Acknowledgments: The authors would like to acknowledge the other members of the MYMeals Study Team, particularly Jisu Jung and Judy Kay for the development of the EaT app. We would like to acknowledge all the participants in the MYMeals Study.

Conflicts of Interest: The authors declare no conflict of interest. Personnel from Cancer Council NSW are on the broader study team, and contributed to the design of the overall MYMeals Study and recruitment of participants. However they were not involved in this validation study. The funders had no role in the analyses, or interpretation of data; writing of the manuscript, or in the decision to publish the results.

References

1. Allman-Farinelli, M.A.; Chey, T.; Bauman, A.E.; Gill, T.; James, W.P. Age, period and birth cohort effects on prevalence of overweight and obesity in Australian adults from 1990 to 2000. *Eur. J. Clin. Nutr.* **2008**, *62*, 898–907. [CrossRef] [PubMed]
2. Laska, M.N.; Hearst, M.O.; Lust, K.; Lytle, L.A.; Story, M. How we eat what we eat: Identifying meal routines and practices most strongly associated with healthy and unhealthy dietary factors among young adults. *Public Health Nutr.* **2015**, *18*, 2135–2145. [CrossRef]
3. Mohr, P.; Wilson, C.; Dunn, K.; Brindal, E.; Wittert, G. Personal and lifestyle characteristics predictive of the consumption of fast foods in Australia. *Public Health Nutr.* **2007**, *10*, 1456–1463. [CrossRef]
4. Australian Bureau of Statistics. 6530.0-Household Expenditure Survey, Australia: Summary of Results, 2015–16. Available online: http://www.webcitation.org/72AkC0CBc (accessed on 9 July 2018).
5. Wellard-Cole, L.; Jung, J.; Kay, J.; Rangan, A.; Chapman, K.; Watson, W.L.; Hughes, C.; Ni Mhurchu, C.; Bauman, A.; Gemming, L.; et al. Examining the Frequency and Contribution of Foods Eaten Away from Home in the Diets of 18- to 30-Year-Old Australians Using Smartphone Dietary Assessment (MYMeals): Protocol for a Cross-Sectional Study. *JMIR Res. Protoc.* **2018**, *7*, e24. [CrossRef] [PubMed]
6. Allman-Farinelli, M. Using digital media to measure diet. *Cab Rev.* **2018**, *13*. [CrossRef]
7. Deloitte Australia. *Smart Everything, Everywhere. Mobile Consumer Survey 2017. The Australian Cut*; Deloitte Australia: Sydney, Australia, 2017.

8.   Rollo, M.E.; Williams, R.L.; Burrows, T.; Kirkpatrick, S.I.; Bucher, T.; Collins, C.E. What Are They Really Eating? A Review on New Approaches to Dietary Intake Assessment and Validation. *Curr. Nutr. Rep.* **2016**, *5*, 307–314. [CrossRef]

9.   Subar, A.F.; Kirkpatrick, S.I.; Mittl, B.; Zimmerman, T.P.; Thompson, F.E.; Bingley, C.; Willis, G.; Islam, N.G.; Baranowski, T.; McNutt, S.; et al. The Automated Self-Administered 24-Hour Dietary Recall (ASA24): A Resource for Researchers, Clinicians and Educators from the National Cancer Institute. *J. Acad. Nutr. Diet.* **2012**, *112*, 1134–1137. [CrossRef] [PubMed]

10.   Chen, J.; Cade, J.E.; Allman-Farinelli, M. The Most Popular Smartphone Apps for Weight Loss: A Quality Assessment. *JMIR Mhealth Uhealth* **2015**, *3*, e104. [CrossRef] [PubMed]

11.   Griffiths, C.; Harnack, L.; Pereira, M.A. Assessment of the accuracy of nutrient calculations of five popular nutrition tracking applications. *Public Health Nutr.* **2018**, *21*, 1495–1502. [CrossRef]

12.   Chen, J.; Berkman, W.; Bardouh, M.; Ng, C.Y.K.; Allman-Farinelli, M. The use of a food logging app in the naturalistic setting fails to provide accurate measurements of nutrients and poses usability challenges. *Nutrition* **2018**. [CrossRef]

13.   Carter, M.C.; Burley, V.J.; Nykjaer, C.; Cade, J.E. 'My Meal Mate' (MMM): Validation of the diet measures captured on a smartphone application to facilitate weight loss. *Br. J. Nutr.* **2013**, *109*, 539–546. [CrossRef] [PubMed]

14.   Rangan, A.M.; O'Connor, S.; Giannelli, V.; Yap, M.L.; Tang, L.M.; Roy, R.; Louie, J.C.; Hebden, L.; Kay, J.; Allman-Farinelli, M. Electronic Dietary Intake Assessment (e-DIA): Comparison of a Mobile Phone Digital Entry App for Dietary Data Collection With 24-Hour Dietary Recalls. *JMIR Mhealth Uhealth* **2015**, *3*, e98. [CrossRef] [PubMed]

15.   Rangan, A.M.; Tieleman, L.; Louie, J.C.; Tang, L.M.; Hebden, L.; Roy, R.; Kay, J.; Allman-Farinelli, M. Electronic Dietary Intake Assessment (e-DIA): Relative validity of a mobile phone application to measure intake of food groups. *Br. J. Nutr.* **2016**, *12*, 2219–2226. [CrossRef]

16.   Ambrosini, G.; Hurworth, M.; Giglia, R.; Trapp, G.; Strauss, P. Feasibility of a commercial smartphone application for dietary assessment in epidemiological research and comparison with 24-h dietary recalls. *Nutr. J.* **2018**, *17*. [CrossRef]

17.   Bucher Della Torre, S.; Carrard, I.; Farina, E.; Danuser, B.; Kruseman, M. Development and Evaluation of e-CA, an Electronic Mobile-Based Food Record. *Nutrients* **2017**, *9*, 76. [CrossRef] [PubMed]

18.   Wellard-Cole, L.; Potter, M.; Jung, J.; Chen, J.; Kay, J.; Allman-Farinelli, M. A Tool to Measure Young Adults' Food Intake: Design and Development of an Australian Database of Foods for the Eat and Track Smartphone App. *JMIR Mhealth Uhealth* **2018**, *6*, e12136. [CrossRef]

19.   Australian Bureau of Statistics. 2033.0.55.001-Census of Population and Housing: Socio-Economic Indexes for Areas (SEIFA), Australia, 2011. Available online: http://www.webcitation.org/6v9YcGNub (accessed on 25 July 2018).

20.   Harris, P.A.; Taylor, R.; Thielke, R.; Payne, J.; Gonzalez, N.; Conde, J.G. Research electronic data capture (REDCap)—A metadata-driven methodology and workflow process for providing translational research informatics support. *J. Biomed. Inform.* **2009**, *42*, 377–381. [CrossRef]

21.   Food Standards Australia New Zealand. AUSNUT 2011–2013. Available online: http://www.webcitation.org/6v9Z4xkkv (accessed on 25 July 2018).

22.   Dunford, E.K.; Wu, J.H.Y.; Wellard-Cole, L.; Watson, W.; Crino, M.; Petersen, K.; Neal, B. A comparison of the Health Star Rating system when used for restaurant fast foods and packaged foods. *Appetite* **2017**, *117*, 1–8. [CrossRef]

23.   Wellard-Cole, L.; Goldsbury, D.; Havill, M.; Hughes, C.; Watson, W.L.; Dunford, E.K.; Chapman, K. Monitoring the changes to the nutrient composition of fast foods following the introduction of menu labelling in New South Wales, Australia: An observational study. *Public Health Nutr.* **2017**, *21*, 1194–1199. [CrossRef]

24.   Australian Bureau of Statistics. *Australian Health Survey Food Model. Booklet*; Australian Bureau of Statistics: Canberra, Australia, 2010.

25.   National Cancer Institute. Epidemiology and Genomics Research Program. Available online: http://www.webcitation.org/6v9ZgmxyQ (accessed on 28 June 2017).

26.   National Cancer Institute. ASA24-Australia. Available online: https://epi.grants.cancer.gov/asa24/respondent/australia.html (accessed on 23 January 2019).

27. National Cancer Institute. Dietary Assessment Primer. Learn More about Energy Adjustment. Available online: https://dietassessmentprimer.cancer.gov/learn/adjustment.html (accessed on 15 January 2019).

28. Altman, D.G.; Bland, J.M. Measurement in Medicine: The Analysis of Method Comparison Studies. *J. R. Stat. Soc.* **1983**, *32*, 307–317. [CrossRef]

29. Schofield, W.N. Predicting basal metabolic rate, new standards and review of previous work. *Hum. Nutr. Clin. Nutr.* **1985**, *39* (Suppl. 1), 5–41.

30. Goldberg, G.R.; Black, A.E.; Jebb, S.A.; Cole, T.J.; Murgatroyd, P.R.; Coward, W.A.; Prentice, A.M. Critical evaluation of energy intake data using fundamental principles of energy physiology: 1. Derivation of cut-off limits to identify under-recording. *Eur. J. Clin. Nutr.* **1991**, *45*, 569–581. [PubMed]

31. Black, A.E. The sensitivity and specificity of the Goldberg cut-off for EI:BMR for identifying diet reports of poor validity. *Eur. J. Clin. Nutr.* **2000**, *54*, 395. [CrossRef]

32. Australian Bureau of Statistics. 2024.0-Census of Population and Housing: Australia Revealed, 2016. Available online: http://www.webcitation.org/6v9Y4HIWI (accessed on 28 June 2017).

33. Subar, A.F.; Freedman, L.S.; Tooze, J.A.; Kirkpatrick, S.I.; Boushey, C.; Neuhouser, M.L.; Thompson, F.E.; Potischman, N.; Guenther, P.M.; Tarasuk, V.; et al. Addressing Current Criticism Regarding the Value of Self-Report Dietary Data. *J. Nutr.* **2015**, *145*, 2639–2645. [CrossRef]

34. Pendergast, F.J.; Ridgers, N.D.; Worsley, A.; McNaughton, S.A. Evaluation of a smartphone food diary application using objectively measured energy expenditure. *Int. J. Behav. Nutr. Phys. Act.* **2017**, *14*. [CrossRef] [PubMed]

35. Australian Bureau of Statistics. Under-reporting in nutrition surveys. In *4363.0.55.001-Australian Health Survey: Users' Guide, 2011–13*; Australian Bureau of Statistics: Canberra, Australia, 2014.

36. Carter, C.M.; Hancock, N.; Albar, A.S.; Brown, H.; Greenwood, C.D.; Hardie, J.L.; Frost, S.G.; Wark, A.P.; Cade, E.J. Development of a New Branded UK Food Composition Database for an Online Dietary Assessment Tool. *Nutrients* **2016**, *8*, 480. [CrossRef] [PubMed]

37. Glanz, K.; Resnicow, K.; Seymour, J.; Hoy, K.; Stewart, H.; Lyons, M.; Goldberg, J. How Major Restaurant Chains Plan Their Menus: The role of profit, demand, and health. *Am. J. Prev. Med.* **2007**, *32*, 383–388. [CrossRef]

Article

Validation of Sensor-Based Food Intake Detection by Multicamera Video Observation in an Unconstrained Environment

Muhammad Farooq [1] , **Abul Doulah** [1], **Jason Parton** [2], **Megan A. McCrory** [3] ,
Janine A. Higgins [4] **and Edward Sazonov** [1,*]

[1] Department of Electrical and Computer Engineering, University of Alabama, Tuscaloosa, AL 35487, USA; mfarooq@crimson.ua.edu (M.F.); adoulah@crimson.ua.edu (A.D.)

[2] Department of Information Systems, Statistics, and Management Sciences, Culverhouse College of Business, University of Alabama, Tuscaloosa, AL 35487, USA; jmparton@cba.ua.edu

[3] Department of Health Sciences, Boston University, Boston, MA 02215, USA; mamccr@bu.edu

[4] Department of Pediatrics, University of Colorado Anschutz Medical Campus Denver, Aurora, CO 80045, USA; janine.higgins@childrenscolorado.org

* Correspondence: esazonov@eng.ua.edu

Received: 4 January 2019; Accepted: 7 March 2019; Published: 13 March 2019

Abstract: Video observations have been widely used for providing ground truth for wearable systems for monitoring food intake in controlled laboratory conditions; however, video observation requires participants be confined to a defined space. The purpose of this analysis was to test an alternative approach for establishing activity types and food intake bouts in a relatively unconstrained environment. The accuracy of a wearable system for assessing food intake was compared with that from video observation, and inter-rater reliability of annotation was also evaluated. Forty participants were enrolled. Multiple participants were simultaneously monitored in a 4-bedroom apartment using six cameras for three days each. Participants could leave the apartment overnight and for short periods of time during the day, during which time monitoring did not take place. A wearable system (Automatic Ingestion Monitor, AIM) was used to detect and monitor participants' food intake at a resolution of 30 s using a neural network classifier. Two different food intake detection models were tested, one trained on the data from an earlier study and the other on current study data using leave-one-out cross validation. Three trained human raters annotated the videos for major activities of daily living including eating, drinking, resting, walking, and talking. They further annotated individual bites and chewing bouts for each food intake bout. Results for inter-rater reliability showed that, for activity annotation, the raters achieved an average (±standard deviation (STD)) kappa value of 0.74 (±0.02) and for food intake annotation the average kappa (Light's kappa) of 0.82 (±0.04). Validity results showed that AIM food intake detection matched human video-annotated food intake with a kappa of 0.77 (±0.10) and 0.78 (±0.12) for activity annotation and for food intake bout annotation, respectively. Results of one-way ANOVA suggest that there are no statistically significant differences among the average eating duration estimated from raters' annotations and AIM predictions (p-value = 0.19). These results suggest that the AIM provides accuracy comparable to video observation and may be used to reliably detect food intake in multi-day observational studies.

Keywords: obesity; dietary assessment; chewing detection; AIM; neural networks; food intake detection; video annotation; sensor validation

1. Introduction

Monitoring and assessment of dietary intake and eating behavior is essential for studying and understanding the factors contributing to obesity and over-weight [1,2]. Traditional approaches of dietary intake assessment utilize self-report methodologies such as 24 h dietary recall [3], food frequency questionnaires [4], and electronic devices for record keeping such as personal data assistants and smart-phones [5]. However, these methods rely heavily on participants' input which results in participant burden and may also result in inaccurate data [6,7]. Over the past decade or so, several automatic food intake detection approaches have been proposed to address the problematic issues associated with self-report by employing different sensing modalities, such as acoustic [8], piezoelectric (e.g., strain gauge) [9–11] and inertial (e.g., accelerometer [11,12]) sensors. Sensor-based approaches require validation for data collection, signal processing, and pattern recognition methods. Many sensors have been validated in laboratory studies; however, validation in unconstrained, free-living or pseudo-free-living environments is required for realistic assessment of sensor performance [13]. For validation, having a robust and objective ground truth metric is essential. Three different methodologies have been widely used for establishment of ground truth data for food intake detection including (1) external observer; (2) push-button by the participant, and (3) video observations of individuals.

External observers have been used extensively to establish ground truth in previous studies. For example, several studies using wearable sensors such as ear-pad microphone [14], acoustic sensor around the neck [15,16] have employed external observers to monitor subjects and manually annotate the collected sensor data. Methods relying on external observers can be labor intensive and may not be accurate for marking the start and end of eating activity as the observers themselves are not involved in the eating activity and mostly rely on visual observation. Another popular approach for ground truth collection is the annotation by the subjects themselves using either pushbutton or mobile apps and have been used in conjunction with a wide variety of sensors such as piezoelectric strain sensor [10,17,18], smart eye-glasses [11,19], and acoustic sensors [20]. The use of push-button by the participants can provide comparatively accurate start and end times of eating activity and therefore could potentially be used for accurate assessment of the developed sensors and related signal processing and pattern recognition methodologies. However, the presence of a push-button can impact the way people would normally eat and interact with their environment (i.e., one hand is always busy with the pushbutton) and could also potentially increase participant burden as well as result in inaccurate labels if the participant is distracted. The accuracy of push-button annotation by participants is also dependent on the participants pushing the button at the correct time (i.e., at the actual start and end times of eating). Therefore, there is a need for assessment methods which do not rely on users.

Another approach for establishing the ground truth data is through video observation of individuals and does not rely on the users. This approach can potentially be used in conjunction with any wearable sensor for monitoring food intake such as chewing and swallowing monitoring systems (piezoelectric strain sensor, swallowing microphones, and electroglottography) [8,21–26], and wrist monitoring systems for tracking bites (for example MEMS gyroscope based system for tracking wrist movements [27], accelerometer present in smart-watches [8]). Video-based annotation methodology has also been utilized in the studies [25,26] for monitoring the feeding behavior of infants in laboratory conditions. A common theme among all the studies which relied on the video observation is the use of a single camera fixated on the participant. This restricts participants to a small defined space, e.g., a dining table, and fails to capture daily activities of the participants. Using a single camera also limits the number of participants that are generally recruited for a study session and usually needs one camera per participant. Video based observations are sensitive to the quality of images/videos taken, orientation of the camera, closeness of the camera to the participant, etc. Another problem associated with video observation is that the results are subjective and dependent on inter- and intra-rater reliability of the human annotators. Therefore, multicamera systems are required which can capture a wide variety of activities performed by the individuals and do not restrict the movements of participants to a designated table/space. At the same time, it is essential to evaluate the inter- and intra-rater reliabilities of the annotation procedure to account for subjectivity of the annotators.

This paper presents results of a study in which multiple participants were monitored simultaneously in a multiroom (4-bedroom) apartment with six cameras installed in different locations. Each participant was wearing a multisensor system called Automatic Ingestion Monitor (AIM [10]) for automatic monitoring of food intake related events. The study was conducted with multiple goals: (1) to establish the reliability of video observations for monitoring food intake bouts using wearable sensors in a pseudo-free-living testing environment; and (2) establish the accuracy of the sensor-based food intake predictions with respect to video observation and evaluate if the AIM sensors can be used as a replacement for video observation in unconstrained environments.

2. Materials and Methods

2.1. Data Collection Protocol

Forty (20 male and 20 female) healthy participants were recruited (aged 24.5 ± 3.4 years; Body Mass Index (BMI) 26.1 ± 5.2 kg/m^2; Mean ± STD). Participants were recruited by advertisements placed around the University of Alabama, Tuscaloosa area and in the University newsletter. Individuals were screened for medical conditions which would impact normal chewing. Those with a history of eating disorders, food allergies or sensitivities, or other conditions which resulted in avoidance of consumption of a wide range of foods (e.g., gluten intolerance, peanut allergy) were excluded from the study. The study protocol was approved by the University of Alabama Institutional Review Board and all individuals provided informed consent before participation in the study.

2.2. Sensor System

Participants were asked to wear a multisensor system AIM (v1.0) [10] comprised of three components: a hand gesture sensor worn on the dominant hand, a piezoelectric strain sensor (LDT0-028K from Measurement Specialties Inc., Hampton, VA, USA) placed on the jaw using medical adhesive, and a data collection module worn around the neck using a lanyard. The hand gesture sensor had an RF transmitter (data sampled at 10 Hz), whereas the data collection module had an RF receiver, and both acted together as proximity sensor to detect characteristic hand to mouth (potential bite) gestures. The data collection module also had preconditioning and signal processing circuitry for the jaw motion sensor (sampled at 1000 Hz). It also included a triaxial accelerometer (ADXL335 from Analog Devices, Norwood, MA, USA) for detecting body acceleration (sampled at 100 Hz). Data from the accelerometer was used for determining physical activity levels. Each participant was also provided with an Android smartphone with a dedicated app to collect data. Data from the data collection module were wirelessly transmitted to the phone via RN-42 Bluetooth module with serial port profile. Details about the sensor system used in this study can be found in [10].

2.3. Experimental Protocol

The observational facility was a 4-bedroom, 3-bathroom apartment with a common living area and kitchen. One of the bedrooms was used by the research staff and therefore, was blocked from access to the participants. Each bedroom had a bed, a study chair and desk; while the living area had a sofa, chairs, dining table, a TV with a game console, and a stationary cycle. The kitchen shelves and refrigerator were fully stocked with daily eating supplies and a variety of different foods (189 items) and the supplies were replenished on regular basis to ensure that none of the items were ever out of stock. A daily inventory was kept of the items consumed. The facility was instrumented with 6 motion-sensitive cameras to capture all the activities performed by the participants. Cameras used in the study were GW-2061IP (GW Security, Inc., El Monte, CA, USA), which provided video recording at fully HD resolution (1080p). The locations of the cameras in the apartment are shown in the Figure 1.

Bathrooms were not monitored due to privacy concerns. Participants were asked to eat only in rooms that were equipped with cameras.

Figure 1. Floorplan of the apartment and placement of the six cameras in the apartment. Cameras were placed such that the area of the coverage is maximized.

Each participant completed the study over three days which were scheduled based on their availability and had an interval of at least three days in between each test day. On any given day, there were no more than three participants in the observational facility. This facilitated interactions among the participants throughout the day, including during meals. On each of the study days, participants reported to the observation facility between 7:00–8:00 a.m. and participated in the experiment until 8:00 p.m. Participants were trained on how to place the piezoelectric strain sensor on the jaw and then the participants self-applied the sensor each study day. For all eating occasions, participants had the option of either eating from the food items available in the apartment's kitchen or to get food on the UA campus at one of the three cafeterias or a food court with multiple fast food vendors. Participants could eat at any time of their choosing, as many times as they wanted, as much as they wanted. They could leave the facility for short periods of time during which they were not monitored. Research assistants kept a record of these times and they were subsequently excluded from the analysis. Upon completion of each study day, participants removed the sensor system and were free to leave.

2.4. Annotation Procedure

To identify the ground truth for each participant's activities, the video recordings were manually annotated by three trained human raters (training described below). The annotation process included two stages—(1) activity annotation and (2) food intake bout annotation. In this case, a food intake bout is defined as a single sitting of eating which involves several bites and chewing bouts and may or may not involve liquid intake. This could be a full meal or a small snack. Figure 2 shows an example of the video screenshot of all six cameras that the raters could see and annotate simultaneously. The activity annotation consisted of identification of six categories: eating food intake bout boundaries, drinking, physically active, physically sedentary, talking, and out of view. Brief definitions of these categories of activities are provided in Table 1a. Some constraints were placed during activity annotation as shown in Table 1b. Out-of-view segments of the videos were not included in the analysis. Start and end time of each activity were recorded.

Figure 2. A snapshot of the software used for video observation and annotation. The annotator can view all six cameras simultaneously and can mark start and end of different activities.

Table 1. (a) Definitions of categories for activity annotation; (b) Constraints placed on activity annotation.

(a)	
Category	**Definition**
Food Intake	Participant was consuming solid food items or solid foods combined with liquids. Eating involved taking bites, chewing, and swallowing of the foods.
Drinking	Participant was consuming just liquids, no bite/chewing were involved.
Physically active	Participant were moving
Physically sedentary	Participant was not in motion, including sitting on the couch/chair, working on the computer or laying down on the bed etc.
Talking	Participant was talking to other participants or talking on the phone.
Out of view	Participant was not in the view of any of the 6 cameras
(b)	
Constraints	**Definition**
1	Participant cannot be physically active and sedentary at the same time.
2	Participant cannot be eating/drinking and talking at the same time.
3	Participant cannot be out of surveillance and physically active at the same time with the exception that when the participant was out with the research assistant getting the food, that was considered as physically active.
4	Restroom use was considered as an out of surveillance category.

After the completion of activity annotation, each food intake bout was further annotated with finer details of individual bites and chewing sequences. Food intake annotations were performed by using a 3-button system and a custom-built software. The 3-button system is shown in Figure 3a, in which button-1 and button-2 were used to indicate bite and chewing events respectively. Additionally, a third button was employed to record potential out of view/frozen video frames. Brief definitions of these categories of events in food intake bout annotation are provided in Table 2. Figure 3b shows an example of the annotation procedure both at activity level and food intake level. For a typical food intake bout, a bite is followed by a sequence of chews and one or more swallows. Swallowing events were difficult to see in the video recordings; therefore, they were not annotated. There were cases where video frames were lost and the transition among the frames was not smooth. This manifested as frozen image frames. Timestamps corresponding to these frames was noted and they were not included in the analysis.

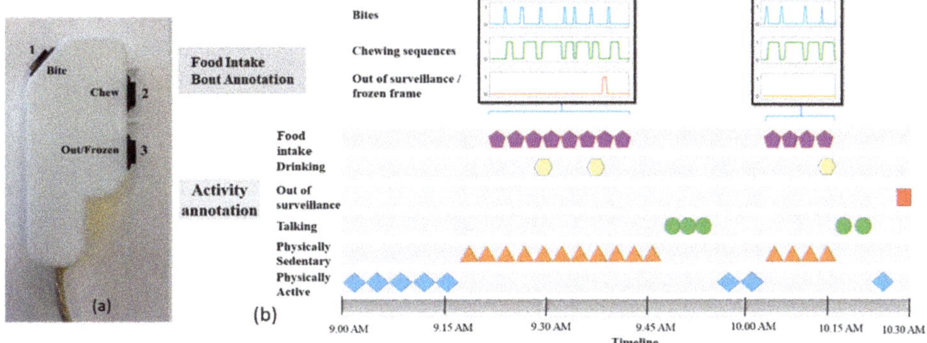

Figure 3. (**a**) The three button systems for annotating the videos of food intake both act activity level as well as meal level; (**b**) Example of the annotation procedure both at the activity and food intake bout level.

Table 2. Definitions of categories for food intake bout annotation.

Category	Definition
Bite	The moment the participant placed the food into mouth and bit down.
Chewing bout	Tracking the jaw movement of the participant immediately after bite until swallowing the food.
Out of view/frozen frame	Frozen video frames or out of camera view (i.e., the participant was not in the selected camera)

2.5. Training of Human Anotators

All the raters were trained before conducting annotation on the full dataset. During training, the raters were provided with specific instructions and supervised by an expert. As a part of activity annotation training, the raters annotated 10 h of video recording. The full day video was played at a high playback speed ($\times 8$) and raters were instructed to pause the video at times when any of the six activities took place. To improve annotation, the raters used rewinding and forwarding of the frames when necessary to identify the start and end times of any category. In addition, raters also used time-stamp information from the research assistant records along with the video observations to annotate videos. Since multiple participants could appear in the camera view, the raters were instructed to complete annotation for one participant at a time and to ignore the other participants who appeared in the video.

Like the activity annotation, raters were given training on use of the 3-button system and custom-built program to annotate food intake bouts. In the training, the raters identified every bite and chewing sequence that took place within a food intake bout. They were instructed to press button-1 once and release immediately each time the participant took a bite. The chewing button was pressed for each entire chewing sequence. The 3rd button was pressed and held for as long as the participant was out of view and for frozen video frames. This process continued until the participant finished the eating event.

2.6. Sensor Signal Processing and Pattern Recognition

One of the goals of the study was to establish the reliability of food intake detection by AIM with respect to the video observations. The same technique can be used for validation of any other sensor for food intake detection. For the validation of the AIM, annotated data was used as reference. Two models for food intake detection were tested. The first model was obtained on an independent dataset trained in a previous study which consisted of a data from 12 participants who wore the AIM device for 24 h [10]. Those participants didn't participate in the current study. Data from the current study were used for testing purposes only. Food intake was detected as 30-s segments labeled as food intake or non-food intake. The data preprocessing and feature computation algorithms were applied to the sensor signals as presented in [10] to ensure that models trained in [10] could be used in this study. The second model utilized the neural network architecture presented in [10], but was trained and validated on data collected during the present study. In this case, a leave-one-participant out cross validation scheme was used, where data from one participant (all days) were used for testing and data collected from the rest of the participants were used for training of the neural networks.

2.7. Statistical Analysis

Statistical comparison was performed to measure the agreement among the raters, and among the video annotation and the AIM-detected food intake. For computing agreement, Cohen's kappa (κ) based inter-rater reliability testing was computed for both activity and food intake bout annotation. The kappa is represented by the following formula:

$$\kappa = \frac{\mathrm{Prob}(a) - \mathrm{Prob}(e)}{1 - \mathrm{Prob}(e)} \tag{1}$$

where $\mathrm{Prob}(a)$ and $\mathrm{Prob}(e)$ represent the probability of observed agreement and expected agreement respectively. The κ can range from -1 to $+1$, where values $\kappa \leq 0$ indicate no agreement, $0.60 < \kappa \leq 0.80$ indicate satisfactory agreement and $\kappa > 0.80$ represent almost perfect agreement.

The inter-rater reliability of the marking of food intake bout boundaries (in the case of activity annotation) and chewing sequences (in the case of food intake annotation) was also evaluated. To evaluate the performance of activity annotation, 1 day of 10 h of video was annotated by each of the three raters after they were trained. For food intake annotation, 10 meals were annotated by each of the three raters.

The following comparisons were performed. To examine inter-rater reliability among the raters, kappa statistics between the three raters were computed and then averaged to obtain Light's kappa. Light's kappa indicates the agreement among the raters when the same day data is annotated by multiple raters. For performance evaluation of the AIM, Light's kappa was used to measure the agreement between the prediction by the AIM and a human rater. For completion, we have also reported the F1-score; which is widely used for performance evaluation of machine learning models. The F1-score is the weighted average of recall and precision. Recall indicates the true positive rate whereas the precision indicates the positive predictive values of the classifier.

Further, a comparison among the average eating duration estimated using the activity level annotation and food intake bout level annotation of the video and AIM prediction is also provided. One-way analysis of variance (ANOVA) was performed with a null hypothesis that average eating duration from all three methods are not statistically different with a p-value of 0.05.

3. Results

For marking food intake events' boundaries in activity annotation, Light's kappa (agreement among the raters) was 0.74. For marking chew sequences in food intake bout annotation, Light's kappa was 0.82. Results of the AIM prediction in comparison to the video annotations are given in Tables 3 and 4. Both activity and meal level predictions from the AIM achieved satisfactory agreement with video annotation (Cohen's kappa of 0.77 and 0.76 respectively, for models trained on the present study dataset). Table 3 also shows the F1-scores achieved by the classifier for both predicting the activity- and meal-level annotations. Table 4 shows the results of AIM prediction when AIM models were trained on the independent dataset from our previous study.

Table 3. Comparison of food intake detection between video based human annotation and AIM predictions based on leave-one-out cross validation.

	Kappa		F1-Score	
	Activity Level	**Food Intake Bout level**	**Activity Level**	**Food Intake Bout Level**
Mean	0.77	0.76	0.8	0.78
STD	0.1	0.12	0.1	0.12

Table 4. Comparison of food intake detection between video based human annotation and AIM predictions based on the model from an earlier study [10].

	Kappa		F1-Score	
	Activity Level	**Food Intake Bout level**	**Activity Level**	**Food Intake Bout Level**
Mean	0.74	0.71	0.77	0.74
STD	0.14	0.11	0.12	0.09

Table 5 shows statistics on the durations of the experiments (from start to end), eating duration marked by the activity level food intake bout annotation, as well as the eating durations predicted by AIM. One-way ANOVA shows that there are no statistically significant differences (p-value 0.19 > 0.05) among the average eating durations (over a day) among activity level annotation, food intake bout level annotation, and the AIM-predicted eating durations.

Table 5. Statistics on Duration of Experiments, Activity, and Food intake bout level eating duration and AIM predicted eating duration. All durations are in minutes.

	Total Duration	**Activity Level (Video)**	**Food Intake Bout Level (Video)**	**AIM Predicted**
Mean	608.6	66.3	37.1	49.4
STD	63.5	30.4	13.4	13.7
25%	589.0	45.9	27.3	40.1
50%	619.3	55.3	35.3	48.3
75%	647.4	78.1	43.4	57.5

4. Discussion

The presented study investigated several issues related to evaluation of wearable sensors for food intake detection in pseudo-free-living environments. Multicamera video observation was used as the gold standard in detection of food intake, instead of relying on pushbuttons which has limitations [10]. As previous research has shown [13], eating behavior varies significantly between strictly controlled laboratory conditions and less restrictive, semi-constrained, or free-living environments. Use of video observation may be a useful tool in establishing the ground truth under the latter conditions.

The use of video-based observation as a means of AIM sensor validation facilitated low participant burden as participants were not required to record their food intake events. Such an approach has multiple advantages. First, not relying on participants to self-report their intake could potentially reduce inaccurate data collection. In addition, presence of multiple cameras did not restrict participants to a confined eating space and they could eat anywhere in the four-bedroom apartment. This approach may have helped the participants mimic their usual daily eating habits, which is desirable in studies of diet and health outcomes.

The inter-rater reliability results for the annotation showed some variability among the raters' perception of eating and not eating. Kappa values of 0.74 (74% agreement) for activity annotation and 0.82 (82% agreement) for food intake bout annotation is good, but not perfect. Although video-based food intake observations have been extensively explored for monitoring in very constrained, laboratory studies, their use in free-living conditions to provide ground truth for wearable sensors may be less reliable as indicated by the kappa for inter-rater reliability metrics. A possible alternative to video-based observation is to use wearable sensors such as the AIM for continuous non-invasive monitoring of eating behavior. Wearable sensors can potentially provide more objective monitoring compared with video-based observations.

Two separate AIM prediction models were tested in this study, and both were compared to video annotation. One model was trained on an independent dataset and the second model was trained on the data collected in the present study. Both models produced results comparable to video annotation, with the first (independent) model resulting in kappa values of 0.74 for activity and 0.71 for meal level annotation. As expected, the recognition model trained on the present dataset had relatively higher agreement (0.77 and 0.76 agreement with raters for activity and food intake bout level annotation respectively) compared to the AIM models trained on independent data. In comparison, inter-rater agreement among raters was 0.74 and 0.82 for activity and food intake bout level, respectively.

One of the possible factors contributing to the strong but not perfect agreement between the AIM detection and the video annotation is the granularity of the epoch size (30 s) used for sensor data processing. This granularity was greatly improved in more recent iterations of the AIM devices [19,28], which were not available at the time of the present experiment. Another source of error is the discrepancy in the observer's ratings which, in turn, affected the fidelity of the AIM predictions. The moderate agreement among raters for video annotation and hence the AIM performance may be attributed to several factors. In some cases, very short snacking events such as eating a small piece of candy may have been missed by the raters. However, such short events were likely captured by the AIM since the AIM is continuously monitoring food intake. Disagreement between the video observation and the AIM could also potentially be explained by constraint # 2 (Table 1b) imposed on the annotation where it was decided that the eating and talking could not be annotated simultaneously. This was because when participants were sitting far from the camera, raters had to zoom in to view the participant, making the view granular and blurry. Raters faced difficulties distinguishing between food intake or talking in such blurred frames. While this could have potentially introduced inaccuracies in the annotation, the AIM would still be likely to capture chewing events during talking if chewing lasts longer than 15 s in a 30-s epoch. A previous study showed that the AIM is able to detect chewing while talking [19]. Another major limitation of identifying ground truth through video observation was the confidence (or lack of thereof) of human raters in their correct identification of the activity shown on video. Many human activities are complex and do not fall easily into predefined categories. Similarly, the raters' expectancy (see what one wants to see) may also have contributed to error. In previous studies ([10,22]), the AIM was to able to distinguish between eating and other activities such as talking and walking etc. and therefore is potentially less prone to the difficulties encountered in video-based observation.

The average experiment duration for all participants was about 10 h out of which about 1 h (66.1 min) (based on the activity level annotation) was spent on eating related activities. The estimated average eating time based on the food intake bout level annotation was 37.1 min, whereas the average estimated eating time based on AIM predictions was 49.4 min. Higher AIM predicted eating durations can be explained by the possibility of raters not being able to mark some chewing events due to occlusion or hard to distinguish eating vs. other activities such as talking. Considering the difficulties in annotating fine level chewing, AIM predicted durations are expected to be more than the fine level chewing (food intake) and less than the activity level eating annotations. However, the one-way ANOVA showed that the differences among the average eating duration are not statistically significant. This shows that estimated eating duration from AIM can provide a good estimate of actual eating duration.

A previous shorter study in free-living conditions that used a push button ground truth reference achieved an average F1-score of 89% when tested on 12 participants using the AIM for 24 h [10] compared to an F1-score of 80% for the present study. However, we would expect similar performance provided more accurate ground truth signal is present. Results of the present study showed that the AIM can provide a reliable prediction of food intake and can potentially be used in place of direct video observations which is labor-intensive and prone to error. Sensors used for passive and automatic detection and identification of food intake have previously been shown to be able to accurately estimate chew counts and chewing rate [22]. Li et al. has shown that increasing number of chews per bite in both obese and healthy participant reduced overall food intake [29]. AIM-like devices can be used for providing near real-time feedback on chewing behavior of individuals and have shown to modify eating behavior to reduce energy intake in a single meal [30]. Similar sensors have been shown to be able to estimate mass of intake only by monitoring chewing behavior [14,31]. While the AIM can accurately detect eating events and can provide information about chewing behavior, in its current form, it does not have the ability to recognize the type of food being consumed which is critical for monitoring caloric intake. Further, integration of computer vision techniques for identification of food type will greatly improve the practical usage of the AIM and similar wearable systems. The present study, together with previous studies in this area, show that wearable systems can be used for not only detecting food intake but also providing other valuable information about eating behavior including quantification of eating rate, duration, and frequency.

5. Conclusions

Human raters achieved an average kappa value of 0.74 and 0.82 for higher level activity annotation and for finer food intake bout level annotation of eating occasions. The AIM predictions were compared with the human raters and achieved a kappa value of 0.8 for detection of food intake. AIM-predicted average eating durations were close to video annotated eating durations. These results indicate that the AIM can potentially be used in studies of food intake in unrestricted environments and provide performance like video annotation without the limitations associated with video annotation.

Author Contributions: Conceptualization, J.P., M.A.M., J.A.H. and E.S.; Data curation, M.F. and A.D.; Funding acquisition, M.A.M., J.A.H. and E.S.; Investigation, M.F.; Methodology, M.F. and E.S.; Project administration, E.S.; Software, M.F.; Supervision, J.P., M.A.M. and E.S.; Validation, M.F. and A.D.; Visualization, M.F. and A.D.; Writing—original draft, M.F. and A.D.; Writing—review & editing, J.P., M.A.M., J.A.H. and E.S.

Funding: Research reported in this publication was supported by the National Institute of Diabetes and Digestive and Kidney Diseases of the National Institute of Health under Award Number R01DK100796. The content is solely the responsibility of the authors and does not necessarily represent the official views of the National Institute of Health.

Conflicts of Interest: The authors declare no conflict of interest.

References

1. Forslund, H.B.; Lindroos, A.K.; Sjöström, L.; Lissner, L. Meal patterns and obesity in Swedish women—A simple instrument describing usual meal types, frequency and temporal distribution. *Eur. J. Clin. Nutr.* **2002**, *56*, 740–747. [CrossRef] [PubMed]
2. Dhurandhar, N.V.; Schoeller, D.; Brown, A.W.; Heymsfield, S.B.; Thomas, D.; Sørensen, T.I.; Speakman, J.R.; Jeansonne, M.; Allison, D.B.; Energy Balance Measurement Working Group. Energy balance measurement: When something is not better than nothing. *Int. J. Obes.* **2015**, *39*, 1109–1113. [CrossRef] [PubMed]
3. Jonnalagadda, S.S.; Mitchell, D.C.; Smiciklas-Wright, H.; Meaker, K.B.; Van Heel, N.A.N.C.Y.; Karmally, W.; Ershow, A.G.; Kris-Etherton, P.M. Accuracy of Energy Intake Data Estimated by a Multiplepass, 24-hour Dietary Recall Technique. *J. Am. Diet. Assoc.* **2000**, *100*, 303–311. [CrossRef]
4. Day, N.; McKeown, N.; Wong, M.; Welch, A.; Bingham, S. Epidemiological assessment of diet: A comparison of a 7-day diary with a food frequency questionnaire using urinary markers of nitrogen, potassium and sodium. *Int. J. Epidemiol.* **2001**, *30*, 309–317. [CrossRef] [PubMed]
5. Beasley, J.M.; Riley, W.T.; Davis, A.; Singh, J. Evaluation of a PDA-based dietary assessment and intervention program: a randomized controlled trial. *J. Am. Coll. Nutr.* **2008**, *27*, 280–286. [CrossRef]

6.   Black, A.E.; Goldberg, G.R.; Jebb, S.A.; Livingstone, M.B.; Cole, T.J.; Prentice, A.M. Critical evaluation of energy intake data using fundamental principles of energy physiology: 2. Evaluating the results of published surveys. *Eur. J. Clin. Nutr.* **1991**, *45*, 583–599. [PubMed]

7.   Livingstone, M.B.E.; Black, A.E. Markers of the Validity of Reported Energy Intake. *J. Nutr.* **2003**, *133*, 895S–920S. [CrossRef]

8.   Sazonov, E.; Schuckers, S.; Lopez-Meyer, P.; Makeyev, O.; Sazonova, N.; Melanson, E.L.; Neuman, M. Non-invasive monitoring of chewing and swallowing for objective quantification of ingestive behavior. *Physiol. Meas.* **2008**, *29*, 525–541. [CrossRef]

9.   Alshurafa, N.; Kalantarian, H.; Pourhomayoun, M.; Liu, J.J.; Sarin, S.; Shahbazi, B.; Sarrafzadeh, M. Recognition of Nutrition Intake Using Time-Frequency Decomposition in a Wearable Necklace Using a Piezoelectric Sensor. *IEEE Sens. J.* **2015**, *15*, 3909–3916. [CrossRef]

10.   Fontana, J.M.; Farooq, M.; Sazonov, E. Automatic Ingestion Monitor: A Novel Wearable Device for Monitoring of Ingestive Behavior. *IEEE Trans. Biomed. Eng.* **2014**, *61*, 1772–1779. [CrossRef]

11.   Farooq, M.; Sazonov, E. A Novel Wearable Device for Food Intake and Physical Activity Recognition. *Sensors* **2016**, *16*, 1067. [CrossRef] [PubMed]

12.   Rahman, S.A.; Merck, C.; Huang, Y.; Kleinberg, S. Unintrusive Eating Recognition Using Google Glass. In Proceedings of the 9th International Conference on Pervasive Computing Technologies for Healthcare, ICST, Istanbul, Turkey, 20–23 May 2015; pp. 108–111.

13.   Doulah, A.; Yang, T.; Parton, J.; Higgins, J.A.; McCrory, M.; Sazonov, E. The Importance of Field Experiments in Testing of Sensors for Dietary Assessment and Eating Behavior Monitoring. In Proceedings of the 2018 40th Annual International Conference of the IEEE Engineering in Medicine and Biology Society (EMBC), Honolulu, HI, USA, 18–21 July 2018.

14.   Amft, O.; Kusserow, M.; Troster, G. Bite Weight Prediction From Acoustic Recognition of Chewing. *IEEE Trans. Biomed. Eng.* **2009**, *56*, 1663–1672. [CrossRef] [PubMed]

15.   Yatani, K.; Truong, K.N. BodyScope: A Wearable Acoustic Sensor for Activity Recognition. In Proceedings of the 2012 ACM Conference on Ubiquitous Computing, New York, NY, USA, 5–8 September 2012; pp. 341–350.

16.   Olubanjo, T.; Ghovanloo, M. Tracheal activity recognition based on acoustic signals. In Proceedings of the 2014 36th Annual International Conference of the IEEE Engineering in Medicine and Biology Society (EMBC), Chicago, IL, USA, 26–30 August 2014; pp. 1436–1439.

17.   Fontana, J.M.; Farooq, M.; Sazonov, E. Estimation of feature importance for food intake detection based on Random Forests classification. In Proceedings of the 2013 35th Annual International Conference of the IEEE Engineering in Medicine and Biology Society (EMBC), Osaka, Japan, 3–7 July 2013; pp. 6756–6759.

18.   Farooq, M.; Sazonov, E. Comparative testing of piezoelectric and printed strain sensors in characterization of chewing. In Proceedings of the 2015 37th Annual International Conference of the IEEE Engineering in Medicine and Biology Society (EMBC), Milan, Italy, 25–29 August 2015; pp. 7538–7541.

19.   Farooq, M.; Sazonov, E. Segmentation and Characterization of Chewing Bouts by Monitoring Temporalis Muscle Using Smart Glasses With Piezoelectric Sensor. *IEEE J. Biomed. Health Inform.* **2017**, *21*, 1495–1503. [CrossRef] [PubMed]

20.   Kalantarian, H.; Alshurafa, N.; Le, T.; Sarrafzadeh, M. Monitoring eating habits using a piezoelectric sensor-based necklace. *Comput. Biol. Med.* **2015**, *58*, 46–55. [CrossRef]

21.   Farooq, M.; Fontana, J.M.; Sazonov, E. A novel approach for food intake detection using electroglottography. *Physiol. Meas.* **2014**, *35*, 739. [CrossRef]

22.   Farooq, M.; Sazonov, E. Automatic Measurement of Chew Count and Chewing Rate during Food Intake. *Electronics* **2016**, *5*, 62. [CrossRef]

23.   Lopez-Meyer, P.; Makeyev, O.; Schuckers, S.; Melanson, E.L.; Neuman, M.R.; Sazonov, E. Detection of Food Intake from Swallowing Sequences by Supervised and Unsupervised Methods. *Ann. Biomed. Eng.* **2010**, *38*, 2766–2774. [CrossRef] [PubMed]

24.   Makeyev, O.; Lopez-Meyer, P.; Schuckers, S.; Besio, W.; Sazonov, E. Automatic food intake detection based on swallowing sounds. *Biomed. Signal Process. Control* **2012**, *7*, 649–656. [CrossRef] [PubMed]

25.   Farooq, M.; Chandler-Laney, P.; Hernandez-Reif, M.; Sazonov, E. A Wireless Sensor System for Quantification of Infant Feeding Behavior. In Proceedings of the Conference on Wireless Health, New York, NY, USA, 14–16 October 2015; pp. 16:1–16:5.

26. Farooq, M.; Chandler-Laney, P.C.; Hernandez-Reif, M.; Sazonov, E. Monitoring of infant feeding behavior using a jaw motion sensor. *J. Healthc. Eng.* **2015**, *6*, 23–40. [CrossRef]

27. Dong, Y.; Scisco, J.; Wilson, M.; Muth, E.; Hoover, A. Detecting periods of eating during free-living by tracking wrist motion. *IEEE J. Biomed. Health Inform.* **2014**, *18*, 1253–1260. [CrossRef]

28. Farooq, M.; Sazonov, E. Accelerometer-Based Detection of Food Intake in Free-Living Individuals. *IEEE Sens. J.* **2018**, *18*, 3752–3758. [CrossRef]

29. Li, J.; Zhang, N.; Hu, L.; Li, Z.; Li, R.; Li, C.; Wang, S. Improvement in chewing activity reduces energy intake in one meal and modulates plasma gut hormone concentrations in obese and lean young Chinese men. *Am. J. Clin. Nutr.* **2011**, *94*, 709–716. [CrossRef] [PubMed]

30. Farooq, M.; McCrory, M.A.; Sazonov, E. Reduction of energy intake using just-in-time feedback from a wearable sensor system. *Obesity* **2017**, *25*, 676–681. [CrossRef] [PubMed]

31. Fontana, J.M.; Higgins, J.A.; Schuckers, S.C.; Bellisle, F.; Pan, Z.; Melanson, E.L.; Neuman, M.R.; Sazonov, E. Energy intake estimation from counts of chews and swallows. *Appetite* **2015**, *85*, 14–21. [CrossRef] [PubMed]

 *nutrients*

Article

Validity and Reproducibility of a Self-Administered Food Frequency Questionnaire for the Assessment of Sugar Intake in Middle-Aged Japanese Adults

Rieko Kanehara [1,2], Atsushi Goto [1,*], Ayaka Kotemori [1], Nagisa Mori [1], Ari Nakamura [1], Norie Sawada [1], Junko Ishihara [1,3], Ribeka Takachi [1,4], Yukari Kawano [5], Motoki Iwasaki [1], Shoichiro Tsugane [1] and for the JPHC FFQ Validation Study Group [1]

[1] Epidemiology and Prevention Group, Centre for Public Health Sciences, National Cancer Centre, Tokyo 104-0045, Japan; rkanehar@ncc.go.jp (R.K.); asunami@ncc.go.jp (A.K.); nagmori@ncc.go.jp (N.M.); ar-nakamura@my-zaidan.or.jp (A.N.); nsawada@ncc.go.jp (N.S.); j-ishihara@azabu-u.ac.jp (J.I.); rtakachi@cc.nara-wu.ac.jp (R.T.); moiwasak@ncc.go.jp (M.I.); stsugane@ncc.go.jp (S.T.)
[2] Department of Food and Nutritional Science, Graduate School of Agriculture, Tokyo University of Agriculture, Tokyo 156-8502, Japan
[3] Department of Food and Life Science, School of Life and Environmental Science, Azabu University, Kanagawa 252-5201, Japan
[4] Department of Food Science and Nutrition, Faculty of Human Life and Environment, Nara Women's University, Nara 630-8506, Japan
[5] Department of Nutritional Science, Faculty of Applied Bioscience, Tokyo University of Agriculture, Tokyo 156-8502, Japan; y1kawano@nodai.ac.jp
* Correspondence: atgoto@ncc.go.jp; Tel.: +81-3-3547-5201 (ext. 3334)

Received: 21 December 2018; Accepted: 26 February 2019; Published: 5 March 2019

Abstract: We evaluated the validity and reproducibility of estimated sugar intakes using a food frequency questionnaire (FFQ) among middle-aged Japanese adults in the Japan Public Health Centre-Based Prospective (JPHC) study. In subsamples of the JPHC study (Cohorts I and II in multiple areas), we computed Spearman's correlations of FFQ results with urine sugar concentrations and dietary records (DR) for validity; we evaluated correlations between two FFQs for reproducibility. During 1994–1998, participants (Cohort I: $n = 27$ [men], $n = 45$ [women]) provided two (spring and fall) 24-h urine samples and completed 7-consecutive-day DR per season (I: $n = 99$, $n = 113$; II: $n = 168$, $n = 171$) and two FFQs (147 food items) at yearly intervals (I: $n = 101$, $n = 108$; II: $n = 143$, $n = 146$). Sugar intakes from FFQ were correlated with urinary sugar (de-attenuated correlations: 0.40; 95%CI: 0.19, 0.58). After adjustment for sociodemographic and lifestyle variables, correlations between FFQ and DR for men and women were 0.57 (0.42, 0.69) and 0.41 (0.24, 0.55) (I) and 0.56 (0.44, 0.65) and 0.34 (0.20, 0.47) (II), respectively. Correlations between FFQs for men and women were 0.63 (0.49, 0.73) and 0.55 (0.41, 0.67) (I) and 0.66 (0.55, 0.74) and 0.63 (0.52, 0.72) (II). In conclusion, our study showed moderate FFQ validity and reproducibility for sugar intake evaluation.

Keywords: food frequency questionnaire; sugar intakes; dietary record; East Asians

---

1. Introduction

The prevalence of obesity and chronic diseases, such as diabetes, is rising steadily worldwide [1,2], leading to increased financial burden from medical expenses and the need to identify preventive measures urgently. The potential role of dietary sugar (especially free or added sugars) consumption in the development of these health conditions has drawn much attention. The World Health Organization (WHO) recommended in the guideline for sugar intake that the intake of free sugars (added or processed sugars, and sugars in honey, syrups, and fruits juices), should be less than 10% of the energy intake [3]. Previous studies, primarily among Westerners, examined associations between the consumption of sugars (mono- and di-saccharide; fructose, glucose, and sucrose) and chronic diseases or conditions, suggesting that the overconsumption of free sugars may lead to chronic diseases [4–6]. Among Japanese populations, however, few previous studies [7,8] have examined the associations between sugars and health conditions. Moreover, according to previous studies on estimations of sugar intakes [9,10], amount and source of dietary sugar consumption among Japanese populations may differ from that among European populations. Owing to these differences between Japanese and Europeans, health impacts of sugar intakes may also differ. Hence, the impact of sugar intakes on the health of the Japanese population merits further investigation.

In Japan, the Ministry of Education, Culture, Sports, Science, and Technology published standard tables of detailed food composition for carbohydrates in 2015 [11]. This further served as a motivation to quantify the dietary intake of sugars among Japanese populations. The food frequency questionnaire (FFQ) is widely used and is less burdensome as a dietary assessment method among study participants than other methods, such as the dietary record (DR). However, it is necessary to verify whether the health impact of nutrient intakes can be accurately estimated using the FFQ [12]. The Japan Public Health Centre-Based Prospective (JPHC) study [13] is a large-scale, nationwide, population-based cohort study with a follow-up period of over 20 years, since its establishment in 1990. In a subsample of the JPHC study, we examined the validity of sugar intakes estimated based on the FFQ, by comparing urinary sugar concentrations as an objective biomarker and DR results for 7 consecutive days per season (28- or 14-d). In addition, the reproducibility was compared using two FFQs completed at a yearly interval.

2. Materials and Methods

2.1. JPHC Validation Study and Participants

The Japan Public Health Centre-Based Prospective (JPHC) study is a prospective cohort study conducted on men and women aged 40 to 69 years. Cohorts I (since 1990) and II (since 1993) were living in five (Ninohe, Yokote, Saku, Ishikawa, and Katsushika) and six (Mito, Kashiwazaki, Chuo-higashi, Kamigoto, Miyako, and Suita) public health centre (PHC) areas, respectively. A 5-year follow-up study was conducted in 1995 (Cohort I) and 1998 (Cohort II) using the FFQ. The FFQ was developed based on weighed 3-d DR survey data from Cohort I participants. Validation studies, for the FFQ, and described previously [14,15], were carried out among a subsample of participants in the JPHC Study Cohorts I and II.

In brief, the Cohort I validation study was performed from February 1994 to February 1996 while Cohort II was performed from May 1996 to February 1998. Participants completed 28 d (14 d for Ishikawa PHC area) DR, they also completed the FFQ twice, while some in Cohort I also collected 24-h stored urine. The FFQ, completed by participants after 3 months of completing the DR, was used for the validation (FFQv). Participants also completed another FFQ (FFQr) at yearly intervals (9-month interval for Mito PHC area) that was used to determine reproducibility. Sample size calculations revealed that approximately 112 participants would be required to detect a CC of 0.25 with $\alpha = 0.05$ and $\beta = 0.20$ separately for men and women, and Cohorts I and II. A total of 120 married couples in Cohort I and 196 married couples in Cohort II were recruited. The participants or their spouses who were out of the age range for the cohorts were excluded. Furthermore, data of the participants without a complete 28 d (14 d for Ishikawa PHC area) DR or FFQv were excluded from validation, while those without a complete FFQr were excluded from reproducibility. Thus, data from a total of 215 participants (102 men and 113 women) from Cohort I and 350 participants (174 men and 176 women) from Cohort II were included for the validation between DR and FFQv. For the calculation of partial correlation coefficients, we further excluded those who had missing data for occupation, smoking status or alcohol intake, leaving a total of 212 participants (99 men and 113 women) from

Cohort I and 339 participants (168 men and 171 women) from Cohort II. Furthermore, 72 participants (27 men and 45 women) were included for the validation between the biomarker and FFQv or DR. From Cohort I, 209 participants (101 men and 108 women) and Cohort II, 289 participants (143 men and 146 women) were included for the reproducibility between FFQv and FFQr [14,15].

All participants gave their oral or written informed consent for participation in the JPHC validation study. The protocol for the current study, including data analysis and the measurement of urinary sugar concentrations, was conducted according to the guidelines laid down in the Declaration of Helsinki and approved by the human ethics review committee of the National Cancer Centre of Japan (No. 2016-428).

2.2. Food Frequency Questionnaire in the 5-Year Follow-Up Survey

The FFQ (which included 147 food items) required information about the usual food consumption during the previous year. Basically, questions about portion size (<0.5 (small)/one (medium)/>1.5 (large) times the reference amount) and frequency (almost never, one to three times per month, one to two times per week, three to four times per week, five to six times per week, once per day, two to three times per day, four to six times per day, and seven or more times per day) were asked. Further questions about consumptions of rice (bowl size/number of bowls per day/consumptions of vitamin reinforced rice and millet), miso soup (number of days eaten per week or month/number of bowls per day/taste intensity), alcohol (number of days drank per week or month/amount per day and types of liquor), supplements (number of tablets per day or week/period), were asked. Additionally, the added sugar and milk for coffee and tea, the usual cooking method, and the amount of noodles soup consumed were also enquired.

2.3. Biomarker for Sugar Intakes

Of 215 participants who completed the DR and the FFQv in Cohort I (the cohort used in developing the FFQ) [16], 72 collected their urine for 24 h. The urine collections were performed for two days (on any day during the 7-d DR period, once in spring and fall). After recording the total volume of the urine collected in a portable device (Urine Mate P, Sumitomo Bakelite, Tokyo, Japan), the urine samples were frozen and stored at $-80\,^{\circ}\text{C}$ [14]. Cohort II participants were not asked to provide their urine samples.

Concentrations of sucrose and fructose in the urine (μg/mL) were measured with a kit (F-kit Sucrose/D-glucose/D-Fructose; Roche/R-Biopharm AG, Darmstadt, Germany) and NanoDrop ND-1000 Spectrophotometer (NanoDrop Technologies, Wilmington, DE, USA). A total of 144 samples (72 participants per season for 2 seasons) were analysed. For quality control, four samples were measured twice, and the quality of the method was assessed. Intra-assay coefficients of variation (CVs) were 4.2% or lower. Other samples were measured once.

2.4. Dietary Records

A total of 565 participants in Cohorts I and II completed the 7 consecutive DR days over each of the four seasons (two seasons [winter and summer], were used for the Ishikawa PHC area because of the subtropical climate during which seasonal variations were likely to be brief). The participants were instructed by the research dietitians to record all foods and beverages prepared and consumed; using a specially developed booklet, they were asked to describe, in as much detail as possible, the methods and recipes used in the preparation. The dietitians checked the records during the survey and reviewed them in a standardized way. Details have been reported elsewhere [14,15].

2.5. Food Composition Table of Carbohydrates, and Nutritional Calculation for FFQ and DR

The 2015 standard tables of food composition in Japan for available carbohydrates include monosaccharides (glucose, fructose, and galactose), disaccharides (sucrose, maltose, lactose, and trehalose), and polysaccharide (starch), and these can be digested and absorbed in the human body [11].

Carbohydrates cover 854 of all 2191 food items in the 2015 standard tables of food composition in Japan [17]. Intakes of glucose, fructose, galactose, sucrose, maltose, lactose, total sugars (sum of these six mono- or disaccharides), and starch from the FFQ and the DR were calculated using the standard tables of food composition for the available carbohydrates [11].

In the FFQ, 75 of 147 food items were covered by the table while 72 were not. Eighteen food items (Table S1) were substituted for by the following methods [18] using different parts of the same species, similar species, or same species with different cooking or purification methods. Among the 54 remaining food items, 48 with <1 g of carbohydrates available per portion size were regarded as containing no carbohydrate. Finally, the remaining six foods items (Table S1) were prepared by dietitians (A.K. and R.K.) using the recipes that were based on the ingredient blending ratio from food manufacturers, cookbooks, and the component values of proteins, lipids, and carbohydrates listed in the appendix in the 2015 Japan standard tables of food composition.

For the FFQ, we also included sugars added to foods during cooking in the calculations by preparing a recipe of the main menus: sugar intakes from table sugar, miso, soy sauce, cooking sake, and sweet cooking rice wine (mirin). First, the main DR menu (e.g., boiled chub mackerel) for each food group (meat, fish, vegetable) and each cooking method (raw, simmered, grilled, fried, stir-fried, others) were selected to cover > 80% of the DR frequent food items. Secondly, selected menu recipes were prepared by dietitians as described above, and sugar intakes were calculated for each of the menu. Thirdly, we calculated the weighted average values of sugar intakes for each of the classifications based on the frequency of occurrence of the menus in the DR because there were multiple menus in the same classification of dishes. For meat (beef, pork, and chicken), the values of sugar intake were calculated using food menus. For fish and vegetable, the values of sugar intake were calculated based on the cooking method (raw, simmered, grilled, fried, stir-fried, and others).

In the DR, a total of 1241 food items were recorded. Of these, 743 were not included in the 2015 standard tables of food compositions for the available carbohydrates. Among food items not included in the table, we substituted 141 foods with different parts of the same species, similar species, same species with different cooking or purification methods (119 food items), or recipes prepared by dietitians (22 food items). The 141 foods included cereals, sugars and sweeteners, pulses, nuts and seeds, vegetables, fruits, milk and milk products, confectionaries, beverages, seasonings and spices, and prepared foods. The remaining food items (602: some vegetables and fruits, mushrooms, algae, fish, meat, eggs, oils and fat, beverages, and seasonings and spices) were not substituted by any other foods. Only twenty-six out of 602 food items contained more than 5 g available carbohydrate, and frequencies of consumption for these foods were extremely low. Therefore, they were considered to have little contribution to the total sugar consumption.

Intakes of energy, protein, fat, and carbohydrate from the FFQ and the DR were calculated using the 2015 Japan standard tables of food composition [17] for reference.

2.6. Statistical Analysis

Major food groups contributing to sugar intakes, by gender, were identified by sugar intakes from the DR. The mean intake and standard deviation (mean ± SD) of glucose, fructose, galactose, sucrose, maltose, lactose, total sugars (sum of these six mono or disaccharides), starch, energy, protein, fat, and carbohydrate from the FFQ and the DR were calculated by gender and by cohort groups. Differences were calculated using the following formula: intakes according to the FFQ—intakes according to the DR. Mean and 95% confidence interval (95%CI) of the differences were calculated. Spearman's rank CCs and Pearson CCs between the FFQv and DR (for validity), and between the FFQv and FFQr (for reproducibility), were calculated for crude and energy-adjusted values of sugar and macronutrient intakes. Correlation coefficients calculated with 95%CI using Fisher's z-transformation.

Energy-adjusted values were estimated using the residual and nutritional density methods. Nutritional density (% energy) was calculated with the following formula: energy intake from sugars/total energy intake × 100. The metabolized energy conversion factor (General Atwater factor) for monosaccharides is 3.75, and the conversion factor from disaccharides to monosaccharides is 1.05 [19]. Moreover, for validity, partial CCs adjusted for age, areas, occupations (primary industry, professionals and office workers, self-employed and others, unemployed), body mass index (BMI), total energy intake, smoking status (never, past, current), and alcohol (nondrinker, ≤ 4 days per week, ≥ 5 days per week) were calculated.

Urinary sugars have been suggested as a useful biomarker to estimate the total sugar intakes, independent of measurement errors from self-reported measures [20]. Spearman's rank CCs and Pearson CCs were calculated to compare between DR or FFQv and the sum of the urinary concentration of sucrose and fructose (urinary sugars). To compare between FFQv and urinary sugars, we used total sugar intakes from the FFQv and the mean urinary concentrations of sugars collected in the spring and fall. For the DR, we compared sugar intakes from the mean 14-d DR in spring and fall with urinary concentrations of sugars, and also compared the mean 7-d DR with urinary sugars separately for the spring and fall. CCs were reported for crude values, energy-adjusted values (for the DR and the FFQv, % energy), and creatinine-adjusted values (for urinary sugars, divided by urinary creatinine concentration (mg/dL)). Furthermore, scatter plots between urinary sugars (creatinine-adjusted) and sugar intakes (%energy) from the FFQ and mean 14-d DR are shown.

Additionally, to correct for within-individual random error, energy-adjusted (% energy) or creatinine-adjusted Spearman's CCs (comparing the FFQv vs. DR; and the urinary sugars vs. the FFQv and DR) were de-attenuated based on the method in SAS macro ("rankcorr_mmer.sas") provided by Dr. Bernard Rosner [12,21,22] using probit transformation and multiplying each with the adjustment factor. The adjustment factors were calculated by using the following formula for FFQv vs. DR and FFQv vs. urinary sugars: $\sqrt{1 + \frac{\lambda}{k}}$, where k is the average DR days for FFQv vs. DR; or the frequency of the urine collection for FFQv vs. urinary sugars; and λ is the ratio of within- to between-subject variance within the 14- or 28-day DR or urinary sugars collected twice, using the random-effects model [12]. For urinary sugars vs. the DR, the adjustment factor was taken into account for the within-individual random errors in measurement of both the DR and urinary sugars. The formula of the adjustment factor was the following: $\sqrt{1 + \frac{\lambda_{dr}}{k_{dr}}} \times \sqrt{1 + \frac{\lambda_{urine}}{k_{urine}}}$, where k_{dr} is the average DR days and k_{urine} is the frequency of the urine collection; and λ_{dr} is the ratio of within- to between-subject variance within the 14- or 28-day DR and λ_{urine} is for urinary sugars [12].

To evaluate intra-subject variations for urinary sugars, the ratio of within- to between-subject variance ($\sigma^2_{ws}/\sigma^2_{bs}$) and intra-class CCs [ICC; $\sigma^2_{bs}/(\sigma^2_{bs} + \sigma^2_{ws})$] of urinary sugars collected in the spring and fall were calculated, using the random-effects model [12]. ICCs of the two FFQs and 28- or 14-d DR were also calculated.

In addition, the proportion of participants, who were classified into the same, adjacent, and extreme categories using the cross classification by quintile [23] for energy-adjusted (% energy) total sugar intakes or creatinine-adjusted urinary sugar, was calculated. The adjacent categories included the proportion of participants who were not in the same category by quintile between the two measurement methods (the FFQv vs. DR or the FFQv vs. urinary sugar), but only in the +1 or −1 difference categories. The extreme categories included the proportion of participants who were misclassified into the opposite side class (for example, the class for the FFQv was the highest, but the DR's was the lowest).

Agreement between total sugar intakes from the FFQv and DR were examined using Bland-Altman analysis. We plotted the mean total sugar intakes from the FFQv and DR on the x-axis, and the difference between them (FFQv and DR) on the y-axis using energy-adjusted (% energy) and log-transformed values. Mean difference $\pm 1.96 \times$ SD was calculated as the limit of agreement [24–26].

For parametric methods such as Pearson CCs and the Bland-Altman analysis, all nutrient intake values were log-transformed to fulfill the assumption of normality. Statistical significance was set at a p value of <0.05. All statistical analyses were implemented in SAS version 9.3.

3. Results

For sugar intakes, contribution proportions by food groups were calculated from the DR (men: $n = 276$, women: $n = 289$). Contribution proportions of fruits, mostly from apples, citrus, bananas, and Japanese persimmons, were the highest for total sugars in both men and women. For women, the proportion of confectioneries in total sugars was higher than that for men (Table 1). In detail, contribution proportions by foods for each of the mono- and di-saccharides and starch are shown in Table S2. The mean (SD) of % energy for total and free sugars was 9.5% (3.3%) and 3.9% (2.3%) for men ($n = 276$), 13.6% (3.2%) and 5.9% (2.3%) for women ($n = 289$). For free sugars, the number of participants who consumed more than 5% was 68 (24.6%) in men and 186 (64.4%) in women. Furthermore, the number of participants who consumed more than 10% was 6 (2.2%) in men and 16 (5.5%) in women.

For validation, participants' characteristics were described in previous studies [14,15]. In short, the mean (SD) age and BMI were 55.6 (5.2) years and 24.3 (3.0) kg/m^2 for men in Cohort I ($n = 102$); 53.3 (5.3) years and 23.9 (3.1) kg/m^2 for women in Cohort I ($n = 113$); 58.9 (7.6) years and 23.7 (2.6) kg/m^2 for men in Cohort II ($n = 174$); and 55.9 (7.1) years and 23.7 (3.2) kg/m^2 for women in Cohort II ($n = 176$). The percentages of participants who had history of diabetes, hypertension, dyslipidaemia, and obesity (BMI $\geq$ 25 kg/m^2) were 7.8%, 18.6%, 5.9%, and 42.2% for men in Cohort I; 3.5%, 22.1%, 8.0%, and 31.0% for women in Cohort I; 8.1%, 20.1%, 4.6%, and 28.2% for men in Cohort II; and 1.1%, 17.6%, 6.3%, and 29.6% for women in Cohort II, respectively.

Table 1. Major food groups contributing to sugar intakes from the dietary records (Cohorts I and II).

Men ($n = 276$)		Women ($n = 289$)	
Food Groups	**(%)**	**Food Groups**	**(%)**
Total sugars			
Fruits	21.3	Fruits	24.4
Apples	(5.0)	Apples	(5.5)
Citrus	(3.9)	Citrus	(5.0)
Bananas	(3.0)	Japanese persimmons	(3.5)
Vegetables	14.3	Confectionaries	19.3
Onions	(2.1)	Traditional fresh and semi-dry confectionery	(10.8)
Carrots	(1.8)	Cake and pastry	(2.7)
Japanese Radishes	(1.8)	Traditional dry confectionery	(1.6)
Sugars and sweeteners	14.2	Milk and milk products	13.1
Sugars	(13.6)	Liquid milk	(8.6)
Honey and syrup	(0.6)	Yogurt	(2.4)
-	-	Ice cream	(1.5)
Confectionaries	13.4	Sugars and sweeteners	12.4
Traditional fresh & semi-dry confectionery	(8.1)	Sugars	(11.8)
Cake & pastry	(1.5)	Honey and syrup	(0.5)
Bun with filling	(1.3)	-	-
Milk and milk products	11.1	Vegetables	12.3
Liquid milk	(8.0)	Onions	(1.7)
Yogurt	(1.7)	Pumpkin and squash	(1.5)
Ice cream	(1.0)	Carrots	(1.5)
Non-alcoholic beverages	8.3	Non-alcoholic beverages	5.9
Carbonated beverage	(3.6)	Carbonated beverage	(1.8)
Coffee	(2.1)	Lactic acid bacteria beverage	(1.4)
Lactic acid bacteria beverage	(1.0)	Fruit drinks	(1.0)
Seasonings	4.8	Seasonings	3.7
Miso	(2.8)	Miso	(2.1)
Japanese Worcester sauce	(0.6)	-	-
Soy sauce	(0.5)	-	-
Alcohol	4.5		
Fermented alcoholic beverage	(2.5)		
Compound alcoholic beverage	(2.0)		
-	-		

Table 1. *Cont.*

Men (*n* = 276)		Women (*n* = 289)	
Food Groups	**(%)**	**Food Groups**	**(%)**
Cereals	3.3		
Bread	(1.9)		
Rice	(0.8)		
Noodles	(0.6)		
Glucose			
Vegetables	26.9	Fruits	30.1
Fruits	20.8	Vegetables	28.0
Alcohol	19.4	Seasonings	14.3
Seasonings	15.1	Non-alcoholic beverages	8.4
Non-alcoholic beverages	8.7	Alcohol	8.4
Cereals	3.9	Cereals	4.1
Fructose			
Fruits	41.4	Fruits	48.2
Vegetables	32.8	Vegetables	28.3
Non-alcoholic beverages	13.1	Non-alcoholic beverages	11.1
Cereals	4.5	Cereals	4.6
Seasonings	3.6		
Galactose			
Seasonings	55.6	Milk and milk products	57.2
Milk and milk products	43.8	Seasonings	42.1
Sucrose			
Sugars and sweeteners	27.8	Confectionaries	34.7
Confectionaries	25.9	Sugars and sweeteners	22.6
Fruits	20.1	Fruits	21.3
Non-alcoholic beverages	7.9	Vegetables	4.9
Vegetables	5.9	Non-alcoholic beverages	4.5
Milk and milk products	3.0	Milk and milk products	4.1
Maltose			
Cereals	41.4	Cereals	32.5
Potatoes	21.5	Potatoes	27.0
Confectionaries	16.8	Confectionaries	22.3
Alcohol	8.8	Alcohol	6.3
Milk and milk products	4.2	Milk and milk products	5.4
Lactose			
Milk and milk products	93.0	Milk and milk products	93.0
Non-alcoholic beverages	3.2	Confectionaries	3.7
Starch			
Cereals	90.7	Cereals	85.0
Confectionaries	3.6	Confectionaries	7.0
		Potatoes	3.7

Food groups contributing to at least 3% of sugars intakes were listed. For total sugars, the top three contributing foods were listed. Non-alcoholic beverages category included 100% fruit juices (including reconstituted fruit juices), fruit drinks (less than 100% fruit juices), lactic acid bacteria beverages, coffee flavoured milk beverages, maccha, coffee, cocoa, and carbonated beverages.

3.1. Validation Using Biomarkers as a Reference

Sugar intake assessed with the DR and FFQv is shown in Table 2. Urinary sugar concentrations were correlated with total sugars (% energy) from the FFQv (de-attenuated Spearman's CC: r = 0.40, 95%CI: 0.19, 0.58) (Table 3; Figure 1a); and total sugars (% energy) from the 14-d DR (r = 0.89, 95%CI: 0.82, 0.93) (Table 3; Figure 1b). The $\sigma^2_{ws}/\sigma^2_{bs}$ ratios and ICCs, as measures of intra-subject variation, were high and low for urinary sugars, respectively ($\sigma^2_{ws}/\sigma^2_{bs}$ ratios: 5.62; ICCs: 0.15, *n* = 72).

For comparisons of the total sugars form the FFQv and urinary sugars based on the joint classification by quintile, 63% of the participants were classified into the same or adjacent categories, and 6.0% were classified into the extreme categories (Table S3-1). For details, 16 out of the participants (n = 72) were classified into the same, 45 were classified into the same or adjacent, while 4 were classified into the extreme categories (Tables S3-1 and S3-2).

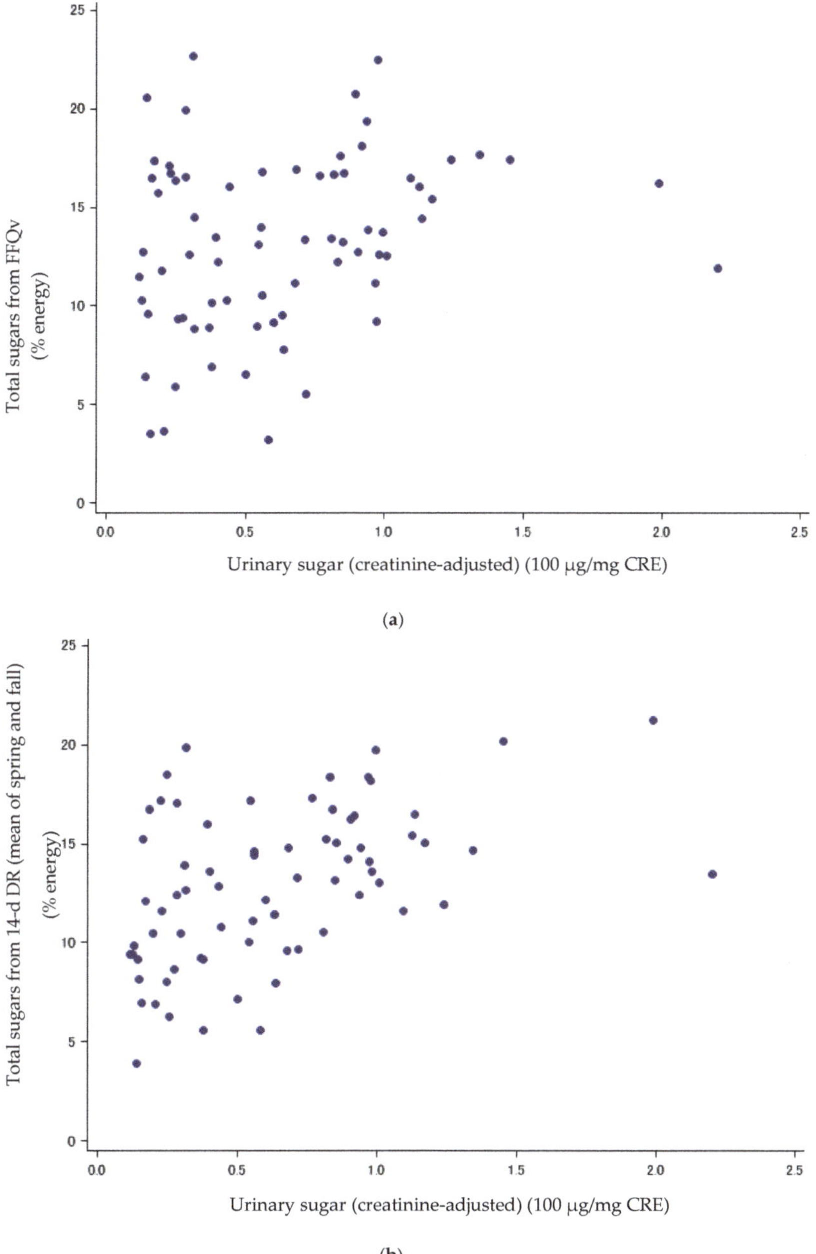

(a)

(b)

Figure 1. (**a**) Scatter plots between urinary sugars and total sugars from FFQv (n = 72, Cohort I). FFQv, food frequency questionnaire for validity. (**b**) Scatter plots between urinary sugars and total sugars from DR (n = 72, Cohort I). DR, dietary record.

Table 2. Sugar intakes assessed with DR for 28 or 14 days and FFQv in Cohort I & II and differences.

	DR			FFQv			Mean of Difference [1]	
	Mean	SD	Median	Mean	SD	Median		(95%CI)
Cohort I								
Men (n = 102)								
Total sugars [2] (g)	56.5	24.7	53.6	69.1	43.0	61.4	12.7	(5.3, 20.0)
Glucose (g)	12.4	5.0	11.8	17.1	9.0	15.5	4.6	(3.2, 6.1)
Fructose (g)	9.7	5.0	8.3	13.9	11.0	11.9	4.2	(2.4, 6.1)
Galactose (g)	0.2	0.3	0.2	0.2	0.4	0.1	0.0	(−0.1, 0.0)
Sucrose (g)	27.7	15.4	24.9	28.1	21.5	21.8	0.5	(−3.4, 4.3)
Maltose (g)	0.9	0.5	0.8	1.1	0.6	0.9	0.1	(0.0, 0.3)
Lactose (g)	5.6	4.6	4.7	8.8	10.2	7.3	3.2	(1.4, 5.0)
Starch (g)	222.0	61.2	204.0	213.1	64.4	203.6	−8.9	(−18.7, 0.9)
Energy (kcal)	2392	435	2372	2408	692	2357	16	(−105, 136)
Protein (g)	91.1	15.6	90.1	88.1	35.6	82.2	−3.0	(−9.1, 3.1)
Fat (g)	59.6	10.9	59.4	61.5	27.1	59.2	1.9	(−3.1, 7.0)
Carbohydrate (g)	323.6	82.4	312.8	325.2	105.0	311.5	1.6	(−14.2, 17.3)
Women (n = 113)								
Total sugars [2] (g)	64.5	22.9	65.2	74.3	52.6	62.3	9.8	(0.7, 19.0)
Glucose (g)	11.4	3.7	11.5	16.4	12.1	13.7	5.0	(2.9, 7.1)
Fructose (g)	10.8	4.4	10.9	16.7	14.9	12.9	5.9	(3.2, 8.5)
Galactose (g)	0.2	0.2	0.2	0.3	0.3	0.1	0.0	(0.0, 0.1)
Sucrose (g)	34.1	14.7	33.5	30.4	24.7	23.4	−3.7	(−8.1, 0.8)
Maltose (g)	1.1	0.6	1.0	1.2	0.8	0.9	0.0	(−0.1, 0.2)
Lactose (g)	6.8	4.0	6.8	9.4	7.7	9.1	2.6	(1.5, 3.7)
Starch (g)	161.6	37.4	160.7	180.1	51.5	172.8	18.5	(11.0, 26.0)
Energy (kcal)	1856	321	1839	2054	846	1916	198	(54, 343)
Protein (g)	75.5	13.0	75.9	82.0	44.1	72.7	6.5	(−1.2, 14.3)
Fat (g)	53.1	10.4	52.5	59.4	32.4	51.7	6.3	(0.3, 12.3)
Carbohydrate (g)	261.9	58.9	264.6	292.0	110.0	278.6	30.1	(12.5, 47.6)

Table 2. *Cont.*

	DR			FFQv			Mean of Difference [1] (95%CI)
	Mean	SD	Median	Mean	SD	Median	
Cohort II							
Men (n = 174)							
Total sugars [2] (g)	57.1	20.2	54.3	67.0	35.9	58.0	9.9 (5.1, 14.7)
Glucose (g)	12.8	4.3	12.5	16.0	8.4	13.7	3.2 (2.1, 4.4)
Fructose (g)	9.6	4.1	9.2	12.5	7.9	10.5	2.9 (1.9, 4.0)
Galactose (g)	0.2	0.2	0.2	0.3	0.4	0.1	0.0 (0.0, 0.1)
Sucrose (g)	27.3	12.9	24.5	27.4	18.5	24.6	0.1 ($-$2.3, 2.5)
Maltose (g)	1.5	1.1	1.2	1.3	0.8	1.0	$-$0.2 ($-$0.4, $-$0.1)
Lactose (g)	5.7	4.4	4.6	9.6	11.9	8.0	3.9 (2.3, 5.4)
Starch (g)	200.6	45.2	192.8	197.0	60.6	184.2	$-$3.7 ($-$11.1, 3.8)
Energy (kcal)	2268	358	2263	2291	690	2215	23 ($-$75, 121)
Protein (g)	86.6	14.9	86.1	82.0	31.6	76.2	$-$4.7 ($-$9.5, 0.1)
Fat (g)	54.8	11.7	54.3	59.6	26.1	54.3	4.9 (0.9, 8.9)
Carbohydrate (g)	306.9	57.5	303.1	304.8	91.7	289.3	$-$2.1 ($-$14.2, 10.0)
Women (*n* = 176)							
Total sugars [2] (g)	63.9	19.7	62.4	69.0	39.4	60.8	5.1 ($-$0.2, 10.5)
Glucose (g)	11.5	3.0	11.4	14.6	8.4	13.2	3.1 (1.9, 4.3)
Fructose (g)	10.5	3.4	10.0	14.2	9.3	12.6	3.7 (2.4, 5.0)
Galactose (g)	0.3	0.2	0.2	0.3	0.5	0.2	0.1 (0.0, 0.1)
Sucrose (g)	32.3	13.6	30.9	28.3	19.3	24.6	$-$4.0 ($-$6.6, $-$1.4)
Maltose (g)	1.8	1.3	1.4	1.2	0.8	1.1	$-$0.6 ($-$0.8, $-$0.4)
Lactose (g)	7.5	4.4	6.9	10.3	10.9	9.3	2.8 (1.4, 4.2)
Starch (g)	150.6	30.9	148.8	166.6	43.6	163.9	16.0 (10.1, 21.9)
Energy (kcal)	1761	263	1753	1929	708	1794	167 (69, 266)
Protein (g)	71.2	11.1	70.9	76.3	34.6	69.9	5.1 (0.2, 10.0)
Fat (g)	48.1	9.9	47.9	57.6	30.2	51.0	9.5 (5.1, 13.9)
Carbohydrate (g)	253.3	45.7	249.0	271.6	87.4	257.8	18.4 (6.6, 30.1)

DR, dietary record; FFQv, food frequency questionnaire for validity; SD, standard deviation; CI, confidence interval. [1] Mean of (intakes from FFQv—intakes from DR). [2] "Total sugars" represents the sum of the crude consumption of the following saccharides: glucose, fructose, galactose, sucrose, maltose, and lactose.

Table 3. The correlations between urinary sugars and dietary sugars by DR or FFQ (*n* = 72, Cohort I).

| | Spearman's Rank Correlation Coefficient | | | | | |
| | Crude [1] | | Adjusted [2] | | De-Attenuated [3] | |
	r	95%CI	r	95%CI	r	95%CI
FFQv vs. urine [4]	0.25	(0.02, 0.45)	0.27	(0.04, 0.47)	0.40	(0.19, 0.58)
14-d DR [5] vs. urine [4]	0.46	(0.25, 0.62)	0.48	(0.28, 0.64)	0.89	(0.82, 0.93)
7-d DR vs. urine (spring)	0.31	(0.08, 0.50)	0.24	(0.01, 0.44)	0.27	(0.04, 0.47)
7-d DR vs. urine (fall)	0.38	(0.16, 0.56)	0.41	(0.19, 0.58)	0.46	(0.26, 0.62)

DR, dietary record; FFQv, food frequency questionnaire for validity; r, correlation coefficient; CI, confidence interval. The $\sigma2ws/\sigma2bs$ ratios: 5.62 and intra-class correlation (ICC): 0.15 for urinary sugars. [1] DR or FFQv (crude) vs. urinary sugars (crude). [2] DR or FFQv (energy-adjusted by nutritional density method (percentage of energy)) vs. urinary sugars (creatinine-adjusted). [3] Adjusted Spearman's correlation coefficients were multiplied using probit transformation with regard to repeats of urinary sugar measures (twice) for FFQv vs. urine, repeats of both urinary sugar measures (twice) and DR measures (14 times) for 14-d DR vs. urine, and repeats of DR measures (7 times) for 7-d DR vs. urine. [4] Urinary sugar was calculated as the mean of the spring and fall values. [5] Dietary sugars from DR were calculated as the mean of the spring and fall values (the same seasons as when the urine was collected).

3.2. Validation Using DR as a Reference

The major sources of total sugars were sucrose, glucose, and fructose. Total sugar intakes were higher among women than men. The SD of sugar intake from the FFQ tended to be larger than in the DR. Overall, total sugar intakes from the FFQv were over-estimated when compared to the DR (Table 2). For the energy-adjusted (% energy) total sugars, Spearman's CCs (95%CI) were 0.64 (0.50, 0.74) for men and 0.48 (0.32, 0.61) for women in Cohort I; 0.62 (0.52, 0.71) for men and 0.37 (0.23, 0.49) for women in Cohort II (Table S4, Figure S1). Results became slightly weaker after adjusting for age, areas, occupations, BMI, total energy intake, smoking status, and alcohol intake; partial Spearman's CCs (95%CI) were 0.57 (0.42, 0.69) for men and 0.41 (0.24, 0.55) for women in Cohort I; 0.56 (0.44, 0.65) for men and 0.34 (0.20, 0.47) for women in Cohort II (Table 4). The CCs were moderate, and higher in Cohort I than in Cohort II, and higher for men than for women. De-attenuated Spearman's CCs based on the probit transformation method were slightly stronger (Table S5). Pearson CCs also showed moderate correlations, and de-attenuated Pearson CCs were slightly stronger (not shown in tables). Furthermore, in any of the cohorts by gender, the differences did not depend on the magnitude of the mean total sugar intakes (Figure S2).

For comparisons of the FFQv and DR sugars based on the cross classification by quintile, about 80% men and 70% women were classified into the same or adjacent categories of sugar intakes (total sugars), and less than 6.0% of men and women were classified into the opposite extreme categories (Table S6).

3.3. Reproducibility

For reproducibility, participants' characteristics were described in previous studies [15,27]. For almost all of the sugars, estimated intakes from the FFQr were neither over- nor under-estimated when compared to the FFQv (Table 5). For total sugars (% energy), Spearman's CCs (95%CI) were 0.63 (0.49, 0.73) for men and 0.55 (0.41, 0.67) for women in Cohort I; and 0.66 (0.55, 0.74) for men and 0.63 (0.52, 0.72) for women in Cohort II. The CCs were moderate and slightly lower for women (Table 6).

Table 4. Partial Correlations between FFQv and DR for 28 or 14 days (Validity).

| | Partial Spearman's Rank Correlation Coefficient [1] | | | | | |
| | Crude | | Residual Method [2] | | Density Method [3] | |
	r	95%CI	r	95%CI	r	95%CI
Cohort I						
Men (*n* = 99)						
Total sugars [4]	0.49	(0.33, 0.63)	0.52	(0.36, 0.65)	0.57	(0.42, 0.69)
Glucose	0.52	(0.36, 0.65)	0.56	(0.41, 0.68)	0.57	(0.42, 0.69)
Fructose	0.54	(0.38, 0.67)	0.54	(0.39, 0.67)	0.55	(0.40, 0.68)
Galactose	0.29	(0.10, 0.46)	0.48	(0.31, 0.62)	0.36	(0.17, 0.52)
Sucrose	0.45	(0.28, 0.60)	0.46	(0.29, 0.60)	0.51	(0.35, 0.64)
Maltose	0.41	(0.23, 0.56)	0.45	(0.27, 0.59)	0.46	(0.29, 0.60)
Lactose	0.61	(0.48, 0.72)	0.54	(0.38, 0.66)	0.56	(0.41, 0.68)
Starch	0.53	(0.38, 0.66)	0.42	(0.24, 0.57)	0.37	(0.19, 0.53)
Protein	0.19	(−0.01, 0.37)	0.25	(0.06, 0.43)	0.25	(0.06, 0.43)
Fat	0.28	(0.09, 0.46)	0.39	(0.21, 0.55)	0.40	(0.22, 0.55)
Carbohydrate	0.37	(0.19, 0.53)	0.44	(0.26, 0.58)	0.46	(0.29, 0.60)
Women (*n* = 113)						
Total sugars [4]	0.26	(0.07, 0.42)	0.36	(0.19, 0.51)	0.41	(0.24, 0.55)
Glucose	0.19	(0.01, 0.37)	0.26	(0.08, 0.42)	0.29	(0.11, 0.45)
Fructose	0.29	(0.11, 0.45)	0.31	(0.13, 0.47)	0.35	(0.17, 0.50)
Galactose	0.51	(0.36, 0.63)	0.49	(0.33, 0.62)	0.51	(0.36, 0.63)
Sucrose	0.23	(0.04, 0.39)	0.38	(0.21, 0.53)	0.35	(0.18, 0.50)
Maltose	0.31	(0.13, 0.47)	0.41	(0.25, 0.55)	0.37	(0.20, 0.52)
Lactose	0.69	(0.57, 0.77)	0.67	(0.56, 0.76)	0.66	(0.54, 0.75)
Starch	0.30	(0.12, 0.46)	0.44	(0.28, 0.58)	0.32	(0.14, 0.48)
Protein	0.18	(0.00, 0.36)	0.28	(0.10, 0.44)	0.27	(0.09, 0.43)
Fat	0.20	(0.01, 0.37)	0.44	(0.28, 0.58)	0.42	(0.25, 0.56)
Carbohydrate	0.20	(0.01, 0.37)	0.44	(0.28, 0.58)	0.39	(0.22, 0.54)
Cohort II						
Men (*n* = 168)						
Total sugars [4]	0.42	(0.28, 0.53)	0.56	(0.45, 0.66)	0.56	(0.44, 0.65)
Glucose	0.27	(0.13, 0.41)	0.44	(0.31, 0.56)	0.44	(0.30, 0.55)
Fructose	0.42	(0.28, 0.53)	0.53	(0.41, 0.63)	0.52	(0.40, 0.62)
Galactose	0.66	(0.57, 0.74)	0.64	(0.55, 0.73)	0.66	(0.56, 0.74)
Sucrose	0.42	(0.29, 0.54)	0.52	(0.40, 0.62)	0.53	(0.41, 0.63)
Maltose	0.17	(0.02, 0.31)	0.26	(0.11, 0.39)	0.29	(0.15, 0.42)
Lactose	0.75	(0.68, 0.81)	0.74	(0.66, 0.80)	0.74	(0.66, 0.80)
Starch	0.43	(0.30, 0.55)	0.55	(0.44, 0.65)	0.46	(0.33, 0.57)
Protein	0.16	(0.01, 0.30)	0.41	(0.27, 0.53)	0.36	(0.23, 0.49)
Fat	0.21	(0.06, 0.35)	0.52	(0.40, 0.62)	0.48	(0.35, 0.59)
Carbohydrate	0.37	(0.23, 0.49)	0.55	(0.43, 0.65)	0.48	(0.36, 0.59)
Women (*n* = 171)						
Total sugars [4]	0.23	(0.08, 0.37)	0.38	(0.25, 0.51)	0.34	(0.20, 0.47)
Glucose	0.26	(0.12, 0.40)	0.30	(0.16, 0.43)	0.30	(0.16, 0.43)
Fructose	0.31	(0.17, 0.44)	0.32	(0.18, 0.45)	0.32	(0.18, 0.45)
Galactose	0.58	(0.47, 0.67)	0.62	(0.52, 0.70)	0.63	(0.53, 0.71)
Sucrose	0.18	(0.03, 0.32)	0.33	(0.18, 0.45)	0.30	(0.16, 0.43)
Maltose	0.17	(0.02, 0.31)	0.18	(0.04, 0.33)	0.19	(0.04, 0.33)
Lactose	0.65	(0.55, 0.73)	0.72	(0.64, 0.78)	0.72	(0.64, 0.78)
Starch	0.27	(0.13, 0.40)	0.39	(0.25, 0.51)	0.34	(0.20, 0.47)
Protein	0.22	(0.07, 0.36)	0.30	(0.16, 0.43)	0.28	(0.13, 0.41)
Fat	0.28	(0.14, 0.42)	0.41	(0.28, 0.53)	0.36	(0.22, 0.49)
Carbohydrate	0.14	(−0.01, 0.28)	0.43	(0.30, 0.55)	0.39	(0.25, 0.51)

DR, dietary record; FFQv, food frequency questionnaire for validity; CC, correlation coefficient; CI, confidence interval. [1] Correlation coefficients were adjusted for age, area, occupation (primary industry, professionals and office workers, self-employed and others, unemployed), body mass index (BMI), total energy intake, smoking status (never, past, current), and alcohol (non-drinker, ≤ 4 days per week, ≥ 5 days per week). [2] Sugar and other nutrients intakes were adjusted for energy intake by residual model. [3] Sugar and other nutrients intakes were energy-adjusted using the density method (percentage of energy). [4] "Total sugars" represents the sum of the crude consumption of following saccharides: glucose, fructose, galactose, sucrose, maltose, and lactose.

Table 5. Sugar intakes assessed with FFQv and FFQr in Cohorts I and II and differences.

	FFQv			FFQr			Mean of Difference [1] (95%CI)	
	Mean	SD	Median	Mean	SD	Median		
Cohort I								
Men (*n* = 101)								
Total sugars [2] (g)	69.3	43.2	61.5	70.4	41.6	61.3	1.1	(−6.7, 9.0)
Glucose (g)	17.1	9.0	15.5	17.9	11.1	15.7	0.8	(−1.1, 2.6)
Fructose (g)	13.9	11.1	12.0	14.1	10.1	11.6	0.2	(−1.9, 2.2)
Galactose (g)	0.2	0.4	0.1	0.2	0.2	0.1	0.0	(−0.1, 0.0)
Sucrose (g)	28.2	21.6	21.9	27.9	20.3	23.5	−0.3	(−3.6, 3.0)
Maltose (g)	1.1	0.6	0.9	1.2	1.4	0.9	0.2	(−0.1, 0.4)
Lactose (g)	8.8	10.2	7.3	9.1	9.7	8.8	0.4	(−2.0, 2.8)
Starch (g)	213.4	64.7	203.8	213.9	71.7	195.4	0.5	(−11.5, 12.6)
Energy (kcal)	2403	694	2354	2418	741	2415	14	(−129, 158)
Protein (g)	87.7	35.5	81.5	87.9	35.0	84.6	0.2	(−6.2, 6.6)
Fat (g)	61.5	27.2	59.1	62.2	29.1	57.1	0.7	(−4.9, 6.3)
Carbohydrate (g)	325.4	105.5	316.3	326.8	107.8	304.3	1.4	(−17.8, 20.6)
Women (*n* = 108)								
Total sugars [2] (g)	74.2	52.9	62.8	74.9	32.6	67.9	0.7	(−8.2, 9.6)
Glucose (g)	16.4	12.2	13.7	16.5	7.7	14.4	0.1	(−1.8, 2.1)
Fructose (g)	16.7	15.0	13.1	16.2	8.8	13.8	−0.4	(−2.9, 2.0)
Galactose (g)	0.2	0.2	0.1	0.3	0.3	0.2	0.0	(0.0, 0.1)
Sucrose (g)	30.7	25.1	23.6	31.1	16.3	29.0	0.5	(−3.8, 4.8)
Maltose (g)	1.2	0.8	0.9	1.2	0.7	1.0	0.0	(−0.1, 0.2)
Lactose (g)	9.1	7.1	9.1	9.6	10.1	8.0	0.5	(−1.3, 2.3)
Starch (g)	179.9	52.4	172.6	182.8	46.8	177.4	2.9	(−4.7, 10.5)
Energy (kcal)	2048	860	1914	2082	601	1959	33	(−109, 176)
Protein (g)	81.5	44.6	71.9	81.9	29.4	76.8	0.4	(−7.4, 8.1)
Fat (g)	59.3	32.9	51.6	61.5	27.2	54.2	2.3	(−3.8, 8.3)
Carbohydrate (g)	291.4	111.8	277.5	295.4	78.9	279.1	4.0	(−13.5, 21.5)
Cohort II								
Men (*n* = 143)								
Total sugars [2] (g)	64.8	31.3	55.9	68.1	34.2	62.1	3.4	(−1.1, 7.8)
Glucose (g)	16.0	8.5	13.7	16.5	9.0	14.9	0.5	(−0.5, 1.6)
Fructose (g)	12.4	7.7	10.4	13.0	8.6	10.3	0.5	(−0.5, 1.6)
Galactose (g)	0.2	0.3	0.1	0.2	0.3	0.1	0.0	(0.0, 0.0)
Sucrose (g)	25.9	14.7	24.1	27.6	16.4	25.3	1.7	(−0.3, 3.7)
Maltose (g)	1.3	0.8	1.0	1.3	0.8	1.2	0.1	(0.0, 0.2)
Lactose (g)	8.9	9.2	8.4	9.5	9.4	8.6	0.5	(−1.3, 2.4)
Starch (g)	191.6	57.4	182.5	202.2	58.9	193.9	10.6	(3.4, 17.8)
Energy (kcal)	2251	650	2167	2387	740	2210	135	(44, 227)
Protein (g)	80.4	30.1	74.9	86.6	33.7	81.2	6.2	(1.9, 10.5)
Fat (g)	59.0	25.2	54.0	64.8	28.9	58.9	5.8	(1.6, 10.0)
Carbohydrate (g)	296.4	85.2	279.6	311.9	90.4	296.5	15.6	(4.9, 26.2)
Women (*n* = 146)								
Total sugars [2] (g)	68.6	36.6	59.1	72.2	42.1	63.2	3.6	(−2.1, 9.4)
Glucose (g)	14.7	8.2	13.2	15.6	9.8	13.8	0.9	(−0.4, 2.1)
Fructose (g)	14.2	9.1	12.5	15.1	12.4	12.8	0.9	(−0.8, 2.6)
Galactose (g)	0.4	0.5	0.2	0.3	0.4	0.2	0.0	(−0.1, 0.1)
Sucrose (g)	27.4	16.9	23.9	29.0	19.0	24.6	1.6	(−1.0, 4.1)
Maltose (g)	1.2	0.7	1.1	1.3	0.9	1.1	0.1	(−0.1, 0.2)
Lactose (g)	10.8	11.6	9.4	11.0	9.6	9.4	0.2	(−1.7, 2.1)
Starch (g)	163.0	38.7	163.5	170.3	40.1	169.5	7.3	(0.9, 13.7)
Energy (kcal)	1911	621	1769	2036	635	1893	125	(36, 214)
Protein (g)	75.5	30.1	69.8	80.8	30.4	75.3	5.3	(1.3, 9.4)
Fat (g)	57.5	27.2	50.8	63.1	27.0	58.0	5.5	(1.4, 9.7)
Carbohydrate (g)	267.9	78.6	254.8	280.4	82.8	266.2	12.4	(1.0, 23.8)

FFQv, food frequency questionnaire for validity; FFQr, food frequency questionnaire for reproducibility; SD, standard deviation; CI, confidence interval. [1] Mean of (intakes from FFQr - intakes from FFQv). [2] "Total sugars" represents the sum of the crude consumption of the following saccharides: glucose, fructose, galactose, sucrose, maltose, and lactose.

Table 6. Correlations between FFQv and FFQr (Reproducibility).

	Spearman's Rank Correlation Coefficient						ICC for FFQ
	Crude		Energy-Adjusted (Residual) [1]		Energy-Adjusted (Density) [2]		
	r	95%CI	r	95%CI	r	95%CI	
Cohort I							
Men (*n* = 101)							
Total sugars [3]	0.61	(0.48, 0.72)	0.53	(0.38, 0.66)	0.63	(0.49, 0.73)	0.66
Glucose	0.65	(0.52, 0.75)	0.55	(0.40, 0.68)	0.64	(0.51, 0.74)	0.63
Fructose	0.65	(0.52, 0.75)	0.56	(0.41, 0.68)	0.60	(0.46, 0.71)	0.61
Galactose	0.73	(0.62, 0.81)	0.42	(0.24, 0.57)	0.66	(0.53, 0.76)	0.67
Sucrose	0.75	(0.65, 0.82)	0.70	(0.59, 0.79)	0.76	(0.67, 0.83)	0.75
Maltose	0.58	(0.43, 0.69)	0.68	(0.56, 0.77)	0.68	(0.56, 0.78)	0.50
Lactose	0.63	(0.50, 0.74)	0.51	(0.35, 0.64)	0.53	(0.37, 0.65)	0.50
Starch	0.69	(0.57, 0.78)	0.62	(0.49, 0.73)	0.51	(0.35, 0.64)	0.46
Energy	0.47	(0.30, 0.61)					
Protein	0.56	(0.40, 0.68)	0.47	(0.30, 0.61)	0.56	(0.41, 0.68)	0.60
Fat	0.51	(0.35, 0.64)	0.60	(0.46, 0.72)	0.59	(0.44, 0.70)	0.59
Carbohydrate	0.66	(0.53, 0.76)	0.65	(0.52, 0.75)	0.60	(0.46, 0.71)	0.54
Women (*n* = 108)							
Total sugars [3]	0.65	(0.52, 0.75)	0.57	(0.43, 0.69)	0.55	(0.41, 0.67)	0.62
Glucose	0.69	(0.57, 0.78)	0.55	(0.41, 0.67)	0.61	(0.48, 0.72)	0.63
Fructose	0.65	(0.53, 0.75)	0.55	(0.40, 0.67)	0.60	(0.47, 0.71)	0.63
Galactose	0.59	(0.45, 0.70)	0.51	(0.36, 0.64)	0.52	(0.37, 0.64)	0.54
Sucrose	0.67	(0.55, 0.76)	0.61	(0.48, 0.72)	0.61	(0.48, 0.72)	0.59
Maltose	0.69	(0.57, 0.77)	0.64	(0.51, 0.74)	0.67	(0.54, 0.76)	0.68
Lactose	0.74	(0.64, 0.82)	0.74	(0.64, 0.82)	0.75	(0.65, 0.82)	0.71
Starch	0.71	(0.60, 0.79)	0.56	(0.42, 0.68)	0.60	(0.46, 0.71)	0.56
Energy	0.69	(0.58, 0.78)					
Protein	0.68	(0.56, 0.77)	0.42	(0.25, 0.56)	0.49	(0.33, 0.62)	0.47
Fat	0.67	(0.55, 0.76)	0.58	(0.44, 0.69)	0.59	(0.46, 0.70)	0.59
Carbohydrate	0.72	(0.62, 0.80)	0.55	(0.40, 0.67)	0.58	(0.44, 0.69)	0.57
Cohort II							
Men (*n* = 143)							
Total sugars [3]	0.63	(0.52, 0.72)	0.65	(0.55, 0.74)	0.66	(0.55, 0.74)	0.64
Glucose	0.71	(0.62, 0.78)	0.63	(0.52, 0.72)	0.66	(0.55, 0.74)	0.63
Fructose	0.64	(0.53, 0.73)	0.61	(0.50, 0.71)	0.62	(0.51, 0.71)	0.63
Galactose	0.72	(0.63, 0.79)	0.68	(0.57, 0.76)	0.69	(0.60, 0.77)	0.76
Sucrose	0.68	(0.58, 0.76)	0.67	(0.57, 0.75)	0.68	(0.57, 0.76)	0.67
Maltose	0.63	(0.51, 0.72)	0.67	(0.57, 0.75)	0.68	(0.58, 0.76)	0.78
Lactose	0.68	(0.58, 0.76)	0.69	(0.60, 0.77)	0.70	(0.60, 0.77)	0.64
Starch	0.69	(0.59, 0.77)	0.64	(0.54, 0.73)	0.62	(0.51, 0.71)	0.64
Energy	0.59	(0.47, 0.69)					
Protein	0.60	(0.49, 0.70)	0.61	(0.50, 0.70)	0.60	(0.49, 0.70)	0.68
Fat	0.56	(0.44, 0.66)	0.63	(0.52, 0.72)	0.61	(0.50, 0.70)	0.64
Carbohydrate	0.64	(0.53, 0.73)	0.69	(0.59, 0.77)	0.66	(0.56, 0.75)	0.70
Women (*n* = 146)							
Total sugars [3]	0.64	(0.53, 0.73)	0.59	(0.47, 0.68)	0.63	(0.52, 0.72)	0.48
Glucose	0.60	(0.49, 0.70)	0.41	(0.26, 0.54)	0.45	(0.32, 0.57)	0.45
Fructose	0.58	(0.46, 0.67)	0.36	(0.21, 0.50)	0.45	(0.31, 0.57)	0.45
Galactose	0.65	(0.55, 0.74)	0.61	(0.50, 0.71)	0.65	(0.55, 0.74)	0.55
Sucrose	0.63	(0.52, 0.72)	0.53	(0.41, 0.64)	0.57	(0.45, 0.67)	0.55
Maltose	0.69	(0.60, 0.77)	0.71	(0.62, 0.78)	0.69	(0.59, 0.76)	0.65
Lactose	0.76	(0.68, 0.82)	0.79	(0.72, 0.84)	0.79	(0.72, 0.85)	0.73
Starch	0.53	(0.40, 0.64)	0.51	(0.38, 0.62)	0.58	(0.46, 0.68)	0.56
Energy	0.60	(0.48, 0.69)					
Protein	0.65	(0.54, 0.73)	0.52	(0.39, 0.63)	0.59	(0.47, 0.69)	0.54
Fat	0.60	(0.48, 0.69)	0.51	(0.38, 0.62)	0.50	(0.36, 0.61)	0.37
Carbohydrate	0.58	(0.46, 0.68)	0.50	(0.37, 0.61)	0.50	(0.37, 0.62)	0.48

DR, dietary record; FFQv, food frequency questionnaire for validity; r, correlation coefficient; CI, confidence interval; ICC, intra-class correlation coefficient. [1] Sugar and other nutrients intakes were adjusted for energy intake by residual model. [2] Sugar and other nutrients intakes were energy-adjusted using the density method (percentage of energy). [3] "Total sugars" represents the sum of the crude consumption of the following saccharides: glucose, fructose, galactose, sucrose, maltose, and lactose.

4. Discussion

We evaluated the validity and reproducibility of sugar intakes assessed by the FFQ in a subsample of the JPHC study. For validity, de-attenuated Spearman's CC was 0.40 between total sugar intake from the FFQv and urinary sugar concentrations. Furthermore, after adjusting for age, areas, occupations, body mass index, total energy intake, smoking status and alcohol, partial correlations of sugar intakes between the FFQv and 28- or 14-d DR ranged from 0.34 to 0.57. These results suggested moderate validity of the FFQ. Compared with the 1-year interval FFQ, correlations ranged from 0.55 to 0.66, indicating moderate reproducibility. Our results for the JPHC study verified that it is possible to use the FFQ for the assessment of the health impacts of sugar intakes. It is expected that future studies will clarify the health impacts of sugar consumption in Japan.

The results of our study are in general agreement with previous studies. Smith et al. [28] assessed the validity of the FFQ by comparing it with three 4-d weighted DRs, and the reproducibility by comparing it with a 12–18-month interval FFQ. Spearman's CCs of sugar intakes were 0.47 (energy-adjusted, 34 men and 45 women) for validity and 0.67 (96 men and 135 women) for reproducibility. Willett et al. [29] evaluated the validity and reproducibility of FFQ comparing it with 28-d DRs and 1-year interval FFQ, respectively, in 173 women. In the study, the Pearson CCs of sucrose intakes were 0.41 (energy-adjusted) and 0.71 for validity and reproducibility, respectively.

The SD of the total sugar intake assessed with the FFQv was almost double that of the DR, indicating that the between-person variation would be overestimated. Therefore, when we examine the association of total sugar intake from the FFQ with disease risks, such misclassification tends to attenuate relative risk estimates.

Except for glucose and galactose, women consumed larger amount of sugars than men according to the DR. The difference was remarkable in sucrose intake, because the contribution proportion of confectionaries (which were one of the main sources of sucrose) was higher in women. Furthermore, the % energy of free sugars in women was also higher than that for men. These characteristics in the source of sugars in women might affect the relationship between sugar intake and health conditions.

In the Bland-Altman plots, differences in total sugar intakes (% energy) between the FFQv and the DR did not differ based on the magnitude of the mean total sugar intakes (% energy). Moreover, FFQ estimates for total sugar intake were overestimated, especially in men.

Correlations between sugar intakes from the FFQv and DR among women in Cohort II were weaker than those in other groups. In previous studies, CCs between carbohydrate intakes estimated from the FFQv and DR among women in Cohort II were lower than for men (Cohort I, men: 0.56; women: 0.37; Cohort II, men: 0.59; women: 0.39) [15,23]. Because men tend to be unconcerned about their daily diets, it might have been easier for men to complete the FFQ, which requires simplified dietary habits [23].

Urinary sugars have been drawing attention as a useful biomarker not affected by measurement errors in self-reported measures [20] and the use of the same food composition table. We evaluated the validity of using urinary concentration of fructose and sucrose as an objective biomarker. In this study, sugar intakes from the FFQv were correlated with the mean concentrations of urinary sugars collected twice (spring and fall) (r = 0.40, 95%CI: 0.19, 0.58). The correlation between sugar intakes from the 7-day DR and urinary sugars, both collected in spring, was weak (r = 0.27, 95%CI: 0.04, 0.47), while the correlation between those collected in the fall was moderate (r = 0.46, 95%CI: 0.26, 0.62). For urinary sugars, $\sigma^2_{ws}/\sigma^2_{bs}$ was high and ICC was low; therefore, the concentrations of sugars in urine were likely to be influenced by within-subject variance and seasonal variations. Furthermore, a previous study [30] showed that participants who consumed higher added sugar resulted in better correlations between dietary sugar intakes and urinary sugar excretions (r = 0.77) than those who consumed lower added sugar (r = 0.15). Thus, the high consumption of sugars might have led to stronger correlations. In our study, total sugar intakes from the DR in the fall were higher than in spring due to increasing fruit intakes (interquartile range of total sugar intakes: 46.0–80.3 g/day in spring; 53.6–90.3 g/day in fall; fruit intakes: 65–192 g/day in spring; 120–273 g/day in fall; n = 72). Therefore, it can be speculated that the correlations between urinary sugars and the 7-d DR were higher in the fall than in spring, as a result of the seasonal variations in total sugar intakes. Of note is the correlation between sugar intakes and urinary sugars in a previous study [31] in which urine was collected daily based on a 30-day diet (r = 0.84, total sugars, n = 13). This seems to suggest that multiple measurements of urinary sugars lead to a high correlation between sugar intakes and urinary sugars and may be more useful than single or double measurements for examining the validity of sugar intakes.

Our study has several strengths. First, we complemented the standard tables of food composition for available carbohydrates with the substitution methods, because the tables do not cover all food items occurring in the FFQ and DR. Furthermore, we also included sugars added to foods during cooking. Accordingly, most of the food items and menus that contain non-negligible amount of sugars were included in the nutrient calculation. Secondly, we examined urinary sugars as an objective biomarker for validation. We found that urinary sugars were useful to some extent in evaluating the validity of the FFQ.

Despite these strengths, our study had some limitations. First, common errors in sugar intake assessments from the FFQ and the DR remained because we used the same food composition table for the nutritional calculation. In both the FFQ and the DR sugar intake estimation, we were unable to consider the heterogeneity of sugar contents in each food since our estimation was based only on the sugar content of foods on the standard tables of food composition. Therefore, the correlations between FFQ and DR might be overestimated. Furthermore, because both of the FFQ and DR are self-reported dietary assessments, the overestimations of the correlations also possibly existed. By contrast, urinary sugars were not affected by this limitation of the food composition tables [12] and the property of these dietary surveys. Our results showed a correlation between sugar intakes from the FFQ and urinary sugars, supporting the validity of the FFQ. Second, some foods in the FFQ and the DR were not assigned sugar contents and were not included in the calculations. However, we believe that this may not have seriously biased our estimates because we evaluated most food items that provide more than 1 g of carbohydrates per portion size. Third, because the dietary data in this study were collected before 2000, they may be different from contemporary dietary habits. Therefore, the results in this study might not be generalizable to studies conducted later. Fourth, correlations of sugar intakes and urinary sugar concentrations may differ by the form of the sugar in the food. Indeed, consistent with a previous report [32], urinary sugar concentrations were more strongly correlated with free sugar intakes measured by the DR than with other sugar intakes in our study (data not shown), suggesting that the validity of sugar intake by form may deserve further investigation.

5. Conclusions

We observed moderate correlations between sugar intakes from the FFQ and urinary sugar, and the DR, as well as between the two FFQs at yearly intervals. The FFQ used in the 5-year follow-up JPHC study may be useful in ranking individuals for sugar intakes in the JPHC study population. These findings suggest that the FFQ may be helpful in assessing the association of sugar intakes with health conditions in Japan.

Supplementary Materials: The following are available online at http://www.mdpi.com/2072-6643/11/3/554/s1, Table S1: Substituted food items in the FFQ, Table S2: Major foods contributing to sugar intake according to the DR (Cohorts I & II), Table S3-1. Comparison of FFQ for sugar intakes with urinary sugars based on cross classification by quintile (%), Table S3-2. Frequency and means of FFQ for sugar intakes with urinary sugars based on cross classification by quintile (Cohort I, *n* = 72), Table S4: Correlations between FFQv and DR for 28 or 14 days, Table S5: Rank correlation coefficients between % energy of sugar intake assessed using the DR for 28 days and FFQv in Cohorts I and II using the probit transformation method with correction for measurement error, Table S6: Comparison of FFQv with DR for sugar intakes based on cross classification by quintile (%), Figure S1: Scatter plots between total sugars from FFQ and DR, Figure S2: Bland-Altman plot for the comparison of FFQ and DR in measuring the total sugar intake.

Author Contributions: Conceptualization, A.G., N.S., J.I., R.T., M.I. and S.T.; Data curation, R.K., A.G., A.K., N.M., A.N., N.S., J.I. and R.T.; Formal analysis, R.K. and A.G.; Funding acquisition, A.G. and S.T.; Investigation, R.K. and A.G.; Methodology, R.K., A.G., N.S., J.I., R.T., Y.K., M.I. and S.T., Project administration, A.G., N.S., M.I., and S.T.; Resources, A.G., N.S., M.I. and S.T.; Software, R.K., A.G. and A.K.; Supervision, A.G., N.S. and M.I.; Validation, R.K. and A.G.; Visualization, R.K. and A.G.; Writing-original draft, R.K.; Writing-review & editing, R.K., A.G., A.K., N.M., A.N., N.S., J.I., R.T., Y.K., M.I., and S.T.

Funding: This research was funded by the JSPS KAKENHI (grant number 15K21389, 18K10095) from Japan Society for the Promotion of Science (JSPS); AMED (grant number JP18ck0106370) from Japan Agency for Medical Research and Development; the National Cancer Center Research and Development Fund (since 2011); and a Grant-in-Aid for Cancer Research from the Ministry of Health, Labour and Welfare of Japan (from 1989 to 2010).

Acknowledgments: We are grateful to Mitsuhiko Noda, Professor, Department of Diabetes Research, Diabetes Research Centre, National Centre for Global Health and Medicine Saitama Medical University and Akiko Nanri, Associate Professor, Department of Food and Health Sciences, International College of Arts and Sciences Fukuoka Women's University, for their helpful comments in the design of this study. We also thank all members in each study area and at the central office, for their efforts in the baseline survey and the follow-up. The investigators in the validation study on the self-administered FFQ in the JPHC Study (the JPHC FFQ Validation Study Group) and their affiliations at the time of the study were: S. Tsugane, S. Sasaki, and M. Kobayashi, Epidemiology and Biostatistics Division, National Cancer Centre Research Institute East, Kashiwa; T. Sobue, S. Yamamoto, and J. Ishihara, Cancer Information and Epidemiology Division, National Cancer Centre Research Institute, Tokyo; M. Akabane, Y. Iitoi, Y. Iwase, and T. Takahashi, Tokyo University of Agriculture, Tokyo; K. Hasegawa and T. Kawabata, Kagawa Nutrition University, Sakado; Y. Tsubono, Tohoku University, Sendai; H. Iso, Tsukuba University, Tsukuba; S. Karita, Teikyo University, Tokyo; the late M. Yamaguchi and Y. Matsumura, National Institute of Health and Nutrition, Tokyo [33].

Conflicts of Interest: The authors declare no conflict of interest.

References

1. World Health Organization. Obesity and Overwight. Available online: http://www.who.int/en/news-room/fact-sheets/detail/obesity-and-overweight (accessed on 3 September 2018).
2. World Health Organization. Diabetes. Available online: http://www.who.int/mediacentre/factsheets/fs312/en/ (accessed on 21 September 2017).
3. World Health Organization. *Guideline: Sugars Intake for Adults and Children*; WHO Document Production Services: Geneva, Switzerland, 2015.
4. Te Morenga, L.; Mallard, S.; Mann, J. Dietary sugars and body weight: Systematic review and meta-analyses of randomised controlled trials and cohort studies. *BMJ* **2012**, *346*. [CrossRef] [PubMed]
5. Sonestedt, E.; Overby, N.C.; Laaksonen, D.E.; Birgisdottir, B.E. Does high sugar consumption exacerbate cardiometabolic risk factors and increase the risk of type 2 diabetes and cardiovascular disease? *Food Nutr. Res.* **2012**, *56*. [CrossRef] [PubMed]
6. Tsilas, C.S.; de Souza, R.J.; Mejia, S.B.; Mirrahimi, A.; Cozma, A.I.; Jayalath, V.H.; Ha, V.; Tawfik, R.; Di Buono, M.; Jenkins, A.L.; et al. Relation of total sugars, fructose and sucrose with incident type 2 diabetes: A systematic review and meta-analysis of prospective cohort studies. *CMAJ* **2017**, *189*. [CrossRef] [PubMed]
7. Kurose, M.M.M.; Watanabe, T. Dental caries and sucrose intake in japanese preschool children. *J. Dent. Health* **1997**, *47*, 683–692. [CrossRef]
8. Saido, M.; Asakura, K.; Masayasu, S.; Sasaki, S. Relationship between dietary sugar intake and dental caries among japanese preschool children with relatively low sugar intake (japan nursery school shokuiku study): A nationwide cross-sectional study. *Matern. Child Health J.* **2016**, *20*, 556–566. [CrossRef] [PubMed]
9. Cust, A.E.; Skilton, M.R.; van Bakel, M.M.; Halkjaer, J.; Olsen, A.; Agnoli, C.; Psaltopoulou, T.; Buurma, E.; Sonestedt, E.; Chirlaque, M.D.; et al. Total dietary carbohydrate, sugar, starch and fibre intakes in the european prospective investigation into cancer and nutrition. *Eur. J. Clin. Nutr.* **2009**, *63*, S37–S60. [CrossRef]
10. Fujiwara, A.; Murakami, K.; Asakura, K.; Uechi, K.; Sugimoto, M.; Wang, H.C.; Masayasu, S.; Sasaki, S. Estimation of starch and sugar intake in a japanese population based on a newly developed food composition database. *Nutrients* **2018**, *10*, 1474. [CrossRef]
11. Ministry of Education, Culture, Sports, Science and Technology, the Government of Japan. *Standard Tables of Food Composition in Japan, Available Carbohydrates, Polyols, and Organic Acids*; Official Gazette Cooperation of Japan: Tokyo, Japan, 2015.

12. Willett, W. *Nutritional Epidemiology*; Oxford University Press: New York, NY, USA, 2012.
13. Watanabe, S.; Tsugane, S.; Sobue, T.; Konishi, M.; Baba, S. Study design and organization of the jphc study. Japan public health center-based prospective study on cancer and cardiovascular diseases. *J. Epidemiol.* **2001**, *11*, S3–S7. [CrossRef]
14. Tsugane, S.; Sasaki, S.; Kobayashi, M.; Tsubono, Y.; Akabane, M. Validity and reproducibility of the self-administered food frequency questionnaire in the JPHC study Cohort I: Study design, conduct and participant profiles. *J. Epidemiol.* **2003**, *13*, S2–S12. [CrossRef]
15. Ishihara, J.; Sobue, T.; Yamamoto, S.; Yoshimi, I.; Sasaki, S.; Kobayashi, M.; Takahashi, T.; Iitoi, Y.; Akabane, M.; Tsugane, S.; et al. Validity and reproducibility of a self-administered food frequency questionnaire in the JPHC study Cohort II: Study design, participant profile and results in comparison with Cohort I. *J. Epidemiol.* **2003**, *13*, S134–S147. [CrossRef]
16. Sasaki, S.; Kobayashi, M.; Ishihara, J.; Tsugane, S. Self-administered food frequency questionnaire used in the 5-year follow-up survey of the JPHC study: Questionnaire structure, computation algorithms, and area-based mean intake. *J. Epidemiol.* **2003**, *13*, S13–S22. [CrossRef] [PubMed]
17. Ministry of Education, Culture, Sports, Science and Technology, the Government of Japan. *Standard Tables of Food Composition in Japan*; Official Gazette Cooperation of Japan: Tokyo, Japan, 2015.
18. Ishihara, J.; Todoriki, H.; Inoue, M.; Tsugane, S. Validity of a self-administered food-frequency questionnaire in the estimation of amino acid intake. *Br. J. Nutr.* **2009**, *101*, 1393–1399. [CrossRef] [PubMed]
19. Charrondière, U.R.S.B.; Haytowitz, D. *Fao/infoods Guidelines for Converting Units, Denominators and Expressions (Version 1.0)*; FAO: Rome, Italy, 2012.
20. Tasevska, N. Urinary sugars—A biomarker of total sugars intake. *Nutrients* **2015**, *7*, 5816–5833. [CrossRef] [PubMed]
21. Rosner, B.; Glynn, R.J. Interval estimation for rank correlation coefficients based on the probit transformation with extension to measurement error correction of correlated ranked data. *Stat. Med.* **2007**, *26*, 633–646. [CrossRef] [PubMed]
22. Bernardrosner Attenuation Point Estimate and Coefficients for Measurement-Error Corrected Rank. Available online: https://sites.google.com/a/channing.harvard.edu/bernardrosner/channing/interval-estimation-for-rank-correlation-coefficients/attenuation-point-estimate-and-coefficients-for-measurement-error-corrected-rank (accessed on 13 April 2018).
23. Tsugane, S.; Kobayashi, M.; Sasaki, S. Validity of the self-administered food frequency questionnaire used in the 5-year follow-up survey of the JPHC study Cohort I: Comparison with dietary records for main nutrients. *J. Epidemiol.* **2003**, *13*, S51–S56. [CrossRef] [PubMed]
24. Bland, J.M.; Altman, D.G. Statistical methods for assessing agreement between two methods of clinical measurement. *Lancet* **1986**, *1*, 307–310. [CrossRef]
25. Bland, J.M.; Altman, D.G. Measuring agreement in method comparison studies. *Stat. Methods Med. Res.* **1999**, *8*, 135–160. [CrossRef] [PubMed]
26. Giavarina, D. Understanding bland altman analysis. *Biochem. Med.* **2015**, *25*, 141–151. [CrossRef] [PubMed]
27. Sasaki, S.; Ishihara, J.; Tsugane, S. Reproducibility of a self-administered food frequency questionnaire used in the 5-year follow-up survey of the JPHC study Cohort I to assess food and nutrient intake. *J. Epidemiol.* **2003**, *13*, S115–S124. [CrossRef]
28. Smith, W.; Mitchell, P.; Reay, E.M.; Webb, K.; Harvey, P.W. Validity and reproducibility of a self-administered food frequency questionnaire in older people. *Aust. N. Z. J. Public Health* **1998**, *22*, 456–463. [CrossRef]
29. Willett, W.C.; Sampson, L.; Stampfer, M.J.; Rosner, B.; Bain, C.; Witschi, J.; Hennekens, C.H.; Speizer, F.E. Reproducibility and validity of a semiquantitative food frequency questionnaire. *Am. J. Epidemiol.* **1985**, *122*, 51–65. [CrossRef] [PubMed]
30. Moore, L.B.; Liu, S.V.; Halliday, T.M.; Neilson, A.P.; Hedrick, V.E.; Davy, B.M. Urinary excretion of sodium, nitrogen, and sugar amounts are valid biomarkers of dietary sodium, protein, and high sugar intake in nonobese adolescents. *J. Nutr.* **2017**, *147*, 2364–2373. [CrossRef] [PubMed]
31. Tasevska, N.; Runswick, S.A.; McTaggart, A.; Bingham, S.A. Urinary sucrose and fructose as biomarkers for sugar consumption. *Cancer Epidemiol. Biomark. Prev.* **2005**, *14*, 1287–1294. [CrossRef] [PubMed]

32. Tasevska, N.; Runswick, S.A.; Welch, A.A.; McTaggart, A.; Bingham, S.A. Urinary sugars biomarker relates better to extrinsic than to intrinsic sugars intake in a metabolic study with volunteers consuming their normal diet. *Eur. J. Clin. Nutr.* **2009**, *63*, 653–659. [CrossRef] [PubMed]
33. National Cancer Center Epidemiology and Prevention Dicision. Members of jphc ffq Validation Study Group. Available online: http://epi.ncc.go.jp/en/jphc/781/7952.html (accessed on 1 June 2018).

 *nutrients*

Article

Requirements of an Application to Monitor Diet, Physical Activity and Glucose Values in Patients with Type 2 Diabetes: The Diameter

Niala den Braber [1,2,*], Miriam M. R. Vollenbroek-Hutten [1,2], Milou M. Oosterwijk [1] , Christina M. Gant [1], Ilse J. M. Hagedoorn [1], Bert-Jan F. van Beijnum [2], Hermie J. Hermens [2,3] and Gozewijn D. Laverman [1]

[1] Department of Internal Medicine/Nephrology, Ziekenhuisgroep Twente (ZGT), 7609 PP Almelo, The Netherlands; m.m.r.hutten@utwente.nl (M.M.R.V.-H.); mi.oosterwijk@zgt.nl (M.M.O.); cm.gant@meandermc.nl (C.M.G.); i.hagedoorn@zgt.nl (I.J.M.H.); g.laverman@zgt.nl (G.D.L.)
[2] Biomedical Signals and Systems (BSS), University of Twente, 7522 NB Enschede, The Netherlands; b.j.f.vanbeijnum@utwente.nl (B.-J.F.v.B.); h.j.hermens@utwente.nl (H.J.H.)
[3] Roessingh Research and Development (RRD), 7522 AH Enschede, The Netherlands
* Correspondence: n.braber@zgt.nl; Tel.: +31-534-896-907

Received: 23 December 2018; Accepted: 11 February 2019; Published: 15 February 2019

Abstract: Adherence to a healthy diet and regular physical activity are two important factors in sufficient type 2 diabetes mellitus management. It is recognized that the traditional treatment of outpatients does not meet the requirements for sufficient lifestyle management. It is hypothesised that a personalized diabetes management mHealth application can help. Such an application ideally measures food intake, physical activity, glucose values, and medication use, and then integrates this to provide patients and healthcare professionals insight in these factors, as well as the effect of lifestyle on glucose values in daily life. The lifestyle data can be used to give tailored coaching to improve adherence to lifestyle recommendations and medication use. This study describes the requirements for such an application: the *Diameter*. An iterative mixed method design approach is used that consists of a cohort study, pilot studies, literature search, and expert meetings. The requirements are defined according to the Function and events, Interactions and usability, Content and structure and Style and aesthetics (FICS) framework. This resulted in 81 requirements for the dietary ($n = 37$), activity and sedentary ($n = 15$), glycaemic ($n = 12$), and general ($n = 17$) parts. Although many applications are currently available, many of these requirements are not implemented. This stresses the need for the *Diameter* as a new personalized diabetes application.

Keywords: Type 2 diabetes mellitus; diabetes management; dietary application; dietary assessment; nutrition; physical activity; blood glucose; mHealth

1. Introduction

Type 2 Diabetes Mellitus (T2DM) is one of the most common chronic diseases, which was recorded to affect 415 million people worldwide in 2015. Its prevalence is still increasing due to the rise of obesity and unhealthy lifestyle with an expected prevalence of 642 million in 2040 [1,2].

The two main components of a sufficient diabetes management are adherence to a healthy diet and regular physical activity. These lifestyle components are important for both glycaemic control, i.e., keeping blood glucose levels within the target range, and in maintaining long-term health, which includes the prevention of micro- and macrovascular complications [3]. The risk of developing complications increases with a poor diet quality and insufficient physical activity [4–6].

For many T2DM patients, the everyday challenge includes the restriction of carbohydrate intake and the performance of sufficient physical activity while reducing sedentary behaviour. Another challenge is to adhere to, often a plethora of, pharmacological agents. In patients that are treated with insulin, there is the additional burden of insulin injections and the need to monitor blood glucose. Multiple finger-pricks per day, in combination with an estimation of the carbohydrate content of the meal, are usually necessary to determine the appropriate pre-meal insulin dose [7,8].

Diabetes healthcare professionals (HCPs) support patients, whereby they provide education and coaching on lifestyle and pharmacological therapy [9]. However, in the current situation, regular contacts between patients and professionals are particularly suited to monitor the pharmacological management of glycaemic control, blood pressure, and dyslipidaemia. It is increasingly recognized that the traditional set-up of outpatients does not meet the requirements for sufficient lifestyle management, and therefore efforts for the improvement of lifestyle and self-management do not reach full potential [10,11]. To illustrate this, we found in the vast majority of T2DM patients that were treated in our hospital, "ZiekenhuisGroep Twente" (ZGT), where adherence to the guidelines of physical activity and healthy diet was not met. This was reflected by an average body mass index (BMI) of 33 kg/m^2 and a sufficient vegetable intake in only 7% of the patients [12,13]. Of note, lifestyle behaviour is not measured routinely and objectively in clinical practice. However, objective measurements are of great importance in adequately mapping lifestyle, because patients grossly overestimate their healthy lifestyle behaviours [14].

Advances in technology have made it easier to monitor lifestyle, e.g., through smartphone applications and wearable technology. Worldwide, several initiatives have arisen to transform healthcare and health support [15–17]. The use of technology can help to incorporate effective lifestyle management in routine clinical care. Mobile health (mHealth) technology allows for lifestyle parameters to be monitored objectively and continuously [15]. Objective data provides insight in actual lifestyle habits, both to the patient and the HCP, while also increasing awareness [18,19]. Additionally, data regarding carbohydrate- and fat intake, physical activity, medication use, and glucose values can be combined, which is of importance because these factors influence each other [19,20]. When compared to traditional methods of obtaining insight in patients' lifestyle, technology-based methods are ultimately less expensive, can be used when it suits the patient, and provide objective data. Moreover, technology can be used for (digital) tailored coaching [21]. Technologies that measure nutrition, physical activity, or provide glucose values are already available, but there are major limitations [18]. The existing applications are usually developed for personal use rather than for clinical use; no application exists that can measure and integrate all of the aspects that are considered to be necessary for an optimal diabetes management, and coaching functionalities remain only rudimentary [22].

Therefore, our research group aims to develop a digital tool that incorporates lifestyle habits and glucose management in one application, the *Diameter*. The core items to be measured by the *Diameter* are food intake, physical activity, glucose values, and medication use, with a built-in possibility to add other relevant items. This information is integrated in such a way that it gives individual patients and HCPs insight in lifestyle, blood glucose levels, as well as into the effect of lifestyle behaviour on glucose values in daily life, e.g., the HCPs can use the data about diet to gain insight in the (macro) nutrients intake and use this in their daily work to be able to treat the patients better. The lifestyle data can subsequently be used to give patients tailored coaching in order to improve adherence to lifestyle guidelines and, if necessary, medication use. The *Diameter* can be implemented as stand-alone application as well as in blended care forms, in which regular doctor visits are combined with online interventions. This paper describes the process of formulating the main requirements of the *Diameter* to measure dietary intake, physical activity, and glucose levels. The requirements for the evaluation of medication adherence and coaching will be addressed later.

2. Materials and Methods

We defined the requirements of the core items of the *Diameter* using an iterative mixed method design approach that combines data derived from several sources (Figure 1). Using the findings from the Diabetes and Lifestyle Cohort Twente (DIALECT), a literature search, and two pilot studies, the initial requirements were assessed. Subsequently, these preliminary requirements were discussed and are further elaborated upon in expert meetings to formulate the final set of requirements. The whole process is based on the approach of "the Function and events, Interactions and usability, Content and structure, and Style and aesthetics" (FICS) framework to enable the proper communication between researchers and the developers [23,24]. These requirements provide the foundation for the technology design of the *Diameter*, including what the system should do, which data is collected, what should be displayed, and what the user will experience. Using the iterative approach, the development runs parallel with the user experience, resulting in continuously made improvements that are based on feedback from both patients and professionals.

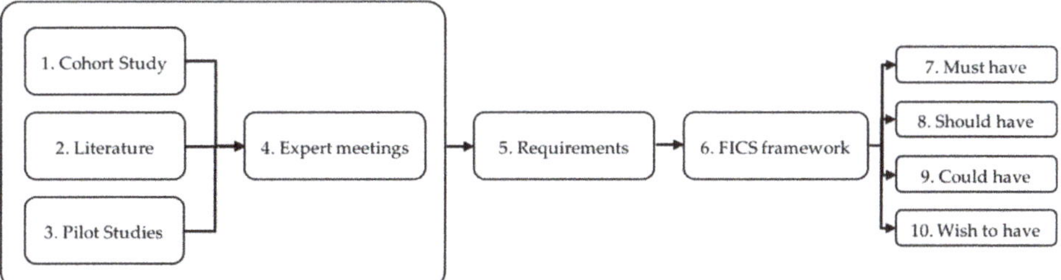

Figure 1. Requirements were formulated from insights gathered in the Diabetes and Lifestyle Cohort Twente (DIALECT) cohort study (**1**), literature research (**2**) and pilot studies (**3**). In expert meetings (**4**) these requirements were discussed and new requirements (**5**) were added. The requirements were formulated according to the Function and events, Interactions and usability, Content and structure and Style and aesthetics (FICS) framework (**6**). The requirements were labelled during expert meetings as "must" (**7**), "should" (**8**), "could" (**9**), and "wish to" (**10**) have.

2.1. Cohort Study

As a first source of requirements, experiences in clinical practice were used to evaluate our methods for the data collection in patients that were included in the DIALECT-2 study. DIALECT-2 is a cohort study that was performed in ZGT, Almelo and Hengelo, the Netherlands, and it is designed to investigate the relationship between lifestyle parameters and long-term outcome in T2DM patients. As such, several parameters of interest for the *Diameter* are measured in DIALECT-2 [12]. The inclusion of DIALECT-2 ($n = 400$) is ongoing and is expected to be completed in 2020. In this study, for the diet, activity, and glucose part, 64, 98, and 60 patients were included from the DIALECT-2 cohort, respectively.

In DIALECT-2, a food diary was used to assess actual dietary intake. Towards this purpose, the patients were instructed to register the timing, amount, and type of all dietary intakes for two consecutive weeks. We have developed software that automatically calculates the intake of food components and macronutrients from these data, using an algorithm that was based on the Dutch Food Composition Table [25]. The entry of nutritional data is important for the *Diameter*, therefore the food diaries of 64 patients included in DIALECT-2 were analysed to evaluate the adherence of registration of the food intake. We evaluated adherence of registration for the respective meals during the day, and whether adherence changes during the two-week registration period. Furthermore, we investigated the daily distribution and between meal variability of registered carbohydrate intake, since this parameter is of particular importance in diabetes. The evaluation of adherence in registering food intake was based on three items: appropriate registration of meal records, appropriate registration of time records, and quantitative and qualitative description of content (e.g., 1 glass of milk), expressed as percentage of total meal moments, with 100% representing full compliance. Carbohydrate intake was calculated for the three main meal moments (i.e., breakfast, lunch, and dinner), the in-between meals, and the total intake in grams per day. Variability in carbohydrate intake for each meal was calculated using the within-person coefficient of variation (COV). Next, we calculated the number of days needed to get a representative estimation of a person's true intake with a specified degree of error using the calculated COV [26].

Second, we assessed the requirements for the collection of data on physical activity. To this end, we evaluated the process of data collection on physical activity in the first 98 patients that were included in DIALECT-2. These patients wore a step counting device for one week (Fitbit® Flex, Fitbit Inc., San Francisco, CA, USA) to measure physical activity and sedentary behaviour. The measurements with the activity trackers were analysed in terms of the use of activity sensors by patients, and activity behaviour in terms of activity bouts, sedentary behaviour, and sedentary bouts. The Fitbit has previously been validated for measuring step count during aerobe activities and for measuring sedentary behaviour [27–29]. Raw Fitbit data (steps/min) were organized into ready variables by an algorithm written in MATLAB® (2016b, The MathWorks, Inc., Natick, MA, USA). We evaluated whether the patients experience limitations in the use of these devices and which are the technical limitations.

We used a flash glucose monitoring (FGM) device, the Freestyle Libre® (Abbott Diabetes Care, Alameda, CA, USA), to evaluate the experienced limitations by the first 60 patients that were included in DIALECT-2 [30]. The Freestyle Libre sensor measures the average subcutaneous interstitial glucose level every 15 min throughout a two-week period. These glucose values, which serve as a surrogate for blood glucose, are transferred to a reader by the patients every 8 h. After the registration period of two weeks, the glucose data was derived from the reader, analysed by a MATLAB script, and then used for further analysis when ≥ 5 days of measurements, each with $\geq 90\%$ available measurements, were available. Again, the experiences with the use of these kind of glucose sensors are translated into requirements for the *Diameter*.

2.2. Literature Search

We reviewed the existing mobile health (mHealth) applications that monitor dietary intake, physical activity, or glucose values to investigate the experiences with those applications and to determine the strength and limitations. There are many basic features in existing applications with undisputable value. We implicitly intend to use current standard features for the *Diameter* and will not formulate these as separate requirements. Rather, for the purpose of this study, we focus on functionalities that are either new or not yet standard.

A PubMed search was performed in the beginning of 2017 using a combination of the following search terms: "application", "diabetes", "diet", "eHealth", "food", "food diary", "mHealth", "nutrition", "nutrition diary", "smartphone", "physical activity", "exercise", "blood glucose", or "continuous glucose monitoring". A selection of relevant studies, which were performed in T2DM, was used to derive requirements for the *Diameter* based on the results.

2.3. Pilot Studies

We performed two pilot studies to gain insight into what people experience or want to experience using a mobile application. The aim of pilot study 1 was to identify the strengths and limitations of the state-of-the-art existing applications for diet registration. Therefore, in the beginning of 2017, the major mobile platforms (Apple iOS App Store, Google Play Store) were searched to find suitable diet registration mHealth applications. The following search terms were used (translated in Dutch): eating, calorie counter, food, food diary, nutrition, nutrition diary, and diabetes. mHealth applications with the following criteria were included: functionality on both Apple iOS and Android, at least fifty user reviews or ratings, capabilities for nutritional monitoring, the presence of a searchable nutritional database, the use of Dutch measurement units (e.g., gram, slices, cups), and the latest version of the application must have been released after 2014. Paid mHealth applications or those that were not available in Dutch were excluded [31]. For applications that met the criteria, all relevant information, such as the developer and the average user rating, was documented. This resulted in eight applications that were downloaded onto a smartphone by the researcher and then evaluated on certain key characteristics, including e.g., the use of the Dutch Food Composition Table as nutritional database [25], the possibility of manual entry of products, foods specificity (e.g., whole grain bread or white bread), and specific time of ingestion. Based on these criteria, the three most suited mHealth applications were selected for further testing by healthy volunteers for three days, in random order. After the test period, the usability of the mHealth applications was determined using a ten item-questionnaire that was based on the System Usability Scale (SUS) and the Unified Theory of Acceptance and Use of Technology (UTAUT) that was specifically designed for this purpose, and by two open questions to appoint the strong and weak characteristics [32–34]. The SUS assesses the usability and the UTAUT predicts the intention to use a technology [35].

In pilot study 2, the aim was to determine disease awareness in T2DM patients. To this end, we evaluated the awareness and knowledge of T2DM patients on the subject of healthy lifestyle and their illness. Additionally, we assessed patients' requirements in future (coaching) technology that supports better diabetes management and healthy lifestyle choices [36,37].

2.4. Expert Meetings

During monthly expert meetings, the progress regarding the development of the *Diameter* was evaluated. The average number of attendants at these meetings was 10, including a wide array of experts, i.e., clinicians, professors in telemedicine, engineers, software developers, and researchers with expertise in (technical) medicine, nutrition, biomedical engineering, and computer science. During these meetings, the preliminary requirements that were derived from the cohort study, literature, and pilot studies were presented and discussed. Additionally, requirements were formulated during these meetings based on the expertise of the participants in the meetings. Taking existing applications regarding diet, physical activity, and glucose values as a starting point, the requirements that were distinctive as compared to the existing applications described in literature were formulated according to the FICS categories by a technical physician and then discussed with the experts. The requirements were prioritized with "must have", "should have", "could have", or "wish to have" during brainstorms with the experts. This approach was used to ensure clear future communication with software developers regarding the required system functionality.

3. Results

The requirements that were derived from the cohort study, literature, pilot studies, and expert meetings are described below for the separate components (diet, physical activity/sedentary behaviour, glucose values) and as shared (i.e., applicable on all categories) requirements. The requirements are noted with identification (ID) in parentheses, coded using an F, I, C, or S to classify the requirements according to the FICS structure. An overview of the requirements, with corresponding ID numbers, can be found in the four Supplementary Tables S1–S4.

3.1. Requirement for Measuring Dietary Intake

3.1.1. Cohort Study

We evaluated two-week food diaries in 64 patients who participated in DIALECT-2. The total carbohydrate intake of breakfast, lunch, dinner, and in-between meals contributed to 20.0%, 22.8%, 28.2%, and 29.0% of total intake, respectively. The COV was for each meal moment 45.0, 49.9, 53.4, and 70.2, respectively. Consequently, as the COV for carbohydrate intake for in-between meals was the highest, the in-between meals require the highest number of days (11 days) for a valid estimate within

30% of the true intake, higher than the number of days that are needed for breakfast (four days), lunch (eight days), and dinner (nine days) [38]. This information is necessary, because the compliance of registering diet decreased with in total 4.2% in the two-week period [26]. This indicates that, without further intervention, these patients will not register their diet properly during longer periods. Patients need to know what the minimum number of days is to enter their diet to be able to help them properly with their diabetes management (I1). This will help in motivating them to adhere. In addition, the *Diameter* should be intelligent and learn from the previously entered food items and use this to show smart options to ease dietary registration (F1). For example, it should inquire whether the patient ate their usual breakfast and, if not, provide the option to enter an alternative breakfast. For dinner, the system must automatically save different frequently used meals, and it should ask the patient whether they had eaten one of the saved meals, and, if not, have the option to change the saved meals (e.g., portion size). Secondly, to prevent the decline and insufficient registration of diet, the system must keep track of a personalized history per type of meal to prevent the patient from having to go over long lists of food options (F2).

In 41% of the patients, the overall description of food intake was not sufficient to draw meaningful conclusions, with the lowest compliance for dinner registration: Of the 64 patients, 26 were excluded due to too low compliance, and in 19 of these, this was due to inadequate description of the dinner [26]. In order to improve the overall adherence for diet registration, dinner entry in particular must be specifically designed to prevent underreporting. Therefore, we conclude that the system should split the dinner into main components, such as rice/potatoes/pasta, vegetables, meat/fish/meat substitute, gravy/sauce, and other (dessert/drinks etc.) (F3). Besides the data entry for dinner, the complete reporting of beverages is an item of concern. To reduce underreporting, we decided that the system must ask whether the patient drank something when food is entered without beverage (I2).

3.1.2. Literature Search

The literature search resulted in a selection of seven articles, which were suitable to derive some important requirements for the *Diameter*. These articles report that the use of mHealth applications for monitoring nutrition results in barriers for both patients and clinicians [39,40]. Existing mHealth applications often only record the number of calories, whereas for diabetic patients, specific nutritional components have particular interest for blood glucose control, i.e., carbohydrates [41]. Specific information is necessary to demonstrate the impact of a specific type of food on the glucose levels [31,42]. This results in the requirements that the amount of carbohydrates must be displayed (F4) and the effect of carbohydrates or a specific type of food on the glucose value should be shown (F5). Patients also appreciate to see, in addition to the amount of consumed calories, the amount of calories that are left to eat that day and the amount of calories burnt by their physical activity (F6) [43]. Finally, healthy recipe suggestions should also be given (C1) [44].

3.1.3. Pilot Studies

In pilot study 1, 20 healthy volunteers tested the usability of three existing mHealth applications; FatSecret, MyFitnessPal, and Virtuagym, for the monitoring of dietary intake. We demonstrated that improvements are necessary to implement a food tool in clinical practice. The usability study led to a number of requirements for the *Diameter*. Firstly, the time and date of ingestion should be registered (F7) before a meal or food product can be entered (S1), and the current date should be shown by default (S2). Furthermore, household measurements must be connected to the food product (F8), it should show pictures of food products (S3), it should be possible change or remove entered data (F9), it should be possible to add food products to a meal category, such as breakfast, lunch, and dinner (F10), a suggestion must appear when typing the first letters of a food product (I3), it must be made clear and easy to find the right food products (C3 and C4), an overview must be given of the consumed calories, amount of fat, carbohydrates, proteins, and percentage dietary reference intakes (F4 and F11), and often used food products must be remembered as is also described as result of the cohort study (F3). As addition to manual entry, having the optional use of a barcode scanner is often desired (F12), it should be possible to use the system on a website on the laptop or personal computer (I4), the application should send reminders or push messages to stimulate complete data input (I5), and it should provide short educational facts, e.g., explain what a calorie is (C5 and S4) [35].

3.1.4. Expert Meetings

A main point, not yet addressed in the pilot, cohort, and literature study, was how to motivate patients to persistently use the application. Therefore, the focus of one expert meeting was on this subject. The attendants were challenged to make a quick design of a food registration tool with particular focus on attraction and ease of use, and this resulted in some requirements that aimed to make the tool interactive and visual. One requirement that followed was to display a graphical image of a plate, bowl, and a glass and to give the option to virtually drag food and drink to this plate and glass (S5). The idea behind this is that the amounts of food can be more easily determined when presented on the virtual plate. Also, there are applications in development that use photography technology to assess diet. This is a promising approach for entry food data, reducing the number of actions that are required to simply taking a picture of the meal [16,45]. The incorporation of nutrient estimation using photos taken with the mobile is desired (F13). People usually have a limited repertoire of meals, hence there must be an option to enter standard meals and to let the system remember earlier registered meals (F14), and there should also be the option to enter and save own recipes for re-use (I6). Fourthly, the system should ask smart questions (F15). For example, after initial inquiries on the regular use of milk and sugar in coffee, the system can ask, 'did you add sugar as you usually do?' when the user registers intake of coffee. This reduces the number of required actions, gives the impression of a personalized approach, and makes it interactive.

Other requirements that followed from the expert meetings are to store the data output per day and present it separately for breakfast, lunch, dinner, in-between meals, and total carbohydrate intake (F16). This gives a clear overview of the carbohydrate distribution over the day. Secondly, the application should be able to give a healthier option of a product that the patient entered (C6), e.g., whole grain bread instead of white bread. This gives the patient insight in healthier decisions. Thirdly, there must be an option to add new products to the database to keep the food database up to date (F17). Food products that are added by patients should be entered in a separate database. The products in this database can be checked on nutritional value and then added to the general database to give all users the ability to register this product. Fourthly, the app must contain a guideline with instructions to monitor food (C7). Fifthly, the data output is presented in gram per day. For each product, the amount of e.g., carbohydrates in gram per day should be calculated, based on nutritional values per 100 g of the food item, as noted in the Dutch Food Composition Table (C8). Finally, it should be possible to use the system independent of internet connection. At any time of the day, patients should have access to the food record to limit memory bias (I7).

3.2. Requirements for Measuring Physical Activity and Sedentary Behaviour

3.2.1. Cohort Study

A Fitbit activity tracker was worn by 98 patients of the DIALECT-2 cohort until then to measure the number of steps per minute during one week. During these measurements we gained general insight in the applicability of such sensors in daily practice, but also the data on physical activity and sedentary behaviour of T2DM patients were helpful.

Regarding the applicability, the battery of the accelerometer needs to be charged by the user approximately every four to five days for 2 h. Some patients forgot to re-attach the sensor after charging. Some activities, like cycling, a common activity in the Netherlands, are not registered while using an accelerometer around the wrist. Based on these findings, the following requirements for the Diameter were formulated: The system must give a notification when the battery is empty (F18) and needs detection when the tracker is not worn to give a reminder to wear it (F19). Activities that a regular accelerometer around the wrist does not detect must also be measured by using an application on the mobile device that detects cycling (F20) or by the manual insertion of non-recordable activities (I8).

Regarding physical activity, only a few patients met the criteria of an activity bout, which is an activity of at least 10 consecutive minutes of moderate to vigorous physical activity (MVPA) with ≥ 95 steps/min [46,47]. About two-thirds of the patients (69%) had no bout of MVPA at all in seven days, whereas ≥ 150 min MVPA per week is recommended, e.g., to reduce the risk of cardiovascular diseases [5,6,48]. Nevertheless, the vast majority (93%) were able to achieve the intensity of moderate activity at one point during the follow-up time, however the duration of the activity was too short [49]. The *Diameter* should provide education for patients regarding activity bouts and physical activity (C9), in order to stimulate patients to achieve MVPA bouts, which are beneficial for their diabetes regulation [42,44]. Also, the application should detect when the intensity of MVPA or duration of an activity bout is not met, in order to provide the basis for the future coaching module to be developed (F21).

Independent of the amount of MVPA, sedentary behaviour increases the risk of morbidity and mortality [50–53], and it is recommended that sedentary bouts are be interrupted every 30 min with an activity of light intensity of more than 10 steps/min for at least 1 min [48,54]. Our data demonstrated that our T2DM patients had no movement for 76% of the total waking hours per day, of which 7 h were spent in prolonged sedentary bouts of at least 30 consecutive minutes [49]. We therefore want the system to create awareness regarding sedentary behaviour and motivate patients to minimize sedentary time (C10).

3.2.2. Literature Search

There are various studies addressing the issue of evaluating physical activity while using mHealth applications. An important finding was that patients prefer the visual demonstration of their activities, by, for example, bar charts over merely numerical presentation (S6) [55]. Also, patients liked to be reminded when they were inactive for a prolonged time period (I9) [43]. Literature also confirmed the notion of paying attention to the management of body weight management, being closely related to physical activity and diet. Weight loss strongly contributes to improved glycaemic control. It is therefore necessary to show patients the burned calories during a performed activity (C11) and to have an option to register body weight (I11) to follow up on weight (loss) in time (C12) [11,44,48,50]. One of the aims in lifestyle management is to reduce sedentary behaviour. The *Diameter* should therefore detect sedentary periods (I11) and it must give educational information regarding sedentary behaviour (C10), as also described in the cohort study [39,56].

3.2.3. Expert Meetings

During the expert meetings, some additional requirements concerning physical activity and sedentary behaviour were formulated: First, there is a strong preference for the option to connect multiple types of activity sensors (I12). Secondly, the system must start with measuring of a baseline to determine the current activity and sedentary behaviour of the patient (F22). When, in the future, a coaching module is added in the *Diameter*, these baseline data should be available.

3.3. Requirements for Measuring Glucose Values

3.3.1. Cohort Study

In the DIALECT-2 cohort, we performed flash continuous glucose measurements in 60 patients during two weeks. Following, we gained insights in the applicability of FGM sensors, the level of glycaemic control, and the daily glucose variability between patients, which are used to formulate requirements.

First, we investigated the usage of the sensor. Of the 60 patients, 12 patients had to be excluded because there were less than five days in which at least 90% of data were available. The most important causes for data loss were the patient forgetting to scan the sensor by the patient and the premature loss of the Freestyle Libre sensor. Of the latter, in 35% of the cases, detachment of the sensor was the problem, 29% of which occurred in the first week. In seven patients, a new sensor was attached. The system must give notifications to reduce data loss by not scanning the sensor (I13) and it should have the option to connect a new Freestyle Libre sensor (I14).

Regarding the glucose level of these patients, the level is, on average, 58.6% of the time that patients are in between the glucose target ranges (≥ 4 mmol/L and ≤ 8.4 mmol/L). The remainder of the time they have hypoglycaemia (4.2%) or hyperglycaemia (37.2%). Patients need the insight in their glucose values, including the percentage of hypo-, hyper-, and normoglycaemic episodes, throughout the day and the progression over time to be better capable in keeping the values between the target ranges (F23). Besides that, they need insight in their glucose variability (F24). The patients with high glucose variability need to have these insights to be able to act better on the glucose values and decrease the variability. To be able to do this, education is needed regarding glucose targets, glucose variability, and on the unfavourable long-term effects of uncontrolled hyperglycaemia (C13).

3.3.2. Literature Search

There are several glucose sensors on the market that are used for continuous glucose monitoring (CGM) in the clinical practice of diabetes management. These so-called real-time CGM sensors are usually used in conjunction with a subcutaneous insulin pump and the use is labour intensive for the patient, because twice daily calibration by finger pricks is still needed, rendering such sensors unsuitable for the purpose of the *Diameter* [57]. We therefore decided to choose the only alternative currently available, i.e., the Freestyle Libre system. It is obvious that using CGM or FGM provides detailed information regarding glycaemic control, with the additional advantage of wireless transfer of the glucose data (I15) [55,58]. For the *Diameter*, these data are to be used to display glucose trends (F23) and to provide the patient and HCP with insight about blood glucose levels and glucose variability (C13), as also described in response to the cohort study. This helps to effectively engage patients and give them the insight that they need [44,58]. The data should be displayed in an easy understandable graph that shows how physical activity and nutrients affect blood glucose (S7) [42].

3.3.3. Expert Meetings

Additional requirements concerning blood glucose were formulated based on the expert meetings. First, the target range for blood glucose should be adjustable to allow for personalized targets (F25). In addition, there is the wish for an option to warn when the glucose values are out of range. (I16). Secondly, it is desired that the *Diameter* can synchronise with insulin pumps and also with other glucose measuring devices than the FGM currently used (F26). This includes the option of manually inserting measured glucose values by a finger-prick (I17). Thirdly, the systems should allow for the development of algorithms that predict glucose levels based on the measured blood glucose data, physical activity data, and food intake data (C14).

3.4. Shared Requirements

3.4.1. Literature Search

Beside requirements for the separate items to be measured, we also derived some general shared requirements from the literature for the *Diameter*. The lack of coordination with the HCPs is a main issue with currently available lifestyle applications. Data that are collected with such applications by an individual are not available for the clinicians, and certainly not in an organized fashion. We want to develop an application that allows for connecting and sharing the data with the care provider (I18) (I19). This can be achieved by the option to automatically generate a report of the data, which is digitally sent to the HCP. The communication and information exchange should contribute to better decision-making [55,58]. Other important requirements are related to the convenience of use.

To this purpose, the language and tone should be accessible, encouraging, and supportive (C15) [59], generally intelligible symbols and terms should be used (S8) [59], and the necessity for scrolling must be minimized (S9) [44]. In addition, medical terminology must be used where needed, but clear explanations should be provided (C16) [44]. Also, keeping the necessary active use to less than 15 min a day is desired (I20) and the applications must be provided in the native language of the user (C17), resulting in the persistent use of the app [60]. Finally, costs appears to be a significant concern, with most people being unwilling to pay anything for apps and discontinuing use when they find that in-app payments are required, therefore the aim is to provide the app for free (I21) [43].

3.4.2. Pilot Studies

In pilot study 2, 19 patients participated in the questionnaires and interviews about awareness and a technology to support diabetes management. Firstly, the patients were asked questions about the following subjects to examine their overall awareness: influence of nutrition and exercise on glycaemic control, diabetes complications, self-management of the respondent, and desirable behaviour on exercise, nutrition, and weight. Approximately 40% of the patients correctly answered half of the questions regarding overall awareness. However, most of the respondents did not have the supposed knowledge regarding the effect of exercise and nutrition on glycaemic regulation. For example, only a few patients were aware of the effect of even a small percentage of weight loss on the improvement of the blood glucose level. The level of awareness and the interviews held with the patient, resulted in requirements to give the patients more knowledge and insight, as noted earlier (F4, F6, F23, C5, C9, C10, C13, S4). New requirements that were mentioned in the interviews were to also incorporate blood pressure measurements (I22) and incorporate an insulin bolus calculator, i.e., a calculator to determine the appropriate insulin dose before the meal, based on the current glucose level and the amount of carbohydrates in the meal (F27) [36].

3.4.3. Expert Meetings

In the expert meetings, it was brought up that patients may have a preference to focus on specific lifestyle items and this may change in time. For example, the aim may be to get more active, instead of focusing on healthy diet. It should therefore become possible to use the application selectively to prevent overload with excessive information (I23). Also, data must be stored according to the European privacy laws (F28). Thirdly, the data must be stored for at least one year (C18). Fourthly, the clocks of all different sensors must be synchronized (F29) to ensure that no discrepancies in time can occur. Fifthly, an overview of the data in the past can be found in an overview per day, per week, and per month (S10). This gives the ability to look at trends. Finally, all of the patients are able to enter all desired data without assistance, independent of education level (I24).

3.5. Overview of the Requirements

The requirements are organised according to the FICS framework and labelled as "must have", "should have", "could have", and "wish to have". In total, 81 requirements were formulated, which are of added value when compared to current applications. Of these, 29 are formulated as functional requirements, 24 as interactive and usability requirements, 18 as substantive and structural, and 10 as style, as can be seen in Table 1. Of these requirements, 74% were labelled as "must" and "should" have and 26% as "could" and "wish to" have. The requirements were labelled during the expert meetings. The "could" and "wish to" have requirements were labelled in this category, because they were valued as requirements that are not of utmost importance in this stage of the development of the app, to use the app, or to receive the desired information and insight. However, these requirements have the potential to increase the ease of use or have the potential to motivate to use the app more.

Table 1. The number of formulated requirements as a result of the cohort study, literature search, pilot study, and expert meetings per FICS category for diet, physical activity (PA), and sedentary behaviour (SB), glucose values and shared.

	Diet (*n*)	PA and SB (*n*)	Glucose Values (*n*)	Shared (*n*)	Total (*n*)
Function & events	17	5	4	3	29
Interaction & usability	7	5	5	7	24
Content & structure	8	4	2	4	18
Style & aesthetics	5	1	1	3	10
	37	15	12	17	81

4. Discussion

We are developing a digital tool, the *Diameter*, with the aim of improving adherence to lifestyle in patients with T2DM. The tool collects and integrates information of diet, physical activity, and glucose values, i.e., items that are pivotal for the management of patients with T2DM. In order to formulate the requirements for this tool, we applied a mixed method design approach in which we used experiences from large scale data collection in a cohort study, performed a literature search, performed pilot studies, and organized expert meetings.

Although many applications are already available to assist patients in monitoring lifestyle behaviour like their diet and physical activity, these applications are not designed for the follow-up of chronic diseases, like diabetes. Therefore, they lack integration with glucose measurements and also do not allow blended care, i.e., function in connection with HCPs [39,40,55,61,62]. Due to the tight relationship between lifestyle and glucose management, an integrated approach using these key elements, together with a blended care setting, were starting points for us. We evaluated existing applications for lifestyle and T2DM to determine the requirements for each part of the *Diameter*. Of these applications, 15 requirements were of value and were therefore adopted. Additionally, 49 new requirements were formulated for the diet, physical activity, and glucose value part during the cohort study, pilot studies, and expert meetings. Besides, shared requirements that are of importance for the integration of the components also had to be formulated, resulting in another 17 shared requirements.

To achieve adherence to lifestyle advice in diabetes, it is important that patients have knowledge on how lifestyle behaviours affect their condition. The pilot study concerning awareness showed there is an urge to improve knowledge of patients on the effects of carbohydrates and activity on the glucose values. We expect that demonstrating the effects that food choices directly have on glucose level will provide a very strong feedback mechanism that may help to stimulate healthy behaviour. The necessity for lifestyle interventions is emphasized by the results of our cohort study that there is a lot of room for improvement on diet, physical activity, sedentary behaviour, and glucose values. Therefore, 14 requirements, mostly labelled as content and structure and style and aesthetics, of the *Diameter* are related to providing insight in counting carbohydrates, glucose values, physical activity, and how these factors mutually influence each other.

Almost half of the requirements, 37 out of 81, are formulated for diet registration. From our findings in the cohort study and the pilot study it was clear that, in order to gain complete and reliable entry of dietary information, the functionality of a new nutrition entry tool needs substantial improvement when compared with existing diet applications. The existing applications require frequent manual data entry and the process of data entry is relatively time consuming. The requirements defined to solve these issues relate to incorporating smart options, mostly labelled as function and events requirements. E.g., the application must learn from past entered data and the use this information to ask personalized questions and to give personalized suggestions in the future. Another functional requirement to increase the ease of use was to enter a main meal by using pre-defined components.

The integration of smart options is important, not only for the dietary entry, but also in the other components of the *Diameter*. For example, the application must recognize when the glucose sensor needs to be scanned, when the patient has not worn the activity tracker for too long, and it must detect activities that the activity tracker cannot measure.

Incorporating the right interaction and usability requirements are also of importance in making the application more user-friendly and to stimulate maintenance of use. Examples of such requirements are the possibility to enter own recipes, the possibility to use the app without access to the internet, to allow own choice of activity tracker and glucose sensor, to enable potential connection with a (digital) healthcare professional, and individual choice for which educational modules are switched on/-off.

As stated above, providing insight in the effect of diet and activity on the glucose values is a key element of the *Diameter*. To this end, the application can generate (e.g., past day, week or month) a graphical presentation of these data and can recognize and show trends for different time periods. Reports of the data can be shared with the diabetes professional, allowing for blended care.

The strengths of this study are that the requirements were developed from an integration of four approaches (cohort study, literature search, pilot studies, and expert meetings), that the requirements are developed in an iterative process, and the main components for T2DM treatment are taken into account. However, there are a few sources of potential bias. There could be selection bias, because the requirements were developed from a complicated T2DM population, which was located in a specific region in the Netherlands. It is possible that this resulted in other requirements than we would have found by researching a population less complicated or with different ethnical background. Also, it is possible that a different composition of the group of experts would have led to other requirements due to differences in personal experience and specific knowledge. However, when expanding to other populations, the iterative approach of the development process should circumvent these potential sources of bias.

With the formulation of the initial requirements that are described in this paper, the first step of the development of the *Diameter* has been taken. From here, we intend to let patients use a first version of the Diameter in a new pilot study. The data generated in this pilot study will serve several purposes. First, an enhanced set of requirements will be derived. We also intend to formulate requirements for the monitoring of adherence to pharmacological therapy. Also, in a parallel process, the data will be used as input for the development of the coaching part of the *Diameter* to be developed. The coaching part will be designed to provide data-driven tailored coaching, based on inter alia, individual patient data, preferences, and comorbidities in order to achieve and maintain adherence to lifestyle recommendations.

In summary, the development of the *Diameter* is an iterative process, using a multi-method approach. Every time after introducing a new version, a pilot study will be performed to evaluate the app in terms of effectiveness, acceptance, and feasibility, and to fine-tune the requirements.

5. Conclusions

The development of a new tool for patients with T2DM is important, because insight in their diet, activity, and glucose values is currently lacking. This study describes the development of the requirements for the first version of the *Diameter*, which are focused on gathering the necessary data and giving patients insight. For the development of future versions of the Diameter, in which a tailored data-driven coaching module will be incorporated inter alia, it is an important step to be able to efficiently collect lifestyle and glucose data.

Future research is needed to develop the desired application for the patients to receive tailored coaching, blended in their healthcare. This future research will focus on development of the tailored coaching module and algorithms that take into account the duration of diabetes, comorbidities, and personal preferences. Also, research is necessary to optimize the usability of the application for patients and HCPs by evaluating experiences of different diabetes populations (e.g., first line and second line healthcare) with the app, evaluating feasibility of the application integrated in clinical care, and by optimizing adoption and implementation of the application.

Supplementary Materials: The following are available online at http://www.mdpi.com/2072-6643/11/2/409/s1, Table S1: requirements of the dietary part of the Diameter, Table S2: requirements of the activity part of the

Diameter, Table S3: requirements of the glucose measurement part of the Diameter, Table S4: requirements of the shared part of the Diameter.

Author Contributions: M.M.R.V.-H., H.J.H., B.-J.F.v.B., G.D.L. and N.B. designed the study. N.B., M.M.O., C.M.G. and I.J.M.H. included patients and performed the measurements. N.B., M.M.R.V.-H. and G.D.L. wrote the paper. All the authors approved the final version of the paper.

Funding: This research was financially supported by an unrestricted research grant from the Pioneers in Health Care Innovation Fund, established by the University of Twente, Medisch Spectrum Twente and ZiekenhuisGroep Twente and by the Dutch Diabetes Research Foundation (grant No. 2017.30.005). N.B. is appointed by the grant Exceptional and Deep Intelligent Coach (EDIC, grant No. 628.011.021) financed by the Netherlands Organisation for Scientific Research (NWO).

Acknowledgments: We thank Michèle H.T. Lankheet, Rosan S. Nobbenhuis, Anouk S. ten Voorde, Nienke F.M. Tuinstra, Anna I. Bouwer and Annis C. Jalving for their contribution to the patient inclusion. We thank David R. de Meij for developing the digital tool for calculation of the nutrients from the paper food diaries.

Conflicts of Interest: The authors declare no conflict of interest.

References

1. International Diabetes Federation. *IDF Diabetes Atlas, 8th Edition 2017*; International Diabetes Federation: Brussels, Belgium, 2017.
2. World Health Organization. *Global Report on Diabetes*; WHO Press: Geneva, Switzerland, 2016; ISBN 978-92-4-156525-7.
3. Marin-Penalver, J.J.; Martin-Timon, I.; Sevillano-Collantes, C.; Del Canizo-Gomez, F.J. Update on the treatment of type 2 diabetes mellitus. *World J. Diabetes* **2016**, *7*, 354–395. [CrossRef] [PubMed]
4. Gambardella, J.; Santulli, G. Integrating diet and inflammation to calculate cardiovascular risk. *Atherosclerosis* **2016**, *253*, 258–261. [CrossRef] [PubMed]
5. Santulli, G.; Ciccarelli, M.; Trimarco, B.; Iaccarino, G. Physical activity ameliorates cardiovascular health in elderly subjects: The functional role of the beta adrenergic system. *Front. Physiol.* **2013**, *4*, 209. [CrossRef] [PubMed]
6. Rutten, G.E.H.M.; De Grauw, W.J.C.; Nijpels, G.; Houweling, S.T.; Van de Laar, F.A.; Bilo, H.J.; Holleman, F.; Burgers, J.S.; Wiersma, T.J.; Janssen, P.G.H. NHG-Standaard Diabetes mellitus type 2 (derde herziening). *Huisarts Wet.* **2013**, *56*, 512–525.
7. Broom, D.; Whittaker, A. Controlling diabetes, controlling diabetics: Moral language in the management of diabetes type 2. *Soc. Sci. Med.* **2004**, *58*, 2371–2382. [CrossRef] [PubMed]
8. Ong, W.M.; Chua, S.S.; Ng, C.J. Barriers and facilitators to self-monitoring of blood glucose in people with type 2 diabetes using insulin: A qualitative study. *Patient Prefer. Adherence* **2014**, *8*, 237–246. [CrossRef] [PubMed]
9. Polonsky, W.H.; Fisher, L.; Schikman, C.H.; Hinnen, D.A.; Parkin, C.G.; Jelsovsky, Z.; Petersen, B.; Schweitzer, M.; Wagner, R.S. Structured self-monitoring of blood glucose significantly reduces A1C levels in poorly controlled, noninsulin-treated type 2 diabetes: Results from the Structured Testing Program study. *Diabetes Care* **2011**, *34*, 262–267. [CrossRef]
10. Wolpert, H.A.; Anderson, B.J. Management of diabetes: Are doctors framing the benefits from the wrong perspective? *BMJ* **2001**, *323*, 994–996. [CrossRef]
11. Kadirvelu, A.; Sadasivan, S.; Ng, S.H. Social support in type II diabetes care: A case of too little, too late. *Diabetes Metab. Syndr. Obes.* **2012**, *5*, 407–417. [CrossRef]
12. Gant, C.M.; Binnenmars, S.H.; Berg, E.V.D.; Bakker, S.J.L.; Navis, G.; Laverman, G.D. Integrated Assessment of Pharmacological and Nutritional Cardiovascular Risk Management: Blood Pressure Control in the DIAbetes and LifEStyle Cohort Twente (DIALECT). *Nutrients* **2017**, *9*, 709. [CrossRef]
13. Gant, C.M.; Binnenmars, S.H.; Harmelink, M.; Soedamah-Muthu, S.S.; Bakker, S.J.L.; Navis, G.; Laverman, G.D. Real-life achievement of lipid-lowering treatment targets in the DIAbetes and LifEstyle Cohort Twente: Systemic assessment of pharmacological and nutritional factors. *Nutr. Diabetes* **2018**, *8*, 24. [CrossRef] [PubMed]
14. Oosterom, N.; Gant, C.M.; Ruiterkamp, N.; van Beijnum, B.F.; Hermens, H.; Bakker, S.J.L.; Navis, G.; Vollenbroek-Hutten, M.M.R.; Laverman, G.D. Physical Activity in Patients with Type 2 Diabetes: The Case for Objective Measurement in Routine Clinical Care. *Diabetes Care* **2018**, *41*, e50–e51. [CrossRef] [PubMed]

15. Nurmi, J.; Knittle, K.; Helf, C.; Zwickl, P.; Lusilla Palacios, P.; Castellano Tejedor, C.; Costa Requena, J.; Myllymäki, T.; Ravaja, N.; Haukkala, A. A Personalised, Sensor-Based Smart Phone Intervention for Physical Activity and Diet – PRECIOUS N-of-1 Trial. *Front. Public Health* **2016**, *4*. [CrossRef]

16. Mezgec, S.; Korousic Seljak, B. NutriNet: A Deep Learning Food and Drink Image Recognition System for Dietary Assessment. *Nutrients* **2017**, *9*, 657. [CrossRef]

17. Helbostad, J.L.; Vereijken, B.; Becker, C.; Todd, C.; Taraldsen, K.; Pijnappels, M.; Aminian, K.; Mellone, S. Mobile Health Applications to Promote Active and Healthy Ageing. *Sensors* **2017**, *17*, 622. [CrossRef] [PubMed]

18. Holtz, B.; Lauckner, C. Diabetes management via mobile phones: A systematic review. *Telemed. J. E-Health* **2012**, *18*, 175–184. [CrossRef] [PubMed]

19. Graffigna, G.; Barello, S.; Bonanomi, A.; Menichetti, J. The Motivating Function of Healthcare Professional in eHealth and mHealth Interventions for Type 2 Diabetes Patients and the Mediating Role of Patient Engagement. *J. Diabetes Res.* **2016**, *2016*, 2974521. [CrossRef] [PubMed]

20. Kumar, S.; Nilsen, W.J.; Abernethy, A.; Atienza, A.; Patrick, K.; Pavel, M.; Riley, W.T.; Shar, A.; Spring, B.; Spruijt-Metz, D.; et al. Mobile health technology evaluation: The mHealth evidence workshop. *Am. J. Prev. Med.* **2013**, *45*, 228–236. [CrossRef]

21. Forster, H.; Walsh, M.C.; Gibney, M.J.; Brennan, L.; Gibney, E.R. Personalised nutrition: The role of new dietary assessment methods. *Proc. Nutr. Soc.* **2016**, *75*, 96–105. [CrossRef]

22. Webb, T.L.; Joseph, J.; Yardley, L.; Michie, S. Using the internet to promote health behavior change: A systematic review and meta-analysis of the impact of theoretical basis, use of behavior change techniques, and mode of delivery on efficacy. *J. Med. Internet Res.* **2010**, *12*, e4. [CrossRef]

23. Widya, I.; Bults, R.; de Wijk, R.; Loke, B.; Koenderink, N.; Batista, R.; Jones, V.; Hermens, H. *Requirements for a Nutrition Education Demonstrator*; Springer: Berlin/Heidelberg, Germany, 2011; pp. 48–53.

24. Jackson, M. The meaning of requirements. *Ann. Softw. Eng.* **1997**, *3*, 5–21. [CrossRef]

25. National Institute for Public Health and the Environment (RIVM). *Dutch Food Composition Database—NEVO Online Version 2016/5.0*; RIVM: Bilthoven, The Netherlands, 2016.

26. Oosterwijk, M.M. Dietary Assessment and Carbohydrate Variability in Patients with Type 2 Diabetes Mellitus. Master's Thesis, Wageningen University & Research, Wageningen, The Netherlands, 2018.

27. Nelson, M.B.; Kaminsky, L.A.; Dickin, D.C.; Montoye, A.H. Validity of Consumer-Based Physical Activity Monitors for Specific Activity Types. *Med. Sci. Sports Exerc.* **2016**, *48*, 1619–1628. [CrossRef] [PubMed]

28. Kooiman, T.J.; Dontje, M.L.; Sprenger, S.R.; Krijnen, W.P.; van der Schans, C.P.; de Groot, M. Reliability and validity of ten consumer activity trackers. *BMC Sports Sci. Med. Rehabil.* **2015**, *7*, 24. [CrossRef] [PubMed]

29. Diaz, K.M.; Krupka, D.J.; Chang, M.J.; Peacock, J.; Ma, Y.; Goldsmith, J.; Schwartz, J.E.; Davidson, K.W. Fitbit(R): An accurate and reliable device for wireless physical activity tracking. *Int. J. Cardiol.* **2015**, *185*, 138–140. [CrossRef]

30. Bailey, T.; Bode, B.W.; Christiansen, M.P.; Klaff, L.J.; Alva, S. The Performance and Usability of a Factory-Calibrated Flash Glucose Monitoring System. *Diabetes Technol. Ther.* **2015**, *17*, 787–794. [CrossRef] [PubMed]

31. Darby, A.; Strum, M.W.; Holmes, E.; Gatwood, J. A Review of Nutritional Tracking Mobile Applications for Diabetes Patient Use. *Diabetes Technol. Ther.* **2016**, *18*, 200–212. [CrossRef] [PubMed]

32. Brooke, J. SUS: A "quick and dirty" usability scale. In *Usability Evaluation in Industry*; Jordan, P.W., Thomas, B., Weerdmeester, B.A., McClelland, A.L., Eds.; Taylor and Francis: London, UK, 1996; pp. 189–194.

33. Bangor, A.; Kortum, P.T.; Miller, J.T. An Empirical Evaluation of the System Usability Scale. *Int. J. Hum. Comput. Interact.* **2008**, *24*, 574–594. [CrossRef]

34. Bangor, A.; Kortum, P.; Miller, J. Determining what individual SUS scores mean: Adding an adjective rating scale. *J. Usability Stud.* **2009**, *4*, 114–123.

35. Nobbenhuis, R. Type 2 Diabetes Mellitus and Lifestyle; Insufficient Adherence to the Dutch Standards of Healthy Physical Activity and Healthy Nutrition: Oppurtunities for Improvement with mHealth? Bachelor's Thesis, Vrije Universiteit Amsterdam, Amsterdam, The Netherlands, 2017.

36. Lankheet, M.H.T. Development of a Coaching Technology for Diabetes Type 2 Patients to Motivate Them into Exercise and Nutrition (Lifestyle) Changes. Master's Thesis, University of Twente, Enschede, The Netherlands, 2017.

37. Vollenbroek-Hutten, M.M.; Lankheet, M.H.T.; Goolkate, T.; van Beijnum, B.J.; Hegeman, H.H. Insufficient Behavioral Change Skill Hampers Adoption of Ehealth Services. In Proceedings of the 4th International Conference on Information and Communication Technologies for Ageing Well and e-Health, Funchal, Madeira, Portugal, 22–23 March 2018. [CrossRef]

38. Palaniappan, U.; Cue, R.I.; Payette, H.; Gray-Donald, K. Implications of day-to-day variability on measurements of usual food and nutrient intakes. *J. Nutr.* **2003**, *133*, 232–235. [CrossRef]

39. Rollo, M.E.; Aguiar, E.J.; Williams, R.L.; Wynne, K.; Kriss, M.; Callister, R.; Collins, C.E. eHealth technologies to support nutrition and physical activity behaviors in diabetes self-management. *Diabetes Metab. Syndr. Obes.* **2016**, *9*, 381–390. [CrossRef]

40. Hoppe, C.D.; Cade, J.E.; Carter, M. An evaluation of diabetes targeted apps for Android smartphone in relation to behaviour change techniques. *J. Hum. Nutr. Diet.* **2017**, *30*, 326–338. [CrossRef] [PubMed]

41. AlEssa, H.B.; Bhupathiraju, S.N.; Malik, V.S.; Wedick, N.M.; Campos, H.; Rosner, B.; Willett, W.C.; Hu, F.B. Carbohydrate quality and quantity and risk of type 2 diabetes in US women. *Am. J. Clin. Nutr.* **2015**, *102*, 1543–1553. [CrossRef] [PubMed]

42. Chomutare, T.; Fernandez-Luque, L.; Arsand, E.; Hartvigsen, G. Features of mobile diabetes applications: Review of the literature and analysis of current applications compared against evidence-based guidelines. *J. Med. Internet Res.* **2011**, *13*, e65. [CrossRef]

43. Krebs, P.; Duncan, D.T. Health App Use Among US Mobile Phone Owners: A National Survey. *JMIR mHealth uHealth* **2015**, *3*, e101. [CrossRef] [PubMed]

44. Pal, K.; Dack, C.; Ross, J.; Michie, S.; May, C.; Stevenson, F.; Farmer, A.; Yardley, L.; Barnard, M.; Murray, E. Digital Health Interventions for Adults with Type 2 Diabetes: Qualitative Study of Patient Perspectives on Diabetes Self-Management Education and Support. *J. Med. Internet Res.* **2018**, *20*, e40. [CrossRef] [PubMed]

45. Rhyner, D.; Loher, H.; Dehais, J.; Anthimopoulos, M.; Shevchik, S.; Botwey, R.H.; Duke, D.; Stettler, C.; Diem, P.; Mougiakakou, S. Carbohydrate Estimation by a Mobile Phone-Based System Versus Self-Estimations of Individuals with Type 1 Diabetes Mellitus: A Comparative Study. *J. Med. Internet Res.* **2016**, *18*, e101. [CrossRef] [PubMed]

46. Tudor-Locke, C.; Sisson, S.B.; Collova, T.; Lee, S.M.; Swan, P.D. Pedometer-Determined Step Count Guidelines for Classifying Walking Intensity in a Young Ostensibly Healthy Population. *Can. J. Appl. Physiol.* **2005**, *30*, 666–676. [CrossRef]

47. Marshall, S.J.; Levy, S.S.; Tudor-Locke, C.E.; Kolkhorst, F.W.; Wooten, K.M.; Ji, M.; Macera, C.A.; Ainsworth, B.E. Translating physical activity recommendations into a pedometer-based step goal: 3000 steps in 30 minutes. *Am. J. Prev. Med.* **2009**, *36*, 410–415. [CrossRef]

48. American Diabetes Association. Standards of Medical Care in Diabetes-2017: Summary of Revisions. *Diabetes Care* **2017**, *40*, S4–S5. [CrossRef]

49. Hagedoorn, I. What is the Actual Daily Movement in Patients with Complicated Type 2 Diabetes Mellitus? Master's Thesis, University of Groningen, Groningen, The Netherlands, 2017.

50. Wijndaele, K.; Brage, S.; Besson, H.; Khaw, K.T.; Sharp, S.J.; Luben, R.; Wareham, N.J.; Ekelund, U. Television viewing time independently predicts all-cause and cardiovascular mortality: The EPIC Norfolk study. *Int. J. Epidemiol.* **2011**, *40*, 150–159. [CrossRef]

51. Thorp, A.A.; Owen, N.; Neuhaus, M.; Dunstan, D.W. Sedentary behaviors and subsequent health outcomes in adults a systematic review of longitudinal studies, 1996-2011. *Am. J. Prev. Med.* **2011**, *41*, 207–215. [CrossRef] [PubMed]

52. Grontved, A.; Hu, F.B. Television viewing and risk of type 2 diabetes, cardiovascular disease, and all-cause mortality: A meta-analysis. *JAMA* **2011**, *305*, 2448–2455. [CrossRef] [PubMed]

53. Dunstan, D.W.; Barr, E.L.; Healy, G.N.; Salmon, J.; Shaw, J.E.; Balkau, B.; Magliano, D.J.; Cameron, A.J.; Zimmet, P.Z.; Owen, N. Television viewing time and mortality: The Australian Diabetes, Obesity and Lifestyle Study (AusDiab). *Circulation* **2010**, *121*, 384–391. [CrossRef] [PubMed]

54. Colberg, S.R.; Sigal, R.J.; Yardley, J.E.; Riddell, M.C.; Dunstan, D.W.; Dempsey, P.C.; Horton, E.S.; Castorino, K.; Tate, D.F. Physical Activity/Exercise and Diabetes: A Position Statement of the American Diabetes Association. *Diabetes Care* **2016**, *39*, 2065–2079. [CrossRef] [PubMed]

55. Arsand, E.; Tatara, N.; Ostengen, G.; Hartvigsen, G. Mobile phone-based self-management tools for type 2 diabetes: The few touch application. *J. Diabetes Sci. Technol.* **2010**, *4*, 328–336. [CrossRef] [PubMed]

56. Schoeppe, S.; Alley, S.; Rebar, A.L.; Hayman, M.; Bray, N.A.; Van Lippevelde, W.; Gnam, J.-P.; Bachert, P.; Direito, A.; Vandelanotte, C. Apps to improve diet, physical activity and sedentary behaviour in children and adolescents: A review of quality, features and behaviour change techniques. *Int. J. Behav. Nutr. Phys. Act.* **2017**, *14*. [CrossRef] [PubMed]

57. Edelman, S.V.; Argento, N.B.; Pettus, J.; Hirsch, I.B. Clinical Implications of Real-time and Intermittently Scanned Continuous Glucose Monitoring. *Diabetes Care* **2018**, *41*, 2265–2274. [CrossRef]

58. Goyal, S.; Cafazzo, J.A. Mobile phone health apps for diabetes management: Current evidence and future developments. *QJM* **2013**, *106*, 1067–1069. [CrossRef]

59. Arnhold, M.; Quade, M.; Kirch, W. Mobile applications for diabetics: A systematic review and expert-based usability evaluation considering the special requirements of diabetes patients age 50 years or older. *J. Med. Internet Res.* **2014**, *16*, e104. [CrossRef]

60. Van Kerkhof, L.; van de Laar, K.; Schooneveldt, B.; Hegger, I. *e-Medication Met Behulp van Apps: Gebruik en Gebruikerservaringen*; Rijksinstituut voor Volksgezondheid en Milieu (RIVM): Bilthoven, The Netherlands, 2015.

61. El-Gayar, O.; Timsina, P.; Nawar, N.; Eid, W. Mobile applications for diabetes self-management: Status and potential. *J. Diabetes Sci. Technol.* **2013**, *7*, 247–262. [CrossRef]

62. McMillan, K.A.; Kirk, A.; Hewitt, A.; MacRury, S. A Systematic and Integrated Review of Mobile-Based Technology to Promote Active Lifestyles in People with Type 2 Diabetes. *J. Diabetes Sci. Technol.* **2017**, *11*, 299–307. [CrossRef] [PubMed]

Article

Evaluation of the Recipe Function in Popular Dietary Smartphone Applications, with Emphasize on Features Relevant for Nutrition Assessment in Large-Scale Studies

Liangzi Zhang [1,2], Eline Nawijn [2], Hendriek Boshuizen [1,2] and Marga Ocké [1,2,*]

1   National Institute for Public Health and the Environment (RIVM), 3721 MA Bilthoven, The Netherlands; liangzi.zhang@wur.nl (L.Z.); hendriek.boshuizen@wur.nl (H.B.)
2   Division of Human Nutrition, Wageningen University & Research, 6708 PB Wageningen, The Netherlands; eline_nawijn@hotmail.com
*   Correspondence: marga.ocke@rivm.nl; Tel.: +31-317488077

Received: 27 December 2018; Accepted: 17 January 2019; Published: 19 January 2019

Abstract: Nutrient estimations from mixed dishes require detailed information collection and should account for nutrient loss during cooking. This study aims to make an inventory of recipe creating features in popular food diary apps from a research perspective and to evaluate their nutrient calculation. A total of 12 out of 57 screened popular dietary assessment apps included a recipe function and were scored based on a pre-defined criteria list. Energy and nutrient content of three recipes calculated by the apps were compared with a reference procedure, which takes nutrient retention due to cooking into account. The quality of the recipe function varies across selected apps with a mean score of 3.0 (out of 5). More relevant differences (larger than 5% of the Daily Reference Intake) between apps and the reference were observed in micronutrients (49%) than in energy and macronutrients (20%). The primary source of these differences lies in the variation in food composition databases underlying each app. Applying retention factors decreased the micronutrient contents from 0% for calcium in all recipes to more than 45% for vitamins B6, B12, and folate in one recipe. Overall, recipe features and their ability to capture true nutrient intake are limited in current apps.

Keywords: diet apps; recipe calculations; nutrient retention; food record; dietary intake assessment; technological innovations

---

1. Introduction

When assessing the dietary intake of a large population, an accurate dietary assessment plays a fundamental role [1]. Self-report dietary assessment methods, such as 24-hour dietary recall (24HDR), dietary record (DR), and food frequency questionnaire (FFQ), are commonly used to assess food consumption at both individual and population level [2]. Since underreporting, overreporting, misreporting, and interviewer bias can occur in those methods [3–5], assessing dietary intake with a high level of accuracy continues to be a major challenge in nutritional epidemiology and monitoring [6,7]. Moreover, cumbersome procedures of collecting details of foods are time-consuming and are associated with a high burden for both the respondent and the researcher [8]. This is especially the case for 24HDR and DR, which are open methods, and for which repeated measurements are needed to estimate usual dietary intake [9]. The burden laid on respondents can also lead to a low response rate, which may lead to bias in the survey results and diminish the representativeness of the sample [10].

Progress in Information and Communication Technology (ICT) in the past few decades has led to investigations into innovative strategies to overcome drawbacks of traditional pen-and-paper and interviewer-based dietary assessment methods [11,12]. One such innovative strategy is the use of mobile applications (apps) on smartphones for a dietary record. In the last decade, an increase in the number of smartphone users has led to a proliferation of mobile applications (apps) [13]. A popular category within all these apps are the health and fitness-related apps [14], mostly aimed at supporting dietary change and weight management [15,16]. Those apps usually include a food diary function, in which users can record the foods consumed and the consumed quantities. Apart from searching in a pre-defined food and beverage list and selecting pre-defined portion sizes [17], various features are available to help identify consumed foods, estimate portion size, and decrease the burden of food entering. Examples of those features are image-based food recognition and barcode scanner. Their potential on reducing the respondents' burden, decreasing the effort of multiple self-administrations and on improving food recording accuracy have been investigated in both experimental and observational epidemiological studies, and have shown some promising results [6,18]. However, the knowledge on the performance of other specific features is still limited [19].

One feature of food diary apps is the recipe function for entering mixed dishes prepared at home. These are dishes consisting of multiple foods, with specific food preparation and often with cooking involved. For user-friendliness, the recipe function should be structured in a way that could easily guide the users in recording necessary information of a recipe. It should be able to assess the recipe intake of an individual, while mixed dishes are often prepared for more than one person [19]. Furthermore, for a better estimation of nutrient intake, an accurate recipe calculation should take nutrient loss of ingredients during cooking and food processing into account [20].

Some food diary apps have introduced a recipe function through the recent years [21,22]. The effectiveness of these recipe functions in capturing the food consumption and nutrient intake has not been fully evaluated. Moreover, the question whether the features of available recipe functions are also appropriate for dietary assessment as part of large-scale studies remains unanswered. Therefore, the aim of this study was to make an inventory of recipe function features in apps that could facilitate the estimation of nutrient intake of a large population. Furthermore, another aim was to evaluate the accuracy of the recipe function in capturing nutrient intake of popular dietary assessment apps by comparing their nutrient calculation with a standard calculation procedure.

2. Materials and Methods

The starting point for app selection was an identification of dietary assessment smartphone apps in the Health & Fitness category of iTunes App Store and Google Play Store in the Netherlands between 15 and 23 October 2016. This selection was performed by Maringer et al. [20] and resulted in the identification of 176 dietary assessment apps. Further screening was performed in August 2017. Inclusion of a subselection of apps for this study required the app to meet the following criteria: (1) user rating >3 in iTunes App Store and Google Play Store, (2) user rating count >500 in iTunes App Store and Google Play Store, (3) >10,000 downloads in the both stores, (4) a recipe function which was freely available, actually present and functional. A recipe function was defined as "a functionality in which the user can create a mixed dish by entering and specifying the amount of each ingredient within the dish" [23,24]. Each app underwent initial screening based on descriptions and associated images in the app stores to check for the presence of a recipe function. Apps were downloaded onto a OnePlus 3T smartphone running Android 7.1.1 and a Huawei Mate 8 running EMUI 5.0.1 for analysis. The apps were checked manually to confirm whether a recipe function was freely available, actually present, and functional. Basic descriptive information about the apps was identified, such as app name, version number, operating platforms, number of installs, ratings, whether they can synchronize with their website, and country of origin. Subsequently, the recipe function of the selected apps was evaluated.

To our knowledge, no widely accepted standard evaluation of the quality of the recipe function of apps exists. Therefore, a criteria list was made for evaluating features in the individual recipe function of apps. For each feature on the criteria list a rubric of assessment was created with a 1 (low)–5 (high) scoring scale. The criteria list and assessment rubric were modified upon findings from a pilot scoring and feedback from two nutritionists and three dietitians with different specializations. The criteria list and assessment include the following aspects of creating an individual recipe: options in searching ingredients, ways to record relevant information of the recipe, whether raw or cooked ingredients could be selected, consumed amount for both ingredients and the whole recipe, energy and nutrient expression, and whether the recipe could be saved and edited later (Table 1). Two researchers scored all the selected apps according to the criteria list independently. Inconsistent scores among the two researchers were discussed to reach agreed final scores. For scoring the criterion whether both raw and cooked foods are available in the food list, nine foods from the three most frequently used Dutch recipes (explained in next paragraph) were entered in each app (kale, potato, milk, mushroom, onion, salami, beef, pepper, and tomato).

To be able to evaluate the accuracy of energy and nutrient content estimations, three recipes were entered into the individual recipe function of each app. The selection of recipes was performed by exploring the most frequent reported recipes in the Dutch diet using the data of the Dutch National Food Consumption Survey (DNFCS) 2007–2010 [25]. Three recipes with different preparation methods, like stewing, baking, and frying, were chosen from the twenty most frequently consumed recipes. The chosen recipes were boerenkool stamppot (mashed potato with kale), pizza with salami, tomato, and mushrooms, and hachee (a traditional Dutch stew based on beef and onions). Raw ingredients of the recipes were entered in the selected apps and a set of rules for entering ingredients were followed, in case the exact match of food items or amount indications could not be found across apps. If available, energy, macro- and micronutrient values of the recipe were obtained based on the displayed nutrient content in the app. For those apps where the nutrient contents were not shown at the recipe level, values from ingredients of a recipe were added up by researchers. Then, nutrient contents from the apps were compared with nutrient contents derived from the Dutch food composition database (NEVO) [26]. To account for nutrient loss due to cooking, retention factors suggested by the European Food Information Resource [27] were applied to the nutrients derived by NEVO, see complete calculation in Supplementary Material (Tables S1–S11). A retention factor larger than 0 and lower than 1 implied nutrient loss due to cooking. A retention factor of 1 was used for energy and macronutrients for all ingredients in all recipes since they were not easily affected by cooking. Next to energy and macronutrient, micronutrients such as sodium, potassium, vitamin A represented as retinol equivalent (RE), vitamin C, calcium, vitamin E, vitamin B1, vitamin B2, vitamin B6, vitamin B12, and folate were selected for comparison between apps and the reference measure. Of these, sodium, potassium, and vitamin E had a retention factor of 1 for all ingredients in the three recipes mentioned above, hence, were deleted from analysis. Calcium also had a retention factor of 1, but was maintained in the analysis as an example.

General characteristics of the 12 evaluated dietary assessment apps with recipe function were summarized. For each app, the mean score and standard deviation over all nine criteria was calculated (see Table 1). The mean and standard deviation of scores across apps were calculated for each criterion. Energy and nutrient content estimations of the three recipes for each app were analyzed using descriptive statistics. For nutrients with retention factor of 1, a direct comparison could be made with the nutrient contents derived from NEVO combining nutrient contents of raw ingredients in the appropriate amounts. For the micronutrients with retention factors below 1, the reference was the NEVO nutrient contents of the raw ingredients after applying the relevant retention factors. For showing the effect of the retention factors, a comparison with NEVO nutrient contents of raw ingredients without applying retention factors was also made. A difference in values between apps and the reference of more than 5% from the Daily Reference Intake (DRI) for adults was considered out of range [28].

Table 1. Rubric for assessment of the individual recipe function in dietary assessment apps, giving a score between 1 (low) and 5 (high) per feature.

Feature	Mark for Feature				
	1	2	3	4	5
Recipe creation options (name, photo, ingredients, servings)	The user can only create a recipe by adding ingredients and amounts	The user can create a recipe by giving it a name and adding ingredients and amounts	The user can create a recipe by giving it a name, adding ingredients and amounts, and number of servings.	The user can create a recipe by giving it a name, add ingredients and amounts, number of servings, and explanation of preparation.	The user can create a recipe by giving it a name, add ingredients and amounts, number of servings, explanation of preparation, and a photo.
Ingredients search options within a recipe	Can only search in one way	Can search in 2 or 3 ways	Can search in 4 or 5 ways	Can search in 6 or 7 ways	Can search in 8 or more ways
Reminders for frequently forgotten ingredients (e.g., olive oil, butter, salt)	App does not give reminders for frequently forgotten ingredients				App gives reminders for frequently forgotten ingredients
Preparation indication of ingredients	It is unclear whether the entered ingredients are prepared or not	The user can only select the prepared or the unprepared ingredient from the food list	The user can select both the prepared and the unprepared ingredients from the food list for some foods	The user can select both the prepared and the unprepared ingredients from the food list for all foods	The user can select an ingredient and indicate the preparation (unprepared, prepared (cooked, grilled, etc.))
Entering consumed amount at recipe level	User cannot indicate consumed amount	User can indicate the consumed amount, but the type of indication of the amount is inappropriate	User can indicate the consumed amount, and the appropriate type(s) of indication is given. However, inappropriate amounts are also given	User can indicate consumed amount and the appropriate type(s) of indication are given	User can indicate the consumed amount and the user can choose from a lot of appropriate types of indications (grams, portion in grams, portion as photo, fraction of recipe) OR can manually add amount indications
Entering prepared amount at ingredient level	User cannot indicate prepared amount	User can indicate the prepared amount, but the type of indication of the amount is limited (1 or 2 options) OR other types of indications (portion in grams, portion as photo, fraction of recipe)	User can indicate prepared amount from more than 2 options.	User can indicate prepared amount from more than 2 options. AND other types of indications (portion in grams, portion as photo, fraction of recipe)	User can indicate prepared amount from more than 2 options, and other types of indications (grams, portion in grams, portion as photo, fraction of recipe), and can manually add amount indications
Save and edit function for recipe	The user can create recipe, but cannot save it to use it later	The user can create a recipe and save it to use it later	The user can create a recipe and save it in a categorized way OR the user can create a recipe and edit it; premium only	The user can create a recipe and edit it later	The user can save the created recipe to use it later, edit it later on, and can save it in a categorized way
Energy and macronutrient information at recipe level	Energy and macronutrient content are not shown	Energy content is shown in kcal (KJ), macronutrient content is not shown	Energy content is shown in kcal (KJ), macronutrient content is shown in grams OR energy is shown in % of Reference Daily Allowance (RDA) *; premium only	Energy content is shown in kcal (KJ) and % of RDA, macronutrient content is shown in grams OR macronutrient content is shown in grams and % of RDA; premium only	Energy content is shown in kcal (KJ) and % of RDA, macronutrient content is shown in grams and % of RDA
Micronutrient information at recipe level	No micronutrient information available	Micronutrient information exists for only premium account	Information on less than 3 micronutrients	Information on 3–6 micronutrients	Information on more than 6 micronutrients

* Reference Daily Allowance (RDA): The average daily dietary intake level sufficient to meet the nutrient requirement (for the specified indicator of adequacy) of nearly all (97% to 98%) healthy individuals in a particular life stage and gender group [29].

To visualize the correlation between apps and nutrients, a principal component analysis (PCA) was conducted for each recipe separately with energy and macronutrients divided by their DRIs being set as variables. The first two principal components represent the most variation. This was done for energy and macronutrients only, since only 3 apps showed information on absolute amounts of micronutrients. The descriptive statistics were calculated using Excel 2016 software and the PCA was conducted in R version 3.5.0 (The R Foundation for Statistical Computing, Vienna, Austria).

3. Results

3.1. App Selection

The starting point was a selection of 176 popular dietary assessment apps with food recording and available in English identified by Maringer et al. [21]. Then, apps were further narrowed down, with inclusion criteria of a user rating >3 in the iTunes App Store and Google Play Store, a user rating count > 500 in iTunes App Store and Google Play Store, >10,000 downloads in the Google Play Store, and a claimed recipe function in the app description. After manually checking for the presence of an individual recipe function in 30 included apps, 17 apps were excluded from further evaluation because of dysfunction of the app, the absence or dysfunctionality of a recipe function, or the inability to use the app due to requirements of a membership. After final exclusion of one app with a non-functioning individual recipe function, a total of 12 apps (21% of 57) were selected for evaluation in detail (Figure 1).

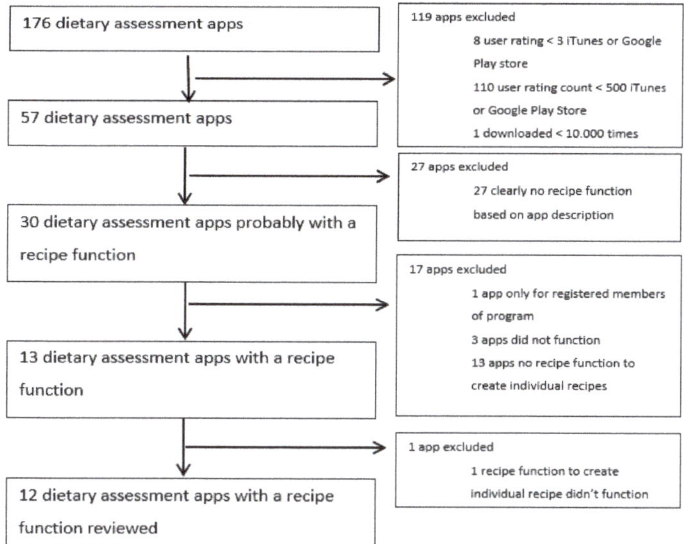

Figure 1. Flow diagram of selection procedure of dietary assessment apps with recipe function showing the number of apps included or excluded.

General characteristics of the remaining 12 apps can be found in Table 2. All apps operated on an Android platform, whereas IOS ranked as the second most-prevalent platform (10 apps). The highest number of installs was 50 million with 1844 thousand ratings for MyFitnessPal, the lowest was 100 thousand installs and 2000 ratings for Nutracheck. The rating scores among the apps ranged from 4.2 to 4.6 with the maximum score of 5.0. Four apps were made by US companies, two apps were made in Germany, and the rest of apps were made in other countries, mostly northwest Europe.

Table 2. General characteristics, such as platforms available, number of installs on Google Play Store, user rating on Google Play Store and country of twelve popular dietary assessment apps with a recipe function (*n* = 12).

	App Name (Version)	Platforms	Installs Google Play Store (Million)	Rating Google Play Store (The number of Ratings/1000)	Country
1	MyFitnessPal (18.6.0)	Android, IOS, Windows Phone	50–100	4.6 (1844)	USA
2	FatSecret (7.8.27)	Android, IOS, Windows Phone, Watch OS, Blackberry OS	10–50	4.4 (223)	Australia
3	YAZIO (4.0.1)	Android, IOS	5–10	4.6 (109)	Germany
4	Lose It! (9.4.5)	Android, IOS	5–10	4.4 (68)	USA
5	Lifesum (6.2.4)	Android, IOS, Watch OS, Android Wear	5–10	4.4 (165)	Sweden
6	MyPlate (3.2.2)	Android, IOS, Watch OS	1–5	4.6 (22)	USA
7	MyNetDiary (6.4.7)	Android, IOS, Watch OS	1–5	4.5 (26)	USA
8	Calories! (8.1.6)	Android	1–5	4.3 (10)	Germany
9	The Secret of Weight (2.4.24)	Android, IOS	1–5	4.3 (14)	France
10	Virtuagym Food (2.4.0)	Android, IOS	1–5	4.5 (28)	The Netherlands
11	Health Infinity (HI) (2.0.58)	Android	0.1–0.5	4.2 (9)	India
12	Nutracheck (5.0.12)	Android, IOS	0.1–0.5	4.3 (2)	UK

3.2. Qualitative Recipe Function Assessment

Agreed scores given to recipe functions of each app are shown in Table 3. Mean overall score of both apps and criteria was 3.0 (out of 5.0). The app Calories! had the highest score for its recipe function with an average score of 3.9 however, in contrast, Calories! had a rating score and number of installations at the lower range compared to other apps (Table 2). MyPlate and Health Infinity, on average, had the lowest scores of 2.2 and 2.3, respectively.

The apps that had relative higher popularity, such as MyFitnessPal, Lose It!, Lifesum, and MyPlate, did not have any criterion that scored 5, while Calories! was achieved a score of 5 three times. Health Infinity scored 1 most often (three times) compared to other apps.

Specifically, most of the evaluated apps could save a self-created recipe and edit it later, hence, this criterion ranked the highest (mean = 4.3) compared to other criteria. None of the apps included reminders for frequently forgotten ingredients, therefore, all apps scored 1 for that criterion. The available options that existed for searching ingredients for recipes included text search, barcode scanning, voice record, recent/frequent/saved food, create new food, choose from categories, and choose from a list of all food in alphabetic order. The number of options ranged from 2 to 6, where half of the apps had only 2 to 3 options, while only Nutracheck had all 6 options. The most frequently adopted options were search in a textbox and barcode scanning. FatSecret and Virtuagym Food had four searching options for food entering, but only two options for adding ingredients to recipes. In terms of options in searching raw or cooked foods, nearly all apps had both raw and cooked options for all or at least some foods in their dataset (mean = 3.3). An exception was The Secret of Weight, where, for the most foods, the text indicated raw while the picture showed cooked foods. In terms of indicating consumed amount in both ingredients and recipes, in Calories!, one could manually add a new serving unit to ingredients but not in recipes whereas, in Virtuagym Food, this was the other way around. Health Infinity had no options to chooe the amount of recipe consumed (scored as 1), and had only one built-in option when choosing the amount of ingredients. In terms of macronutrient information, Calories! was the only app that had energy and macronutrients expressed as both absolute amounts (mg, μg, etc.) and % of Recommended Daily Allowance (RDA). Most apps had energy and macronutrients shown only in absolute amounts. Since only four apps showed micronutrient for recipes, the average score for micronutrient availability ranked the second lowest with a score of 2.7. Among the apps with micronutrients, Calories! and MyNetDiary had both absolute amounts and % RDA for more than six micronutrients, while Virtuagym Food had only actual amounts. MyFitnessPal had only % RDA of less than six micronutrients.

Table 3. Agreed scores for the recipe function of 12 popular dietary assessment apps using the criteria list based on a 1(low)–5 (high) scale.

App Name (Version) Criteria List	MyFitnessPal (18.6.0)	FatSecret (7.8.27)	YAZIO (4.0.1)	Lose It! (9.4.5)	Lifesum (6.2.4)	MyPlate (3.2.2)	MyNetDiary (6.4.7)	Calories! (8.1.6)	The Secret of Weight (2.4.24)	Virtuagym Food (2.4.0)	Health Infinity (HI) (2.0.58)	Nutracheck (5.0.12)	Mean	SD
Options (name, photo, ingredients, servings)	3	5	5	3	4	2	5	4	5	2	2	3	3.6	1.2
Options to search ingredients	2	2	3	3	3	2	3	3	2	2	2	4	2.6	0.6
Reminders for frequently forgotten ingredients (e.g., oil, spices, salt)	1	1	1	1	1	1	1	1	1	1	1	1	1.0	0.0
Entering ingredients—preparation indication	4	3	3	4	3	3	3	4	2	4	4	3	3.3	0.6
Consumed amount recipe level	4	4	4	4	4	2	4	4	4	5	1	4	3.7	1.0
Consumed amount ingredient level	3	3	3	3	3	3	3	5	2	3	2	3	3.0	0.7
Save and edit	4	5	5	4	4	4	3	4	4	4	5	5	4.3	0.6
Energy and macronutrient expression at recipe level	4	4	3	3	3	2	3	5	2	3	3	3	3.2	0.8
Micronutrient availability at recipe level	4	3	3	1	2	1	5	5	1	5	1	1	2.7	1.6
Mean	3.2	3.3	3.3	2.9	3.0	2.2	3.3	3.9	2.6	3.2	2.3	3.0	3.0	0.5
SD	1.0	1.2	1.2	1.1	0.9	0.9	1.2	1.2	1.3	1.3	1.3	1.2	0.9	-

3.3. Accuracy of Energy and Macronutrient Content Estimations

The differences in energy and macronutrient content estimations of the three recipes between the 12 popular dietary assessment apps and the value derived from NEVO are presented in Table 4. Macronutrient contents for both recipes and ingredients were not available in The Secret of Weight. Heterogeneity in differences was observed between recipes and between nutrients. Pizza had fewer differences >5% ($n = 7$) in the DRI as compared to boerenkool stamppot ($n = 10$) and hachee ($n = 12$). Carbohydrates ($n = 2$) and energy ($n = 3$) contents had fewer differences >5% in the DRI than protein ($n = 13$) and fat ($n = 11$). In total, around 20% of the differences were >5% DRI. Most apps underestimated the macronutrient content in boerenkool stamppot and pizza, while this was not observed in hachee.

With 7 out of 12, Nutracheck had the most discrepancies >5% in the DRI compared to the reference, mainly caused by a discrepancy in fat and protein contents. YAZIO and Lifesum only had one difference of more than 5%. Health Infinity had lower protein contents in all three recipes, whereas Lose It! had lower fat in all three recipes. Virtuagym Food and YAZIO had similar patterns in all recipes, and both had lower fat in hachee as outliers. MyNetDiary had all macronutrients being out of range once, including a lower carbohydrate, lower protein, and higher fat in three recipes, respectively. In Figure 2, apps are plotted against the first and second principal component of all differences in macronutrient contents. Macronutrients plotted further from the center indicate a larger variance. Apps situated in the same direction with a certain nutrient indicate an overestimation of the nutrient and vice versa. Nutracheck laid outside compared to other apps for all three recipes. MyFitnessPal was the only app without discrepancies of more than 5%. Therefore, it was located around the center of the graph in all three recipes.

Table 4. Difference in energy (kcal) and macronutrient content (gram) estimations for one portion of each of three recipes between 12 dietary assessment apps and reference values using NEVO.

Recipes	Macronutrients	NEVO [a]	MyFitness Pal	FatSecret	YAZIO	Lose It!	Lifesum	MyPlate	MyNet Diary	Calories!	The Secret of Weight	Virtuagym Food	Nutra Check	HI	Mean	SD
Boerenkool stamppot	Energy (kcal)	472	4	−42	10	−16	−69	−28	−53	−93	−116 *	−62	59	−44	−38	46
	Fat (g)	10.9	−0.2	−5.1 *	−0.4	−3.7 *	−1.0	−0.6	−0.2	−0.9	-	0.9	6.6 *	−2.9	−0.7	2.9
	Protein (g)	17.0	0.1	0.4	0.8	0.9	−5.2 *	−1.7	−0.2	−5.3 *	-	−1.9	−11.1 *	−17.0 *	−3.6 *	5.4
	Carbohydrate (g)	70.4	−0.1	0.3	11.8	1.2	−2.9	10.2	−15.1 *	−14.1 *	-	−11.4	−9.0	−6.1	−3.2	8.6
Pizza with salami, tomato, and mushroom	Energy (kcal)	483	−36	−5	−2	−42	−5	−35	0	−24	−7	−8	−47	−41	−21	17
	Fat (g)	25.9	−2.6	−0.3	0.3	−2.9	−0.3	−4.4 *	−0.7	−1.6	-	−0.1	−5.4 *	−2.9	−1.9	1.8
	Protein (g)	22.1	−2.3	−1.2	−0.2	−2.7 *	−0.8	−2.6 *	−5.1 *	−1.0	-	−0.9	−2.6 *	−3.8 *	−2.1	1.4
	Carbohydrate (g)	38.8	0.1	0.6	0.3	1.9	1.9	11.8	−0.8	−2.8	-	−0.4	4.2	−2.8	1.3	3.9
Hachee	Energy (kcal)	316	15	−43	−47	−119 *	7	12	75	32	58	−46	142 *	19	9	65
	Fat (g)	17.9	2.2	−4.3 *	−4.5 *	−8.8 *	2.5	1.7	8.4 *	−0.3	-	−5.1 *	10.8 *	2.4	0.4	5.6
	Protein (g)	23.3	−0.9	−0.6	−1.0	−11.2 *	−0.8	−21.3 *	−1.3	12.5 *	-	1.3	9.0 *	−0.1	−1.3	8.5
	Carbohydrate (g)	13.7	1.7	3.8	3.7	3.8	−0.9	2.3	−4.7	−4.1	-	−0.5	3.1	−1.7	0.6	3.0

[a] Energy and macronutrient contents of one recipe portion by adding nutrient contents of raw ingredients derived from Dutch food composition database (NEVO); retention factors were all 1. * Discrepancy with reference >5% of the Dietary Reference Intakes (DRI), which is 100 kcal out of 2000 kcal for energy, 3.5 g out of 70 g for fat, 2.5 g out of 50 g for protein, and 13 g out of 260 g for carbohydrate.

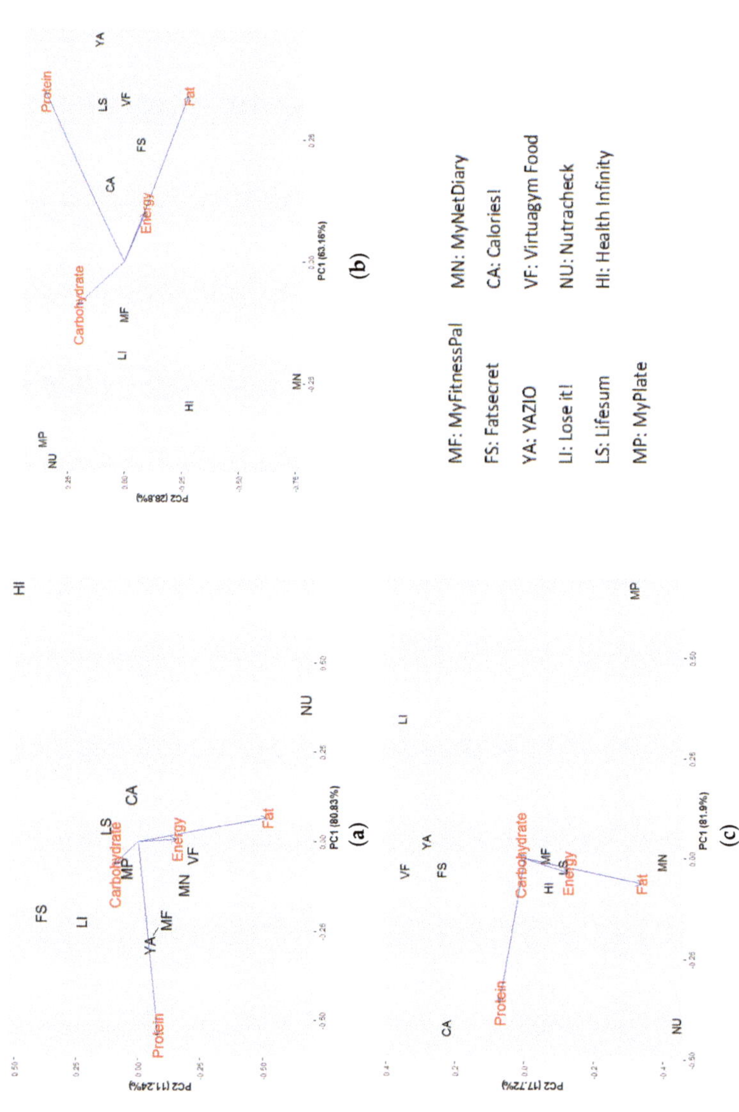

Figure 2. Principal Component Analysis (PCA) showing the variation strength and trend of macronutrient difference compared with the reference contents from different apps in (**a**) Boerenkool stamppot, (**b**) Pizza with salami, tomato, and mushrooms, (**c**) Hachee.

3.4. Accuracy of Micronutrient Content Estimations

The micronutrient contents were analyzed for MyNetDiary, Calories! and Virtuagym in which it was available. The differences in micronutrient content estimations of the three recipes between the three popular dietary assessment apps, the micronutrient calculated from NEVO values in raw foods and the reference where retention factors was applied to NEVO are presented in Table 5. For most micronutrients except calcium, applying retention factors resulted in lower micronutrient levels than micronutrient levels in raw ingredients. The relative differences between the reference and using NEVO without applying retention factors ranged from 0% for calcium in all recipes, vitB12 in stamppot and vitB2 in hachee to more than 45% for vitamins B6, B12 and folate in hachee. Over the 3 recipes, 8 out of 24 differences (33%) were relevant (>5% of DRI) in case of a high content and high vulnerability of these nutrients of raw ingredients in a certain recipe. The relatively large difference in vitamin B6 and B12 in Hachee can be explained by the sensitivity to heat and the two cooking procedures in this recipe, i.e. frying and stewing. Whereas, boerenkool stamppot (n = 5) had more relevant differences than the other two recipes (n = 1 and 2 respectively), due to its high contents of vitamin C, vitamin A, vitamin B1, vitamin B6 and folate even if the retention factor was not so different from 1 (for example, vitamin A with a retention factor of 0.9).

A larger proportion of difference >5% DRI was found in micronutrients (49%) than in energy and macronutrients (20%) when compared with the reference values. Among the three apps, MyNetDiary showed more differences > 5% DRI (n = 14 out of 24) than the other two apps (Virtuagym n =10, Calories! n =11) when comparing micronutrient values with the reference. In contrast to macronutrient comparisons, apps more often overestimated the contents of micronutrient in the recipes. The number and extent of overestimations were slightly larger when comparing with the reference than comparing with NEVO without applying retention factors, since the retention factors resulted in lower micronutrient contents in the reference values. The proportions of relevant differences found after comparing the apps to NEVO with or without applying retention factors were rather similar (49% vs. 51%), illustrating that in many cases the effects of differences in nutrient databases were much larger than differences due to applying retention factors.

Table 5. Comparison of micronutrient contents between recipes added by raw ingredients from three apps with recipes added by raw ingredients from the Dutch food composition database (NEVO), with NEVO multiplied by retention factors.

Recipes	Micronutrients	NEVO [a]	R [b]	NEVO-R	MyNetDiary			Calories!			Virtuagym		
					App	App-NEVO	App-R	App	App-NEVO	App-R	App	App-NEVO	App-R
Boerenkool stamppot	Calcium(mg)	494	494	0	431	−63 *	−63 *	573	80 *	80 *	391	−102 *	−102 *
	Vitamin C(mg)	294	187	107 *	327	33 *	140 *	327	33 *	140 *	362	68 *	174 *
	Vitamin A(μg)	1774	1606	168 *	2557	783 *	951 *	2320	546 *	714 *	74	−1701 *	−1532 *
	Vitamin B1(mg)	0.66	0.60	0.06 *	0.32	−0.34 *	−0.28 *	0.57	−0.09 *	−0.03	0.49	−0.17 *	−0.11 *
	Vitamin B2(mg)	0.43	0.41	0.02	0.56	0.13 *	0.15 *	0.79	0.36 *	0.38 *	0.48	0.05	0.07
	Vitamin B6(mg)	1.49	1.34	0.15 *	1.38	−0.11 *	0.04	1.70	0.21 *	0.36 *	1.30	−0.19 *	−0.04
	Vitamin B12(μg)	0.11	0.11	0.00	0.43	0.32 *	0.32 *	-	-	-	0.19	0.08	0.08
	Folate(μg)	198	142	56 *	407	208 *	265 *	94	−104 *	−48 *	-	-	-
Pizza	Calcium(mg)	339	339	0	293	−46	−46	290	−48	−48	293	−46	−46
	Vitamin C(mg)	6	5	1	5	−1	3	8	2	3	5	−1	0
	Vitamin A(μg)	188	183	5	205	17	22	204	17	22	97	−91 *	−86 *
	Vitamin B1(mg)	0.21	0.18	0.03	0.75	0.54 *	0.57 *	0.29	0.08 *	0.11 *	0.77	0.56 *	0.59 *
	Vitamin B2(mg)	0.31	0.30	0.01	0.62	0.31 *	0.32 *	0.38	0.07	0.08 *	0.62	0.31 *	0.32 *
	Vitamin B6(mg)	0.26	0.24	0.02	0.27	0.01	0.03	0.31	0.05	0.07	0.26	0.00	0.02
	Vitamin B12(μg)	1.10	1.01	0.09	1.00	−0.10	−0.01	-	-	-	1.02	−0.08	0.01
	Folate(μg)	92	67	24 *	129	38 *	62 *	45	−47 *	−23 *	77	−15	10
Hachee	Calcium(mg)	51	51	0	66	15	15	48	−3	−3	67	16	16
	Vitamin C(mg)	6	5	1	8	2	3	7	1	3	9	3	4
	Vitamin A(μg)	136	129	7	108	−28	−21	123	−13	−6	94	−42 *	−34
	Vitamin B1(mg)	0.10	0.06	0.04	0.19	0.09 *	0.13 *	0.16	0.06 *	0.10 *	0.19	0.09 *	0.13 *
	Vitamin B2(mg)	0.19	0.19	0.00	0.21	0.02	0.02	0.24	0.05	0.05	0.25	0.06	0.06
	Vitamin B6(mg)	0.39	0.20	0.19 *	0.73	0.34 *	0.53 *	0.34	−0.05	0.14 *	0.73	0.34 *	0.53 *
	Vitamin B12(μg)	2.95	1.46	1.49 *	2.70	−0.25 *	1.24 *	-	-	-	2.69	−0.26 *	1.23 *
	Folate(μg)	28	15	13	57	29 *	42 *	29	1	14	13	−15	−2
The number of differences >5% DRI				8		15	14		10	11		12	10
The number of positive differences				8		11	12		7	9		5	7

[a] Micronutrient contents of one recipe portion by adding nutrient contents of raw ingredients derived from Dutch food composition database (NEVO). [b] The reference measure where retention factors (RF) were multiplied by each micronutrient content derived from NEVO. * Discrepancy with reference >5% of the Dietary Reference Intake (DRI) which is 5 mg for vitamin C, 49 mg for calcium, 35 μg for vitamin A, 0.06 mg for vitamin B1, 0.08 mg for vitamin B2, 0.08 mg for vitamin B6, 0.20 μg for vitamin B12, and 17 μg for folate.

4. Discussion

The current study evaluated the recipe function that was available in only one-fifth of the popular available food diary apps. We found a varying quality of recipe features across selected apps which were, on average, judged as suboptimal from research perspectives. Furthermore, capturing the true nutrient intake of mixed dishes is a challenge for this innovative dietary assessment method. A comparison of energy, macro-, and micronutrient contents of recipes between apps with a reference standard recipe calculation showed variation in terms of their ability to accurately estimate nutrient contents. In only three apps was micronutrient information available for recipes, and none of these apps included a procedure to take nutrient losses due to recipe processing into account, and the variability in micronutrient content databases was large.

This is the first study to evaluate the recipe function of current popular dietary assessment apps in a standardized way in which the quality assessment was performed using a rubric of assessment which was made prior to the evaluation. The scores of recipe function were discussed by two researchers, which has eliminated mistakes and the bias of scoring. From the quality assessment of the recipe functions, apps were given a mean overall score of 3.0 (out of 5.0) where the highest score was 3.9 and the lowest 2.2. No correlations were found between the scores given in this study and the popularity and user ratings in app stores. This could illustrate that the recipe function was not the main aspect contributing to users' overall app-experiences, or that researchers and users have different needs for dietary apps [9]. Some simplified features might be favored by users since it was observed that the user's time invested for understanding and learning about an app should be small to sustain long-term app usage [30], whereas researchers are more concerned with features that could enable detailed and accurate data collection. This preference gap between the app users and researchers is important to select suitable features to be included in dietary assessment tools for large nutrition monitoring studies.

Although the quality of recipe function in popular apps was not investigated before [13], several features of a recipe function were investigated by others since they are also relevant for recording food intake. In terms of options for searching ingredients in apps from the current study, all apps had a text searching option and the majority of the apps had a barcode function. Barcode scanning has been shown to save time and was favored by users in recording branded food items, however, the resulting nutrient intake estimation depends largely on the quality of the underlying food composition database within the app [31]. An aspect in which these apps differ from many web-based tools is that most of them do not have portion images, which may due to limited space in the user interface. Previous research has found that the incorporation of portion images was preferred by all age groups [9]. However the overall advantage of using portion images remains unknown [17]. In terms of nutrient information, the energy and macronutrient information was more complete in apps than micronutrient information, and this complied with the fact that energy and macronutrients were more closely correlated with weight change, which was the aim for most apps.

Features specific for creating recipes were evaluated. For instance, in addition to other basic features for entering recipes (i.e., add a name, ingredients, and serving number of the recipe), half of the evaluated apps had the capability to enter a photo and cooking explanation. However, this information was not used by the app to estimate nutrient intake. A photo of the recipe could help identify and estimate the amount of food consumed by participants, and could also reduce the extent of underreporting, especially for people with low literacy levels [17], while a cooking explanation provided information relevant for nutrient retention estimation. However, with the extra efforts required in using these features, they might be practical only in small-scale studies. Unlike computer/web-based dietary assessment tools for research purposes [32], all apps lack reminders for frequently forgotten ingredients when creating recipes (e.g., oil, spices, sugar, etc.), which may have partly contributed to the systematic underestimation of macronutrients in most apps found in other studies [33]. Also, current apps did not have pre-defined recipes that could be adapted by users whereas, in some computer-based software, standard recipes could be adapted by switching ingredients or changing the amount of ingredients [32]. However, the practicality of above features to be included in apps or to be used by participants, without the help of researchers, remains questionable. As a simpler alternative, the feature for saving frequently consumed or favorite foods in current apps was shown to save the efforts of users from entering the same recipes repeatedly and searching for food in a comprehensive food list [34].

In the present study, differences in energy, macro-, and micronutrient contents were found between the apps and the reference measure, which could be explained by several reasons. There were substantial differences in the nutrient contents of the recipe ingredients between apps, showing the differences in underlying nutrient databases. Apps were made by companies from different countries and they might have incorporated a nutrient database from their own countries which might have varying nutrient contents for certain foods, due to different cultivating environments [35]. Another source of nutrient values might be input from the app users. This has the benefit of customization of food consumed, however, has shortcomings in the accuracy of nutrients and can lead to quality losses in the food database [14].

The inability to enter exactly the same ingredients across the apps and the limited choice of food amounts may additionally explain part of the variation in nutrient estimation [33]. For example, it was difficult to find an exact match of beef steak in hachee, since there was a large variety of beef steak in different apps, and food amounts in grams were not available in some apps. However, for most other recipe ingredients, this problem did not occur. For micronutrients, the difference was also due to applying retention factors to the reference nutrient values, whereas all apps came up with the nutrient content of recipes by simply adding up the nutrient content of each ingredient without taking nutrient retention into account.

Variations of nutrient content of three recipes between apps and the reference measure were observed in the present study, with fewer variations in energy and macronutrient than in micronutrient contents. Similarly, comparable energy contents across apps were also observed in a study where nutrient contents from the barcode scanning of 100 food products in apps were compared with product labels [31]. Likewise, Griffiths et al. compared the results of five commercial apps with thirty 24 h dietary recalls collected using the Nutrition Data System for Research (NDSR), and found a better validity of energy estimation than nutrients [33]. The mean difference of 22 kcal in energy across all apps and recipes in this study was similar with the 30 kcal mean energy difference of 23 apps compared with the three days' weighed food record in the study of Chen et al. [14]. The wider range of energy difference (-167 to 262 kcal) in Chen's study compared to the energy difference in our study (-118 to 141 kcal) is possibly due to a higher number of apps evaluated, and a larger amount of foods being entered in apps in Chen's study. These findings indicated a relatively reliable energy estimation for both generic and branded food items in the current apps. Still, it was noteworthy that the largest difference of around 345 kcal between apps from both studies could impact the accuracy on both individual and population nutrient intake estimations. A trend of underestimation of energy and macronutrient contents in apps compared to reference in our study was consistent with the study by Griffiths et al. The reason in the study of Griffiths was because the food preparation details were captured by the reference (NDSR), but not in the apps. By contrast, in our study, the food details were equally captured by both the reference and apps, and the reporting bias by participants did not exist since the foods were being entered by researchers. Hence, the main reason of underestimation is the inaccuracy of the nutrition databases within the apps.

A proper way of calculating the nutrient contents within a recipe requires the consideration of nutrient loss during cooking. Currently, the nutrient retention for foods based on different cooking processes is not calculated automatically in any dietary assessment tools, and none of the apps had instructions on using the recipe function. Although existing recipes in food composition tables take the nutrient loss into account, none of the food composition databases cover all the variations on recipes made individually [14]. Alternatively, cooked ingredients could be chosen from the food list. However, the availability of cooked ingredients was incomplete, and this would also require participants to know the amount of the prepared ingredients (which might be smaller due to shrinkage during preparation). Hence, we entered ingredients as raw ingredients, as that is the most logical option for a user.

This is the first study to investigate the discrepancies of nutrient content between raw ingredients in different apps, compared to a more accurate estimation that takes the nutrient loss into account. Only three out of twelve apps had comprehensive micronutrient information, with both actual amounts and percentage of RDA. The large variation in micronutrient content found in this study implied the importance of choosing the right nutrient database, especially when micronutrient intake estimation is part of the study purposes. The input of raw ingredients potentially leads to overestimation of several heat-sensitive micronutrients, which was shown in the micronutrient comparison between NEVO with the reference method in this study. Moreover, the results showed that the extent of difference depends largely on the nutrient contents in the recipe. Therefore, it was suggested that retention factors are most influential when applied to recipes with high micronutrient contents (e.g., boerenkool stamppot).

NEVO was chosen as the reference measure for nutrient estimations, which was a well-maintained food composition database that had all the data on the nutrition values that were assessed and has a standardized food-compiling procedure that follows the guidelines set by EuroFIR [36,37]. Retention factors applied in this study were the most up-to-date values from the harmonization of retention factors provided by 17 EuroFIR partners [38]. However, the results of nutrient differences may lack representativeness in this study, due to a limited recipe selection. To develop a full picture of the importance of recipe calculation, additional studies, that include more recipes and an evaluation on their contribution to population nutrient intake, will be needed. Furthermore, the evaluation was done only from a research perspective in this study, while user perspective was not analyzed for the apps. Especially factors that could affect the individual's ability to accurately enter the recipe consumed were not examined. Further development of an app for large nutrition monitoring studies would benefit from an evaluation on app users' perspectives.

5. Conclusions

In popular food diary apps, the quality of recipe functions is suboptimal from a research perspective. All apps follow a basic nutrition-calculating algorithm, without taking nutrient retention into consideration. This leads to inaccurate nutrient intake estimations in the case that recipes are an important source of micronutrients which are vulnerable to the effects of food processing. Moreover, across apps, there is large variability in nutrient databases. From a research perspective and out of interest regarding micronutrient intake, a balance between user-friendliness and completeness of the recipe function is important. In order to obtain more insight into the need for more complex recipe functionalities, further studies on their potential impact on the nutrient intake estimations in large nutrition-monitoring studies and users' perspective are needed.

Supplementary Materials: The following are available online at http://www.mdpi.com/2072-6643/11/1/200/s1, Tables S1–S3: energy and macronutrient in NEVO and apps, differences between app and NEVO for three recipes, Tables S4–S11: recipe calculation of vitamin C, calcium, vitamin A, vitamin B1, vitamin B2, vitamin B6, vitamin B12, and folate.

Author Contributions: L.Z. led the writing, evaluated the apps, conducted micronutrient analysis and has primary responsibility for the final content; E.N. developed the criteria list, evaluated the apps, conducted macronutrient data analysis and made the first draft; H.B. provided guidance on methodology and data analysis; M.O. designed the research and supervised the research progress. All authors assisted with interpretation of data, read and edited each draft of the manuscript. All authors read and approved the final manuscript.

Funding: This research received no specific grant from any funding agency in the public, commercial or not-for-profit sectors.

Acknowledgments: The authors would like to acknowledge the personnel provided guidance on developing the criteria list. Also, the authors appreciate Ido Toxopeus and Ceciel Dinnissen for providing feedback to the manuscript.

Conflicts of Interest: The authors declare no conflict of interest.

References

1. Cade, J.E. Measuring Diet in the 21st Century: Use of New Technologies. *Proc. Nutr. Soc.* **2017**, *76*, 276–282. [CrossRef] [PubMed]
2. Shim, J.-S.; Oh, K.; Kim, H.C. Dietary Assessment Methods in Epidemiologic Studies. *Epidemiol. Health* **2014**, *36*, e2014009. [CrossRef] [PubMed]
3. Klesges, R.C.; Eck, L.H.; Ray, J.W. Who Underreports Dietary Intake in a Dietary Recall? Evidence from the Second National Health and Nutrition Examination Survey. *J. Consult. Clin. Psychol.* **1995**, *63*, 438–444. [CrossRef] [PubMed]
4. Livingstone, M.B.; Robson, P.J.; Wallace, J.M. Issues in Dietary Intake Assessment of Children and Adolescents. *Br. J. Nutr.* **2004**, *92* (Suppl. 2), S213–S222. [CrossRef] [PubMed]
5. Schober, M.F. Asking Questions and Influencing Answers. In *Questions about Questions: Inquiries into the Cognitive Bases of Surveys*; Russell Sage Foundation: New York, NY, USA, 1992; pp. 15–48.
6. Boushey, C.J.; Spoden, M.; Zhu, F.M.; Delp, E.J.; Kerr, D.A. New Mobile Methods for Dietary Assessment: Review of Image-Assisted and Image-Based Dietary Assessment Methods. *Proc. Nutr. Soc.* **2017**, *76*, 283–294. [CrossRef] [PubMed]
7. Ramachandran, K. Nutrition Monitoring and Surveillance. *NFI Bull.* **2006**, *27*, 5. [CrossRef]
8. Ocke, M.C.; Van Rossum, C.T.M.; de Boer, E.J. Dietary Surveys: National Food Intake. In *The Encyclopedia of Food and Health*; Caballero, B., Finglas, P., Toldrà, F., Eds.; Academic Press: Oxford, UK, 2016; Volume 2, pp. 432–438.
9. Carter, M.C.; Albar, S.A.; Morris, M.A.; Mulla, U.Z.; Hancock, N.; Evans, C.E.; Alwan, N.A.; Greenwood, D.C.; Hardie, L.J.; Frost, G.S.; et al. Development of a UK Online 24-H Dietary Assessment Tool: Myfood24. *Nutrients* **2015**, *7*, 4016–4032. [CrossRef]
10. Boshuizen, H.C.; Viet, A.L.; Picavet, H.S.; Botterweck, A.; van Loon, A.J. Non-Response in a Survey of Cardiovascular Risk Factors in the Dutch Population: Determinants and Resulting Biases. *Public Health* **2006**, *120*, 297–308. [CrossRef]
11. Conrad, J.; Nothlings, U. Innovative Approaches to Estimate Individual Usual Dietary Intake in Large-Scale Epidemiological Studies. *Proc. Nutr. Soc.* **2017**, *76*, 213–219. [CrossRef]
12. Carter, M.C.; Burley, V.J.; Nykjaer, C.; Cade, J.E. 'My Meal Mate' (Mmm): Validation of the Diet Measures Captured on a Smartphone Application to Facilitate Weight Loss. *Br. J. Nutr.* **2013**, *109*, 539–546. [CrossRef]
13. Xu, W.; Liu, Y. Mhealthapps: A Repository and Database of Mobile Health Apps. *JMIR Mhealth Uhealth* **2015**, *3*, e28. [CrossRef] [PubMed]
14. Chen, J.; Cade, J.E.; Allman-Farinelli, M. The Most Popular Smartphone Apps for Weight Loss: A Quality Assessment. *JMIR Mhealth Uhealth* **2015**, *3*, e104. [CrossRef] [PubMed]
15. Aguilar-Martinez, A.; Sole-Sedeno, J.M.; Mancebo-Moreno, G.; Medina, F.X.; Carreras-Collado, R.; Saigi-Rubio, F. Use of Mobile Phones as a Tool for Weight Loss: A Systematic Review. *J. Telemed. Telecare* **2014**, *20*, 339–349. [CrossRef] [PubMed]
16. Dennison, L.; Morrison, L.; Conway, G.; Yardley, L. Opportunities and Challenges for Smartphone Applications in Supporting Health Behavior Change: Qualitative Study. *J. Med. Internet Res.* **2013**, *15*, e86. [CrossRef] [PubMed]
17. Lieffers, J.R.L.; Hanning, R.M. Dietary Assessment and Self-Monitoring: With Nutrition Applications for Mobile Devices. *Can. J. Diet. Pract. Res.* **2012**, *73*, e253–e260. [CrossRef] [PubMed]
18. *Dietary Assessment: A Resource Guide to Method Selection and Application in Low Resource Settings*; FAO: Rome, Italy, 2018.
19. Illner, A.K.; Freisling, H.; Boeing, H.; Huybrechts, I.; Crispim, S.P.; Slimani, N. Review and Evaluation of Innovative Technologies for Measuring Diet in Nutritional Epidemiology. *Int. J. Epidemiol.* **2012**, *41*, 1187–1203. [CrossRef] [PubMed]
20. Reinivuo, H.; Bell, S.; Ovaskainen, M.-L. Harmonisation of Recipe Calculation Procedures in European Food Composition Databases. *J. Food Compos. Anal.* **2009**, *22*, 410–413. [CrossRef]
21. Maringer, M.; van't Veer, P.; Klepacz, N.; Verain, M.C.D.; Normann, A.; Ekman, S.; Timotijevic, L.; Raats, M.M.; Geelen, A. User-Documented Food Consumption Data from Publicly Available Apps: An Analysis of Opportunities and Challenges for Nutrition Research. *Nutr. J.* **2018**, *17*, 59. [CrossRef]

22. Franco, Z.R.; Fallaize, R.; Lovegrove, A.J.; Hwang, F. Popular Nutrition-Related Mobile Apps: A Feature Assessment. *JMIR Mhealth Uhealth* **2016**, *4*, e85. [CrossRef]
23. Severi, S.; Bedogni, G.; Zoboli, G.P.; Manzieri, A.M.; Poli, M.; Gatti, G.; Battistini, N. Effects of Home-Based Food Preparation Practices on the Micronutrient Content of Foods. *Eur. J. Cancer Prev. Off. J. Eur. Cancer Prev. Organ.* **1998**, *7*, 331–335. [CrossRef]
24. Slimani, N.; Deharveng, G.; Charrondiere, R.U.; van Kappel, A.L.; Ocke, M.C.; Welch, A.; Lagiou, A.; van Liere, M.; Agudo, A.; Pala, V.; et al. Structure of the Standardized Computerized 24-H Diet Recall Interview Used as Reference Method in the 22 Centers Participating in the Epic Project. European Prospective Investigation into Cancer and Nutrition. *Comput. Methods Programs Biomed.* **1999**, *58*, 251–266. [CrossRef]
25. Van Rossum, C.T.M.; Fransen, H.P.; Verkaik-Kloosterman, J.; Buurma, E.M.; Ocke, M.C. *Dutch National Food Consumption Survey 2007–2010: Diet of Children and Adults Aged 7 to 69 Years*; RIVM-report 350070006; RIVM: Bilthoven, The Netherlands, 2011.
26. RIVM. Nevo-Table. Available online: https://nevo-online.rivm.nl/ (accessed on 23 July 2018).
27. Bognár, A. *Tables on Weight Yield of Food and Retention Factors of Food Constituents for the Calculation of Nutrient Composition of Cooked Foods (Dishes)*; BFE Karlsruhe: Karlsruhe, Germany, 2002.
28. Reference Intake. Available online: https://www.voedingscentrum.nl/encyclopedie/referentie-inname.aspx (accessed on 26 December 2018).
29. Barr, S.I.; Murphy, S.P.; Poos, M.I. Interpreting and Using the Dietary References Intakes in Dietary Assessment of Individuals and Groups. *J. Am. Diet. Assoc.* **2002**, *102*, 780–788. [CrossRef]
30. Spriestersbach, A.; Springer, T. Quality Attributes in Mobile Web Application Development. In *Product Focused Software Process Improvement*; Lecture Notes in Computer Science; Springer: Berlin/Heidelberg, Germany, 2004; Volume 3009, pp. 120–130.
31. Maringer, M.; Wisse-Voorwinden, N.; Veer, P.V.; Geelen, A. Food Identification by Barcode Scanning in the Netherlands: A Quality Assessment of Labelled Food Product Databases Underlying Popular Nutrition Applications. *Public Health Nutr.* **2018**, 1–8. [CrossRef] [PubMed]
32. Slimani, N.; Casagrande, C.; Nicolas, G.; Freisling, H.; Huybrechts, I.; Ocke, M.C.; Niekerk, E.M.; van Rossum, C.; Bellemans, M.; De Maeyer, M.; et al. The Standardized Computerized 24-H Dietary Recall Method Epic-Soft Adapted for Pan-European Dietary Monitoring. *Eur. J. Clin. Nutr.* **2011**, *65* (Suppl. 1), S5–S15. [CrossRef]
33. Griffiths, C.; Harnack, L.; Pereira, M.A. Assessment of the Accuracy of Nutrient Calculations of Five Popular Nutrition Tracking Applications. *Public Health Nutr.* **2018**, *21*, 1495–1502. [CrossRef]
34. Rusin, M.; Årsand, E.; Hartvigsen, G. Functionalities and Input Methods for Recording Food Intake: A Systematic Review. *Int. J. Med. Inform.* **2013**, *82*, 653–664. [CrossRef] [PubMed]
35. Marconi, S.; Durazzo, A.; Camilli, E.; Lisciani, S.; Gabrielli, P.; Aguzzi, A.; Gambelli, L.; Lucarini, M.; Marletta, L. Food Composition Databases: Considerations About Complex Food Matrices. *Foods* **2018**, *7*, 2. [CrossRef]
36. Castanheira, I.; Roe, M.; Westenbrink, S.; Ireland, J.; Møller, A.; Salvini, S.; Beernaert, H.; Oseredczuk, M.; Calhau, M.A. Establishing Quality Management Systems for European Food Composition Databases. *Food Chem.* **2009**, *113*, 776–780. [CrossRef]
37. Westenbrink, S.; Oseredczuk, M.; Castanheira, I.; Roe, M. Food Composition Databases: The Eurofir Approach to Develop Tools to Assure the Quality of the Data Compilation Process. *Food Chem.* **2009**, *113*, 759–767. [CrossRef]
38. Vasquez-Caicedo, A.L.; Bell, S.; Hartmann, B. *Report on Collection of Rules on Use of Recipe Calculation Procedures Including the Use of Yield and Retention Factors for Imputing Nutrient Values for Composite Foods*; European Food Information Resource: Brussels, Belgium, 2014.

 *nutrients*

Article

Assessing the Usability of the Automated Self-Administered Dietary Assessment Tool (ASA24) among Low-Income Adults

Julia Kupis [1], Sydney Johnson [1], Gregory Hallihan [1] and Dana Lee Olstad [2,*] 

[1] W21C Research and Innovation Centre, Cumming School of Medicine, University of Calgary, Calgary,
    AB T2N 4Z6, Canada; jnkupis@ucalgary.ca (J.K.); sydney.johnson@ucalgary.ca (S.J.);
    gmhallih@ucalgary.ca (G.H.)
[2] Department of Community Health Sciences, Cumming School of Medicine, University of Calgary, Calgary,
    AB T2N 4Z6, Canada
*   Correspondence: dana.olstad@ucalgary.ca

Received: 9 November 2018; Accepted: 8 January 2019; Published: 10 January 2019

Abstract: The Automated Self-Administered Dietary Assessment Tool (ASA24) is a web-based tool that guides participants through completion of a 24-h dietary recall and automatically codes the data. Despite the advantages of automation, eliminating interviewer contact may diminish data quality. Usability testing can assess the extent to which individuals can use the ASA24 to report dietary intake with efficiency, effectiveness, and satisfaction. This mixed-methods study evaluated the usability of the ASA24 to quantify user performance and to examine qualitatively usability issues in a sample of low-income adults (85% female, 48.2 years on average) participating in a nutrition coupon program. Thirty-nine participants completed a 24-h dietary recall using the ASA24. Audio and screen recordings, and survey responses were analyzed to calculate task times, success rates, and usability issue frequency. Qualitative data were analyzed thematically to characterize usability issues. Only one participant was able to complete a dietary recall unassisted. We identified 286 usability issues within 22 general usability categories, including difficulties using the search function, misunderstanding questions, and uncertainty regarding how to proceed to the next step; 71.4% of participants knowingly misentered dietary information at least once. Usability issues may diminish participation rates and compromise the quality of ASA24 dietary intake data. Researchers should provide on-demand technical support and designers should improve the intelligence and flexibility of the ASA24's search functionality.

Keywords: usability; human factors; dietary assessment; Automated Self-Administered Dietary Assessment Tool (ASA24); 24-h dietary recall; low socioeconomic status

1. Introduction

Accurate and detailed characterization of dietary intake is essential in nutrition research. However, assessment of intake is challenging because researchers must typically rely on self-reported methods, including 24-h dietary recalls, food-frequency questionnaires, and food records [1]. Dietary intake data collected in this way are subject to bias stemming from systematic measurement error, which may result in inaccurate and imprecise estimates of dietary intake [2–4]. Compared to other self-reporting instruments, 24-h dietary recalls capture intake with less bias, and have, therefore, emerged as a preferred means of dietary assessment [3,4]. Until recently, dietary recalls were typically conducted in person or via telephone using the computer-assisted Automated Multiple Pass Method. In this method, trained interviewers use a five-step multiple-pass approach to obtain details of all foods consumed from midnight to midnight the previous day, with manual data entry and auto-coding for most foods [1]. The time and expense of collecting and analyzing data gathered in this way made 24-h recalls impractical for use in most large, community-based studies.

In 2009, the National Cancer Institute released the Automated Self-Administered 24-h Dietary Assessment Tool (ASA24): an automated, self-administered web-based tool that guides participants through completion of a 24-h dietary recall and automatically codes the data [5]. The ASA24-Canada was released in 2014 and updated in 2016 [6]. The self-administered and automated nature of the ASA24 has made collection of 24-h dietary recall data feasible in large studies. Nevertheless, despite its apparent advantages, eliminating contact with an interviewer may introduce additional challenges and different sources of error, with potential implications for the quantity and quality of the data that are collected.

A limited number of studies have discussed the usability of the ASA24 in relation to the quality of dietary intake data collected [2,7]. Although intakes were on average underreported on the ASA24 compared to more objective measures [2,7], the ASA24 nevertheless appeared to provide reasonable estimates of dietary intake comparable to, or better than, those derived from other self-report methods [2,7–9]. Usability related to the acceptability and feasibility of use of the ASA24 has been examined through retrospective questionnaires and by examining completion rates and intake data. Findings suggest that lack of internet access and/or lower levels of computer literacy may limit participation by some populations such as older adults, racial minorities, and those with lower levels of education [8]. Participant workload may also pose a barrier with some participants finding it simpler and faster to interact with an interviewer rather than to search for and select foods themselves [8]. Other limitations include difficulties finding exact matches for foods entered in search bars and a resulting tendency to select items that appear near the top of the list [7,10]. Overall, varying levels of receptivity to using the tool have been reported [8–10].

Comparing intake data generated by the ASA24 to data generated by other measures can allow quantification of measurement error. However, such assessments cannot identify critical points in the reporting process where errors often originate. Similarly, although retrospective reports can identify socio-demographic characteristics of participants who find the ASA24 challenging to complete, the comprehensiveness, accuracy, and ultimate usefulness of these data is diminished by a reliance on participants' ability to recall specific details of difficulties they encountered during a lengthy reporting process (~40 min [9,10]) and by the questionable assumption that participants can accurately pinpoint the errors they made as well as their cause. Moreover, the common use of closed-ended questions to query participants limits the ability to describe specific qualitative aspects of usability in detail.

In this respect, the science of Human Factors may offer new avenues for understanding how human–system interactions contribute to dietary measurement error, particularly those associated with novel technology-enabled assessments such as the ASA24. Human Factors as a science is concerned with the interaction between people and designed systems, and how human limitations and capabilities (e.g., limitations in human working memory), and the design of those systems (e.g., the number of digits an individual has to keep in working memory to complete a task), interact. The study of usability is a sub-discipline within the Human Factors field that seeks to understand the extent to which individuals can interact with a system to achieve a desired outcome with effectiveness (i.e., the accuracy and completeness with which specified users can achieve specified goals in particular environments), efficiency (i.e., the resources expended in relation to the accuracy and completeness of goals achieved), and satisfaction (i.e., the comfort and acceptability of a system to its users and other people affected by its use) [11]. The construct of usability can also be extended to include concepts such as learnability, legibility, readability, and comprehension. A simpler way to frame the usability paradigm is to examine the relationship between usability, utility, and usefulness: a system may be technically able to deliver utility (i.e., it is technically functional) when used under perfect circumstances, but usability defines the required effort, and experience of, an individual to access that utility and ultimately interact with the system to produce a useful outcome [12].

There are a variety of methods available to evaluate software usability. The present study follows the guidelines established by the International Organization for Standardization (ISO), where usability testing of a system engages representative users to complete representative tasks within the system in order to calculate measures of efficiency, effectiveness, and satisfaction [13]. Usability testing also includes a methodology to identify usability issues and inform system design through the collection and analysis of qualitative data regarding users' perspectives [14]. Techniques such as the "think-aloud" method encourage users to verbalize cognitive processes while interacting with the system of interest [15].

To our knowledge, no previous studies have involved structured usability testing of the ASA24 to obtain quantitative measures of efficiency and effectiveness, or to describe qualitative aspects of usability in detail. Therefore, the purpose of this study was to conduct a structured usability test of the ASA24 to generate quantitative measures of user performance (i.e., task success, task time, food item count, and usability issue frequency) and to examine qualitative aspects of usability (i.e., describe usability issues and user preferences) within a specific user population. The results of the usability test can provide insights that can be applied to configure and administer the ASA24 in a manner that makes the tool more usable for individuals completing a dietary recall, thereby increasing the quantity and quality of the dietary intake data that are collected.

2. Materials and Methods

2.1. Study Design

This was a cross-sectional, mixed-methods study in which qualitative and quantitative data were collected concurrently and integrated during analysis. Mixed methods were used for purposes of complementarity to provide a comprehensive and rich understanding of usability issues [16]. This study was conducted as a pre-cursor to a larger funded study designed to investigate the impact of the British Columbia Farmers' Market Nutrition Coupon Program (FMNCP) on the dietary intake and mental and social well-being of program participants. The FMNCP provides low-income households with 16 weeks' worth of vouchers that can be used to purchase selected healthy foods (e.g., fruits, vegetables, nuts, seeds, legumes, meats) at participating farmers' markets in the province of British Columbia, Canada [17]. The study was conducted in accordance with the Declaration of Helsinki and received ethical approval from the Conjoint Health Research Ethics Board at the University of Calgary (REB17-1076).

2.2. Screening Criteria and Recruitment

Study participants were recruited from households participating in the British Columbia FMNCP. Coupons are normally distributed via local community partner organizations in each community. Given their existing relationships with FMNCP participants, community partner organizations were asked to facilitate participant recruitment for the current study. The FMNCP manager invited approximately 100 community partner organizations to recruit participants through individual conversations as well as two broadcast emails, of which 13 agreed, with 6 ultimately enrolling participants spanning 6 different communities. Attempts were made to specifically recruit organizations working with older adults ($\geq$60 years) as well as recent immigrants to ensure the study population was reflective of FMNCP participants who would participate in the subsequent larger study. All participating community organizations were asked to recruit study participants via posters posted on-site, broadcast or direct emails, announcements during programming, and/or in-person requests. Interested individuals were asked to sign a preliminary consent form granting the research team permission to contact them directly and conduct eligibility screening.

Participants were deemed eligible to participate if they met the following inclusion criteria: adults ($\geq$18 years of age), not pregnant or breastfeeding, not reporting having a cognitive disability, and able to speak, read, and write in English. Individuals without home internet/computer access were offered access to both at community partner organization sites. A total of 80 participants were screened for eligibility across the 6 participating community partner sites. Of these, 67 individuals were eligible and were invited to participate in the study, of which 51 agreed. Of those 51 individuals, 11 either cancelled or did not show up to their session and were not able to be rescheduled. One individual attempted to participate but was excluded due to technical difficulties. In total, 39 individuals participated in the study. Usability testing typically gathers in-depth data from a small number of participants (i.e., n = 6–8) and, therefore, our sample size of 39 divided across three groups was deemed sufficient [18].

2.3. Participant Consent and Compensation

Individuals who agreed to participate in the study were sent an e-mail from the research team containing a link to a web-based data collection platform and a username/login and password for the ASA24. Participants read through an online consent form embedded within the web-based data collection platform, and indicated their agreement to participate. Participants were offered $20 worth of FMNCP coupons in appreciation of their time.

2.4. Online Survey Instruments

As the ASA24 is typically administered along with other multi-component questionnaires, participants were asked to complete a socio-demographic and health-related survey prior to using the ASA24 in order to approximate real-world conditions. The web-based data collection platform (SurveyMonkey 2017; San Mateo, CA, USA) guided participants through completion of a socio-demographic questionnaire and multi-item scales to assess social connectedness, perceived stress, and mental well-being, followed by the ASA24 for Canada, 2016 (National Cancer Institute 2016, Rockville, MD). Briefly, the data collection platform consisted of the following:

- *Socio-demographics:* questions to assess date of birth, gender, race/ethnicity, citizenship status, country of birth, marital status, pregnancy, perceived health status, household composition, employment status, education level, household income, receipt of social assistance, food insecurity status, language ability, and usual fruit and vegetable intake. Items were adapted from existing surveys [19,20].
- *Social connectedness:* the 20-item Social Connectedness Scale-Revised, a valid and reliable scale that reflects individuals' perception of their closeness with others [21].
- *Perceived stress:* the 10-item version of the Perceived Stress Scale, a valid and reliable measure of the degree to which respondents perceive their lives to be unpredictable, uncontrollable, or overwhelming [22].
- *Mental well-being:* the 14-item Warwick-Edinburgh Mental Well-being Scale, a valid and reliable measure of both feeling and functioning aspects of mental well-being [23].
- *ASA24:* the ASA24 guides participants through the process of reporting dietary intake for the previous day [6]. The version used by the participants asked for a single "midnight to midnight" recall and included the "location" and "source" modules. In an initial quick list, participants select an eating occasion, indicate time of consumption, and report all foods and beverages consumed at that time. Foods and beverages can be entered by typing in specific search terms and selecting items from a returned list. Details of food types, preparation methods, portion sizes, and additions are subsequently queried during a detailed pass. The system later prompts users to recall frequently omitted/forgotten foods and to complete a final review of all items consumed.

2.5. Test Moderation and Procedure

A usability test moderator is a trained researcher who observes and directly interacts with participants during a usability test. Like all other researcher/participant interactions, moderators attempt to remain unbiased and neutral in their interactions. However, they also seek to elicit information from participants regarding their experiences and may provide participants with assistance to complete tasks [24]. This interaction generates valuable qualitative data, but creates a scenario that is less representative of how individuals typically interact with software under real-use conditions. In the present study, participants were sequentially assigned to complete the ASA24 in one of three "moderation groups" with varying levels of moderator involvement to account for the strengths and limitations of moderator interaction. Sequential assignment was chosen to avoid creating groups of unequal size. Table 1 provides a summary of the differences between the moderation groups that participants were assigned to.

Table 1. Moderator-participant interactions and data sources in the three session types.

	Moderator Probing	Participant Think-Aloud	Audio and Screen Recording	Survey Completion Data
Unmoderated	×	×	×	✓
Semi-Moderated	×	×	✓	✓
Moderated	✓	✓	✓	✓

2.5.1. Moderated and Semi-Moderated Procedure

Participants assigned to the moderated ($n = 10$) and semi-moderated ($n = 12$) groups scheduled a session to join an online meeting with one of two trained moderators who adhered to the same protocol. Participants were encouraged to participate from their computing environment of choice. During the session, participants completed the previously described survey instruments while moderators used Adobe Connect Meeting software (Adobe Systems Incorporated 2017; San Jose, CA, USA) to capture audio and screen recordings. For participants who could not provide digital audio, a telephone recording was used to capture audio data. Moderators also maintained detailed written notes of all sessions.

The process of providing assistance to participants was formalized a priori to ensure consistent moderator–participant interaction. Participants in the moderated and semi-moderated groups were informed that they should use the ASA24 as they normally would, and that moderators could answer questions that they had while trying to complete each task. Moderators encouraged participants to resolve their own problems independently before providing assistance. (For example, if a participant stated "I do not know what to do now," the moderator responded "Where do you think you would click on the page to proceed?"). The moderator allowed participants to experience and express difficulty and frustration until it was determined that the participant was likely to fail the task and/or withdraw from the study, at which point the moderator offered to assist the participant to continue using the platform.

2.5.2. Unmoderated Procedure

Participants assigned to the unmoderated group were split into two sub-groups. The first group of participants ($n = 5$) scheduled a time, at their convenience, to participate in the study and complete the surveys. The second group of participants ($n = 12$) received an unannounced email inviting them to complete the surveys and were given a 36-h time frame in which to do so. Limiting the time frame for survey completion was intended to minimize reactivity, where participants change their dietary intake in anticipation of having to report it [25,26]. Participants in the unmoderated group ($n = 17$) completed data collection entirely independently and had no contact with the study team. Audio and screen recordings were not collected for participants in the unmoderated group and only their survey responses were available for analysis.

2.5.3. Data Collection

The data collection procedures for each of the three moderation groups (as presented in Table 1) are described in detail below:

- Participant think-aloud: the think-aloud procedure involves asking participants to verbalize their thoughts and feelings as they use a system [15]. Participants were asked to "think aloud" as they completed the surveys, were encouraged to elaborate on difficulties encountered (both conceptual and technical in nature), and asked to suggest remedies (where applicable).
- Moderator probing: moderators encouraged participants to expand on their thoughts as they worked through the platform and posed clarifying questions. This probing provided a more comprehensive and in-depth perspective of participants' experiences using the ASA24.
- Audio and screen recording: participants' screens were recorded while they completed the survey instruments. Their verbalizations (think-aloud comments and responses to probing questions) were audio recorded using the Adobe Connect Meeting platform or audio recording over the telephone.
- Survey data: a subset of the information entered into the online survey instruments (as described in Section 2.4) was relevant to the usability test, including participant socio-demographic information and administrative and food intake data from the ASA24 to measure task completion and food item count.

2.5.4. Participant Tasks

Usability testing requires participants to complete pre-determined and standardized "tasks" that are representative of how they would typically interact with a system. Within the context of this study, the participant's objective was to complete a dietary recall using the ASA24. The research team defined four tasks and eight subtasks to meet this objective (see Figure 1). All participants had to complete the tasks of reading the introduction (which includes the ASA24 User Orientation), reporting meals (including snacks and drinks), adding details to those meals, and reviewing and completing their entries. The subtasks varied depending on whether or not the individual had consumed the meal in question (i.e., breakfast, brunch, lunch, dinner, supper, snack, just a drink, just a supplement).

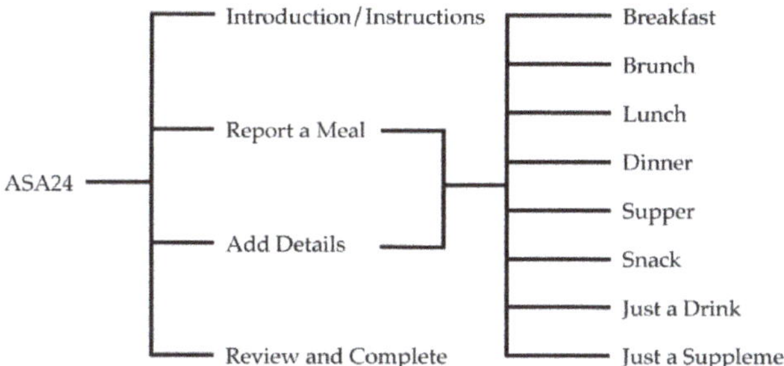

Figure 1. Participant tasks and sub-tasks while completing a dietary recall with the Automated Self-Administered Dietary Assessment Tool (ASA24).

2.5.5. Measuring Usability

Measuring usability requires specifying relevant usability metrics. The research team identified four quantitative metrics relevant to the performance of participants using the ASA24. Task success was measured for all three moderation groups, while task time, food item count, and usability issue count could only be quantified for the moderated and semi-moderated groups.

1. Task success, a nominal variable, was defined in three categories: failure to complete the task, completion of the task with moderator assistance, and completion of the task without moderator assistance. The task in this case was completing a dietary recall using the ASA24.
2. Task time, a continuous variable, was defined as the time between a clearly defined start and stop point for each task.

a. Introduction/Instructions: start time when webpage has loaded, stop time when participant clicks "report a meal" button

b. Report a meal: start time when participant has selected a meal from dropdown menu, stop time when participant clicks "finish with this meal" button

c. Add details: start time when participant clicks "add details" button, stop time when ASA24 system loads "review and complete" section

d. Review and complete: start time when webpage has loaded, stop time when ASA24 system dialogue box appears indicating participant has finished their dietary recall

3. Food item count, a nominal variable, was included to account for variability in participants' dietary intake, which directly affects task performance (i.e., participants who had eaten more food items had more data to enter). This measure was defined as the number of ASA24 output line items in the analytic file for "Items" per meal. For example, if the separate line items for a lunch meal were "fruit, Not Specified (NS) as to type" and "water, municipal" then that meal was assigned a food item count of 2. If there were three separate line items for a meal such as "fruit, NS as to type", "water, municipal", and "cheese, cheddar", then that meal was assigned a Food Item Count of 3.

4. Usability issue count, a nominal variable, was defined as the total number of individual usability issues observed per task. Usability issue identification is defined in detail Section 2.6.2. Generally usability issues are observable events or actions associated with difficulty completing a task.

2.6. Data Analysis

Data were collected from the online survey described in Section 2.4, as well as audio and screen recordings from participants in the moderated and semi-moderated groups. The focus of the quantitative and qualitative analysis presented here is only on the data directly relevant to the usability test of the ASA24 (i.e., audio/screen recordings, participant characteristics as reported in the socio-demographic questionnaire, and user data from the ASA24). This section reports the procedures that were used to transform, aggregate, and analyze these data. Data were originally stratified by age (i.e., seniors, non-seniors) and session type (i.e., moderated, semi-moderated, and unmoderated because procedures differed by group). However, performance did not differ by age and, therefore, data are stratified by session type only.

Two members of the research team worked together to first analyze the audio and screen recordings to extract relevant qualitative (i.e., usability issues) and quantitative (i.e., usability metrics) data for subsequent analysis. This audiovisual analysis process is described in Section 2.6.1 and produced a time-stamped record of events and actions for each participant, which was the foundation for subsequent analyses. Qualitative data (e.g., a participant verbalizing that they were unable to find a food item they were searching for) were aggregated and analyzed using thematic analysis, described in Section 2.6.2. Quantitative data (i.e., participant task times, task success, food item count, and usability issue count) were analyzed using descriptive and correlational statistics, described in Section 2.6.3.

2.6.1. Audiovisual Analysis

Audio and screen recordings from moderated and semi-moderated sessions were analyzed using a software package designed for the behavioural analysis of observational data (Noldus Observer XT (v.14, Noldus Information Technologies, Wageningen, Netherlands). The methodological analysis of audiovisual recordings of users interacting with a "health system" to derive quantitative and qualitative data is described in detail by Mackenzie and Xiao [27]. Central to the process are clear and consistent operational definitions, which analysts assign to events and actions of interest. An operational definition is the formalization of an observable phenomenon so that a researcher can consistently and independently detect it. In this case, the phenomena of interest were usability issues and usability metrics. The two analysts systematically reviewed the audio/screen recordings to identify usability issues (described in Section 2.6.2.) and calculate task times and usability issue counts (described in Section 2.6.3). The two research analysts worked through this process collaboratively and relied on clear and consistent operational definitions to ensure accuracy in the analysis process.

2.6.2. Qualitative Analyses to Identify and Categorize Usability Issues

Usability issues are observable participant behaviours reflecting inefficiency, ineffectiveness, dissatisfaction, or confusion during the use of a system, and an interpretation of the cause of those behaviours relative to the participant's attempt to complete a task [14]. The research team first identified individual usability issues specific to each participant through audiovisual analysis. These individual usability issues were then subject to thematic analysis, which involved identifying, analyzing, and reporting patterns (themes) within the data [28] to aggregate individual issues into general usability issues. For example, two users may have separately expressed confusion on how to complete an aspect of a task, so these two individual issues would be categorized as the general issue "How to Complete Task Unclear." To aid in the interpretation of the significance and impact of these general issues, they were then thematically organized based on their relationship to the usability constructs of efficiency, effectiveness, satisfaction, and comprehension.

The two research analysts met regularly to review the data and ensure consistency in interpretations of usability issues. The analysts determined that a point of thematic saturation was reached during the analysis as few to no new usability issues were being observed in the data [29]. In addition, the frequency distribution of usability issues (presented in Section 3.5.) reveals that usability issues in the tails of the distribution were being identified.

2.6.3. Quantitative Analyses of Usability Metrics

Descriptive statistics were calculated for task time, task success, usability issue count, and food item count. Task success was calculated for all three moderation groups and nominally measured as either failure to complete the ASA24, completion of the ASA24 with assistance, or completion of the ASA24 without assistance. Average task time and individual usability issue count were calculated for the moderated and semi-moderated conditions. The proportional frequency of general usability issues (%) was calculated by dividing the number individual issues counted within a general usability issue category, divided by the total individual usability issue count. For example, 33 out of the 286 individual usability issues were classified as "Question Not Understood", the proportional frequency of the general usability issue "Question Not Understood" was, therefore, 11.5%. The proportion (%) of participants affected by a general usability issue was calculated by dividing the number of participants that encountered each general usability issue by the total number of participants. Correlations between task time, individual usability issue count, and food item count (from the add details task) were calculated for the report a meal and/or add details tasks using Pearson's product-moment correlation, with $p < 0.05$ indicating statistically significant correlations. These quantitative analyses were conducted in Microsoft Excel, (2013, Microsoft Corporation, Redmond Washington, WA, USA).

3. Results

3.1. Participant Characteristics

Across all moderation groups, participants were mostly female (85.3%) and non-seniors (55.9%). The majority of participants had either only a high school diploma or below (41.1%), or a certificate or diploma below a Bachelor's degree (including trade certificates and diplomas) (53.7%). While 73.5% of participants were severely or moderately food insecure, the majority still self-reported being in good health or better (79.4%). All participants were from low-income households, which is a criterion for participation in the FMNCP. All participants in the moderated and semi-moderated groups completed the survey using a laptop or desktop computer. It is not known how participants in the unmoderated group accessed the ASA24. Further details can be found in Table 2.

Table 2. Participant characteristics according to moderation group.

	Moderated ($n = 10$)	Semi-Moderated ($n = 12$)	Unmoderated [1] ($n = 12$)	Total ($n = 34$)
Age (years), mean $\pm$ standard deviation (SD)	43.9 $\pm$ 20.3	43.5$\pm$ 17.7	57.2 $\pm$ 18.4	48.2 $\pm$ 19.0
Seniors, *n* (%)	4 (40.0)	4 (33.3)	7 (58.3)	15 (44.1)
Non-seniors, *n* (%)	6 (60.0)	8 (66.7)	5 (41.7)	19 (55.9)
Gender, *n* (%)				
Women	7 (70.0)	10 (83.3)	12 (100.0)	29 (85.3)
Men	3 (30.0)	2 (16.7)	0 (0.0)	5 (14.7)
Self-reported health, *n* (%)				
Excellent	0 (0.0)	1 (8.3)	0 (0.0)	1 (2.9)
Very Good	0 (0.0)	1 (8.3)	3 (25.0)	4 (11.8)
Good	8 (80.0)	7 (58.3)	7 (58.3)	22 (64.7)
Fair	1 (10.0)	3 (25.0)	2 (16.7)	6 (17.7)
Poor	1 (10.0)	0 (0.0)	0 (0.0)	1 (2.9)
Education level, *n* (%)				
Less than high school	1 (10.0)	3 (25.0)	4 (33.3)	8 (23.5)
High school	5 (50.0)	1 (8.3)	0 (0.0)	6 (17.6)
Trade certificate/diploma	1 (10.0)	3 (25.0)	0 (0.0)	4 (11.8)
Certificate or diploma below Bachelor's	3 (30.0)	4 (33.3)	7 (58.3)	14 (41.9)
Bachelor's degree	0 (0.0)	0 (0.0)	1 (8.3)	1 (2.9)
Above Bachelor's degree	0 (0.0)	1 (8.3)	0 (0.0)	1 (2.9)
Employment Status, *n* (%)				
Full-time employment	0 (0.0)	2 (16.7)	2 (16.7)	4 (11.8)
Part-time employment	1 (10.0)	1 (8.3)	1 (8.3)	3 (8.8)
Unemployed: Looking for work	1 (10.0)	0 (0.0)	1 (8.3)	2 (5.9)
Unemployed: Not looking for work	0 (0.0)	4 (33.3)	2 (16.7)	6 (17.7)
Student	1 (10.0)	0 (0.0)	0 (0.0)	1 (2.9)
Retired	3 (30.0)	1 (8.3)	5 (41.7)	9 (26.5)
Other	4 (40.0)	4 (33.3)	1 (8.3)	9 (26.5)
Receiving Social Assistance or Welfare, *n* (%)	5 (50.0)	7 (58.3)	5 (41.7)	17 (50.0)
Immigration Status, *n* (%)				
Native Canadian	9 (90.0)	12 (100)	10 (83.3)	31 (91.2)
Immigrant: English-speaking country	1 (10.0)	0 (0.0)	2 (16.7)	3 (8.8)
Immigrant: Non-English-speaking country	0 (0.0)	0 (0.0)	0 (0.0)	0 (0.0)
Marital Status, *n* (%)				
Single, never married	1 (10.0)	4 (33.3)	4 (3.3)	9 (26.5)
Living common-law	3 (30.0)	1 (8.3)	0 (0.0)	4 (11.8)
Married	1 (10.0)	2 (16.7)	1 (8.3)	4 (11.8)
Separated	2 (20.0)	2 (16.7)	0 (0.0)	4 (11.8)
Divorced	3 (30.0)	1 (8.3)	6 (50.0)	10 (29.4)
Widowed	0 (0.0)	1 (8.3)	1 (8.3)	2 (5.9)
Did not answer	0 (0.0)	1 (8.3)	0 (0.0)	1 (2.9)
Food Insecurity Status, *n* (%)				
Food secure	1 (10.0)	1 (8.3)	3 (25.0)	5 (14.7)
Marginally food insecure	2 (20.0)	2 (16.7)	0 (0.0)	4 (11.8)
Moderately food insecure	3 (30.0)	3 (25.0)	7 (58.3)	13 (38.2)
Severely food insecure	4 (40.0)	6 (50.0)	2 (16.7)	12 (35.3)
Access to Internet at Home, *n* (%)				
Yes	8 (80.0)	11 (91.7)	12 (100.0)	31 (91.2)
No	2 (20.0)	0 (0.0)	0 (0.0)	2 (5.9)
Did not answer	0 (0.0)	1 (8.3)	0 (0.0)	1 (2.9)
Preferred Internet Access Method [2], *n* (%)				
Computer		7 (58.3)		7 (58.3)
Smartphone		3 (25.0)		3 (25.0)
No Preference		2 (16.7)		2 (16.7)
Did not answer		1 (8.3)		1 (8.3)

[1] Not all participants in the unmoderated group completed the sociodemographic questionnaire; [2] Some participants chose both computer and smartphone as preferred methods to access the internet.

3.2. Successful Completion of Dietary Recall Using the Automated Self-Administered Dietary Assessment Tool (ASA24)

In the unmoderated group 94.1% of participants (n = 16) failed to fully complete the ASA24. These 16 failures can be characterized as follows:

- Three participants did not initiate either the socio-demographic/health-related survey or the ASA24.
- Two participants initiated the socio-demographic/health-related survey after the 36-h time window had passed, but did not fully complete it or initiate the ASA24.
- Eight participants completed the socio-demographic/health related survey within the 36-h time window, but failed to initiate the ASA24.
- Three participants completed the socio-demographic/health-related survey and initiated the ASA24 within the 36-h time window, but did not fully complete the ASA24 (two of the three participants quit before reporting a meal, and the third participant quit during the add details section).

Participants in the unmoderated condition were not observed and there are insufficient data to determine the specific causes of these failures. In the semi-moderated group, 91.7% (n = 11) of participants completed the ASA24 with assistance from the moderator (one failed to complete the ASA24), while 100% of the participants (n = 10) in the moderated group completed the ASA24 with the assistance of a moderator. In total, only one participant was able to complete the ASA24 without the assistance of a moderator (Figure 2). Because one participant in the semi-moderated group failed to complete the ASA24, quantitative and qualitative data (task completion times, food item count, and usability issues) from this participant were not available for analysis. Additionally, one of the 22 participants in the moderated and semi-moderated group accessed the ASA24 help feature.

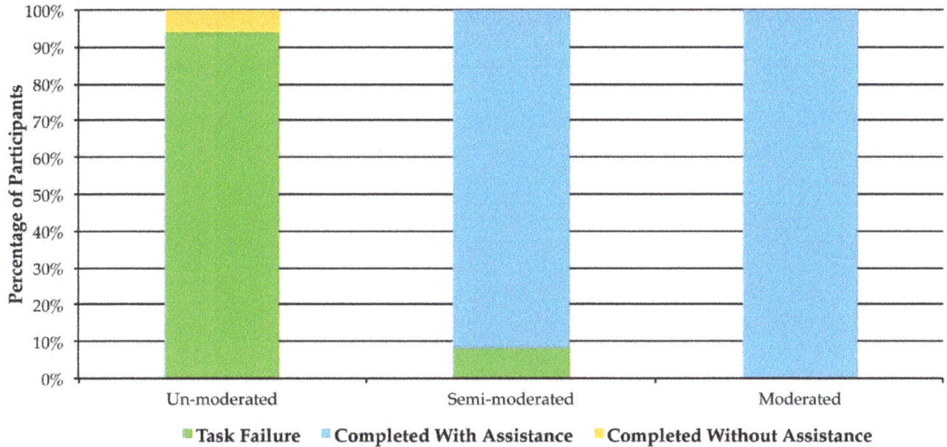

Figure 2. Task success rates for the three groups: unmoderated (n = 17), semi-moderated (n = 12), and moderated (n = 10).

3.3. Time to Complete ASA24 Tasks

Across both the moderated and semi-moderated groups, the average time it took participants to complete the entire ASA24 (i.e., all four tasks) was 27.4 min (standard deviation (SD) = 12.9). For the moderated group, the average time to complete all tasks was 33.4 min (SD = 14.1). Average task completion time for all four tasks in the semi-moderated group was 21.9 min (SD = 8.7). The report a meal and add details tasks took the longest to complete (Figure 3).

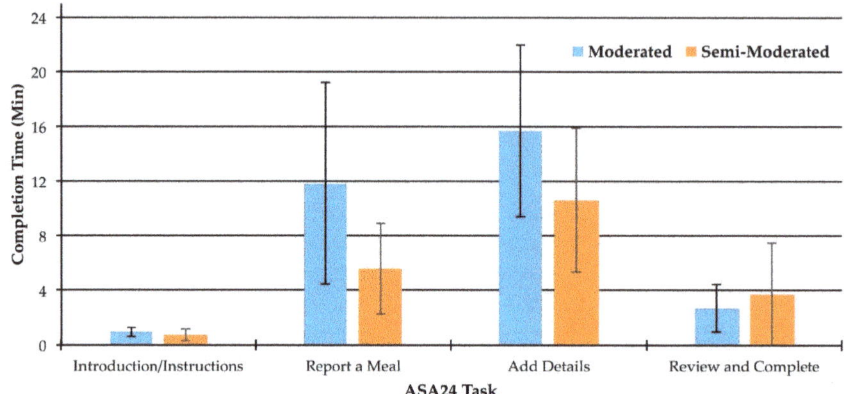

Figure 3. Average task completion times for each task in the moderated ($n = 10$) and semi-moderated ($n = 11$) groups. Error bars represent one standard deviation above and below the mean.

3.4. Food Item Count

Food item count provides an indication of differences in the number of steps a participant had to go through to complete the add details task. The average number of food items reported in the add details section in the moderated and semi-moderated groups was 11.5 items (SD = 5.7). The average number of items reported by participants in the moderated group was 12.1 items (SD = 5.6). The average number of items reported by participants in the semi-moderated group was 10.9 items (SD = 5.6).

3.5. Usability Issue Frequency

There were a total of 286 individual usability issues observed in the moderated and semi-moderated groups across all four tasks. There were an average of 13.7 individual usability issues (SD = 8.0) identified per participant. The average number of individual usability issues that participants in the moderated group encountered was 16.6 (SD = 8.7), while those in the semi-moderated group encountered an average of 11.1 (SD = 6.2) usability issues. The majority of individual usability issues were encountered during the report a meal and add details tasks (Figure 4).

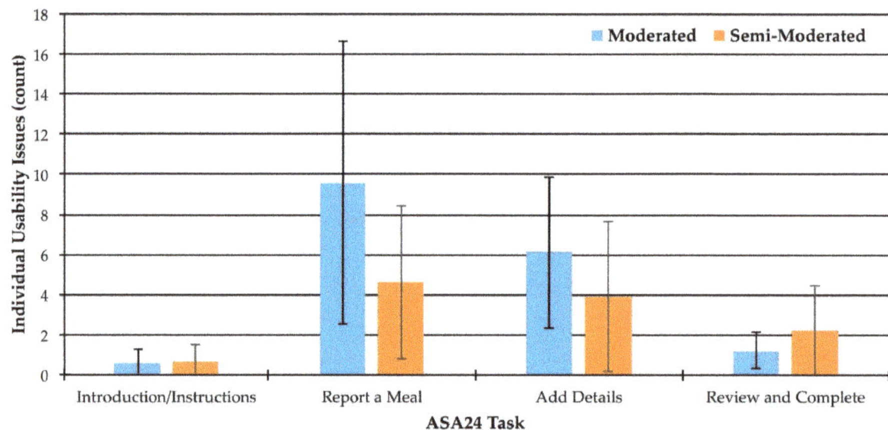

Figure 4. Average number of individual usability issues encountered by participants in the moderated ($n = 10$) and semi-moderated ($n = 11$) groups across all four tasks in the ASA24. Error bars represent one standard deviation above and below the mean.

The 286 individual usability issues identified were classified within 22 general usability issue categories (see Section 3.7 for definitions and examples). For example, 33 individual issues were identified in which participants did not understand a question asked in the ASA24, all of which corresponded with the general usability issue category of "Question Not Understood". The proportional frequency of these general usability issues is presented in Figure 5. This analysis revealed that the five most frequent general usability issues were:

1. Search Item Missing/Inaccurate (13.6%)
2. Question Not Understood (11.5%)
3. Next Step Unclear (9.9%)
4. Submits Incorrect Information—Known to User (9.4%)
5. Misclick (8.7%)

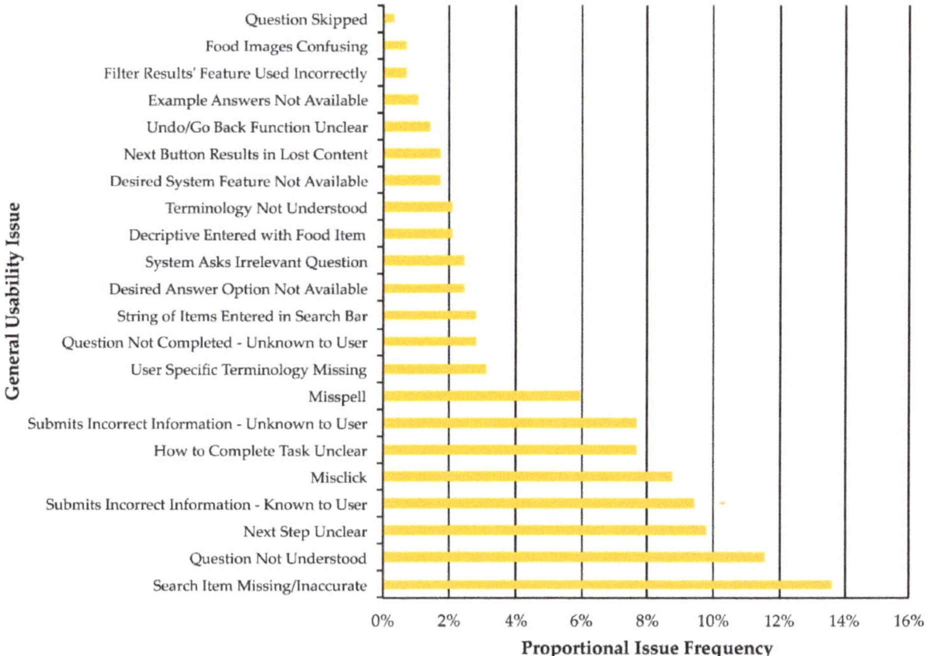

Figure 5. Proportional frequency (%) of each general usability issue across all ASA24 tasks for participants in the moderated ($n = 10$) and semi-moderated ($n = 11$) groups.

The frequency with which participants in the semi-moderated and moderated groups encountered each of the 22 general usability issues is presented in Figure 6. These data indicate the prevalence of each general usability issue within the sample of participants (e.g., out of 21 participants, 14 (66.7%) did not understand an ASA24 question at least once). The five most common general usability issues across all participants were:

1. Search Item Missing/Inaccurate (76.2% of participants, $n = 16$)
2. Submits Incorrect Information—Known to User (71.4% of participants, $n = 15$)
3. Next Step Unclear (71.4% of participants, $n = 15$)
4. Question Not Understood (66.7% of participants, $n = 14$)
5. Misclick (61.9% of participants, $n = 13$)

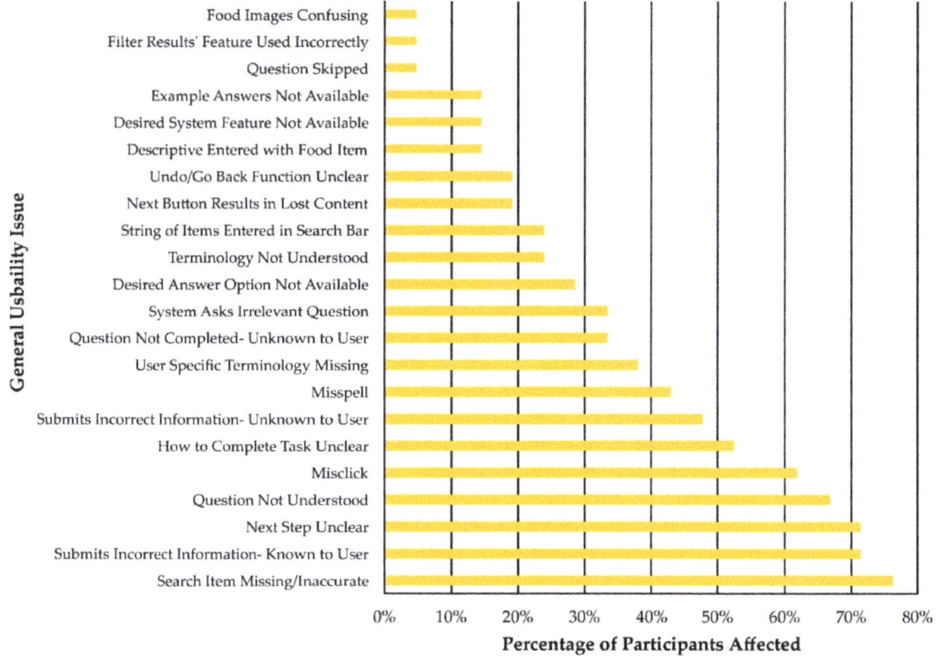

Figure 6. Percentage (%) of participants in the semi-moderated (*n* = 11) and moderated (*n* = 10) groups that were observed being affected by each general usability issue at least once.

3.6. Correlated Measures

A strong and statistically significant positive correlation was observed between task time and individual usability issue count for the report a meal task, r (91) = 0.81, p < 0.01. A moderate and statistically significant positive correlation was observed between task time and individual usability issue count for the add details task, r (89) = 0.54, p < 0.01. A moderate and statistically significant positive correlation was observed between task time and food item count for the add details task, r (89) = 0.53, p < 0.01. Finally, a moderate and statistically significant positive correlation was observed between food item count and individual usability issue count for the add details task, r (89) = 0.45, p < 0.01.

3.7. Usability Issue Definitions and Examples

The 22 general usability issues and their definitions have been thematically organized based on their relationship to four typical components of usability: Effectiveness, Efficiency, Satisfaction, and Comprehension.

3.7.1. Usability Issues Related to Effectiveness

Effectiveness refers to a user's ability to perform a task [30]. Usability issue-related impediments to successful completion of the ASA24 are presented in Table 3. A total of 99 out of the 286 (34.6%) individual usability issues identified relate to the Effectiveness of the ASA24.

Table 3. Descriptions and examples of usability issues related to Effectiveness.

Usability Issue	Definition	Example/Application
Question not completed—unknown to user	User does not answer questions because they do not realize they are mandatory.	Occurred most often in the "frequently forgotten foods" section in which some participants did not realize they had to respond to each question and not just the questions they found applicable (although the system later forced them to provide an answer).
Undo/go back function unclear	It is unclear to user how to undo an action or revert back to a previous screen.	Occurred when participants wanted to edit a meal that had already been submitted as well as general confusion whilst navigating through the tool.
Next step unclear	User is confused about how to proceed, particularly when transitioning from one task to the next.	The language on some buttons did not match the participant's expectations (e.g., the button to begin adding details read "add details" but participants expected to click a button labeled "next"). Buttons were also often located "below the fold" meaning that participants had to scroll down to see them.
How to complete task unclear	User knows what goal they want to accomplish but are unsure how to do so.	Many participants were unsure how to begin entering their meals or had other questions about how to do so. For example, "I made pizza last night so do they want me to put in pizza dough, sauce, cheese?"
Submits incorrect information—unknown to user	User misinterprets task and enters incorrect information but do not realize that they have made an error.	Occurred most often when participants were entering food items. Most of these issues were clear to the moderators. However, when moderators suspected that participants had entered something incorrectly without realizing it, they asked the participant to clarify what they had intended to do. For example, entering "coconut" instead of "coconut sugar", but believing they had entered coconut sugar.
Question skipped	User chooses not to answer a question.	Occurred when participants decided not to answer a question that was asked (although the system later forced them to provide an answer).
Descriptive entered with food item	User enters either the size, amount, number of food items, or other adjectives in the search bar along with the item itself.	Occurred when entering foods in the search bar (the ASA24 does not recognize adjectives). For example, searching for "cold cereal" instead of "cereal".
String of items entered in search bar	User enters several food items into the search bar, not realizing that items must be entered individually.	Occurred when entering foods in the search bar. For example, entering "eggs and toast and water and coffee" into the search bar.

3.7.2. Usability Issues Related to Efficiency

Efficiency refers to the effort required by a user to complete a task [30]. In this study, efficiency related to all of the factors that influenced a participant's ability to complete tasks in a timely and logical way. As opposed to Effectiveness, which refers to whether the user can accomplish a task, Efficiency relates to the effort required to do so. Table 4 presents definitions and examples of usability issues that relate to the ASA24's Efficiency. A total of 131 out of the 286 (45.8%) individual usability issues related to Efficiency. Table 5 lists food items users were unable to find using the ASA24 search function. Although missing food items reduced the speed of completion, users were still able to complete the task, albeit inefficiently and with inaccurate data.

Table 4. Descriptions and examples of usability issues related to Efficiency.

Usability Issue	Definition	Example/Application
User-specific terminology missing	Users search for food items in terms familiar to them that the system did not recognize.	Occurred when entering foods in the search bar. For example, searching for "Palm Bay" instead of the system recognized "vodka cooler".
System asks irrelevant question	User is asked questions that are not relevant based on answers they have previously given.	Occurred most frequently during "add details" tasks. For example, asking participants where they obtained the ingredients for their tap water.
Search item missing/inaccurate	Users search for food items in common language but the system returns either no results or inaccurate results.	Occurred when entering foods in the search bar. See Table 5 for a list of items not found in the system and items participants substituted when this occurred.

Table 4. *Cont.*

Usability Issue	Definition	Example/Application
Next button results in lost content	Users proceed to the next task without correctly submitting the previous task. After realizing the error and reverting back, content previously entered has been lost.	Occurred most often when participants selected "finish with this meal" (similar to the function of a "next" button) before correctly adding all food items to the meal.
Submits incorrect information—known to user	Users deliberately enter incorrect information as determined via the participant's verbalization that they were doing so.	Occurred in situations where the participant felt it was too much work to enter the information accurately, because they weren't sure what the correct information was, or because they wanted to enter the information accurately but did not know how to do so. For example, "This is assuming I only had one burger so I am just going to say 3 patties because realistically I had 3 burgers."
Misclick	Users click in a location that is different from where they need to click to accomplish a task.	Occurred often when the participants clicked "next", which took them to the "add details" task when they had not finished reporting all of their meals.
Misspell	User makes a spelling mistake when searching for or entering a food item.	Occurred when entering foods in the search bar or via free text.
Filter results feature used incorrectly	User misinterprets the list of food items returned from search and/or uses it incorrectly.	Occurred when participants reviewed the list of food items returned from search. For example, the ASA24 offers functionality to filter the search results. One participant interpreted those filter options as ingredients for the food items they were entering, which resulted in confusion.

Table 5. Food items searched by the user that did not return the desired result, and the resulting food item selected by the user.

Food Item Searched by User	Food Item(s) Selected by User
Apricots	Could not/did not enter item
Greek yogurt	Fruit-flavoured yogurt
Coconut cookie	Sugar cookie
English muffin	Multigrain bread
Cherries	Fruit salad
Coconut sugar	Coconut
Green smoothie	Fruit smoothie
Spring water	Bottled water
Gluten free bread	Whole wheat bread
Garlic powder	Could not/did not enter item
French vanilla cream	Half and half cream
Rice bowl	Could not/did not enter item
Chocolate covered almonds	Candy
Salt	Could not/did not enter item
Pepper	Could not/did not enter item
Chipotle steak sandwich	Steak
Tacos	Taco
Onion (as addition to hamburger)	Could not/did not enter item
Tomato (as addition to hamburger)	Could not/did not enter item
Dill pickle (as addition to hamburger)	Could not/did not enter item

3.7.3. Usability Issues Related to Satisfaction

Satisfaction refers to a user's subjective impression of how well a system meets their personal expectations, and can include the desires of the individual with respect to how they would like to use or interact with a system [30]. Table 6 presents a description of the general usability issues that related to user Satisfaction in using the ASA24. A total of 12 out of the 286 (4.2%) individual usability issues related to Satisfaction.

Table 6. Descriptions and examples of usability issues related to Satisfaction.

Usability Issue	Definition	Example/Application
Desired answer option not available	Response options did not reflect the desired response options of the user	Occurred most frequently when answering whether the amount of food eaten was "usual", "much more than usual", or "much less than usual". Many participants felt that "more than usual" or "less than usual" would have provided a more accurate picture of the quantity eaten.
Desired system feature not available	System features do not allow users to report intake accurately in a convenient way	Many participants wanted an easier way to accurately represent their diet than what was offered by the system. For example, many wanted to report water consumed intermittently throughout the day as opposed to entering each instance of water consumption individually.

3.7.4. Usability Issues Related to Comprehension

Comprehension refers to whether a user can understand the intended meaning of, and draw accurate conclusions from, the information presented [31]. Comprehension is relevant to text as well as images and diagrams [32]. Usability issues related to a participant's difficulty or inability to understand questions throughout the ASA24 are presented in Table 7. A total of 44 out of the 286 (15.4%) individual usability issues related to user Comprehension.

Table 7. Descriptions and examples of usability issues related to Comprehension.

Usability Issue	Definition	Example/Application
Question not understood	Users do not understand the meaning of a question	Participants often misunderstood questions. This occurred most frequently during the "add details" task, when the system switched from requesting details about one item, to requesting details about another, but participants did not notice that they were now being asked to add details of a new food item.
Terminology not understood	Users do not understand the terminology contained within a question	Participants often did not understand specific words used in certain questions. For example, when users were asked to report meals, both "dinner" and "supper" were response options which led to user confusion.
Example answers not available	Users are not presented with the same answer options as presented in the question	There was a mismatch between the way some questions were phrased and the available response options. For example, one participant was asked "String cheese: was it regular, reduced fat, low fat, non-fat, or something else?" and the answer options in the dropdown menu were "part skim", "other", and "don't know".
Food images confusing	Food images are not representative of items actually consumed	Occurred when food images did not resemble the items that participants had eaten. For example, one participant who had sliced their zucchini lengthwise became confused when the system showed a zucchini that was sliced width-wise.

4. Discussion

This study presents the results of a structured usability test of the ASA24 within a group of low-income, non-University and food insecure adults (85% female, average age 48.2 years) living in British Columbia, Canada. Detailed quantitative data describing how effectively (i.e., task success) and efficiently (i.e., completion time and usability issue count) participants were able to use the ASA24 were collected in real-time and analyzed. In addition, qualitative data were collected and analyzed to explain why individuals performed the way they did. This approach represents a considerable advance over previous studies that evaluated the usability of the ASA24 by relying on users to identify and report retrospectively problematic issues themselves. When considering why researchers, developers, and users of the ASA24 should be concerned about usability, it is helpful to consider the relationship between usability, utility, and usefulness [12]. In the context of the ASA24, utility is the system's technical capability to enable participants to independently search for, and accurately enter, dietary intake data in a manner that provides the information that researchers require. Usability refers to quantitative and qualitative aspects reflecting how individuals feel and behave while interacting with the ASA24 to access that utility. Usefulness can be considered from two perspectives: (1) how useful is the ASA24 to users who need to enter dietary intake data and (2) how useful is the ASA24 for nutrition researchers who rely on the data it captures. A system with "high usability" makes it easy for users to access its utility and will, therefore, provide data that are more useful for researchers. The purpose of performing a usability test of the ASA24 was to quantify user performance such that usability issues could be identified, quantified and qualitatively described in order to identify opportunities to improve the tool.

4.1. Findings and Recommendations to Enhance Effectiveness of the ASA24

Only one of 17 participants in the unmoderated group successfully completed a dietary recall using the ASA24. Thirteen of the 17 failed to initiate the ASA24 altogether while the other three began the ASA24 but did not complete it. This task failure rate is concerning considering that the unmoderated group was the most representative of how individuals use the ASA24 outside the context of a usability test (i.e. without the support of a moderator). Time pressures may have prevented some participants from initiating the survey within the 36-h time window provided. Others completed the initial socio-demographic/health-related survey, but failed to initiate the ASA24, suggesting that problems logging into the ASA24 and/or participant fatigue led to task failure. Because these participants were not recorded or observed by the moderator it is not possible to identify specific factors that contributed to task failure. Notably, 58% of unmoderated participants were seniors, whereas less than half of those in the other groups were seniors. Among those who initiated, but did not fully complete the ASA24, general usability issues described in Section 3.7.1 can provide a sense of why they may have struggled successfully to complete the ASA24 independently. Of particular note was the common issue of participants being uncertain of the sequence of steps needed to complete a task (Next Step Unclear = 71.4% of participants (15 of 21), How to Complete Task Unclear = 52.4% of participants (11 of 21)). When participants were unclear on how to proceed with a task, it is likely that without the support of a moderator they may have exited the system or potentially entered data incorrectly.

Further study is needed to understand task failure in individuals using the ASA24 without the assistance of a moderator. Website analytics can provide a useful means of determining problematic sections of the ASA24. For example, a website's Exit Rate is the percentage of individuals who leave a website from an individual page [14]. The pages of the ASA24 with the highest Exit Rates will likely be associated with task failures. These data could be analyzed by ASA24 developers.

In the moderated and semi-moderated groups, 21 of 22 participants completed the dietary recall process successfully; however, all of these participants relied on assistance from a moderator at some point to assist them in navigating through usability issues. Whether or not this assistance was necessary for participants to complete a dietary recall is unclear; however, it is clear that the majority of participants benefitted from the support of a moderator. These quantitative data indicate that although the self-administered nature of the ASA24 facilitates data collection, it may ultimately result in low participation rates for groups who encounter frequent usability issues. As the complexity of interaction with a software system increases, it is reasonable to expect that a degree of technical support might be required to support individuals in using that system. None of the participants indicated that they had used the ASA24 before and, therefore, they can be considered novices, demonstrating performance typical of a population of untrained users. Based on these findings, researchers relying on the ASA24 to assess dietary intake in similar populations might expect novice users to require technical support to effectively use the tool for the first time. Thus, availability of on-demand technical assistance may be important to maximize the quantity and quality of data that are collected via the ASA24 and support participant retention.

4.2. Findings and Recommendations to Enhance Efficiency of Use of the ASA24

Participants in the present study completed the ASA24 more quickly (~27 min) than has been reported in other studies (~35 min [9,10]) despite the think-aloud procedure and interaction with a moderator. However, these times are comparable to the range reported on the ASA24 website (17–34 min). A strong correlation was observed between task time and food item count in the moderated and semi-moderated groups, with participants consuming an average of 11.5 items daily. Although other studies have not reported item count, an average of 11.5 items daily appears low.

Given that the majority of participants had experienced moderate or severe food insecurity during the previous month, this low item count may be related to diminished food access and dietary diversity in our sample, potentially leading to lower ASA24 completion times relative to previous studies. Other characteristics, such as living alone or aging, may also be implicated due to reduced interest in, or capacity to cook. Individuals who are less inclined to cook may consume relatively simple meals with few ingredients. In addition to participant characteristics, the context of completion may also have been influential. Moderators assisted all participants in the moderated and semi-moderated groups at least once. Therefore, it is possible that the task completion times we observed are simply what can be expected when participants have support to use the tool.

Task time itself is a particularly meaningful usability metric when evaluating repetitive tasks [14]. If task time can be reduced for tasks that most participants must perform repeatedly, gains in efficiency will be achieved. The add details task was the most lengthy to complete for both the moderated and semi-moderated groups. Therefore, supporting users during the add details task will have the greatest impact with respect to enhancing efficiency of use of the ASA24.

Another aspect of how efficiently the ASA24 can be used relates to the overall number and types of usability issues that participants encountered. Participants encountered an average of 13.7 individual usability issues per session. This is concerning considering that usability issues can diminish data quality and present opportunities for task failure or study dropout. The most common general usability issue, experienced by 76.2% of participants (16 of 21), was the inability to find a specific food item. Considering that one of the primary functions of the ASA24 is to enable users to record food intake independently, the prevalence of this usability issue is noteworthy. The inability to locate items was a key source of dietary measurement error. For instance, in the moderated and semi-moderated groups, 71.4% of participants (15 of 21) indicated that they knowingly entered incorrect information at least once, primarily because they were unable to find a specific food item using the search function. Table 5 demonstrates that when this occurred users often selected other items (e.g., substituting steak for a chipotle steak sandwich), used the "I can't find what I'm looking for" function (the impact of which was having to answer supplemental questions, some of which were irrelevant), or omitted the item entirely. The quantitative impact of this error is unclear, as the current study was not intended to quantify measurement error, but rather to examine its source in order to understand how to mitigate it. However, others have shown that energy intake in adults aged 50–74 years is underestimated by 15–17% on the ASA24 compared to recovery biomarkers, with no difference in mean protein and sodium densities [2]. The current findings can provide complementary data to understand factors that contribute to misreporting of energy intake using the ASA24.

Additional search-related usability issues concern findings that participants often entered a string of items (e.g., "eggs and toast and water and coffee") or entered additional descriptive information (e.g., "cold cereal") into the search bar. Users will increasingly expect any web-based platform they interact with to provide them with "Google-Like" performance, likely with little appreciation for the investment that providing this functionality requires of the developer. Therefore, in addition to adding new food items to the ASA24 database, the intelligence of the ASA24 search algorithm could be improved to recognize plural forms of food items (e.g., the system returned a result for "taco" but not "tacos") and to suggest potential matches when multiple words or descriptors are entered into the search bar.

Usability issues related to misclicking and misspelling were also prevalent. This could be a reflection of users not knowing what to do in the system, contributing to data errors or task failures (e.g., exiting the system). Additionally, entering an incorrectly spelled food item into the system can lead the user through additional irrelevant questions. This contributes to increased task times, and potentially decreased user satisfaction. This inefficiency could be quantified in subsequent usability tests using the metric of "lostness" [33] by comparing the number of steps an individual performs to the minimum number of steps possible. This calculation would enable the impact of misclicks to be more thoroughly understood from an efficiency perspective. Some degree of user error (whether typos or accidental clicks) is outside the control of designers or researchers, however others can be addressed through relatively simple design changes (e.g., the visual differentiation or clarification of a button or text field's function).

4.3. Findings and Recommendations to Enhance Satisfaction with the ASA24

Satisfaction in use is perhaps one of the most easily conceptualized aspects of usability; presumably, if someone is happy using a system then it is likely usable. Previous studies have asked participants to self-report their satisfaction with the ASA24, finding that the majority of participants had a favourable view of the system [8,34–36]. The current study examined specific system features that contributed to user satisfaction or dissatisfaction. Given these different outcomes, ability to compare our findings with those of others was limited. However, similar themes emerged in our study compared to others', including frustrations about the time involved in completing a recall and how to proceed to the next step, not understanding how to use the search function, and not being able to find food items [10]. Users who feel that a system is not designed to allow them to use it intuitively are unlikely to want to continue to use that system. One design approach to address satisfaction is to provide users with shortcuts that allow them to duplicate repetitive actions [37]. Future usability tests could compare how easy individuals expect a task to be before attempting it (expectation score) to how easy they found it after completing it (experience score). When users expect a task to be easy to complete but then find it difficult (i.e., expectation score is much higher than the experience score), it is very likely to lead to dissatisfaction [38]. This approach would help prioritize potential design changes specifically to improve satisfaction.

4.4. Findings and Recommendations to Enhance Comprehension of the ASA24

The second most frequent general usability issue identified was Question Not Understood, which accounted for 11.5% of all issues and was experienced by 66.7% of participants (14 of 21). Often, when participants were confused, it was observed that they had only partially processed the elements of the question being asked (e.g., mistaking which particular food item they were being asked about) or that they had not understood the specific words or phrasing of the question itself (e.g., general comments such as "I don't know what this means."). An individual who does not understand what the ASA24 is asking them to do, or who becomes confused by terminology or imagery, is likely to have difficulty using the tool to provide accurate information. A general approach to enhancing comprehension is to match the user's mental model (e.g., ensuring language doesn't exceed the user's reading level, presenting visuals in a way that match the way the user consumes the food, requesting measurements of quantity that align with the user's method of measurement, etc.). This can be challenging given the wide variability in individuals' mental models, particularly for a tool such as the ASA24 that is intended to be used broadly across multiple populations. Two alternate design strategies that might be considered would be to: (1) allow users to customize the ASA24 to match their mental model (e.g., switching units from metric to imperial, alternative visual depictions of foods, customizable reading levels for text presentation), or (2) standardize the tool but ensure that it has been optimized to meet the needs of the majority of users through comparative comprehension testing.

4.5. User Characteristics and Usability Testing

Participants were recruited from a population participating in a nutrition coupon program and as such were primarily low-income, low- to mid-educated, and food insecure adults, many of whom were seniors. Older adults and those of a lower socioeconomic status may have lower computer literacy compared to the general population [39] and it is, therefore, possible that some of the usability issues identified here may be particular to this sample. As we did not assess computer literacy, it is unclear whether this was an issue in our sample. Darajeh and Singh [40] have summarized design recommendations to enhance usability for those with lower computer literacy, including creating simple layouts with limited clutter, providing user guides, reducing the use of complex terminology, creating simple navigation paths, using similar functions for different tasks, and including descriptive text for tool use. Dietary inequities and strategies to support optimal dietary patterns among older adults are significant concerns worldwide, and thus our findings can inform nutrition studies among these priority populations.

4.6. Strengths and Limitations

One of the strengths of the mixed methods approach was the ability to both quantify and qualitatively describe usability issues, providing a much more comprehensive and in-depth perspective of the usability of the ASA24. Moreover, the think-aloud procedures and inductive nature of the analyses enlarged the scope of investigation beyond researchers' pre-determined questions and response options to uncover novel usability issues. The qualitative analysis had a sufficient sample size to reach a point of thematic saturation and we are, therefore, confident that the analysis uncovered the most salient usability issues in this particular sample.

The validity of a usability test is partially dependent on creating test conditions that reflect the actual conditions under which a user interacts with a system. The presence of a moderator may create a Hawthorne or Observer Effect in which participant behavior changes due to being observed, participants are overly reliant on assistance from a moderator to complete a task, or experience heightened sensitivity to usability issues. Having participants think-aloud may also create additional cognitive demand and thereby alter task performance. These factors could all contribute to a test scenario in which reported usability issues and performance metrics are not perfect representations of those that would have been encountered during actual use.

In addition, the definition of task failure as used in this study is specific to our methodological design. Participants were not given multiple attempts to complete the ASA24; if they failed to complete it upon their first attempt this was recorded as a task failure. However, the tool does allow participants to complete a dietary recall in multiple attempts. This study also looked at the ASA24 when used in combination with a socio-demographic/health-related survey. This survey was administered before the ASA24, which may have influenced participant behavior. For example, completing surveys prior to the ASA24 may create additional fatigue, affecting motivation to complete the ASA24. Finally, participants in the unmoderated group may have failed to complete the ASA24 for reasons unrelated to its usability (e.g., interruptions, variability in motivation). Researchers interested in using the ASA24 in a similar population should be aware of these potentially high participant drop-out rates.

4.7. Help Documentation and Training in Relation to Usability.

The ASA24 does provide a help guide for users and Best Practices information for researchers, in addition to the help feature embedded in the ASA24 [41]. However, just one of the 22 participants observed in the moderated and semi-moderated groups accessed the ASA24's help feature. Help functions do not, however, improve usability because they place the onus of efficient and effective system use on the user, increasing their workload rather than making system changes to enhance usability. Moreover, users often do not read support materials [42]. One of the main benefits of conducting usability testing, or designing usable systems, is that the process will reduce costs associated with training and customer support [43].

5. Conclusions

This study demonstrates how the usability of the ASA24 affects its usefulness for a particular group of users entering dietary intake data as well as for researchers studying that information. One of the primary benefits of the self-administered nature of the ASA24 is the relative ease (for researchers) with which dietary intake data can be collected. However, the results of this study highlight important limitations of this self-administered approach. Task success data reveal that the vast majority of individuals in our sample had difficulty independently using the ASA24 without the support of a moderator. Moreover, the frequency and nature of usability issues identified suggest that information was often entered inaccurately, as 71.4% of participants knowingly misentered dietary information at least once. Other key usability issues encountered were related to difficulties using the search function, not understanding certain questions, uncertainty regarding how to proceed to the next step, and misclicks. It is not clear to what extent our findings are specific to our sample of primarily non-university educated adult females (average age 48.2 years), with a low household income. We expect that other groups may encounter similar challenges, albeit perhaps at a lower frequency. Our findingscan help to understand how the ASA24 can be improved to make it more intuitive and simple for individuals from a wider variety of populations to use, thereby enhancing the accuracy of dietary intake reporting. The following recommendations are intended to address key usability issues users encountered.

5.1. Key Recommendations for Designers of the ASA24

1. Improve the intelligence of the ASA24 search algorithm such that it is responsive to user search behaviors including pluralization, synonyms, multi-item searches and misspelling.
2. Remove irrelevant questions (e.g., asking the user where they obtained the ingredients for their tap water).
3. Use intuitive language and terminology (e.g., instead of a button that says "add details", change so it reads "next").
4. Create a demonstration video that shows participants how to use the ASA24 and that addresses key usability issues they may encounter (i.e., those that developers cannot remedy through design changes).

5.2. Key Recommendations for Researchers Using the ASA24

1. Technical assistance should be available on demand to assist participants who are using the ASA24, particularly for first time users.
2. Strategies for effective use of the ASA24 should be explained in advance.
3. Request and allow participants to report retrospectively that they entered information that was not accurate (e.g., if they could not find an exact match for a particular food) and allow them to report what the actual food was.

Author Contributions: Study design: G.H. and D.L.O.; Data collection and analysis: J.K., G.H., and S.J.; Manuscript preparation: J.K., S.J., G.H. and D.L.O.; Funding Acquisition, D.L.O. All authors reviewed and approved the final version of the manuscript.

Funding: This research was funded by the Calgary Centre for Clinical Research.

Acknowledgments: The research team would like to thank W21C Senior Research Associate Julie Babione for her support in developing the protocol for this research study.

Conflicts of Interest: The authors declare no conflict of interest. The funders had no role in the design of the study; in the collection, analyses, or interpretation of data; in the writing of the manuscript; and in the decision to publish the results.

References

1. Thompson, F.E.; Subar, A.F. Dietary assessment methodology. In *Nutrition in the Prevention and Treatment of Disease*, 3rd ed.; Coulston, A.M., Boushey, C.J., Ferruzzi, M.G., Eds.; Elsevier: London, UK, 2007; pp. 5–46. ISBN 9780123918840.
2. Park, Y.; Dodd, K.W.; Kipnis, V.; Thompson, F.E.; Potischman, N.; Schoeller, D.A.; Baer, D.J.; Midthune, D.; Troiano, R.P.; Bowles, H. Comparison of self-reported dietary intakes from the automated self-administered 24-h recall, 4-d food records, and food-frequency questionnaires against recovery biomarkers. *Am. J. Clin. Nutr.* **2018**, *107*, 80–93. [CrossRef] [PubMed]
3. Subar, A.F.; Kipnis, V.; Troiano, R.P.; Midthune, D.; Schoeller, D.A.; Bingham, S.; Sharbaugh, C.O.; Trabulsi, J.; Runswick, S.; Ballard-Barbash, R. Using intake biomarkers to evaluate the extent of dietary misreporting in a large sample of adults: the OPEN study. *Am. J. Epidemiol.* **2003**, *158*, 1–13. [CrossRef] [PubMed]

4. Schatzkin, A.; Kipnis, V.; Carroll, R.J.; Midthune, D.; Subar, A.F.; Bingham, S.; Schoeller, D.A.; Troiano, R.P.; Freedman, L.S. A comparison of a food frequency questionnaire with a 24-h recall for use in an epidemiological cohort study: Results from the biomarker-based Observing Protein and Energy Nutrition (OPEN) study. *Int. J. Epidemiol.* **2003**, *32*, 1054–1062. [CrossRef] [PubMed]

5. Subar, A.F.; Kirkpatrick, S.I.; Mitti, B.; Zimmerman, T.P.; Thompson, F.E.; Bingley, C.; Willis, G.; Islam, N.G.; Baranowski, T.; McNutt, S.; et al. The automated self-administered 24-h recall (ASA24): A resource for researchers, clinicians, and educators from the National Cancer Institute. *J. Acad. Nutr. Diet* **2012**, *112*, 1134–1137. [CrossRef] [PubMed]

6. ASA24-Canada 2017. Available online: www.asa24.ca (accessed on 1 July 2017).

7. Kirkpatrick, S.; Subar, A.F.; Douglass, D.; Zimmerman, T.P.; Thompson, F.E.; Kahle, L.L.; George, S.M.; Dodd, K.W.; Potischman, N. Performance of the automated self-administered 24-h recall relative to a measure of true intakes and to an interviewer- administered 24-h recall. *Am. J. Clin. Nutr.* **2014**, *100*, 233–240. [CrossRef]

8. Ettienne-Gittens, R.; Boushey, C.J.; Au, D.; Murphy, S.P.; Lim, U.; Wilkens, L. Evaluating the feasibility of utilizing the automated self-administered 24-h (ASA24) dietary recall in a sample of multiethnic older adults. *Procedia. Food Sci.* **2013**, *2*, 134–144. [CrossRef] [PubMed]

9. Thompson, F.E.; Dixit-Joshi, S.; Potischman, N.; Dodd, K.W.; Kirkpatrick, S.I.; Lawrence, H.; Kushi, G.L.; Alexander, L.A.; Coleman, T.P.; Zimmerman, M.E.; et al. Comparison of interviewer-administered and automated self-administered 24-h dietary recalls in 3 diverse integrated health systems. *Am. J. Epidemiol.* **2015**, *181*, 970–978. [CrossRef] [PubMed]

10. Kirkpatrick, S.I.; Gilsing, A.M.; Hobin, E.; Solbak, N.M.; Wallace, A.; Haines, J.; Mayhew, A.J.; Orr, S.K.; Raina, P.; Robson, P.J.; et al. Lessons from studies to evaluate an online 24- hour recall for use with children and adults in Canada. *Nutrients* **2017**, *9*, 100. [CrossRef]

11. International Organization for Standardization. Ergonomics of Human-System Interaction—Part 11: Usability Definitions and Concepts. ISO Standard No. 9241-11:2018. Available online: https://www.iso.org/standard/63500.html (accessed on 1 August 2017).

12. Nielsen Norman Group. Available online: https://www.nngroup.com/articles/usability-101-introduction-to-usability/ (accessed on 23 October 2018).

13. International Organization for Standardization. Software engineering—Software product Quality Requirements and Evaluation (SQuaRE)—Common Industry Format (CIF) for usability test reports. ISO/IEC Standard No. 25062:2006. Available online: https://www.iso.org/standard/43046.html (accessed on 1 August 2017).

14. Tullis, A.; Albert, B. *Measuring the User Experience: Collecting, Analyzing, and Presenting Usability Metrics*, 2nd ed.; Elsevier: Waltham, MA, USA, 2013; ISBN 978-0-12-415781-1.

15. Van Someren, M.W.; Barnard, Y.F.; Sandberg, J.A. *The Think Aloud Method: A Practical Approach To Modelling Cognitive Processes*, 1st ed.; Academic Press: London, UK, 1994; pp. 27–28. ISBN 0-12-714270-3.

16. Greene, J.C.; Caracelli, V.J.; Graham, W.F. Toward a conceptual framework for mixed-method evaluation designs. *Educ. Eval. Pol. Anal.* **1989**, *11*, 255–274. [CrossRef]

17. British Columbia Association of Farmers' Markets. Farmers' Market Nutrition Coupon Program. Available online: http://www.bcfarmersmarket.org/nutrition-coupon-program (accessed on 14 February 2017).

18. Israelski, E.W. Testing and evaluation. In *Handbook of Human Factors in Medical Device Design*; Weinger, M.B., Gardner-Bonneau, D.J., Wiklund, M.E., Eds.; CRC Press Taylor & Francis Group: Boca Raton, FL, USA; ISBN 978-0-8058-5627-9.

19. Canadian Community Health Survey. Available online: http://www23.statcan.gc.ca/imdb/p2SV.pl?Function=getSurvey&Id=259374 (accessed on 1 July 2017).

20. Canadian Community Health Survey, Cycle 2.2, Nutrition (2004): Income-Related Household Food Security in Canada. Available online: https://www.canada.ca/en/health-canada/services/food-nutrition/food-nutrition-surveillance/health-nutrition-surveys/canadian-community-health-survey-cchs/canadian-community-health-survey-cycle-2-2-nutrition-2004-income-related-household-food-security-canada-health-canada-2007.html#appa (accessed on 1 July 2017).

21. Lee, R.M.; Draper, M.; Lee, S. Social connectedness, dysfunctional interpersonal behaviours, and psychological distress: Testing a mediator model. *J. Couns. Psychol.* **2001**, *48*, 310–318. [CrossRef]

22. Cohen, S.; Kamarck, T.; Mermelstein, R. A global measure of perceived stress. *J. Health Soc. Behav.* **1983**, *24*, 385–396. [CrossRef] [PubMed]

23. Tennant, R.; Hiller, L.; Fishwick, R.; Platt, S.; Joseph, S.; Weich, S.; Parkinson, J.; Secker, J.; Stewart-Brown, S. The warwick-edinburgh mental well-being scale (WEMWBS): development and UK validation. *Health Qual Life Outcomes* **2007**, *5*, 63. [CrossRef] [PubMed]

24. Dumas, J.S.; Loring, B.A. *Moderating Usability Tests: Principles and Practices for Interacting*, 1st ed.; Elsevier: San Franciso, CA, USA, 2008; chapter 2; ISBN 9780123739339.

25. Olsho, L.E.; Klerman, J.A.; Wilde, P.E.; Bartlett, S. Financial incentives increase fruit and vegetable intake among Supplemental Nutrition Assistance Program participants: A randomized controlled trial of the USDA Healthy Incentives Pilot. *Am. J. Clin. Nutr.* **2016**, *104*, 423–435. [CrossRef]

26. National Institutes of Health, National cancer institute: 24-h Dietary Recall (24HR) at a Glance. Available online: https://dietassessmentprimer.cancer.gov/profiles/recall/ (accessed on 14 February 2017).

27. Mackenzie, C.F.; Xiao, Y. Video analysis: an approach for use in healthcare. In *Handbook of Human Factors and Ergonomics in Health Care and Patient Safety*, 2nd ed.; Carayon, P., Ed.; CRC Press, Taylor & Francis Group: Boca Raton, FL, USA, 2012; pp. 526–527. ISBN 978-1138074590.

28. Braun, V.; Clarke, V. Using thematic analysis in psychology. *Qual. Res. Psychol.* **2006**, *3*, 77–101. [CrossRef]

29. Guest, G.; Bruce, A.; Johnson, L. How many interviews are enough? An experiment with data saturation and variability. *Field Methods* **2006**, *18*, 59–82. [CrossRef]

30. Kortum, P. *Usability Assessment: How to Measure the Usability of Products, Services, and Systems*; SAGE Publication Ltd: London, UK, 2008; pp. 39–40. ISBN 978-0-945289-49-4.

31. Nielsen Norman Group. Available online: https://www.nngroup.com/articles/testing-content-websites/ (accessed on 1 October 2017).

32. Nielsen Norman Group. Available online: https://www.nngroup.com/articles/legibility-readability-comprehension/ (accessed on 1 October 2017).

33. Smith, P.A. Towards a practical measure of hypertext usability. *Interact. Comput.* **1996**, *8*, 365–381. [CrossRef]

34. Boushey, C.J.; Spoden, M.; Edward, J.D.; Fengging, Z.; Bosch, M.; Ahmad, Z.; Shvetsiv, Y.B.; DeLany, J.P.; Kerr, D.A. Reported energy intake accuracy compared to doubly labeled water and usability of the mobile food record among common dwelling adults. *Nutrients* **2017**, *9*, 312. [CrossRef] [PubMed]

35. Hongu, N.; Pope, B.T.; Bilgic, P.; Suzuki, A.; Kim, A.S.; Merchant, N.C.; Roe, D.J. Usability of a smartphone food picture app for assisting 24-h dietary recall: a pilot study. *Nutr. Res. Pract.* **2015**, *9*, 207–212. [CrossRef] [PubMed]

36. Jacques, S.; Lemieux, S.; Lamarche, B.; Laramee, C.; Corneau, L.; Lapointe, A.; Tessier-Grenier, M.; Robataille, J. Development of a web-based 24-h dietary recall for a French-canadian population. *Nutrients* **2016**, *8*, 724. [CrossRef]

37. Molich, R.; Nielsen, J. Improving a human-computer dialogue. *Commun. ACM* **1990**, *33*, 338–348. [CrossRef]

38. Rich, A.; McGee, M. Expected usability magnitude estimation. In Proceedings of the Human Factors and Ergonomics Society Annual Meeting 2004, New Orleans, LA, USA, 20–24 September 2004; Sage: Los Angeles, CA, USA, 2004; pp. 912–916.

39. Choi, N.G.; Dinitto, D.M. The digital divide among low-income homebound older adults: Internet use patterns, eHealth literacy, and attitudes toward computer/Internet use. *J. Med. Internet. Res.* **2013**, *15*, 1–12. [CrossRef] [PubMed]

40. Darejeh, A.; Sungh, D. A review on user interface design principles to increase software usability for users with less computer literacy. *J. Comput. Sci.* **2013**, *9*, 1143. [CrossRef]

41. ASA24: Instructions for Study Staff & Respondents. Available online: https://epi.grants.cancer.gov/asa24/resources/instructions.html (accessed on 5 December 2018).

42. Nielsen, J. *Usability Engineering (Interactive Technologies)*; Morgan Kaufmann Publishers Inc.: San Francisco, CA, USA, 1993; ISBN 978-0-12-518406-9.

43. Bias, R.; Mayhew, D. *Cost-Justifying: An Update for the Internet Age*, 2nd ed.; Morgan Kaufman: San Francisco, CA, USA, 2005; ISBN 9780080455457.

Article

Effectiveness of the Nutritional App "MyNutriCart" on Food Choices Related to Purchase and Dietary Behavior: A Pilot Randomized Controlled Trial

Cristina Palacios [1,*], Michelle Torres [2], Desiree López [2], Maria A. Trak-Fellermeier [3], Catherine Coccia [1] and Cynthia M. Pérez [4]

1   Department of Dietetics and Nutrition, Robert Stempel College of Public Health & Social Work, Florida International University, Miami, FL 33199, USA; ccoccia@fiu.edu
2   Nutrition Program, Graduate School of Public Health, Medical Sciences Campus, University of Puerto Rico, San Juan 00935, Puerto Rico; michelle.torres13@upr.edu (M.T.); desire.lopeztravieso@upr.edu (D.L.)
3   Center for Clinical Research and Health Promotion, School of Dental Medicine, Medical Sciences Campus, University of Puerto Rico, San Juan 00935, Puerto Rico; maria.trak@upr.edu
4   Department of Biostatistics and Epidemiology, Graduate School of Public Health, Medical Sciences Campus, University of Puerto Rico, San Juan 00935, Puerto Rico; cynthia.perez1@upr.edu
*   Correspondence: cristina.palacios@fiu.edu; Tel.: +1-305-348-3235

Received: 2 October 2018; Accepted: 6 December 2018; Published: 12 December 2018

Abstract: Objective: To pilot test the effectiveness of "MyNutriCart", a smartphone application (app) that generates healthy grocery lists, on diet and weight. Methods: A pilot randomized trial was conducted to test the efficacy of using the "MyNutriCart" app compared to one face-to-face counseling session (Traditional group) in Hispanic overweight and obese adults. Household food purchasing behavior, three 24-h food recalls, Tucker's semi-quantitative food frequency questionnaire (FFQ), and weight were assessed at baseline and after 8 weeks. Statistical analyses included t tests, a Poisson regression model, and analysis of covariance (ANCOVA) using STATA. Results: 24 participants in the Traditional group and 27 in the App group completed the study. Most participants were women (>88%), with a mean age of 35.3 years, more than a high school education (>80%), a family composition of at least three members, and a mean baseline body mass index (BMI) of 34.5 kg/m^2. There were significant improvements in household purchasing of vegetables and whole grains, in individual intakes of refined grains, healthy proteins, whole-fat dairies, legumes, 100% fruit juices, and sweets and snacks; and in the individual frequency of intake of fruits and cold cuts/cured meats within the intervention group ($p < 0.05$). However, no significant differences were found between groups. No changes were detected in weight. Conclusions: "MyNutriCart" app use led to significant improvements in food-related behaviors compared to baseline, with no significant differences when compared to the Traditional group. Cost and resource savings of using the app compared to face-to-face counseling may make it a good option for interventionists.

Keywords: nutritional application; smartphone; DGA; dietary behaviors; household food purchase behavior; obesity; overweight weight control

---

1. Introduction

Diet-mediated chronic conditions affect half of the US adult population [1]. These could be prevented by following the science-based Dietary Guidelines for Americans (DGA) [2]. However, adherence to these guidelines is suboptimal [3]. In fact, several task forces have pinpointed the gap in translating the DGA recommendations into positive dietary changes [4–6], noting that the main barrier is the translation of the guidelines into practical, food-based recommendations and as such, new approaches are needed to implement these guidelines. In particular, innovative approaches should aim to improve grocery shopping, a critical moment when individuals need assistance for purchasing healthy foods [7]. With the huge food variety available in supermarkets, together with a large amount of nutritional information, and the limited time to read and understand nutrition labels, individuals may feel overwhelmed [8]. In fact, it has been reported that the main barrier to healthful shopping is a lack of self-efficacy in choosing healthy foods [9]. Therefore, interventions aimed at guiding individuals to choose healthy foods when grocery shopping may increase DGA adherence. This could be achieved by leveraging technology to help people make better choices at the point of purchase.

The use of tablets or smartphones for accessing the Internet is widespread, offering a unique platform for interventions. In 2016, 68% of all US adults owned a smartphone and 77% of them downloaded applications (apps) [10]. A myriad of nutrition and fitness apps have become extremely accessible via portable electronic devices with the capacity to calculate caloric requirements, track food intake and physical activity, and access healthy cooking information. In fact, studies have found better self-monitoring adherence and changes in dietary behaviors and/or weight control from using smartphone apps compared to traditional methods [11,12]. In addition to self-monitoring, a study found that "nudging" people to make healthy food purchases from local vendors resulted in improved awareness and consumption of healthy foods [11]. However, there are no available apps that translate the DGA into a healthy grocery list.

Therefore, in collaboration with technology experts, we developed the "MyNutriCart" app to help individuals make smart and healthy choices when purchasing foods at grocery stores [13]. This app automatically generates a healthy grocery list following DGA recommendations and accounts for the family's nutritional needs, within a pre-specified budget [13]. The purpose of this study is to report on the pilot test of this app for improving household food purchase behavior and for improving individual dietary behaviors, compared to a traditional nutritional counseling face-to-face session in a convenience sample of overweight and obese Hispanic adults. As a secondary aim, we examined the potential effect of the intervention on weight control. We hypothesized that the use of the app would improve household food purchasing behavior when grocery shopping, which in turn would positively influence the individual frequency and intake of healthy foods and weight control compared to a traditional nutritional counseling session.

2. Materials and Methods

2.1. Study Design

We conducted a pilot randomized clinical trial to test the effectiveness of the "MyNutriCart" app on household food purchase behavior, individual dietary behaviors, and individual weight control. Recruited participants were randomly assigned to either the App group or the Traditional group for 8 weeks. Diet and weight were assessed at baseline and after 8 weeks of intervention. This time frame was chosen for this pilot study as this is the time frame used in similar studies using apps with significant changes in diet and/or weight. Recruitment was conducted between December 2015 and March 2016 and all study visits were conducted at the Medical Sciences Campus, University of Puerto Rico. The Institutional Review Board at this institution approved this study. Prior to the study, all recruited participants provided written consent.

2.2. Participants, Eligibility, and Recruitment

For this pilot trial, a convenience sample of participants was recruited between January–March of 2016 using flyers posted on the university intranet and around campus, shopping malls, clinics/medical offices, and by word of mouth. Overweight and obese adults aged 21–45 years were invited to participate in a study to test an app that helps individuals select healthier food, which could impact their dietary behaviors and weight control. Additional inclusion criteria were: being the main household shopper (i.e., responsible for >50% of the household grocery acquisition), shopping at a grocery store at least once weekly, owning a smartphone (iPhone or Android) with internet access, and willingness to be randomized into one of the two groups. We excluded those already using apps to monitor diet and/or physical activity or those enrolled in weight loss programs. Pregnant women, individuals with chronic health conditions (i.e., diabetes, kidney disease), or with reported food allergies were deemed ineligible.

2.3. Intervention Groups

Participants were equally randomized to either the App or Traditional groups using a simple computerized randomization scheme. Participants were assigned their allocation following a sequentially numbered container mechanism. Randomization was done by the statistician.

2.3.1. "MyNutriCart" (App Group)

Participants allocated to the App group were guided by the research assistant in how to download and navigate the app. The MyNutriCart app was developed to guide individuals to make smart and healthy choices when purchasing foods at grocery stores, as recently published [13]. Briefly, the app provided a healthy grocery list based on the daily nutritional recommendations of the individuals that constitute the participant's household. This list took into consideration a pre-defined budget, which was maximized by connecting to supermarkets' discounts. It also integrated the following aspects:

- Estimation of energy requirements for each family member based on age, sex, and physical activity using the equations from the Dietary Reference Intakes [14]. The app automatically subtracted 500 kcals from the total calculated energy requirement for study participants only (not family members) to allow for a weight loss of about 1 pound/week [15];
- General food recommendations from the DGA [2], such as consumption of half of the grains as whole grains and low-fat dairy products, in addition to a variety of protein foods (beans, eggs, poultry, fish, and seashells);
- Number of servings per food group, based on the caloric level of each member, as recommended by the DGA [2]. Servings of each food group from each member were added to get a total of each food group per day;
- Intended number of days of the shopping event to multiply the servings per food group to get a total of foods to purchase;
- Participant's pre-specified budget and weekly discounts offered by the largest local supermarkets (which was retrieved from an independent and free website service) to maximize the budget;
- Sample menus for each caloric level of the household based on local preferences, which were previously designed by a registered dietitian (RD).

The primary goal of the app use is the establishment of a healthy eating pattern; hence, energy dense items or foods containing added sugar (i.e., sweetened beverages, juices of any type, alcoholic beverages, sweets and desserts, and non-healthy snacks) were excluded from the grocery list. Only healthy versions for the following main food groups were included: fruits, vegetables, dairy products, cereals and grains, and protein foods. Participants were informed that this list would cover most of their energy requirements, but not all, as it excluded items not purchased weekly, such as fats and oils, and other items such as condiments, sauces, spices, coffee or tea, and bottled water. Participants were also instructed to use the app every time they went grocery shopping or at least once per week. To generate the list, the user had to open the app before each grocery event, select a budget amount, and a time frame for that grocery event (i.e., $100 for 7 days). The app then generates a grocery list for each supermarket included in the app based on their weekly specials. Therefore, each list generated was unique. Participants were free to choose from the supermarkets included in the app to do their grocery shopping, based on the convenience of its location, the total amount estimated to pay if all the foods were purchased, and the discounts offered that week. The app did not include notifications or reminders to use the app.

2.3.2. Traditional Nutritional Counseling (Traditional Group)

This group received one face-to-face counseling session with an RD at the beginning of the study. The RD calculated the participant's energy requirements using the Dietary Reference Intakes [14] and subtracted 500 kcals to allow for weight loss [15], similar to the App group. The RD provided the participant with the MyPlate Tip sheets [16], which are based on the DGA recommendations and contain the recommended food groups' servings per caloric level. Also, participants received a sample menu, similar to the menu included in the app. There were no follow-up calls or additional sessions during the study.

During the study, all participants were instructed to maintain their usual physical activity level and to avoid partaking in other programs or sessions related to weight loss or promoting healthy dietary behaviors. Compliance with these study requirements was verified through a brief questionnaire at post-intervention.

2.4. Instruments and Measures

Trained research assistants conducted measures and interviews, as described below:

- Socio-demographics

A short questionnaire was completed at baseline with information about age, sex, educational level, and family composition (number, age, and sex of family members).

- Household food purchase frequency

This was evaluated from grocery receipts collected at baseline and post-intervention. This method has been previously validated to assess household food purchasing behavior [17]. In particular, the purchasing frequency of the following key DGA food groups was evaluated from each grocery receipt: fruits, vegetables, whole grains, 100% fruit juices, and sugar-sweetened beverages (SSB). These were the only food groups selected as they were easily identifiable by name from the grocery receipts. Participants were asked to provide all the grocery receipts available from their grocery events near the baseline visit and all of their grocery receipts during the study, either by uploading a picture of the grocery receipt in the app, sending scanned copies by email, or submitting hard copies. We reminded participants throughout the study to keep all their grocery receipts. Each time the food group was identified in the receipt, it was counted as a frequency of one. For example, if a receipt showed: grapes $1.05, oranges $2.33, and bananas $0.99, this was counted as 3 fruits. It was not possible to evaluate amount purchased as this information was not readily available from all grocery receipts. Results were averaged for each food group from the available receipts collected at baseline and at post-intervention.

- Dietary behaviors at the individual level

Participants were interviewed by trained research staff to complete the following questionnaires at baseline and at post-intervention:

- Food frequency questionnaire (FFQ). We used a short version of the Tucker's semi-quantitative FFQ, which was validated in Puerto Rican adults [18]. The questionnaire was interview-administered, and respondents were asked to estimate the frequency of food consumption from 10 categories (daily, weekly, monthly), using the preceding eight weeks as the reference period. Summary questions for the frequency of consumption of the following food groups: fruits, vegetables, starchy vegetables, refined and whole grains, legumes, healthy proteins, red meats, cold cuts and cured meats, whole-fat and low-fat dairy products, 100% fruit juices, and SSB were conducted.
- Intake of foods using three 24-h dietary recalls. These were conducted during 2 non-consecutive weekdays and one weekend day using the Nutrition Data System for Research multi-pass method (5 steps) (Version 25, 2014) [19]. The baseline 24-h recalls were done before participants were

informed about their group assignment; one was done in person at the baseline visit and the other 2 recalls were done by phone in the following 2–3 days. For the post-intervention recalls, we completed the first 2 by phone and the last one when they came to the post-intervention visit. For the first recall, we used a portion size booklet displaying standardized food servings as a visual aid for participants to estimate their usual portion sizes. A copy of this booklet was provided to each participant to take home to help in estimating portion sizes when we called them to complete the other recalls by phone. Intake (in servings) from the following food groups were averaged for the 3 days for both baseline and post-intervention recalls: fruits, vegetables, starchy vegetables, refined and whole grains, legumes, healthy proteins, red meats, cold cuts and cured meats, whole-fat and low-fat dairy products, 100% fruit juices, SSB, and snacks and sweets.

- Weight control at the individual level

Weight and height were assessed at baseline and at post-intervention (only weight). These measurements were taken with participants wearing light clothing, no shoes, hats, or any other objects that could cause interference. Weight was determined in kg using a calibrated scale (BF-350 TANITA, Arlington Heights, IL, USA) with a 0.1 kg accuracy. Height was measured in cm using a portable stadiometer, with a 0.1 cm accuracy (Charder HM200P Portable Stadiometer, Taichung, Taiwan). Measurements were taken in duplicates and averaged. Body mass index (BMI) was calculated as kg/m^2.

2.5. Data Analysis

Descriptive statistics (frequency and percentage for categorical variables and mean (standard deviation) for continuous variables) were reported. Comparison between the App and Traditional groups at baseline and within group changes were performed using Student *t* tests. Analysis of covariance (ANCOVA) was used to assess differences between intervention groups for each outcome assessed, in which intake of foods (in servings) or frequency of food intake or weight/BMI were used as the dependent variables, group assignment as the fixed factor, and baseline value of the dependent variables as covariates. The effect sizes were calculated using the partial eta-squared, and the values 0.01, 0.06, and 0.14 were considered small, moderate, and large effects, respectively [20,21]. Due to the substantial proportion of zeroes in food purchase behavior data, a Poisson regression model was used to assess the effect of the intervention on the food selection after 8 weeks controlling for baseline values. All analyses were computed using Stata version 15 (StataCorp, College Station, TX, USA), and did not adjust for multiplicity nor missing value imputations.

3. Results

A total of 37 participants were randomized to the App group and 38 to the Traditional group, as shown in Figure 1. Not all participants completed all aspects of the study. Within the Traditional group, 18 completed the FFQ, 17 completed the 24-h recalls, and 18 completed the grocery receipt collection. Within the App group, 25 participants completed the FFQ, 15 completed the 24-h recalls, and 13 completed the grocery receipt collection. A total of 17 (8 in the Traditional group and 9 in the App group) completed all aspects of data collection (three 24-h recalls, at least two receipts, the FFQ, and weight measurements, both at baseline and post-intervention). Table 1 summarizes the characteristics of those who completed at least one aspect of the study. No differences were observed in baseline characteristics between intervention groups. Most participants were women (>88%), mean age was 35.3 years, most had more than high school education (>80%), a family composition of at least three members, and a mean baseline BMI of 34.5 kg/m^2. Also, no differences were observed in any of the baseline characteristics between those who completed or fail to complete the study (data not shown).

Table 1. Baseline characteristics of study participants by intervention groups (n = 51).

Variable	Traditional group (n = 24)	App group (n = 27)	p Value *
	Mean (SD) or %		
Age, years	36.8 (5.86)	33.8 (7.30)	0.12
Female sex, %	91.7	88.9	0.56
More than high school education, %	83.3	81.5	0.58
Number of family members in household	3.17 (1.24)	3.11 (1.34)	0.88
Weight (kg)	83.3 (14.9)	93.3 (20.4)	0.09
Height (m)	1.58 (0.06)	1.62 (0.08)	0.12
BMI, kg/m^2	33.3 (5.81)	35.6 (7.50)	0.29
Overweight, %	31.6	30.0	0.92
Obese, %	68.4	70.0	

SD: standard deviation; BMI: body mass index. * t test. Level of significance was $p < 0.05$.

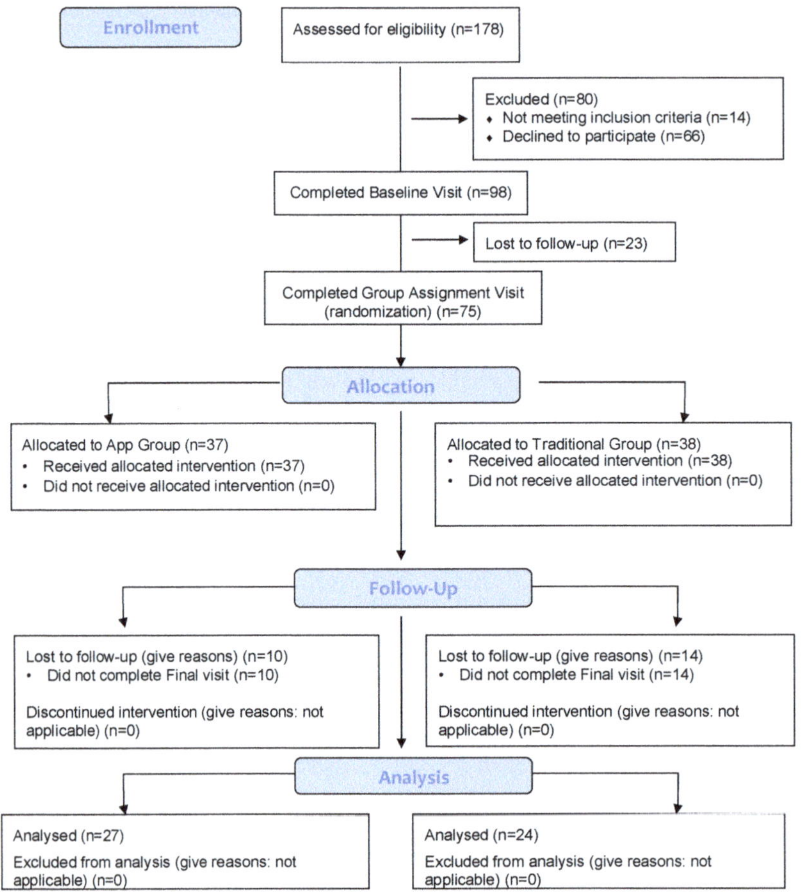

Figure 1. Participant flow chart.

Results for household food purchase frequency are shown in Table 2. Compliance with grocery receipts submission was low, therefore, the analysis included participants that had submitted at least two grocery receipts at baseline and post-intervention. No differences were observed at baseline between intervention groups. Within groups, we observed a significant increase in the frequency of purchase of vegetables and whole grains in the App group ($p < 0.05$) from baseline to post-intervention. We also analyzed the change of household food purchase frequency during the 8 weeks of the study using Poisson regression, adjusting for food purchase behavior at baseline. The coefficient associated with the intervention (App vs. Traditional) is the expected difference in log count between the App group and the Traditional group. Compared to the Traditional group, the estimated Poisson regression coefficient was 0.27 for fruits (standard error [SE] = 0.26; $p = 0.29$), 0.05 for vegetables (SE = 0.19; $p = 0.79$), 0.46 for whole grains (SE = 0.46; $p = 0.41$), 1.36 for 100% fruit juices (SE = 0.78; $p = 0.08$), and 0.51 for SSB (SE = 0.51; $p = 0.09$).

Individual food intake, as assessed from three 24-h recalls, is shown in Table 3. At baseline, the App group consumed significantly fewer servings of whole-fat dairy foods compared to the Traditional group. Within groups, we observed a decrease in the intake of refined grains, healthy proteins, and whole-fat dairy products in the Traditional group ($p < 0.05$) and a significant decrease in the intake of refined grains, legumes, 100% fruit juices, and sweets and snacks in the App group ($p < 0.05$) from baseline to post-intervention. However, when analyzing the change in food intake using ANCOVA to adjust for baseline data, as shown in Table 4, only a trend for a significant decrease in the intake of legumes in the App group compared to the Traditional group ($p = 0.06$) was observed. We also assessed individual food frequency from the FFQ, as shown in Supplementary Table S1, and found that at baseline, the App group consumed low-fat dairy foods with less frequency compared to the Traditional group ($p < 0.05$). Within groups, we observed a decrease in the frequency of intake of cold cuts and cured meats in the Traditional group ($p = 0.05$) and a significant increase in the frequency of intake of fruits in the App group ($p < 0.05$) from baseline to post-intervention. However, when analyzing changes in frequency of food intake using ANCOVA to adjust for baseline data, only a trend for an increase in the frequency of consumption of whole grains ($p = 0.08$) and a significant increase in the frequency of consumption of cold cuts and cured meats in the App group compared to the Traditional group ($p = 0.01$) was observed.

For weight and BMI, there were no differences between or within groups, as shown in Supplementary Table S2. No harm or unintended effects were observed in either of the allocation groups.

Results on the evaluation of the app have been previously published [13]. Briefly, the exit interview at post-intervention showed that most (>50%) considered the app to be feasible, acceptable, usable at least once in the last month and they were satisfied; the short survey completed by participants at the end of their grocery shopping ($n = 23$) showed that 73.1% used the app every time they went grocery shopping and that 26.1% purchased $\geq$70% of the recommended products in the list.

Table 2. Household food purchase behavior at baseline and post-intervention by groups (*n* = 31) [†].

| Variable | Baseline | | | | Post-Intervention | | | | Difference between Groups at Baseline | Difference between Baseline and Post-Intervention | |
| | Traditional Group (*n* = 18) | | App Group (*n* = 13) | | Traditional Group (*n* = 18) | | App Group (*n* = 13) | | | Traditional Group | App Group |
	Mean	SD	Mean	SD	Mean	SD	Mean	SD	*p* Value *	*p* Value *	
Fruits	1.03	1.02	1.27	1.05	1.78	1.63	2.34	2.31	0.53	0.05	0.08
Vegetables	1.75	1.76	1.31	1.03	3.42	4.07	3.71	3.68	0.43	0.07	0.02
Whole grains	0.31	0.55	0.23	0.39	0.61	0.49	0.98	1.37	0.68	0.06	0.04
100% fruit juices	0.14	0.33	0.08	0.19	0.13	0.28	0.49	0.94	0.55	0.45	0.07
SSB [††]	1.14	1.19	0.65	0.85	1.10	1.25	2.13	3.46	0.22	0.46	0.10

[†] Data estimated from two shopping receipts at baseline and at least two shopping receipts during the last 2 weeks of the intervention; [††] SSB: sugar sweetened beverages; SD: standard deviation. * *t* test. Level of significance was *p* < 0.05.

Table 3. Individual intake of food (servings/day) at baseline and post-intervention by groups (*n* = 32) [†].

| Variable | Baseline | | | | Post-Intervention | | | | Difference between Groups at Baseline | Difference between Baseline and Post-Intervention | |
| | Traditional Group (*n* = 18) | | App Group (*n* = 13) | | Traditional Group (*n* = 18) | | App Group (*n* = 13) | | | Traditional Group | App Group |
	Mean	SD	Mean	SD	Mean	SD	Mean	SD		*p* Value *	
Fruits	0.87	0.96	1.13	1.16	1.09	0.99	1.37	1.06	0.49	0.19	0.18
Vegetables	0.65	0.69	0.76	0.69	0.69	0.69	1.45	1.55	0.66	0.43	0.06
Starchy vegetables	1.18	0.94	1.10	0.65	1.10	0.93	1.60	1.28	0.78	0.39	0.10
Refined grains	3.59	2.06	3.36	1.38	2.64	1.65	2.18	1.51	0.71	0.02	0.01
Whole grains	1.12	0.73	1.32	1.03	1.41	0.79	1.78	1.17	0.52	0.06	0.09
Legumes	0.13	0.16	0.23	0.24	0.18	0.18	0.07	0.12	0.14	0.16	0.02
Healthy proteins	1.26	1.20	0.82	0.84	0.77	1.04	0.42	0.59	0.24	0.05	0.10
Red meats	3.90	1.94	4.25	1.98	3.49	1.92	4.22	1.76	0.62	0.23	0.48
Cold cuts & cured meats	0.40	0.47	0.38	0.51	0.50	0.47	0.40	0.44	0.91	0.22	0.46
Whole-fat dairies	1.01	0.58	0.42	0.31	0.63	0.35	0.38	0.28	0.00	0.00	0.34
Low-fat dairies	0.36	0.25	0.75	0.75	0.43	0.60	0.75	0.87	0.06	0.33	0.49
100% fruit juices	0.23	0.41	0.35	0.47	0.09	0.16	0.11	0.28	0.47	0.11	0.01
SSB [††]	0.45	1.29	0.07	0.15	0.24	0.46	0.08	0.19	0.27	0.28	0.41
Sweets and snacks	1.32	1.57	1.36	1.05	0.98	1.16	0.58	0.74	0.94	0.21	0.03

[†] Data collected from three 24-hrs food recalls at baseline and three 24-hrs foods recalls at the end of the study; includes nuts, fish, and poultry; [††] SSB: sugar sweetened beverages; SD: standard deviation. * *t* test. Level of significance was *p* < 0.05.

Table 4. Analysis of covariance for individual food intake (servings/day) at 8 weeks (Traditional group $n = 17$; App group $n = 15$).

Variable	Adjusted Mean Difference	95% CI	p Value *	Partial Eta-Squared
Fruits	0.13	−0.50, 0.77	0.67	0.006
Vegetables	0.74	−0.12, 1.60	0.09	0.10
Starchy vegetables	0.52	−0.29, 1.32	0.20	0.06
Refined grains	−0.35	−1.34, 0.64	0.47	0.02
Whole grains	0.27	−0.38, 0.93	0.40	0.02
Legumes	−0.11	−0.23, 0.004	0.06	0.12
Healthy proteins	−0.25	−0.88, 0.37	0.41	0.02
Red meats	0.68	−0.67, 2.04	0.31	0.04
Cold cuts and cured meats	−0.10	−0.42, 0.23	0.55	0.01
Regular dairies	−0.09	−0.34, 0.17	0.49	0.02
Low-fat dairies	0.35	−0.23, 0.93	0.22	0.05
100% fruit juices	−0.005	−0.16, 0.15	0.94	0.0002
SSB [†]	−0.17	−0.44, 0.10	0.21	0.05
Snacks and sweets	−0.40	−1.13, 0.32	0.26	0.04

Includes nuts, fish, and poultry; [†] SSB: sugar sweetened beverages. * Analysis of covariance (ANCOVA) was used to assess differences between intervention groups, with food intake as the dependent variable, group assignment as the fixed factor, adjusting for food intake at baseline. Level of significance was $p < 0.05$.

4. Discussion

This is the first study to test an app that generates a shopping list based on energy requirements, following the DGA and accounting for budget and supermarkets' discounts. Those using "MyNutriCart" purchased vegetables and whole grains significantly more frequently at the household level, while at the individual level they significantly consumed more servings of refined grains, legumes, 100% fruit juices, and sweets and snacks and significantly consumed fruits more frequently at post-intervention compared to baseline. However, the Traditional group also had some improvements, so when analyzing changes in these behaviors during the study between groups using Poisson regression or ANCOVA, the App group only had a significantly greater frequency of consumption of whole grains and cold cuts and cured meats with a lower intake of legumes compared to the Traditional group. No effects on weight control were detected.

As hypothesized, "MyNutriCart" improved some aspects of household food purchasing behavior (i.e., higher vegetables and whole grains purchase), which translated into a lower intake of refined grains at the individual level. However, it is interesting to note that purchasing vegetables more frequently at the household level did not translate into a greater intake of vegetables at the individual level, although intake did improve somewhat compared to baseline. Since this is a measure of the household food purchase frequency, it may explain why it did not specifically translate to greater vegetable consumption at the individual level. However, compared to the Traditional group, none of the changes regarding household food purchases were significant. This was not expected as the app considered the household budget, the supermarkets' weekly discounts, and only included in the shopping list only those fruits and vegetables offered at a reduced price, to maximize the budget as the price of fresh produce varies considerably depending on the season. Therefore, the app showed participants that healthy foods could be purchased even within a tight budget. Certain food purchasing behaviors are easier to be influenced, such as purchasing whole grains as they are readily available in all supermarkets and most refined grains (i.e., white rice, white bread, white tortillas), have a healthier whole grain option (i.e., brown rice, whole multigrain bread, whole-wheat tortillas). In fact, other studies aiming to improve diet quality have found improvements in whole grains [22], therefore, switching from refined to whole grains seems to be easier than introducing new foods, such as fruits or vegetables. In particular, consumption of fruits and vegetables was low in both groups and improvements were observed in the App group including a greater frequency of consumption of fruits ($p = 0.02$) and a trend in a greater number of servings of vegetable consumed ($p = 0.06$) at post-intervention compared to baseline, which is consistent with other studies conducted in similar groups [23–26]. Other trials targeting fruits and vegetables among populations with traditionally low intakes have also found significant improvements [27,28]. These products are often perceived as expensive [29,30]; which is the main reason our app only included in the shopping list the fresh produce that was on sale that week. However, more intensive interventions may be needed to increase household purchasing of fruits and vegetables and to translate this into a higher individual intake of these foods.

Currently, there are a limited number of studies investigating the purchase of healthy foods in grocery stores and improvement in dietary behaviors using a smartphone app, although there are a few trials that are currently ongoing. A study among 208 adults in Canada testing the "SmartAPPetite" app for 8-10 weeks found a significant decrease in the intake of soft drinks, sugary and fast foods and an increase in homemade meals and fruits, particularly among those using the app more frequently [11]. Also, 46% of participants believed that the messaging changed their food purchasing habits [11]. A study testing an app to improve vegetables among 135 overweight adults for 8 weeks found a significantly greater vegetable intake among the intervention group compared to the control group [31]. Another trial testing the effect of a "SaltSwitch" app among 66 adults with cardiovascular disease for four weeks found a significant reduction in salt purchase, which resulted in a reduction of 0.7 g of salt/day per person, compared to the usual care group [32]. Most of these trials showed that compliance was reduced over time and that those that were more compliant with the intervention (i.e., greater use of the app) had greater outcome effects. However, results from these trials provide evidence on the effects of such apps in improving food selection and purchase, although more studies are needed to understand how individuals use the apps.

Although there are only a few trials testing apps to improve household food purchasing behavior and dietary behaviors, they have the potential to support/reinforce adherence to the DGA. However, this technology may be insufficient for helping individuals make the necessary behavioral changes. As found in our study, only one intervention session without follow-ups or app notifications to remind participants to use the app, led to only a few significant improvements in dietary behaviors. More intensive follow-ups with app notifications may be needed to facilitate behavioral change. Some participants may need more counseling than others, therefore, sessions should be personalized depending on the level of behavioral change needed by each participant. Also, follow-up sessions may be necessary to keep participants motivated in using the app, as we previously reported that only 26% purchased more than 70% of the items recommended on the grocery list at each shopping event [13]. Others have reported that greater app interactions led to greater dietary changes [11]. The low app use in the present study could also explain the lack of greater changes in the study and also on the lack of effects on weight; as evidenced by others [33]. Studies testing apps for weight control have found significant effects [34,35], particularly those using more intensive approaches [12,35] and even among short-term studies [34,35]. Therefore, interventions using smartphone apps may require several sessions/calls/follow-ups during the study to maintain motivation towards using the app.

The present study helped identify the potential of the "MyNutriCart" app to improve household food purchase and individual dietary behaviors using a randomized clinical trial design; with trained research assistants and using validated tools. It also helped identify limitations that should be considered in future investigations, such as its short duration and small sample size. Another limitation was the lack of follow-up messages or app notifications to remind participants to use the app. We did not assess their prior experience with healthy eating, which could have affected our results. The information from grocery receipts was limited to frequency, as amount purchased was not readily available from all grocery receipts. We did not assess if the frequency of grocery shopping changed with the study, which could have affected the selection of foods. We also did not assess how many members of the household each shopping event was intended for; however, the app did take into account the number of members in the household when coming up with the list. In addition, due to low compliance with grocery receipts submission, the analysis was based on at least two receipts at each time, which may not be representative of usual household purchase. We also did not ask if other household members did complementary grocery shopping during the study, which should be accounted for in future studies. "MyNutriCart" was tested among Hispanics and integrated elements of Hispanic diets, but its conception is based on the DGA, hence its applicability does not exclude other ethnic groups. Future studies should also integrate all family members, as the app provides a healthy grocery list for the entire family and we learned at the end of the study that some family members disliked some of the recommended foods, as previously reported [13].

5. Conclusions

In conclusion, the use of the "MyNutriCart" app led to small improvements in household food purchase and individual food intake over the 8-week period compared to the initial assessment but there were basically no significant improvements compared to the Traditional group. Therefore, these results may suggest that the "MyNutriCart" app is as good as the traditional method for improving these behaviors. Using such tools could reduce costs and resources for improving household food purchase and dietary quality. Also, these tools may help reach out to other target groups, that may not reach out to health professionals for improving their diet. However, neither of the interventions led to changes in weight control. More intense interventions with greater follow-up visits, app notifications, calls or messages are needed to achieve greater changes in food-related behaviors and weight outcomes. In the future, larger and longer trials with more intensive follow-ups may be needed to detect changes in the desired outcomes.

Supplementary Materials: The following are available online at http://www.mdpi.com/2072-6643/10/12/1967/s1, Table S1: Frequency of food consumption at baseline and post intervention by groups (n = 43), Table S2: Weight and BMI of study participants at baseline and post intervention by groups (n = 37).

Author Contributions: Conceptualization, C.P. and C.M.P.; Methodology, C.P., C.M.P., M.T., and D.L.; Software, C.M.P.; Validation, C.P. and C.M.P.; Formal Analysis, C.M.P.; Investigation, C.P., M.T., and D.L.; Resources, C.P.; Data Curation, C.P., M.T. and D.L.; Writing—Original Draft Preparation, M.T. and C.P.; Writing—Review & Editing, C.P., C.M.P., M.A.T.F., and C.C.; Supervision, C.P.; Project Administration, C.P.; Funding Acquisition, C.P.

Funding: This study was conducted with funding support in part by the Research Centers in Minority Institutions Program [grant number G12 MD007600] from National Institute on Minority Health and Health Disparities, National Institutes of Health.

Acknowledgments: The authors would like to thank the Nutrition Program, School of Public Health of the University of Puerto Rico and the Medical Sciences Campus for their support and providing their facilities to conduct the research.

Conflicts of Interest: The authors declare no interest conflict related to this publication.

References

1. Ward, B.W.; Schiller, J.S.; Goodman, R.A. Multiple Chronic Conditions Among US Adults: A 2012 Update. *Prev. Chronic Dis.* **2014**, *11*, 130389. [CrossRef] [PubMed]
2. US Department of Agriculture Dietary Guidelines for Americans. 2015. Available online: Health.gov (accessed on 2 June 2017).
3. Guenther, P.M.; Kirkpatrick, S.I.; Reedy, J.; Krebs-Smith, S.M.; Buckman, D.W.; Dodd, K.W.; Casavale, K.O.; Carroll, R.J. The Healthy Eating Index-2010 is a valid and reliable measure of diet quality according to the 2010 Dietary Guidelines for Americans. *J. Nutr.* **2014**, *144*, 399–407. [CrossRef]
4. Rowe, S.; Alexander, N.; Almeida, N.G.; Black, R.; Burns, R.; Bush, L.; Crawford, P.; Keim, N.; Kris-Etherton, P.; Weaver, C. Translating the Dietary Guidelines for Americans 2010 to Bring about Real Behavior Change. *J. Am. Diet. Assoc.* **2011**, *111*, 28–39. [CrossRef]
5. Ivens, B.J.; Smith Edge, M. Food and Nutrition Science Solutions Joint Task Force Translating the Dietary Guidelines to Promote Behavior Change: Perspectives from the Food and Nutrition Science Solutions Joint Task Force. *J. Acad. Nutr. Diet.* **2016**, *116*, 1697–1702. [CrossRef]
6. Lewis, K.D.; Burton-Freeman, B.M. The role of innovation and technology in meeting individual nutritional needs. *J. Nutr.* **2010**, *140*, 426S–436S. [CrossRef] [PubMed]

7. Escaron, A.L.; Meinen, A.M.; Nitzke, S.A.; Martinez-Donate, A.P. Supermarket and Grocery Store–Based Interventions to Promote Healthful Food Choices and Eating Practices: A Systematic Review. *Prev. Chronic Dis.* **2013**, *10*, 120156. [CrossRef] [PubMed]

8. Kalnikaitė, V.; Bird, J.; Rogers, Y. Decision-making in the aisles: Informing, overwhelming or nudging supermarket shoppers? *Pers. Ubiquitous Comput.* **2013**, *17*, 1247–1259. [CrossRef]

9. Hollywood, L.E.; Cuskelly, G.J.; O'Brien, M.; McConnon, A.; Barnett, J.; Raats, M.M.; Dean, M. Healthful grocery shopping. Perceptions and barriers. *Appetite* **2013**, *70*, 119–126. [CrossRef]

10. Kelley, P.G.; Consolvo, S.; Cranor, L.F.; Jung, J.; Sadeh, N.; Wetherall, D. A Conundrum of Permissions: Installing Applications on an Android Smartphone. In *Financial Cryptography and Data Security, FC2012*; Blyth, J., Dietrich, S., Camp, L.J., Eds.; Lecture Notes in Computer Science; Springer: Berlin/Heidelberg, Germany, 2012; Volume 7398, pp. 68–79.

11. Gilliland, J.; Sadler, R.; Clark, A.; O'Connor, C.; Milczarek, M.; Doherty, S.; Gilliland, J.; Sadler, R.; Clark, A.; O'Connor, C.; et al. Using a Smartphone Application to Promote Healthy Dietary Behaviours and Local Food Consumption. *BioMed Res. Int.* **2015**, *2015*, 841368. [CrossRef]

12. Carter, M.C.; Burley, V.J.; Nykjaer, C.; Cade, J.E. Adherence to a smartphone application for weight loss compared to website and paper diary: Pilot randomized controlled trial. *J. Med. Internet Res.* **2013**, *15*, e32. [CrossRef]

13. Lopez, D.; Torres, M.; Velez, J.; Grullon, J.; Negron, E.; Perez, C.; Palacios, C. Development and Evaluation of a Nutritional Smartphone Application for Making Smart and Healthy Choices in Grocery Shopping. *Healthc. Inform. Res.* **2017**, *23*, 16–24. [CrossRef] [PubMed]

14. Standing Committee on the Scientific Evaluation of Dietary Reference Intakes, Institute of Medicine, Food and Nutrition Board. *Dietary Reference Intakes for Energy, Carbohydrate, Fiber, Fat, Fatty Acids, Cholesterol, Protein, and Amino Acids*; National Academy Press: Washington, DC, USA, 2005.

15. Academy of Nutrition and Dietetics Adult Weight Management (AWM) Guideline. Available online: http://www.andeal.org/topic.cfm?cat=2798 (accessed on 20 February 2017).

16. US Department of Agriculture (USDA) Choose MyPlate. Available online: https://www.choosemyplate.gov/MyPlate (accessed on 21 February 2017).

17. French, S.A.; Wall, M.; Mitchell, N.R.; Shimotsu, S.T.; Welsh, E. Annotated receipts capture household food purchases from a broad range of sources. *Int. J. Behav. Nutr. Phys. Act.* **2009**, *6*, 37. [CrossRef] [PubMed]

18. Palacios, C.; Trak, M.A.; Betancourt, J.; Joshipura, K.; Tucker, K.L. Validation and reproducibility of a semi-quantitative FFQ as a measure of dietary intake in adults from Puerto Rico. *Public Health Nutr.* **2015**, *18*, 2550–2558. [CrossRef] [PubMed]

19. Feskanich, D.; Sielaff, B.H.; Chong, K.; Buzzard, I.M. Computerized collection and analysis of dietary intake information. *Comput. Methods Programs Biomed.* **1989**, *30*, 47–57. [CrossRef]

20. Cohen, J. *Statistical Power Analysis for the Behavioral Sciences*; Lawrence Erlbaum Associates: Hillsdale, MI, USA, 1988; Volume 2.

21. Levine, T.R.; Hullett, C.R. Eta Squared, Partial Eta Squared, and Misreporting of Effect Size in Communication Research. *Hum. Commun. Res.* **2002**, *28*, 612–625. [CrossRef]

22. Petrogianni, M.; Kanellakis, S.; Kallianioti, K.; Argyropoulou, D.; Pitsavos, C.; Manios, Y. A multicomponent lifestyle intervention produces favourable changes in diet quality and cardiometabolic risk indices in hypercholesterolaemic adults. *J. Hum. Nutr. Diet.* **2013**, *26*, 596–605. [CrossRef]

23. Torres, R.; Santos, E.; Orraca, L.; Elias, A.; Palacios, C. Diet quality, social determinants, and weight status in Puerto Rican children aged 12 years. *J. Acad. Nutr. Diet.* **2014**, *114*, 1230–1235. [CrossRef] [PubMed]

24. Serrano, M.; Torres, R.; Perez, C.M.; Palacios, C. Social environment factors, diet quality, and body weight in 12-year-old children from four public schools in Puerto Rico. *P. R. Health Sci. J.* **2014**, *33*, 80–87.

25. Torres, R.; Serrano, M.; Perez, C.M.; Palacios, C. Physical environment, diet quality, and body weight in a group of 12-year-old children from four public schools in Puerto Rico. *P. R. Health Sci. J.* **2014**, *33*, 14–21.

26. Guilloty, N.I.; Soto, R.; Anzalota, L.; Rosario, Z.; Cordero, J.F.; Palacios, C. Diet, Pre-pregnancy BMI, and Gestational Weight Gain in Puerto Rican Women. *Matern. Child Health J.* **2015**, *19*, 2453–2461. [CrossRef]

27. Neville, C.E.; McKinley, M.C.; Draffin, C.R.; Gallagher, N.E.; Appleton, K.M.; Young, I.S.; Edgar, J.D.; Woodside, J.V. Participating in a fruit and vegetable intervention trial improves longer term fruit and vegetable consumption and barriers to fruit and vegetable consumption: A follow-up of the ADIT study. *Int. J. Behav. Nutr. Phys. Act.* **2015**, *12*, 158. [CrossRef] [PubMed]

28. Pearson, N.; Atkin, A.J.; Biddle, S.J.H.; Gorely, T. A family-based intervention to increase fruit and vegetable consumption in adolescents: A pilot study. *Public Health Nutr.* **2010**, *13*, 876–885. [CrossRef]

29. Erinosho, T.O.; Pinard, C.A.; Nebeling, L.C.; Moser, R.P.; Shaikh, A.R.; Resnicow, K.; Oh, A.Y.; Yaroch, A.L. Development and implementation of the National Cancer Institute's Food Attitudes and Behaviors Survey to assess correlates of fruit and vegetable intake in adults. *PLoS ONE* **2015**, *10*, e0115017. [CrossRef]

30. McLaughlin, C.; Tarasuk, V.; Kreiger, N. An examination of at-home food preparation activity among low-income, food-insecure women. *J. Am. Diet. Assoc.* **2003**, *103*, 1506–1512. [CrossRef] [PubMed]

31. Mummah, S.; Robinson, T.N.; Mathur, M.; Farzinkhou, S.; Sutton, S.; Gardner, C.D. Effect of a mobile app intervention on vegetable consumption in overweight adults: A randomized controlled trial. *Int. J. Behav. Nutr. Phys. Act.* **2017**, *14*, 125. [CrossRef] [PubMed]

32. Eyles, H.; McLean, R.; Neal, B.; Jiang, Y.; Doughty, R.N.; McLean, R.; Ni Mhurchu, C. A salt-reduction smartphone app supports lower-salt food purchases for people with cardiovascular disease: Findings from the SaltSwitch randomised controlled trial. *Eur. J. Prev. Cardiol.* **2017**, *24*, 1435–1444. [CrossRef] [PubMed]

33. Laing, B.Y.; Mangione, C.M.; Tseng, C.-H.; Leng, M.; Vaisberg, E.; Mahida, M.; Bholat, M.; Glazier, E.; Morisky, D.E.; Bell, D.S. Effectiveness of a smartphone application for weight loss compared with usual care in overweight primary care patients: A randomized, controlled trial. *Ann. Intern. Med.* **2014**, *161*, S5–S12. [CrossRef]

34. Scholtz, C. *A Novel 'Food Lists' App to Promote Weight Loss, Improve Diet Quality, and Strengthen Diet Adherence: The Foodmindr Study*; Arizona State University: Tempe, AZ, USA, 2016.

35. Wharton, C.M.; Johnston, C.S.; Cunningham, B.K.; Sterner, D. Dietary Self-Monitoring, But Not Dietary Quality, Improves With Use of Smartphone App Technology in an 8-Week Weight Loss Trial. *J. Nutr. Educ. Behav.* **2014**, *46*, 440–444. [CrossRef]

 *nutrients*

 MDPI

Article

Active Image-Assisted Food Records in Comparison to Regular Food Records: A Validation Study against Doubly Labeled Water in 12-Month-Old Infants

Ulrica Johansson [1,*], Michelle Venables [2], Inger Öhlund [1] and Torbjörn Lind [1]

[1] Department of Clinical Sciences, Pediatrics, Umeå University, SE 901 85 Umeå, Sweden; inger.ohlund@umu.se (I.Ö.); torbjorn.lind@umu.se (T.L.)
[2] MRC Elsie Widdowson Laboratory, Cambridge CB I 9NL, UK; Michelle.Venables@mrc-epid.cam.ac.uk
* Correspondence: ulrica.johansson@umu.se

Received: 30 October 2018; Accepted: 29 November 2018; Published: 4 December 2018

Abstract: Overreporting of dietary intake in infants is a problem when using food records (FR), distorting possible relationships between diet and health outcomes. Image-assisted dietary assessment may improve the accuracy, but to date, evaluation in the pediatric setting is limited. The aim of the study was to compare macronutrient and energy intake by using an active image-assisted five-day FR against a regular five-day FR, and to validate image-assistance with total energy expenditure (TEE), was measured using doubly labeled water. Participants in this validation study were 22 healthy infants randomly selected from the control group of a larger, randomized intervention trial. The parents reported the infants' dietary intake, and supplied images of main course meals taken from standardized flat-surfaced plates before and after eating episodes. Energy and nutrient intakes were calculated separately using regular FR and image-assisted FRs. The mean ($\pm$ standard deviations) energy intake (EI) was 3902 $\pm$ 476 kJ/day from the regular FR, and 3905 $\pm$ 476 kJ/day from the FR using active image-assistance. The mean EI from main-course meals when image-assistance was used did not differ (1.7 $\pm$ 55 kJ, p = 0.89) compared to regular FRs nor did the intake of macronutrients. Compared to TEE, image-assisted FR overestimated EI by 10%. Without validation, commercially available software to aid in the volume estimations, food item identification, and automation of the image processing, image-assisted methods remain a more costly and burdensome alternative to regular FRs in infants. The image-assisted method did, however, identify leftovers better than did regular FR, where such information is usually not readily available.

Keywords: energy intake; dietary assessment; image-assisted method; infant; food record; doubly labeled water

1. Introduction

Dietary assessments from food records (FR) are commonly used to assess children's food and nutrient intake, and possible relationships between dietary intake and health outcomes [1]. In infants and young children, parents are asked to record everything that the child has eaten and drunk during a predefined time period [2]. However, achieving accurate and reliable dietary intake data can be difficult and demanding: for the parents, the process may be tedious and time consuming [3], and for the clinician or researcher, the generated data may be subject to bias, making interpretation difficult [4]. Meals with complex content, such as main course meals with several ingredients, are challenging to remember, record, and to determine the amounts of the various ingredients [5]. In young children, food records tend to overestimate energy intake, e.g., parents may misreport the child's intake by failing to omit food leftovers and spillage from the FRs [4–6]. In order to better understand the complex relationships between diet and health in young children, it is important to develop dietary assessment methods with higher accuracy and precision [7].

Mobile phone applications and cameras have been shown to improve self-reported dietary intake, and they have likewise increased participants' user satisfaction, compared to conventional methods [8]. Moreover, information obtained from images seems to reduce random and measurement errors for energy intake (EI), especially when it comes from complex and diverse foods. In adults, EI is often underestimated, but this can be corrected with image-assisted dietary assessments [9]. This was also found among overweight and obese children by using digital camera FRs [10]. Previous research in pre-school children has shown that EI assessment using images was not significantly different, compared to measuring total energy expenditure (TEE) with doubly labeled water (DLW) [11]. However, no study to date has investigated methods of active image-assisted FRs in infants [9,12].

The aim of this study was to compare total and main course meal energy and macronutrient intake in 12-month-old, healthy infants using an active image-assisted five-day FR against a regular five-day FR, and to validate the total energy intake measured with the image-assisted food record method against TEE using DLW.

2. Materials and Methods

2.1. Participants and Study Design

The infants in the present study were taking part in an optimized complementary feeding study (OTIS; ClinicalTrials.gov registration number NCT02634749, (n = 250) among 4–6 months old, healthy, full-term infants in Umeå, Sweden, measuring the effects of different complementary diets on various health outcomes and food acceptance. In the present validation study, all infants (n = 27) belonging to the control group in the OTIS trial from September 2016 until July 2017 were selected at 12 months of age. In the control arm of the study (n = 125), the participants were advised to follow the current, Swedish dietary recommendations, but they were otherwise not subject to any intervention [13]. In the present study, as well as in the larger OTIS trial, the inclusion criteria were healthy, singleton infants, 4–6 months of age, born after >37 weeks of gestation and birth weight > 2500 g, living in Umeå municipality. The exclusion criteria were infants with chronic illnesses, iron deficiency, or any other biochemical abnormality, or infants been having started feeding with complementary foods at the time of recruitment.

2.2. Anthropometry

Within two weeks of the participants' 12-month birthday, the infants were invited to the Pediatric research facility at Umeå University Hospital for information on the study procedures, measurements and administration of DLW. Anthropometric data were collected according to standardized procedures [14]: nude weight was measured to the nearest 5 g using electronic scales (Seca 727, Seca, Hamburg, Germany), recumbent length was measured to the nearest 0.1 cm using an infantometer (Seca 416, Seca, Hamburg, Germany), and the head circumference was measured to the nearest 0.1 cm by using a non-stretchable measuring tape (Seca 212, Seca, Hamburg, Germany).

2.3. Doubly Labeled Water

On the same day as the anthropometrical measurements, a pre-dose urine sample was collected by placing an absorbent pad (Bastos Viegas, Penafiel, Portugal) in the diaper of the infant. Each infant was then given an oral weighed dose of DLW consisting of 100 mg/kg ^2H_2O and 280 mg/kg $H_2^{18}O$. Post-dose urine samples were collected at home once daily for 10 consecutive days, with dates and times recorded for all samples by using absorbent pads as described above, omitting the first urine portion of the day. The first post-dose urine sample was collected approximately 24 h after the DLW dose was given and the subsequent pads were collected once daily after that. The parents were asked to remove the pad once it was wet from urine. Each collected pad was stored at −18 °C. The pads were then taken to the Pediatric research facility at Umeå University Hospital and thawed, and the urine content was extracted using a press, collecting the urine in glass bottles. The glass bottles were stored at $-20\ °C$ until transportation to MRC Elsie Widdowson Laboratory, Cambridge, UK for analysis.

2.4. Water Isotope Analysis

Urine samples were prepared for ^{18}O enrichment using the CO_2 equilibration method [15]. The samples were then analyzed using a continuous flow isotope ratio mass spectrometer (IRMS) (AP2003, Analytical Precision Ltd., Northwich, Cheshire, UK). For ^2H, the samples were analyzed using a continuous-flow IRMS (Sercon, ABCA-Hydra 20–22, Sercon Ltd., Crewe, UK). All samples were measured alongside secondary reference standards previously calibrated against the primary international standards Vienna-Standard Mean Ocean Water (vSMOW) and Vienna-Standard Light Antarctic Precipitate (vSLAP) (International Atomic Energy Agency, Vienna, Austria). Sample enrichments were corrected for interference according to Craig [16], and expressed relative to vSMOW. Analytical precisions (SD) were better than ± 0.4 ppm for ^{18}O and ± 1.3 ppm for ^2H. The rate of CO_2 production (R_{CO_2}) was calculated according to Schoeller et al. [17], R_{CO_2} was then converted to TEE using the equation of Elia and Livesey [18], with the food quotient (FQ) calculated according to Jéquier et al. [19]. From TEE, metabolizable energy (ME) was calculated according to Wells and Davies [20].

2.5. Food Record and Dietary Assessment

Parents were asked to record everything that their child ate and drank, including breastmilk and food supplements, e.g., vitamins, using a pre-printed five-day FR. Of these five days, we asked that at least one day was a Saturday or Sunday. The parents started the recording the day after the administration of DLW. Each day, the parents recorded the meal type, time of day, and which foods and drinks the infants were offered, including amounts and brand names. Amounts of foods and drinks were documented using household measures and for bread etc., in slices. Homemade recipes were documented separately, including ingredients, quantities and detailed descriptions of preparation. Unfamiliar dishes were reported in detail with brand name and amounts. Breastmilk was recorded as 'meals' (more than five minutes of breastfeeding) or 'snacks' (less than five minutes of breastfeeding), estimated as 102 or 25 g of milk, respectively [21,22]. The reported food and drink intake was converted to grams using standardized weights for consumed foods from the Swedish Food Agency Database [23]. To calculate the mean daily EI (kJ/day) and macronutrients sub-classes (g/day) from the five day FR, we used the software Dietist Net Pro (Kost och Näringsdata AB, Bromma, Sweden) and the food composition database (version 17 February 2016) from the Swedish National Food Administration. The database was complemented with special products for infants used in the OTIS study, with nutrient contents analyzed and supplied from Semper AB.

2.6. Food Record with an Active Image-Assisted Method

An active image-assisted FR method is a system that captures images, usually photographs, during eating episodes, and is used to enhance or supplement traditional written or electronic FRs [24]. The images provide objective information such as food type, volume, and leftovers, and may even record foods that were forgotten and not reported in the food registration [24]. In this study, we decided to capture two main meals, i.e., the noon (lunch) and late afternoon (dinner) meals, for the image-assisted part. These two meals were expected to represent 30% of the total daily EI, and they included more complex and diverse dishes with a larger amount of ingredients mixed together, which makes assessing the composition and estimating leftovers more challenging [8]. Given the meal frequency of 12-month-old infants and without specific smart phone applications to facilitate this task, we also assumed that the workload for the parents would be too great if they would have to record all of the meals that the child consumed by using the image-assisted method.

During the five-day FR, parents were instructed to serve the two main meals on standardized flat-surfaced plates, which were provided by the researchers to the participants, and then to capture mobile phone images of the plates before and after each meal. The plate served as a reference marker [25]. The written instructions had four examples of main meals served on the standardized flat-surfaced plate before and after intake, with images captured at a 90° angle from a mobile phone camera according to Stumbo (Figure 1) [26]. All 20 images from the 10 main meals during the five-day FR were sent by the parents, usually by directly sharing the images from the participants' mobile phones to the study e-mail account. The participants used their own mobile phones for the photos. If no images were received within five days, a reminder was sent by e-mail to the participant. First, a trained pediatric dietician calculated the mean, daily energy, and macronutrient intakes from the FRs without access to the images. In a second step, the images of the main meals were made available for the dietician, who analyzed the images, taking food leftovers, spillage, etc., into consideration [25]. To assist the dietician in estimating the food items and food volumes on the plate before and after the meal, the dietician was provided with images of the standardized flat-surfaced plate with different quantities of commonly used baby foods, either from glass jars as used in the study, or home-cooked food comparable to an infant's normal portion size. These reference images were similar to the images received from the participants [24]. Finally, the initial calculations from the FR were, if needed, adjusted depending on the results of the analyses of the main meal images, for example, subtracting undocumented leftovers and spillages from the initially estimated intake. This generated two sets of data, one from the regular FR, and one record that included the image-adjusted dietary intakes. The latter also contained specific information on leftovers that was unaccounted for in the regular FR.

(A)

(B)

Figure 1. Images of a main course meal with leftovers before (**A**) and after (**B**) an eating session.

2.7. Pilot Testing

Before embarking on the present study, we performed an unpublished pilot study to assess the feasibility of using image-assisted FR in infants. Parents of fourteen 8–12 months old, healthy, free-living infants were asked to do a five-day food record in a similar way as in the present study, taking mobile phone images of the main course meal together with a regular FR. During the course of five days, 78 meals were recorded by the participants. When comparing the food records with and without image assistance, we found that 23 meals (29%) had to be adjusted, since the meal images contained additional information to correctly estimate the intake from the meals, and 22 of the 23 adjusted meals were overestimations, i.e., the parents had omitted to exclude foods left over or spilled on or around the plate. This resulted in a mean difference in daily EI between the two estimations of 169 ± 146.4 kJ.

2.8. Group Size Calculation

We based the sample size calculation on the pilot study described above. We estimated that 45% of participants would have their FR adjusted when image-assistance was added, and that the difference in EI measured with FRs against TEE with DLW would be 238 ± 193.7 kJ [4]. Given these circumstances, and allowing for a 30% attrition rate, we calculated that we would need 25–30 participants (power 90%, alpha = 0.05) in order to show a significant difference in the measurement error in EI between FRs with and without image-assistance, compared to ME.

2.9. Ethical Considerations

The study was approved by the Regional Ethical Review Board at Umeå University, Sweden (dnr 2016-134-32M).

2.10. Statistical Analyses

Statistical analyses were performed using SPSS 24.0 (SPSS, Chicago, IL, USA). For continuous variables, results are presented as means (± standard deviations, SD or ± 95% confidence intervals, CI) and for categorical variables as numbers and percentages. Normal distribution for continuous variables was assessed with the Shapiro–Wilk test. The energy and macronutrient intake were calculated as kilojoules (kJ) and grams (g) per day, respectively. The significance level was set at $p < 0.05$. Differences between image-assisted FR and regular FR and image-assisted FR and ME were analyzed separately with paired sample *t*-tests. The Bland and Altman method [27] was used to assess the agreement between regular and image-assisted FRs, and between the image-assisted FR and ME calculated from DLW. Reliability between ME and the image-assisted FR method was quantified using a two-way mixed absolute agreement intra-class correlation coefficient (ICC).

3. Results

Of the 27 selected infants, 82% completed the study with a majority being boys (Table 1). Five infants were excluded; three with missing FR information and images, and two infants had insufficient urine samples to allow for the analysis of TEE.

3.1. Energy and Macronutrient Intake

Five of the 22 infants were breastfed, usually 2–3 times per day; in the morning, in the evening, and/or at night. None of the infants were breastfed at the time of any of the main course meals. The average number of meals per day was 7.1 ± 1.1, and image-assistance was used in 29% of these meals. Mean EI and macronutrient intakes, both overall and from the main course meal, were normally distributed. Overall, these intakes were not significantly different between the regular FRs and the image-assisted FRs (Table 2). In particular, EI from the main course meals were a mix of equal numbers of over- and underestimated meals (Table 3) and therefore the errors were balanced out and had no effect on the average energy or macronutrient intake (Table 2). Average ME, calculated from DLW was 3538 ± 428 kJ/day. Bland–Altman plots were used to assess agreement between image-assisted FRs and ME. The mean bias between the methods was 366 kJ/day, with limits of agreement of ±712 kJ/day (Figure 2). There was no significant association between the mean and the difference of EI and ME ($p = 0.53$), indicating no systematic bias across the different levels of EI. The intra-class correlation (ICC) coefficient was 0.81, indicating high reliability between the two methods.

Table 1. Anthropometric and demographic data of the study infants ($n = 22$) and their parents.

	Mean ± SD
Age infants (months)	11.9 ± 0.34
Body weight (kg)	10.5 ± 1.2
Body length (cm)	76.9 ± 3.1
Head circumference (cm)	47.1 ± 0.9
Mothers age (year)	31 ± 5.1
Fathers age (year)	32 ± 5.3
	n (%)
Girls/boys	6 (27)/16(73)
Breastfeeding at 12 mo	5 (23)
≥1 sibling	10 (45)
Education level Mother	
Elementary school	1 (4.5)
High school	5 (22.7)
University	16 (72.7)
Education level Father	
Elementary school	1 (4.5)
High school	7 (31.8)
University	14 (63.6)
Born in Sweden	
Infant	22 (100)
Mother	21 (95.5)
Father	17 (77.3)

SD: standard deviations

Table 2. Total daily energy and macronutrient intake, and the daily energy and macronutrient intake from the main course meals (lunch and dinner combined), estimated by regular five-day food records, and food records with image-assistance in the study infants ($n = 22$).

	Food Record [1]	Food Record with Image-Assistance [1]	Difference [1]	p for Difference [2]
Total intake				
Energy (kJ)	3901 ± 476	3905 ± 476	3.9 ± 48.0	0.71
Protein (g)	29.8 ± 5.7	30.1 ± 6.0	0.2 ± 1.1	0.30
Fat (g)	35.4 ± 6.7	35.4 ± 6.5	0.0 ± 0.6	0.89
Carbohydrate (g)	118.3 ± 17.7	118.4 ± 17.2	0.1 ± 1.5	0.80
Main course meals				
Energy (kJ)	1348 ± 388	1350 ± 377	1.7 ± 55	0.89
Protein (g)	13.0 ± 3.9	13.4 ± 4.1	0.4 ± 1.5	0.19
Fat (g)	12.5 ± 4.2	12.7 ± 3.9	0.2 ± 1.3	0.37
Carbohydrate (g)	38.0 ± 13.3	38.5 ± 12.9	0.5 ± 4.2	0.63

[1] Values are mean ± SD, [2] Paired sample *t*-test.

Table 3. Numbers of non-adjusted and adjusted meals (corrected by dietician after review of meal images) with or without leftovers assessed with five-day food records with active image-assistance.

Meals (*n* = 210)	Leftovers *n* (%)	No Leftovers *n* (%)	Total *n* (%)
Non-adjusted meals	67 (53)	74 (89)	141 (67)
Adjusted meals	60 (47)	9 (11)	69 (33)
Underestimated [1]	34 (27)	2 (3)	36 (52)
Overestimated [2]	26 (20)	7 (8)	33 (48)
Total	127 (60)	83 (40)	210

[1] Underestimated: the recorded amount of food consumed is less than what is estimated from the meal images.
[2] Overestimated: the recorded amount of food consumed is more than what is estimated from the meal images.

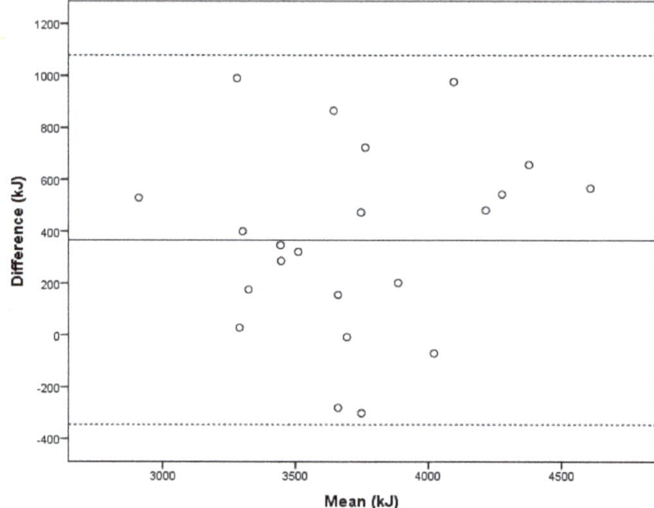

Figure 2. Bland–Altman plot showing the mean versus the difference in energy intake estimated from food records with active image-assistance and metabolizable energy calculated from doubly labeled water in 22 healthy, 12-month-old infants. The *x*-axis shows the mean energy intake (EI) per day (kJ) from FR with image-assistance and metabolizable energy. The solid line (-) shows the mean difference of 366 kJ, and the dashed lines (—) show the 95% limits of agreement (±1.96 SD) of 712 kJ.

When estimating EI from only the main course meals representing 35% of the total daily EI, i.e., when the image-assisted method was used, there was no statistically significant difference between the image-assisted and regular FRs, and no significant differences for any of the macronutrients (Table 2).

3.2. Main Course Meals with an Active Image-Assisted Method

In the five-day FR, 220 main course meals were recorded. Of these, 210 meals (96%) were assessed with both regular FRs, and the active image-assisted method. Ten meals (4%) were excluded because of missing images after the eating episode (Table 3). For the majority of meals, the dietician did no adjustment of the amounts of food consumed from that particular meal after taking the meal images into account. However, for a third of the meals, some adjustments were made. Out of these adjusted meals, about half were underestimations on the part of the regular FR, and leftovers were more common, compared to no leftovers. Of the 22 infants, 17 (77%) had at least one main meal adjusted by the active image-assisted method. The average number of main meals with leftovers over five days were 5.8 ± 3.3 per infant. Three infants had leftovers from all 10 meals, and one infant had no leftovers from any of the eating occasions.

4. Discussion

Previous studies in infants have shown the non-random error to be 5–15% when comparing energy expenditure, measured with doubly labeled water to recorded dietary intake [1]. Such bias increases the risk of type II-error and diminishes the power of the study. Image-assistance, where FRs are complemented with images before and after eating sessions is one possibility to reduce this bias [9].

In the present study, comparing total EI in regular and image-assisted FRs, the difference between methods was only 3.9 kJ and not statistically different. The reliability between image-assisted FR and ME, and the golden standard to assess EI, using ICC was excellent. The bias of 366 kJ means that image-assisted FRs overestimated the EI compared to ME by 10%. This bias was higher but the limits of agreement were narrower than reported in similar studies [4,11]. Previous studies have shown that image assistance has been successful in reducing underreporting, but to our knowledge no study has used the technique in settings when over reporting is an issue, as was the case in the present study [9,10].

Unaccounted leftovers, i.e., the parts of the meal that are left on the plate or that are lost due to spillage, are possible sources of systematic error if they are not subtracted from the estimated intake. In the present study, 60% of the meals that were recorded with active image assistance showed leftovers. However, in more than half of these eating episodes, the parents correctly modified the FR to include the leftovers, and in the other half, where the research dietician did adjust the recording, it was equally common for the dietician to increase the recording as it was to decrease the estimated EI from that meal. In our earlier pilot study, we found a similar proportion of meals (23/78 meals in the pilot vs 69/210 meals in the present study) had been adjusted, but the regular FR overestimated the energy intake by 169 kJ per day, compared to the image-assisted method. Also, in the pilot, the majority of the adjusted meals (22/23) were overestimations. In the present study, underestimations were equally common to overestimations. A possible explanation to this discrepancy may be that the participants in the present study, being part of a large trial, were more experienced in completing food recordings, this being their third in six months compared to the participants in the pilot. We speculate that greater experience explains some of the lower bias in the regular FRs, compared to the image-assisted method found in our study [28,29]. We do not know to what extent the parents in the present study used the images to corroborate their recordings, but we did notice that the parents were skilled at allowing for leftovers in the FRs, which is indicated by the fact that half of the meals where there were leftovers had been already adjusted by the parent. From the present study, we have no information on the amount of leftovers or spillage from the eating episodes where image assistance was not used. We can only speculate that the overestimation of EI compared to ME can be found in unaccounted leftovers, from for example breakfast and snacks. From the FRs, we know that these meals contained large amounts of energy dense foods, such as porridge and milk cereal drinks. Another possible error could have been the preparation of the porridge and milk cereals, where adding too much water could have made the meals more diluted, reducing the actual EI compared to the recorded EI.

The overall energy and macronutrient intake, and its variations, were similar to other studies in the same age group [30–32]. Also, the parents adhered well to submitting the FRs and images, with less than 5% of the main meals having missing images. Breast milk intakes were estimated from feeding episodes and not by direct observation, i.e., test weighing. On average, breast milk contributed to <6% of the mean total daily energy intake. The energy and nutrient content of the products used in the present study (porridge, formula, milk cereal drink, baby food in glass jars) were based on high quality analyzed data supplied from Semper AB.

A strength of the study was that all parents used the same reference marker [25], i.e., the plate estimated the true area of the food portions, and we omitted images of meals, from which other types of plates were used. Also, the same dietician managed all FRs, calculated the dietary intake, and assessed the images, but was blinded as to the outcome, and did not participate in the final analysis, i.e., when the regular FRs was compared to the image-assisted FRs. In the DLW analyses, we used the more accurate FQ [19], instead of the generic RQ suggested by Schoeller et al. [17]. The proportion of parents with university education was higher than people of the same age and gender in the general, Swedish population, and higher than that reported in a recent iron supplementation study in the same geographical area [33]. Despite this, we believe that it is possible to generalize the results to other populations as well.

A limitation in the study was that we did not use the image-assisted method for all meals and snacks. Our hypothesis was to focus on the main meals, i.e., lunch and dinner, which are more complex and diverse in terms of ingredients and nutrient value, and to leave breakfast and snacks, which in this age are less diverse. It is likely that unaccounted leftovers and spillage from energy- and nutrient-dense foods, such as milk cereal drinks or baby porridge, could have been identified through image-assistance. However, we believe that the task of providing before and after photos of every meal and snack, considering the frequency of feedings, including night meals in this age group would have been almost insurmountable for both the parents and researchers.

Variations among different mobile cameras, ambient light conditions, etc., may have contributed to some of the subjectivity of the image analysis [34]. All images were taken at 90° to the plate, which is the most favorable for capturing which ingredients the meal contained. However, to optimize volume calculations, another photo at 45° would have been preferable [9,24,35]. To improve the quality of the images, and to aid in the volume estimation, some kind of mobile phone application would have been desirable, but to the best of our knowledge no such product validated for use in infants is commercially available [34].

5. Conclusions

In conclusion, in 12-month-old infants, the image-assisted method identifies leftovers better than regular FR, where such information is usually not readily available, and it may thereby improve the accuracy of EI and macronutrients. But as seen in this study, parents with earlier experience of food recording were, in many cases, capable of including leftovers in their records, reducing this source of systematic bias. It is possible that this compensation was facilitated by the availability of the images themselves. Also, FRs with or without image-assistance overestimate EI compared to ME. With these caveats, and without validated commercially available software to aid in the volume estimations, food item identification, and automation of the image processing, the image-assisted method remains more costly and burdensome, but possibly a more accurate alternative to regular FRs in infants [34]. In future validation studies, technical solutions for smartphones are required to better identify food items and food volumes from images. Such future software applications would make it possible to estimate more cost-effectively the entire energy intake in infants. Future research should also include training sessions, both for the participants using the technique, and for professionals involved in the dietary assessment with images [36].

Author Contributions: T.L. is the Principal Investigator for the OTIS study. T.L. designed the study, together with U.J., I.Ö., and M.V. contributed to the DLW analysis, T.L., U.J., and I.Ö. contributed to the design of the image-assisted method and dietary assessment. T.L. and U.J. was responsible for the data collection and the data analysis of the image-assisted method and the dietary assessment. All authors approved the final version.

Funding: The study is funded by grants from the Oskar Foundation, the Regional agreement between Umeå University and Västerbotten county council on cooperation in the field of Medicine, Odontology, and Health (ALF, grants VLL-644531, VLL-488901, VLL, 677921, VLL-761381) and Semper AB.

Acknowledgments: The authors thank participating families, Anna Lundman, Charlotta Strömberg, and Mariana Åhfeldt for participant recruitment and data collection, Carina Lagerqvist, and Catharina Lundell for assistance in the DLW-procedure, and Lisbeth Nordström for food record calculations and Lina Johansson for image processing. We also thank the stable isotope team, Priya Singh, Elise Orford, and Kevin Donkers at the MRC Elsie Widdowson Laboratory, UK for their provision of the DLW and sample analysis.

Conflicts of Interest: U.J. is a doctoral student at the Umeå University Industrial Doctoral School for Research and Innovation with Semper AB as the industrial sponsor. None of the other authors had financial or personal interests in any of the organizations sponsoring this research. The sponsors had no role in the design and conduct of the study; in the collection, analysis, and interpretation of the data, or in the preparation or contents of the manuscript.

Abbreviation

FR	Food record
DLW	Doubly labeled water
EI	Energy intake
ICC	Intra-class correlation coefficient
ME	Metabolizable energy
SD	Standard deviation
TEE	Total energy expenditure
OTIS	Optimized complementary feeding study

References

1. Burrows, T.L.; Martin, R.J.; Collins, C.E. A systematic review of the validity of dietary assessment methods in children when compared with the method of doubly labeled water. *J. Am. Diet. Assoc.* **2010**, *110*, 1501–1510. [CrossRef] [PubMed]

2. Lanigan, J.A.; Wells, J.C.K.; Lawson, M.S.; Cole, T.J.; Lucas, A. Number of days needed to assess energy and nutrient intake in infants and young children between 6 months and 2 years of age. *Eur. J. Clin. Nutr.* **2004**, *58*, 745–750. [CrossRef] [PubMed]

3. Moreno, L.A.; Kersting, M.; de Henauw, S.; González-Gross, M.; Sichert-Hellert, W.; Matthys, C.; Mesana, M.I.; Ross, N. How to measure dietary intake and food habits in adolescence: The European perspective. *Int. J. Obes.* **2005**, *29*, 66–77. [CrossRef]

4. Lanigan, J.A.; Wells, J.C.K.; Lawson, M.S.; Lucas, A. Validation of food diary method for assessment of dietary energy and macronutrient intake in infants and children aged 6–24 months. *Eur. J. Clin. Nutr.* **2001**, *55*, 124–129. [CrossRef] [PubMed]

5. Kong, K.; Zhang, L.; Huang, L.; Tao, Y. Validity and practicability of smartphone-based photographic food records for estimating energy and nutrient intake. *Asia Pac. J. Clin. Nutr.* **2017**, *26*, 396–401. [PubMed]

6. Tennefors, C.; Coward, W.A.; Hernell, O.; Wright, A.; Forsum, E. Total energy expenditure and physical activity level in healthy young Swedish children 9 or 14 months of age. *Eur. J. Clin. Nutr.* **2003**, *57*, 647–653. [CrossRef] [PubMed]

7. Buday, R.; Tapia, R.; Maze, G.R. Technology-driven dietary assessment: A software developer's perspective. *J. Hum. Nutr. Diet.* **2014**, *27*, 10–17. [CrossRef]

8. Boushey, C.; Kerr, D.; Wright, J.; Lutes, K.; Ebert, D.; Delp, E. Use of technology in children's dietary assessment. *Eur. J. Clin. Nutr.* **2009**, *63*, 50–57. [CrossRef]

9. Boushey, C.J.; Spoden, M.; Zhu, F.M.; Delp, E.J.; Kerr, D.A. New mobile methods for dietary assessment: Review of image-assisted and image-based dietary assessment methods. *Proc. Nutr. Soc.* **2016**, *76*, 283–294. [CrossRef]

10. Svensson, Å.; Waling, M.; Bäcklund, C.; Larsson, C. Overweight and obese children's ability to report energy intake using digital camera food records during a 2-year study. *J. Nutr. Metab.* **2012**, *2012*, 247389. [CrossRef]

11. Delisle Nyström, C.; Forsum, E.; Henriksson, H.; Trolle-Lagerros, Y.; Larsson, C.; Maddison, R.; Timpka, T.; Löf, M. A mobile phone based method to assess energy and food intake in young children: A validation study against the doubly labelled water method and 24 h dietary recalls. *Nutrients* **2016**, *8*, 50. [CrossRef] [PubMed]

12. Rollo, M.E.; Williams, R.L.; Burrows, T.; Kirkpatrick, S.I.; Bucher, T.; Collins, C.E. What are they really eating? A review on new approaches to dietary intake assessment and validation. *Curr. Nutr. Rep.* **2016**, *5*, 307–314. [CrossRef]

13. National Food Administration. *Good Foods for Infants Under One Year*; Sweden Food Agency: Uppsala, Sweden, 2011.

14. Lohman, T.G.; Hingle, M.; Going, S.B. Body composition in children. *Pediatr. Exerc. Sci.* **2013**, *25*, 573–590. [CrossRef]

15. Roether, W. Water-CO_2 exchange set-up for the routine 18oxygen assay of natural waters. *Int. J. Appl. Radiat. Isot.* **1970**, *21*, 379–387. [CrossRef]

16. Craig, H. Isotopic standards for carbon and oxygen and correction factors for mass-spectrometric analysis of carbon dioxide. *Geochim. Cosmochim. Acta* **1957**, *12*, 133–149. [CrossRef]

17. Schoeller, D.A.; Ravussin, E.; Schutz, Y.; Acheson, K.J.; Baertschi, P.; Jequier, E. Energy expenditure by doubly labeled water: Validation in humans and proposed calculation. *Am. J. Physiol. Regul. Integr. Comp. Physiol.* **1986**, *250*, 823–830. [CrossRef]
18. Elia, M.; Livesey, G. Theory and validity of indirect calorimetry during net lipid synthesis. *Am. J. Clin. Nutr.* **1988**, *47*, 591–607. [CrossRef]
19. Jéquier, E.; Acheson, K.J.; Schutz, Y. Assessment of energy expenditure and fuel utilization in man. *Annu. Rev. Nutr.* **1987**, *7*, 187–208. [CrossRef]
20. Wells, J.C.; Davies, P.S. Estimation of the energy cost of physical activity in infancy. *Arch. Dis. Child.* **1998**, *78*, 131–136. [CrossRef]
21. Paul, A.A.; Black, A.E.; Evans, J.; Cole, T.J.; Whitehead, R.G. Breastmilk intake and growth in infants from two to ten months. *J. Hum. Nutr. Diet.* **1988**, *1*, 437–450. [CrossRef]
22. Michaelsen, K.M.; Skafte, L.; Badsberg, J.H.; Jorgensen, M. Variation in Macronutrients in Human Bank Milk: Influencing Factors and Implications for Human Milk Banking. *J. Pediatr. Gastroenterol. Nutr.* **1990**, *11*, 229–239. [CrossRef] [PubMed]
23. National Food Administration. Available online: https://slv.se (accessed on 1 January 2017).
24. Gemming, L.; Utter, J.; Ni Mhurchu, C. Image-assisted dietary assessment: A systematic review of the evidence. *J. Acad. Nutr. Diet.* **2015**, *115*, 64–77. [CrossRef] [PubMed]
25. Martin, C.K.; Kaya, S.; Gunturk, B.K. Quantification of food intake using food image analysis. *Conf. Proc. IEEE Eng. Med. Biol. Soc.* **2009**, *2009*, 6869–6872. [PubMed]
26. Stumbo, P.J. New technology in dietary assessment: A review of digital methods in improving food record accuracy. *Proc. Nutr. Soc.* **2013**, *72*, 70–76. [CrossRef]
27. Bland, J.M.; Altman, D.G. Statistical methods for assessing agreement between two methods of clinical measurement. *Lancet* **1986**, *1*, 307–310. [CrossRef]
28. Söderberg, L.; Lind, T.; Karlsland Åkeson, P.; Sandström, A.-K.; Hernell, O.; Öhlund, I. A validation study of an interviewer-administered short food frequency questionnaire in assessing dietary vitamin D and calcium intake in Swedish children. *Nutrients* **2017**, *9*, 682. [CrossRef] [PubMed]
29. Öhlund, I.; Lind, T.; Hörnell, A.; Hernell, O. Predictors of iron status in well-nourished 4-y-old children. *Am. J. Clin. Nutr.* **2008**, *87*, 839–845. [CrossRef] [PubMed]
30. Öhlund, I.; Hernell, O.; Hörnell, A.; Stenlund, H.; Lind, T. BMI at 4 years of age is associated with previous and current protein intake and with paternal BMI. *Eur. J. Clin. Nutr.* **2010**, *64*, 138–145. [CrossRef] [PubMed]
31. Huysentruyt, K.; Laire, D.; Avondt, T.V.; Schepper, J.D.; Vandenplas, Y. Energy and macronutrient intakes and adherence to dietary guidelines of infants and toddlers in Belgium. *Eur. J. Nutr.* **2016**, *55*, 1595–1604. [CrossRef] [PubMed]
32. Grote, V.; Verduci, E.; Scaglioni, S.; Vecchi, F.; Contarini, G.; Giovannini, M.; Koletzko, B.; Agostoni, C. Breast milk composition and infant nutrient intakes during the first 12 months of life. *Eur. J. Clin. Nutr.* **2015**, *70*, 250–256. [CrossRef] [PubMed]
33. Szymlek-Gay, E.A.; Domellöf, M.; Hernell, O.; Hurrell, R.F.; Lind, T.; Lönnerdal, B.; Zeder, C.; Egli, I.M. Mode of oral iron administration and the amount of iron habitually consumed do not affect iron absorption, systemic iron utilisation or zinc absorption in iron-sufficient infants: A randomised trial. *Br. J. Nutr.* **2016**, *116*, 1046–1060. [CrossRef] [PubMed]
34. Sharp, D.B.; Allman-Farinelli, M. Feasibility and validity of mobile phones to assess dietary intake. *Nutrition* **2014**, *30*, 1257–1266. [CrossRef] [PubMed]
35. Olafsdottir, A.S.; Hörnell, A.; Hedelin, M.; Waling, M.; Gunnarsdottir, I.; Olsson, C. Development and validation of a photographic method to use for dietary assessment in school settings. *PLoS ONE* **2016**, *11*, e0163970. [CrossRef] [PubMed]
36. Fatehah, A.A.; Poh, B.K.; Shanita, S.N.; Wong, J.E. Feasibility of reviewing digital food images for dietary assessment among nutrition professionals. *Nutrients* **2018**, *10*, 984. [CrossRef] [PubMed]

Article

Technology-Based Dietary Assessment in Youth with and Without Developmental Disabilities

Michele Polfuss [1,2,*], Andrea Moosreiner [3], Carol J. Boushey [4], Edward J. Delp [5] and Fengqing Zhu [5]

1　University of Wisconsin-Milwaukee, College of Nursing, Milwaukee, WI 53211, USA
2　Children's Hospital of Wisconsin, Department of Nursing Research, Milwaukee, WI 53226, USA
3　Medical College of Wisconsin-Clinical and Translational Science Institute, Milwaukee, WI 53226, USA; amoosreiner@mcw.edu
4　Epidemiology Program, University of Hawaii Cancer Center, Honolulu, HI 96789, USA; cjboushey@cc.hawaii.edu
5　School of Electrical and Computer Engineering, Purdue University, West Lafayette, IN 47907, USA; ace@ecn.purdue.edu (E.J.D.); zhu0@ecn.purdue.edu (F.Z.)
*　Correspondence: mpolfuss@uwm.edu; Tel.: +1-414-229-2609

Received: 12 September 2018; Accepted: 8 October 2018; Published: 11 October 2018

Abstract: Obesity prevalence is higher in children with developmental disabilities as compared to their typically developing peers. Research on dietary intake assessment methods in this vulnerable population is lacking. The objectives of this study were to assess the feasibility, acceptability, and compare the nutrient intakes of two technology-based dietary assessment methods in children with-and-without developmental disabilities. This cross-sectional feasibility study was an added aim to a larger pilot study. Children (n = 12; 8–18 years) diagnosed with spina bifida, Down syndrome, or without disability were recruited from the larger study sample, stratified by diagnosis. Participants were asked to complete six days of a mobile food record (mFR™), a 24-h dietary recall via FaceTime® (24 HR-FT), and a post-study survey. Analysis included descriptive statistics for survey results and a paired samples t-test for nutrient intakes. All participants successfully completed six days of dietary assessment using both methods and acceptability was high. Energy (kcal) and protein (g) intake was significantly higher for the mFR™ as compared to the 24 HR-FT (p = 0.041; p = 0.014, respectively). Each method had strengths and weaknesses. The two technology-based dietary assessment tools were well accepted and when combined could increase accuracy of self-reported dietary assessment in children with-and-without disability.

Keywords: dietary assessment; mobile food record; 24-h recall; developmental disabilities; children; spina bifida; down syndrome; technology; pediatrics

---

1. Introduction

Assessment of an individual's dietary intake is an essential component of the prevention and treatment of an abnormal weight status [1]. Details of dietary intake provide valuable information on an individual's nutritional balance and dietary habits [2,3]. The interest in dietary assessment has heightened as the prevalence of obesity has increased. However, there is a lack of testing and development of tools focusing on children with developmental disabilities [4]. This is a critical oversight as the prevalence of obesity is often higher in children with developmental disabilities as compared to children who are typically developing [5].

Recommended assessment methods for dietary intake in children vary based on the child's age and who is reporting [2,4]. Conclusions from a systematic review identified that the 24-h dietary recall reported by the parent for 4 to 11 year olds and dietary history reported by adolescents 16–21 years of age had the highest level of accuracy when compared to doubly labeled water [2]. Challenges to obtaining an accurate dietary assessment include social bias, the burden of time to complete, and the inability of the reporter to estimate portion sizes, identify food preparation methods, and recall foods consumed [1,6]. Currently there is no recommendation for dietary assessment in children with developmental disabilities.

Incorporating technology is thought to improve dietary intake accuracy, appeal to a younger generation, and reduce the burden placed on the reporter [7,8]. One option is the Technology Assisted Dietary Assessment™ (TADA™) system, an image-based dietary assessment system which uses the mobile Food Record™ (mFR™) app to collect images of eating occasions [9–12]. The app can be downloaded onto smart devices (e.g., mobile phone or iPad). The app allows individuals to record images before and after eating occasions and the images upload in real time to a cloud-based server along with contextual information, e.g., time.

The collection of dietary intake in real time is thought to reduce recall bias, provide additional information related to the individual's eating behaviors, and increase convenience for the reporter [11]. Among 41 adolescents (11–15 years of age), use of the mFR™ was accepted by the majority [10]. Bathgate et al. [13] examined the feasibility of using the mFR™ in 59 adolescents and young adults (12–30 years of age; M = 21.5 (SD 4.6)) with Down syndrome. In this sample, 86% of the participants successfully recorded dietary intake using the mFR™ for a minimum of two days [13].

The objectives of this study were to assess the feasibility and acceptability of the mFR™ and a 24-h dietary recall conducted via FaceTime (24 HR-FT) among children with-and-without developmental disabilities. FaceTime is an app available on Apple® products that allows individuals to use WiFi or cellular data to perform a call with video and audio capability. The estimates for total energy and macronutrient intakes were hypothesized to be similar between the methods. Results from this exploratory study can inform future studies to better assess dietary intakes among a vulnerable and understudied population.

2. Materials and Methods

2.1. Study Design and Sample

This study was part of a larger cross-sectional study examining energy expenditure assessment in 36 children with-and-without developmental disabilities [14]. This feasibility study was conducted as an added aim to the original study through an additional funding mechanism. Institutional Review Board approval was granted through a Midwestern Children's Hospital and parents and children provided written informed consent and assent. Study visits were conducted within a Translational Research Unit funded by the Clinical and Translational Science Institute of Southeast Wisconsin.

Participants included 12 of the original 36 children diagnosed with Down syndrome, spina bifida, or no developmental disability. A sample of 12 participants was determined based on funding and feasibility design. Based on a completed permission to contact form from the parent study, participants stratified by diagnosis were randomly recruited for this study.

2.2. Measures

2.2.1. Dietary Assessment (mFR™)

Study participants were provided a mini iPad® (iOS version 9.3, Apple Inc., Cupertino, CA, USA) with the mFR™ and FaceTime app. These community dwelling children were asked to obtain images before and after all meals/snacks for a 24-h period for a total of six days (4 weekdays and 2 weekend days) of their choice over a two-week period. Data collection occurred during late summer and fall seasons. Participants were instructed to eat as usual. The child and parent were provided training and practiced using the mFR™ with a cafeteria meal. Training focused on technical issues, such as the need to incorporate the provided checkered fiducial marker in the eating scene to aid volume estimation (Figure 1), and problem solving for common mealtime issues such as having seconds or placement of food labels within the eating scene to assist the intake analysis. Parental assistance was recommended to be used as needed. Pre- and post-eating occasion images were automatically uploaded to a secure cloud-based server. A trained team member used the images to enter the food intake and amounts using Nutrition Data Systems for Research, a computer-based software application [15].

Figure 1. Before and after Mobile Food Record™ images with the fiducial marker.

2.2.2. Dietary Assessment (24 HR-FT)

Participants were instructed that each subsequent day following the mFR™, they would be asked to complete a 24-h dietary recall conducted via the FaceTime app on the provided mini iPad. Scheduling of the FaceTime calls were predetermined with the family. The 24 HR-FT was conducted by a dietitian trained to use a multiple-pass method which included extracting forgotten foods and detailed portion sizes. During training, participants were provided with a set of standard measuring cups and spoons, a deck of cards, and 2-dimensional portion size tools for use as a reference during the recalls. Parental assistance was recommended to be used as needed. At the time of the 24 HR-FT, the interviewer did not access or preview the mFR™ images.

Following the six days of dietary intake recording by the mFR™ and the 24 HR-FT, the child and parent were asked to complete a post-study survey. The survey included questions on use of parental assistance and details specific to each method.

2.3. Statistical Analysis

The dietary intake data collected using the mFR™ and 24 HR-FT were entered and analyzed using the Nutrition Data System for Research software version 2015 developed by the Nutrition Coordinating Center (NCC), University of Minnesota, Minneapolis, MN [15]. The survey responses were analyzed using descriptive statistics. Daily intake of energy (kcal), carbohydrates, fats, and proteins were compared between the methods with a paired samples t-test. Statistical analyses were performed using SPSS (IBM SPSS Statistics Version 25; Chicago, IL, USA). Statistical significance was set at a p-value < 0.05.

3. Results

3.1. Sample Characteristics

The cohort (n = 12) equally represented the three groups (spina bifida (n = 4), Down syndrome (n = 4), and no disability (n = 4)) with ages between 8 and 18 years old (M = 13.17; SD 3.35) and included six boys and six girls.

3.2. Feasibility and Acceptability

The six days of recording dietary intake with the mFR™ and 24 HR-FT were successfully completed by 12 of the 12 study participants. All 12 children were willing to use the mFR™ and participate in multiple 24 HR-FT in a future study. See Table 1 for additional results. All parents who completed the survey (*n* = 11) were women. Six parents reported assisting their child with the mFR™ and eight assisted with the 24 HR-FT.

Table 1. Child post-study survey result.

Mobile Food Record (mFR™)	
Willing to use TADA mFR™ in future	12/12 (100%)
Ease of use (1-very easy; 10-very difficult)	1 (75%); 3 (8.3%); 4 (8.3%); 9 (8.3%)
Screen easy to read	10 (83.3%) strongly agree and 2 (16.7%) agree
Easy to enter information	8 (66.7%) strongly agree and 4 (33.3%) agree
Information provided was accurate	10 (83.3%) strongly agree and 2 (16.7%) agree
Interfered with daily activities	4 (33.3%) strongly disagree; 7 (33.3%) disagree and 1 (8.3%) agree
24-h Recall by FaceTime (24 HR-FT)	
Willing to use 24 HR-FT in future	12/12 (100%)
Ease of use (1-very easy; 10 very difficult)	1 (83.3%); 2 (8.3%); 4 (8.3%)
Easy to recall food	4 (33.3%) strongly agree and 8 (66.7%) agree
Information provided was accurate	7 (58.3%) strongly agree and 5 (41.7%) agree
Interfered with daily activities	5 (41.7%) strongly disagree; 6 (50%) disagree and 1 (8.3%) agree

TADA™: Technology Assisted Dietary Assessment.

3.3. Energy and Dietary Macronutrients

Significant differences were identified for kcals per day from 24 HR-FT (M = 2020, SD = 626) as compared to mFR™ (M = 1855, SD = 508), t (11) = 2.32, p = 0.041 and for protein (g/day) from 24 HR-FT (M = 80, SD = 27) as compared to mFR™ (M = 69, SD = 19), t (11) = 2.92, p = 0.014 with the 24 HR-FT assessment being higher for both. No significant differences were reported for dietary fats (g/day) between the 24 HR-FT (M = 81, SD = 32) and mFR™ (M = 75, SD = 25), t (11) = 1.29, p = 0.223. Similarly, no significant differences were identified for dietary carbohydrates (g/day) when comparing the 24 HR-FT (M = 250, SD = 73) and mFR™ (M = 233, SD = 70), t = 2.0, p = 0.071.

3.4. Post Hoc Observations

Strengths of the mFR™ included the ability to capture intake not identified by the 24 HR-FT, which was commonly either a snack or non-nutritive item. Weaknesses included the limited ability to extract details from the images, e.g., preparation and food density. Strengths of the 24 HR-FT included the ability to probe and expand on questions related to types of foods and meal components. Weaknesses of the 24 HR-FT included the child's inability to accurately remember intake, identify food preparation details, and estimate portion sizes. Parental involvement was highest among children with Down syndrome and all groups in the age range of 8 to 12 years. Of the parents who assisted their children, there was a generalized reduced awareness of complete dietary intake for the child.

4. Discussion

In this feasibility study, the mFR™ and 24 HR-FT dietary assessment methods were both well accepted by children with and without developmental disabilities. Requesting the use of both methods for a total of six days within a two-week timeframe was feasible for both child and parent schedules. This expanded on what was reported by Bathgate and colleagues [13] who tested the feasibility of using the TADA mFR™ in a slightly older sample of individuals with Down syndrome. In their study, 86% (51/59) of the sample successfully recorded nutritional intake with the mFR™ for a minimum of two days [13], whereas 100% of the sample in the current study successfully collected both the image recordings for the mFR™ and the 24 HR-FT for a total of six days. Notable differences between these studies were that the sample in the current study was smaller, younger, and included children with spina bifida and without developmental disabilities. In addition, the current study was able to provide information on parental assistance with the assessment tools. Benefits and limitations of each of the dietary assessment methods became evident following execution of this study protocol.

A notable strength of the mFR™ was the ability to capture snacks or non-nutritive food choices that were often not reported in the 24 HR-FT. This finding was similar across all participant groups and ages. The omission of this intake in the recalls may have been due to issues of memory, mindless eating, or social desirability bias.

Challenges related to the mFR™ included difficulty in identifying food items from the uploaded images on the web server. Having a single 2-dimensional image did not consistently provide sufficient details regarding the food item, portion size, or the preparation methods. A dietitian completing a brief review with study participants regarding items needing additional information as done by Kerr [11] and Bathgate [13] could address these issues. The mFR™ used a fiducial marker to assist the human eye to estimate volume but potential for error was still present. These challenges are not specific to the mFR™. Food supplies and systems have produced an infinite number of possible nutrient compositions per food item creating challenges for any assessment method.

Completion of the 24 HR-FT was well accepted by the study participants. The use of FaceTime to complete the recall proved to be convenient and offered the investigator and reporter face-to-face interview benefits. Recall appointments were able to take place anywhere there was an internet connection decreasing the burden to participants. The face-to-face interview potentially reduced misreporting by allowing the investigator to observe social cues including eye movements and facial expressions, which assisted in the determination of when to probe for further information.

A common limitation when using the recall method is the inability for the reporter to remember all food consumed. When recalling independently, participants sometimes did not remember intake that they had documented with an image the day before. These image confirmed differences might have contributed to the larger amount of inaccuracies in the children between 8 and 15 years of age. In addition to difficulties recalling consumed items, all participants struggled to describe how food items were prepared, provide food details (e.g., low-fat), and estimate portion sizes. However, the option of using the provided measuring cups and spoons lessened this problem. When given the option of having parental assistance with recalls, children with Down syndrome and all children between 8 and 12 years of age employed this. This may be related to Down syndrome having a higher potential of cognitive impairment and a poor working memory or it may be indicative of this age group. When used, parental assistance was not always useful. Parents were often unaware of specifics related to what their child ate throughout the day. This is not unexpected as food is often consumed outside of the home or can be eaten independently within the home.

When comparing energy and macronutrient intakes between the two methods, dietary fats and carbohydrates were consistent with each other, but energy (kcals) and protein intake were significantly different between the methods with the 24 HR-FT measuring higher for both. The rationale for this difference is uncertain but may stem from the challenges related to extracting details from the TADA™ images or the added benefit of being face-to-face for the 24 HR-FT. As noted above a review process after collecting the images might address this [11,13]. Further study would be needed with larger sample sizes to confirm if these remain consistent findings.

The intent of the study was to compare two novel methods of dietary assessment in children with and without developmental disabilities. Having the ability to perform each method back-to-back not only allowed the authors to compare the methods but it also highlighted how unique attributes of each method could be synergistic if used together. During analysis, it became evident that the TADA™ images captured intake that was not identified by the child during the 24 HR-FT, which may alleviate issues related to the inability to recall food consumed the previous day. In addition, the 24 HR-FT could provide the trained interviewer the opportunity to ask questions or to use props to gain valuable details related to the food in the TADA™ images. While our team did not preview the TADA™ images prior to the subsequent 24 HR-FT, deliberately replicating the sequence of these two methods and using the 24 HR-FT to complement the mFR™ could be extremely valuable and is recommended for future studies.

Study strengths were the inclusion of children with disabilities, the use of the same food composition table, and that a single team member entered all data for analysis. Particular limitations include the small sample size and cross-sectional design that limits the generalizability of study findings. Also, the errors inherent with interpreting dietary information for data entry to a food composition table and lack of an objective biomarker.

5. Conclusions

This feasibility study provided valuable information in a vulnerable subset of children who have a higher prevalence of obesity and could be applied to all children regardless of disabilities. The mFR™ and conducting multiple pass 24-h dietary recalls over FaceTime are two novel methods of assessing dietary intake. The use of technology appeared to benefit acceptance and willingness to complete the tools in a sample of children with-and-without developmental disabilities and their parents. Each tool had its own strengths and weaknesses that could leverage the other. The combination of methods may increase the accuracy of self-reported dietary assessment in children and is recommended for further study in larger samples.

Author Contributions: This manuscript represents the collaborative work of the five authors. The study was conceptualized by A.M., M.P., C.B., E.I.D. and F.M.Z. M.P. acquired funding support. A.M., M.P. and C.B. developed the methodology, C.B., E.I.D., F.M.Z. and A.M. worked with the software. C.B., A.M. and M.P. provided resources and A.M. and M.P. conducted the investigation, were responsible for project administration and overall project supervision. Data was curated by A.M., E.I.D. and F.M.Z.; analysis was conducted by M.P. and A.M. validated the study. The original draft was prepared by A.M. and M.P. and the entire team (A.M., M.P., C.B., F.M.Z. and E.I.D.) reviewed and edited the manuscript. A.M. and M.P. were responsible for visualization of the final product. All authors have read and approved the final manuscript.

Funding: This research was funded by the Clinical and Translational Science Award program of the National Center for Research Resources and the National Center for Advancing Translational Sciences (UL1TR001436). The study was an added aim to the Pilot Project supported by the National Institutes of Nursing Research (P20NR015339) and the Clinical and Translational Science Institute of Southeastern Wisconsin through the Advancing a Healthier Wisconsin endowment of the Medical College of Wisconsin (8UL1TR000055).

Conflicts of Interest: The authors declare no conflicts of interest. The funders had no role in the design of the study; in the collection, analyses, or interpretation of data; in the writing of the manuscript, or in the decision to publish the results.

References

1. Walker, J.; Ardouin, S.; Burrows, T. The validity of dietary assessment methods to accurately measure energy intake in children and adolescents who are overweight or obese: A systematic review. *Eur. J. Clin. Nutr.* **2017**, *72*, 185–197. [CrossRef] [PubMed]

2. Burrows, T.L.; Martin, R.J.; Collins, C.E. A systematic review of the validity of dietary assessment methods in children when compared with the method of doubly labeled water. *J. Am. Diet. Assoc.* **2010**, *110*, 1501–1510. [CrossRef] [PubMed]

3. Naska, A.; Lagiou, A.; Lagiou, P. Dietary assessment methods in epidemiological research: Current state of the art and future prospects. *F1000 Res.* **2017**, *6*, 926. [CrossRef] [PubMed]

4. Ptomey, L.; Willis, E.; Goetz, J.; Lee, J.; Sullivan, D.; Donnelly, J. Digital photography improves estimates of dietary intake in adolescents with intellectual and developmental disabilities. *Disabil. Health J.* **2015**, *8*, 146–150. [CrossRef] [PubMed]

5. Bandini, L.; Danielson, M.; Esposito, L.E.; Foley, J.T.; Fox, M.H.; Frey, G.C.; Fleming, R.K.; Krahn, G.L.; Must, A.; Porretta, D.L.; et al. Obesity in children with developmental and/or physical disabilities. *Disabil. Health J.* **2015**, *8*, 309–316. [CrossRef] [PubMed]

6. Shim, J.-S.; Oh, K.; Kim, H. Dietary assessment methods in epidemiologic studies. *Epidemiol. Health* **2014**, *36*, e2014009. [CrossRef] [PubMed]

7. Boushey, C.J.; Kerr, D.A.; Wright, J.; Lutes, K.D.; Ebert, D.S.; Delp, E.J. Use of technology in children's dietary assessment. *Eur. J. Clin. Nutr.* **2009**, *63*, S50–S57. [CrossRef] [PubMed]

8. Thompson, F.; Subar, A.; Loria, C.M.; Reedy, J.; Baranowski, T. Need for technological innovation in dietary assessment. *J. Am. Diet. Assoc.* **2010**, *110*, 48–51. [CrossRef] [PubMed]

9. Boushey, C.; Spoden, M.; Delp, E.; Zhu, F.; Bosch, M.; Ahmad, Z.; Shvetsov, Y.; DeLany, J.; Kerr, D. Reported energy intake accuracy compared to doubly labeled water and usability of the mobile food record among community dwelling adults. *Nutrients* **2017**, *9*, 312. [CrossRef] [PubMed]

10. Boushey, C.J.; Harray, A.J.; Kerr, D.A.; Schap, T.E.; Paterson, S.; Aflague, T.; Bosch Ruiz, M.; Ahmad, Z.; Delp, E.J. How willing are adolescents to record their dietary intake? The mobile food record. *JMIR mHealth uHealth* **2015**, *3*, e47. [CrossRef] [PubMed]

11. Kerr, D.; Dhaliwal, S.; Pollard, C.; Norman, R.; Wright, J.; Harray, A.; Shoneye, C.; Solah, V.; Hunt, W.; Zhu, F.; et al. BMI is associated with the willingness to record diet with a mobile food.record among adults participating in dietary interventions. *Nutrients* **2017**, *9*, 244. [CrossRef] [PubMed]

12. Zhu, F.; Bosch, M.; Woo, I.; Kim, S.; Boushey, C.; Ebert, D.; Delp, E. The use of mobile devices in aiding dietary assessment and evaluation. *IEEE J. Sel. Top. Signal Process.* **2010**, *4*, 756–766. [PubMed]

13. Bathgate, K.; Sherriff, J.; Leonard, H.; Dhaliwal, S.; Delp, E.; Boushey, C.; Kerr, D. Feasibility of assessing diet with a mobile food record for adolescents and young adults with down syndrome. *Nutrients* **2017**, *9*, 273. [CrossRef] [PubMed]

14. Polfuss, M.; Sawin, K.J.; Papanek, P.E.; Bandini, L.; Forseth, B.; Moosreiner, A.; Zvara, K.; Schoeller, D.A. Total energy expenditure and body composition of children with developmental disabilities. *Disabil. Health J.* **2017**, *3*, 442–446. [CrossRef] [PubMed]

15. University of Minnesota. Nutrition Data System for Research-Nutritional Analysis Software. Available online: http://license.umn.edu/technologies/ndsr87072_nutrition-data-system-for-research-nutritional-analysis-software (accessed on 26 July 2017).

 *nutrients*

Article

A Qualitative Evaluation of the eaTracker® Mobile App

Jessica R. L. Lieffers [1] , Renata F. Valaitis [2], Tessy George [3], Mark Wilson [4], Janice Macdonald [4,5] and Rhona M. Hanning [2,*]

[1] College of Pharmacy and Nutrition, University of Saskatchewan, Saskatoon, SK S7N 2Z4, Canada; jessica.lieffers@usask.ca

[2] School of Public Health and Health Systems, University of Waterloo, Waterloo, ON N2L 3G1, Canada; rfvalait@uwaterloo.ca

[3] Hospital for Sick Children, Toronto, ON M5G 1X8, Canada; t.george@mail.utoronto.ca

[4] Dietitians of Canada, Toronto, ON M5R 1C1, Canada; mark@wilson5.ca (M.W.); janice@childhoodobesityfoundation.ca (J.M.)

[5] Childhood Obesity Foundation, Vancouver, BC V5Z 1M9, Canada

* Correspondence: rhona.hanning@uwaterloo.ca; Tel.: +1-519-888-4567 (ext. 35685)

Received: 29 August 2018; Accepted: 4 October 2018; Published: 9 October 2018

Abstract: Background: eaTracker® is Dietitians of Canada's online nutrition/activity self-monitoring tool accessible via website and mobile app. The purpose of this research was to evaluate the eaTracker® mobile app based on user perspectives. Methods: One-on-one semi-structured interviews were conducted with adult eaTracker® mobile app users who had used the app for $\geq$ 1 week within the past 90 days. Participants (n = 26; 89% female, 73% 18–50 years) were recruited via email. Interview transcripts were coded using first level coding and pattern coding, where first level codes were grouped according to common themes. Results: Participants mentioned several positive aspects of the mobile app which included: (a) Dashboard displays; (b) backed by dietitians; (c) convenience and ease of use; (d) portion size entry; (e) inclusion of food and physical activity recording; and (f) ability to access more comprehensive information via the eaTracker® website. Challenges with the mobile app included: (a) Search feature; (b) limited food database; (c) differences in mobile app versus website; and (d) inability to customize dashboard displayed information. Suggestions were provided to enhance the app. Conclusion: This evaluation provides useful information to improve the eaTracker® mobile app and also for those looking to develop apps to facilitate positive nutrition/physical activity behavior change.

Keywords: mobile applications; adults; nutritional science; qualitative research

1. Introduction

Over the past decade, mobile devices and their applications ("apps") have become an integral part of the everyday lives of many Canadians. The Canadian Radio-television and Telecommunications Commission reported that in 2016, 87%, 77%, and 54% of Canadian adults owned cellphones, smartphones, and tablets, respectively, which is up from 80%, 51%, and 26%, respectively, in 2012 [1]. Use of mobile apps for health-related purposes has also become popular amongst the general public. A recent survey study of mobile phone owners from the United States found that almost 60% of respondents had downloaded a mobile app with health-related content; the authors also found that use of apps for nutrition and fitness was common [2]. In addition, some apps for monitoring eating and activity behaviors (e.g., MyFitnessPal (Under Armour Inc., Baltimore, MD, USA)) have had millions of downloads. The availability of health apps is also expanding daily—a recent report found that about

200 new health apps are added each day [3]. Dietitians are also now commonly encountering clients interested in using nutrition apps in their practice [4,5]. However, despite the popularity of nutrition (and activity) apps, many may have been developed without visible health professional input [6].

Dietitians of Canada's eaTracker® (http://www.eatracker.ca/) is a free, publicly available bilingual (English and French) web-based tool that allows members of the public to track their eating and/or physical activity behaviors and compare them to recommendations (including those set by Health Canada). Users create an account and enter demographic information including their month and year of birth, sex, height, weight, self-reported activity level, pregnancy/breastfeeding status, postal code, province of residence, and country of residence, which allows personalized recommendations to be determined. Users then have the ability to enter eating and physical activity behaviors via a database of available choices (~4500 food items from the Canadian Nutrient File (version 2010) which contains information on average nutrient values for foods available in Canada [7], and ~159 activities). Following data entry, users are able to receive feedback on consumption of calories, Canada's Food Guide [8] servings, 22 nutrients as well as information on physical activity including minutes of activity, minutes of low effort, moderate effort, high effort, and muscle and bone strengthening exercise, and number of calories burned through exercise. eaTracker® also contains other tools such as a recipe analyzer (which allows users to enter ingredients to obtain a nutritional analysis of their recipe, and save this recipe to expedite future entry of this recipe into eaTracker®), a goal setting and tracking tool (My Goals) (described elsewhere) [9,10], and the ability for a dietitian coach to view intake and activity patterns of a group of clients, and provide comments.

In 2014, Dietitians of Canada released free iOS™ and Android™ eaTracker® mobile apps available via the Apple App Store® (Cupertino, CA, USA), and Google Play™ store (Mountain View, CA, USA). The mobile app can be used either by itself or in conjunction with the eaTracker® website. The mobile app allows users to create/access their account, to log and receive feedback on eating and physical activity behaviors, and to set and track goals using the My Goals feature. Users are able to receive feedback on intakes of energy, macronutrients, and number of servings from the four Canada's Food Guide food groups as well as activity behaviors via the mobile app. Users also have the option to visit the eaTracker® website to obtain the more comprehensive assessment of their eating and activity behaviors and use the recipe analyzer. Screenshots of the eaTracker® mobile app are shown in Figure 1.

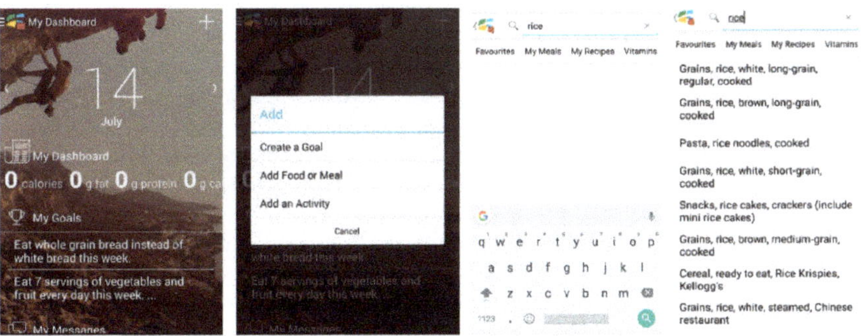

Figure 1. eaTracker® mobile app screenshots.

The purpose of this research was to conduct an evaluation of the eaTracker® mobile app using qualitative one-on-one semi-structured interviews with users. This evaluation will support future modifications to the eaTracker® mobile app and the development of other credible, user-friendly, and effective nutrition and physical activity behavior change mobile apps to optimize the nutritional status of Canadians.

2. Materials and Methods

An advisory committee (*n* = 5 Dietitians of Canada staff members; researchers from University of Waterloo) oversaw the project design, methods (including interview protocol), and analyses. The Dietitians of Canada information technology team also provided expertise and support. The University of Waterloo Office of Research Ethics provided ethics approval (ORE: #20671; approval date: April 30, 2015). The Consolidated Criteria for Reporting Qualitative Research (COREQ) checklist [11] guided study reporting.

Approximately 30 participants were desired a priori for this study; a convenience sampling strategy was used. eaTracker® users were approached about the study through an email invite sent via FluidSurveys (FluidSurveys, Ottawa, ON, Canada). Dietitians of Canada sent an email invite on two occasions (May 12, 2015; June 15, 2015) to eaTracker® users who: (a) Were ≥18 years, (b) Southern Ontario residents (based on self-reported postal code in their eaTracker® account), (c) had used eaTracker® within the past 90 days and, and (d) previously provided permission to be contacted by Dietitians of Canada. The email invite provided a link to an online survey where users were asked whether they had specifically used the eaTracker® mobile app within the past 90 days; if they responded "yes", they were invited to leave their name and contact information. A research assistant (TG) then contacted individuals via email or phone to provide more information about the study. Interested users who had used the eaTracker® mobile app for ≥1 week were invited to complete the interview.

One-on-one, semi-structured interviews were conducted by a TG, a female research assistant who had recently completed a Bachelors degree in Health Studies (with a minor in Nutrition) and was trained in qualitative research methods. The researcher who conducted the interviews was not using eaTracker® at the time of the interview, and was at an arms-length from Dietitians of Canada in order to prevent bias. Participants did not know anything about the researchers except information provided via the information letter and consent forms. Participants were told that the purpose of the study was to evaluate the eaTracker® mobile app by obtaining information on user perspectives regarding app content, service, and functionality. No relationships with participants were established prior to study commencement. Interviews were conducted in-person (in public locations e.g., coffee shops), by telephone, or Skype™ (Microsoft Corporation, Redmond, WA, USA). Interviews were conducted at a time that was convenient for both individuals, and no other individuals were present at the interview except the participant and researcher. Several interviews were conducted by telephone, as participants were located across Southern Ontario; however, if participants were located within close proximity to researchers, efforts were made to conduct in-person interviews. All participants provided written (in-person interviews) or verbal (telephone or online interviews) informed consent. The semi-structured interviews were guided by an interview protocol with open-ended questions designed to address study objectives (Supplementary S1). Both clarifying and elaborating probes were used to gather additional data [12]. The interview protocol was pilot tested with two individuals from the target population prior to data collection. Field notes were taken during interviews. Participants were provided with a free Dietitians of Canada cookbook following the interview as a thank you gift. No participants dropped out after completing the interview. Interviews were completed until data saturation was achieved; no repeat interviews were conducted. All interviews took place between May 2015 and August 2015.

All interviews were audio-recorded and transcribed verbatim. Detailed notes were taken for participants who did not consent to the audio-recording. Any identifying information in the transcripts was removed to maintain confidentiality. Transcripts were reviewed to correct any errors; transcripts were not returned to participants for comment and/or correction, and participants did not provide feedback on study findings.

Data were analyzed using content analysis [13,14]. The interviews were coded by a single trained coder (TG) using NVivo 10 for Mac (QSR International Pty Ltd., Doncaster, Australia). First level coding and pattern coding were used, where first level codes were grouped according to common themes which were derived from the data [15]. To ensure reliability of coding, a second experienced coder reviewed a subset of transcripts (~10%) and associated codes, and themes. Any disagreements were discussed until consensus was reached [16].

3. Results

In total, $n = 4135$ and $n = 1082$ eaTracker® users were sent an email message on May 12, 2015 and June 15, 2015, respectively inviting them to participate in the study. In total, $n = 129$ users completed information on the recruitment survey. Of those, $n = 67$ eaTracker® users who had used the mobile app in the past 90 days provided contact information and were contacted by TG, and in total, $n = 26$ users participated in the one-on-one semi-structured interview (average length: 36 min; range: 20–64 min); one interview was not recorded because of participant request (detailed notes were taken instead for this interview).

Table 1 shows participant demographics and interview methods. Overall, 88.5% of participants were female, and 73.1% were 18–50 years of age; just under 60% of interviews were done by phone. The distribution of age and sex for study participants generally reflects the overall population of eaTracker® users in Canada from July 5, 2015 to September 2, 2015 ($n = 2265$) (79.8% female; 75.7% 18–50 years).

Table 1. Participant demographics and interview method.

	Number (%)
Sex	
Female	23 (88.5)
Male	3 (11.5)
Age (years)	
18–30	9 (34.6)
31–50	10 (38.5)
51–70	7 (26.9)
Interview Method	
Phone	15 (57.7)
Online (e.g., Skype)	2 (7.7)
In-Person	9 (34.6)

Participants reported various nutrition and physical activity goals including: Improving specific eating habits (e.g., follow food guide, consume a balanced diet, plenty of vegetables, decrease saturated fat intake, reduce sodium intake, decrease meat intake, decrease intake of "bad foods," meet iron requirements) ($n = 21$ participants), weight management ($n = 9$), and attaining recommended activity levels ($n = 9$).

In total, $n = 18$ and $n = 11$ participants were using the iOS™ and Android™ eaTracker® mobile apps, respectively. Participants used various devices to access the app; in total, $n = 13$, $n = 9$, $n = 5$, and $n = 2$ participants accessed the mobile app via an iPhone®, Android™ phone, iPad® tablet, and BlackBerry® phone, respectively. Participants mentioned finding out about the eaTracker® mobile app through different channels; the most common ways were via the eaTracker® website, app stores, and through their dietitian or a dietitian they followed on social media. Some participants were also informed about the mobile app through school, the EatRight Ontario website (now rebranded as http://www.unlockfood.ca/), family and friends, trainers, and media.

Information on duration of eaTracker® mobile app use was obtained for $n = 23$ participants. In total, $n = 9$ participants had used the eaTracker® mobile app between one and three weeks, $n = 4$ for about two months, $n = 7$ between three and six months, and $n = 3$ for a year. Most participants described themselves as 'daily users' which could include using the app multiple times/day, using the app after every meal, or using the app at specific times (e.g., morning to enter foods eaten the day before). In addition, $n = 5$ users mentioned that they used the app on a weekly basis, and $n = 4$ used the app less often than weekly (e.g., every few months). Of the $n = 26$ participants, $n = 13$ said that they would continue to use the eaTracker® mobile app in the future, $n = 8$ said that they would not continue to use the mobile app, with the remaining being unsure if they would continue or discontinue eaTracker® mobile app use. Reasons for discontinuation varied and included, for example, preference for other apps, and the food search feature being difficult to use.

In total, $n = 10$ participants mentioned using the eaTracker® mobile app at home. Participant 011 stated, "If I'm out for a meal, I'll do it when I get home; always at home. It just takes too long to do it when I'm out . . . you have to type in the whole word." However, many participants also reported using the mobile app in any setting and even found that it could be "a discussion point." Typically, these participants used the mobile app as soon as they consumed food or beverages. A few participants also mentioned that the mobile app helped them to increase awareness of the foods they were eating throughout the day. Participant 026 mentioned that the use of the mobile app allowed her to " . . . see where I was adding the most calories, cause a lot of these calories were hidden to me. Once I put them in eaTracker I noticed where I was adding too much."

3.1. Positive Aspects of the eaTracker® Mobile App

Participants mentioned several positive aspects of the eaTracker® mobile app which included dashboard displays, backing by dietitians, convenience and ease of use, portion size entry, and inclusion of both food and activity recording components. These findings are described in detail below.

3.1.1. Dashboard Displays

The dashboard display's information about quantities of calories (kcal), macronutrients (g), Canada's Food Guide servings consumed as well as goals set and physical activity behaviors (screenshot in Figure 1). Participants felt positively about how the information was organized on the dashboard and found it visually appealing. Participant 020 mentioned, "I do like the dashboard format where it's like a summary at a glance." Participant 006 mentioned, "I like that (the dashboard is) not too overly physical, like there's not too many graphics or pictures and all that, things that typically take longer to load too that I don't need." Participants also liked other aspects of the dashboard including the ability to change the background picture and to use a swiping motion to view information on behaviors logged for previous days.

3.1.2. App Backed by Dietitians

Several participants liked that the mobile app was developed by a reputable organization (Dietitians of Canada), backed by dietitians, and contained Canadian content. Some participants reported trusting the validity and accuracy of the information presented in the mobile app. For example, Participant 006 mentioned: " . . . the Canadian focus of it too-right? There's a lot of US based stuff; I wanted something that was in Canadian metrics and Canadian context." The fact that Dietitians of Canada developed the mobile app was a motivating factor to continue use for some participants such as Participant 016: "I don't know I guess I just haven't given up on it yet. Because it's gotta be worth it if the Dietitians of Canada suggest it."

3.1.3. Convenience and Ease of Use

Most participants found the mobile app was convenient, easy to use, and an easy way to record both their eating and activity behaviors. Participants also described the mobile app as convenient because having it on their mobile devices allowed them to record their eating and physical activity behaviors during leisure time or soon after they had a meal. The perspective that mobile apps are easier to access versus computers was also described by several participants. Participant 018 explained, "it's a lot easier to pick up a phone like off a counter than if you go onto a computer and login and everything." In addition, participants that used the website and the mobile app together enjoyed the flexibility of being able to access their eaTracker® account in different ways and felt that the two methods to access the tool complimented one another. Participant 001 explained, "I like how they're integrated, (when I) input something on the mobile app, I don't have to go back to the website to change it or make sure it's correct and put something on the website."

3.1.4. Portion Size Entry

The eaTracker® mobile app provides users with several options for entering portion size information for the foods they consume (e.g., volume, weight, and count-based units). Participants felt positively about having access to several units to enter food portion sizes, and felt this was unique to this mobile app. Participant 008 stated, "I like all the quantities that they provide, that's really helpful." This participant also went on to compare the number of measurements provided in the eaTracker® mobile app vs. other non-Canadian commercial apps. They explained, "(name of other commercial app) do(es) have that as well but sometimes they don't have (the units) that you want and I found that (the eaTracker® mobile app) was actually better."

3.1.5. Includes Both Food and Physical Activity Recording

A couple of participants liked that the mobile app included both food and activity recording components. They felt positively about the fact that the activity and food entry were separate distinct components with similar layouts all housed within the same mobile app. Participant 022 stated, "I like that you can enter your physical activity as well as your, your food in there because you know they both go hand in hand when you're worrying about your health." Even participants like Participant 001 who had never reported using the physical activity feature appreciated that this functionality was available. They stated, "(what) I like about the app is that you can add how much physical activity you've done, I've never used that in the app, but you could, so I like that cause with a lot of the other apps it's strictly a food app or strictly health."

3.1.6. Ability to Access More Comprehensive Information via the eaTracker® Website

A few participants also liked that they could visit the eaTracker® website to retrieve a "more detailed description" of their food intake through nutritional reports that were not available via the mobile app. Participant 005 explained that "(the website) gave more information in terms of what vitamins, the nutritional concepts. (Other mobile apps are) strictly limited to calories, fats, sodium and a hand full of nutrients but the eaTracker app was more accurate in terms of stuff you wouldn't regularly think about."

3.2. Challenges with the eaTracker® Mobile App

Participants mentioned some challenges with the eaTracker® mobile app which included difficulties with the search feature, limited food database, differences between the mobile app and the website, and inability to customize nutrition variables displayed on the mobile app dashboard; these four findings will be discussed in more detail. Participants also mentioned other challenges, which included having lunch as the default meal, inability to use the mobile app without Wi-Fi or cellular data, and finding the mobile app background distracting.

3.2.1. Search Feature

Over half of participants described various difficulties with the eaTracker® food search feature which made the mobile app tedious to use. There were also a few concerns mentioned with the activity search feature, although these comments were less common. One reported difficulty was that the search did not bring up relevant items. Participant 025 explained the challenge: "After I started using it, I found it really almost too detailed . . . like when I was looking for like a thing I have eaten, I got 50 options that came up. But I can't scroll through and I wasn't able to find a way to filter them in order to find what I was looking for-something that was close enough." Another challenge was that the search provided unrelated items. Participants who had used other mobile apps for food data entry also discussed the benefit of entering foods using a barcode scanner; participants described this entry as quick as it did not require extra time to search for individual foods. Another participant provided a suggestion that the mobile app bring up common foods eaten at the same time each day when the app is opened (instead of having to search for foods) to simplify data entry.

3.2.2. Limited Food Database

Another concern was the limited food database. Almost 50% of participants had concerns about missing food items from the database, some of which they believed to be common, such as Greek yogurt. Participant 006 mentioned, "it's limited, they don't have Greek yogurt for instance." Participant 020 also mentioned: "I didn't feel like the database of food and nutritional information had represented what I had actually eaten." Participants often compared the eaTracker® mobile app food database to other commercial mobile apps which have much larger food databases, which, according to participants, contained everything from restaurant foods to packaged foods, and trendy foods (e.g., gluten free foods, Goji berries).

Several participants suggested efforts should be made to improve the food database. Participant 021 mentioned: "I would say that it would be good to work on the database of the food. That would be my biggest comment-just to have more foods available to input would be perfect." Participants suggested different ways to improve the food database. One suggestion was to add more restaurant foods. Participant 003 mentioned: "having restaurants-that would be handy, including Canadian restaurants, because nowadays a lot of people are eating out at Tim Hortons or Starbucks or whatever … " Another suggestion was allowing users to update the database on their own. Participant 025 suggested, "I wonder if there might be an opportunity instead of looking for an option in the list that would represent what you just had-if you could take a picture or do something with the nutritional label which is just standard and input it saying this is what I had, right?" Another suggestion was having a community-based database where users could add in nutrition information from the products they typically consume which would allow for a greater number of options to be available for users.

3.2.3. Mobile App Differs from the Website

While most participants felt the eaTracker® mobile app was simple to use, some participants who used the eaTracker® website previously did not enjoy the layout and interface of the mobile app because it did not resemble the website. Participant 012 mentioned: "I didn't really understand (the eaTracker® mobile app). It was just the interface; I wasn't used to it cause it did look different from the website … so yeah, but I definitely didn't start using it immediately." In addition, some participants mentioned concerns that some feedback information available on the website was not available via the mobile app.

Over half of participants wanted more eaTracker® website features available via the mobile app. Several participants wanted to be able to access progress graphs available on the eaTracker® website via the mobile app. Participant 012 mentioned: "Well, personally, I would definitely like to see graphs of everything, so everything that I can see on the website, I should be able to see on eaTracker mobile app in terms of that analysis portion-so both the like micronutrients, macros as well as the like food groups." Others discussed wanting to be able to access the recipe analyzer via the mobile app. Participant 022, suggested, "I would like to see the recipe analyzer feature added to the app, maybe that's way too complicated and there's a reason why they haven't put it on there but I would certainly use it."

3.2.4. Inability to Customize Dashboard Displayed Information

As mentioned previously, participants liked the concept of the dashboard. However, a few participants mentioned a limitation of the eaTracker® mobile app was the inability to customize the specific nutritional variables displayed. Participant 022 mentioned, "I would like to see them add sugar to that-like grams of sugar, to that little part that goes across it (i.e., dashboard) . . . I think a lot of people are watching their sugar these days and it would be handy to know how much sugar is in the foods that you're eating." Participants had varied preferences on the variables that they wanted to be included or excluded depending on their dietary pattern (e.g., vegetarian) or disease state (e.g., diabetes) and wanted the app to be able to accommodate those preferences.

Participants also suggested making the mobile app more interactive by providing customized recommendations based on user-entered data (e.g., goals, nutrient intake, common foods). Participant 008 stated, "if I could just click on a button, say on the day, when I'm low on something I hit a button and it tells you to 'try eating this' and choose a snack that had more of that in it."

4. Discussion

eaTracker® is one of only a few Canadian mobile apps to support nutrition (and activity) behavior change. The current evaluation highlights the enthusiasm of users and provides rich feedback to enhance this mobile app as well as other electronic health tools to optimize nutrition (and activity) behaviors to prevent and manage chronic diseases such as cardiovascular disease, diabetes, and cancer.

One key finding from this study was that participants liked that the mobile app was developed by a reputable organization and for some participants, this was a motivating factor to continue use. Concerns about mobile app credibility and accuracy were also mentioned in a related study by Dennison et al. [17]. This finding is important as many health, nutrition, and weight management apps are not developed with input from health care professionals and professional organizations; Nikolaou and Lean [6] recently found that <1% of weight management apps were developed with visible professional input. Professionals and professional organizations should consider becoming involved in app development and making this involvement clearly visible to users. In addition, implementation of easy ways for consumers to identify nutrition apps developed by dietitians (and other reputable professionals or organizations) may be helpful. One strategy may be to have a list of apps and a badge for those developed with this type of input, which is similar to what Dietitians of Canada does to identify blogs written by dietitians (Dietitians of Canada Member Blogs) [18].

Similar to other qualitative data on user experiences with nutrition mobile apps [17,19–23], this study found that food data entry is a key topic that affects satisfaction with these types of tools. Previous qualitative studies with related mobile apps have also identified that users have a strong desire to record the foods eaten as precisely as possible [17,23], which may be a reason for the high frequency of this type of comment among participants in the current study. Participants in this study liked that multiple units were available for food data entry with the eaTracker® mobile app (which is a limitation of many other similar mobile apps) [19]; however, they encountered challenges with the search feature, as well as frustrations with the small food database. Generally, other studies have found that large databases are convenient and well-liked because of the large variety of foods available which allows an exact item to be found [19,21]. However, difficulties finding correct foods in large databases have also been reported [19,20]. Difficulties with both large and small food databases suggests that issues with food databases are present regardless of database size. While the eaTracker® food database is smaller and does not offer the ability to self-enter foods compared to databases used in other publicly available apps, it has the potential to provide feedback on a larger selection of dietary variables because the Canadian Nutrient File database is used rather than relying on information only available on food labels. Barcode scanners to streamline data entry have been previously reported to be well-liked [19,23]; however, they are only useful for entry of packaged foods and errors with these tools have been reported [19]. Future modifications to eaTracker® and other mobile apps will need to weigh the pros and cons of different options for food entry. Additionally, when the method(s) are chosen, it is important that users are educated on the rationale and pros and cons of the chosen food entry option and strategies to ensure success with data entry for the chosen option. For example, if smaller databases are chosen to be used, users should be provided with strategies on how to find the correct foods (e.g., education on searching for the type of food instead of the brand name, e.g., searching for hazelnut spread instead of Nutella®) to help promote success and satisfaction with the tool. In addition, regardless of database size, implementing strategies to streamline data entry (e.g., via favorites, commonly entered items, recently entered items, optimizing the search feature) would be a worthwhile endeavor.

A notable finding from this study was that participants who had been previous users of the eaTracker® website initially found the eaTracker® mobile app difficult to use. To our knowledge, this is the first time that this qualitative finding has been reported in users of a mobile nutrition (and activity) self-monitoring tool. This finding suggests that users may find transitioning to a different format to be challenging and that it may be necessary to implement strategies to help improve their success.

Participants mentioned that one of their most liked features of eaTracker® is the in-depth feedback provided on several dietary variables; this feedback is provided in different ways including charts and graphs. Some participants wanted mobile app access to include more of this personalized information currently available only on the eaTracker® website. This finding aligns with previous studies in this area which have found that having access to numbers and graphs about progress in a nutrition and/or physical activity behavior change mobile app is generally well-liked and can be motivational [21,23–27]. In addition to having the ability to personalize which eaTracker® feedback information is displayed on the mobile app, personalization of other aspects of the app (e.g., automated recall of favorite foods) (as has also been found in other related studies [28]) is also desired. This should be considered in future eaTracker® mobile app modifications and development of future apps.

Strengths and Limitations

This study has several strengths. Sampling occurred until data saturation was reached. In addition, a variety of participant types were chosen without exclusively focusing on individuals using the app for weight management. In addition, we captured information on real-world experiences as opposed to experiences of use as part of a research trial.

While this study has several strengths, there are some limitations that should be mentioned. It should be noted that only a small subset of eaTracker® users responded to the email invitation which is common for these types of invitations. In addition, participants who completed the interview may be more motivated and willing to contribute feedback compared to the general user group. Participants were also primarily female and 18–50 years of age; however, this distribution generally reflects the overall population of eaTracker® users. In addition, information on education level, income, and ethnicity was not collected from participants.

5. Conclusions

This evaluation of the eaTracker® mobile app provides important insight on real-world user experiences with mobile apps for nutrition (and activity) behavior change. Users liked that the app was developed by a reputable organization, and had multiple ways to enter food data. Professionals and developers should keep in mind that users may have difficulty transitioning between a website and mobile app, and that finding ways to streamline data entry should be a priority. In addition, allowing users to personalize mobile apps would likely help to increase satisfaction. Ultimately, higher user satisfaction may result in improved app adherence which may help to improve nutrition (and activity) behaviors to decrease the burden of chronic disease.

Supplementary Materials: The following are available online at http://www.mdpi.com/2072-6643/10/10/1462/s1, S1: Interview Protocol for eaTracker® Mobile App Users.

Author Contributions: Conceptualization, J.R.L.L., J.M., M.W., R.M.H.; Methodology, J.R.L.L., R.F.V., T.G., R.M.H.; Formal Analysis, J.R.L.L., T.G., R.F.V., R.M.H.; Supervision, R.M.H.; Writing–Original Draft Preparation, J.R.L.L.; Writing–Review & Editing, all authors.

Funding: This research was funded by Dietitians of Canada/Public Health Agency of Canada. JL was funded by a Cancer Care Ontario/Canadian Institutes for Health Research Training Grant in Population Intervention for Chronic Disease Prevention: A Pan-Canadian Program (Grant #53893) (2014-15).

Acknowledgments: The authors would like to thank study participants, Dietitians of Canada/Public Health Agency of Canada for funding, Dietitians of Canada for donating cookbooks for interview participants, members of the advisory committee, Dietitians of Canada Information Technology team, and Corin Schneider for his assistance with transcription.

Conflicts of Interest: The authors declare no conflict of interest.

References

1. Canadian Radio-television and Telecommunications Commission (CRTC)—Communications Monitoring Report 2017. Available online: https://crtc.gc.ca/eng/publications/reports/PolicyMonitoring/2017/cmr2017.pdf (accessed on 14 June 2018).
2. Krebs, P.; Duncan, D.T. Health app use among US mobile phone owners: A national survey. *JMIR Mhealth Uhealth* **2015**, *3*. [CrossRef]
3. IQVIA Institute for Human Data Science–The Growing Value of Digital Health. Available online: https://www.iqvia.com/institute/reports/the-growing-value-of-digital-health (accessed on 14 June 2018).
4. Lieffers, J.R.L.; Vance, V.A.; Hanning, R.M. Use of mobile device applications in Canadian dietetic practice. *Can. J. Diet. Pract. Res.* **2014**, *75*, 41–47. [CrossRef] [PubMed]
5. Chen, J.; Lieffers, J.; Bauman, A.; Hanning, R.; Allman-Farinelli, M. The use of smartphone health apps and other mobile health (mHealth) technologies in dietetic practice: A three country study. *J. Hum. Nutr. Diet* **2017**, *30*, 439–452. [CrossRef] [PubMed]
6. Nikolaou, C.K.; Lean, M.E. Mobile applications for obesity and weight management: Current market characteristics. *Int. J. Obes.* **2017**, *41*, 200–202. [CrossRef] [PubMed]
7. Health Canada—Answers to Frequently Asked Questions. Available online: https://www.canada.ca/en/health-canada/services/food-nutrition/healthy-eating/nutrient-data/frequently-asked-questions-canadian-nutrient-file.html (accessed on 14 June 2018).
8. Health Canada. *Eating Well with Canada's Food Guide*; Health Canada: Ottawa, ON, Canada, 2007.
9. Lieffers, J.R.L.; Haresign, H.; Mehling, C.; Hanning, R.M. A retrospective analysis of real-world use of the eaTracker® My Goals website by adults from Ontario and Alberta, Canada. *BMC Public Health* **2016**, *16*. [CrossRef] [PubMed]
10. Lieffers, J.R.L.; Haresign, H.; Mehling, C.; Arocha, J.F.; Hanning, R.M. The website-based eaTracker® 'My Goals' feature: A qualitative evaluation. *Public Health Nutr.* **2017**, *20*, 859–869. [CrossRef] [PubMed]
11. Tong, A.; Sainsbury, P.; Craig, J. Consolidated criteria for reporting qualitative research (COREQ): A 32-item checklist for interviews and focus groups. *Int. J. Qual. Health Care* **2007**, *19*, 349–357. [CrossRef] [PubMed]
12. Creswell, J. *Educational Research: Planning, Conducting, Evaluating Quantitative and Qualitative Research*, 3rd ed.; Pearson Education: Upper Saddle River, NJ, USA, 2008.
13. Hsieh, H.F.; Shannon, S.E. Three approaches to qualitative content analysis. *Qual. Health Res.* **2005**, *15*, 1277–1288. [CrossRef] [PubMed]
14. Patton, M.Q. *Qualitative Research & Evaluation Methods*, 3rd ed.; Sage Publications: Thousand Oaks, CA, USA, 2002.
15. Miles, M.B.; Huberman, A.M.; Saldana, J. *Qualitative Data Analysis: A Methods Sourcebook*; Sage Publications: Thousand Oaks, CA, USA, 2013.
16. Creswell, J.W. *Qualitative Inquiry and Research Design: Choosing Among Five Approaches*, 2nd ed.; Sage Publications: Thousand Oaks, CA, USA, 2007.
17. Dennison, L.; Morrison, L.; Conway, G.; Yardley, L. Opportunities and challenges for smartphone applications in supporting health behavior change: Qualitative study. *J. Med. Internet Res.* **2013**, *15*. [CrossRef] [PubMed]
18. Dietitians of Canada. Member Blogs. Available online: https://www.dietitians.ca/Media/Member-Blogs.aspx (accessed on 13 June 2018).
19. Lieffers, J.R.L.; Arocha, J.F.; Grindrod, K.; Hanning, R.M. Experiences and perceptions of adults accessing publicly available nutrition behavior-change mobile apps for weight management. *J. Acad. Nutr. Diet* **2018**, *118*, 229–239. [CrossRef] [PubMed]

20. Lister, C.; West, J.; Richards, R.; Crookston, B.; Cougar Hall, P.; Redelfs, A.H. Technology for health: A qualitative study on barriers to using the iPad for diet change. *Health* **2013**, *5*, 761–768. [CrossRef]

21. Shigaki, C.L.; Koopman, R.J.; Kabel, A.; Canfield, S. Successful weight loss: How information technology is used to lose. *Telemed. E-Health* **2014**, *20*, 144–151. [CrossRef] [PubMed]

22. Gowin, M.; Cheney, M.; Gwin, S.; Franklin Wann, T. Health and fitness app use in college students: A qualitative study. *Am. J. Heal. Educ.* **2015**, *46*, 223–230. [CrossRef]

23. Tang, J.; Abraham, C.; Stamp, E.; Greaves, C. How can weight-loss app designers' best engage and support users? A qualitative investigation. *Br. J. Health Psychol.* **2015**, *20*, 151–171. [CrossRef] [PubMed]

24. Casey, M.; Hayes, P.S.; Glynn, F.; ÓLaighin, G.; Heaney, D.; Murphy, A.W.; Glynn, L.G. Patients' experiences of using a smartphone application to increase physical activity: The SMART MOVE qualitative study in primary care. *Br. J. Gen. Pract.* **2014**, *64*, e500–e508. [CrossRef] [PubMed]

25. Fukuoka, Y.; Lindgren, T.; Jong, S. Qualitative exploration of the acceptability of a mobile phone and pedometer-based physical activity program in a diverse sample of sedentary women. *Public Health Nurs.* **2012**, *29*, 232–240. [CrossRef] [PubMed]

26. Fritz, T.; Huang, E.M.; Murphy, G.C.; Zimmermann, T. Persuasive technology in the real world: A study of long-term use of activity sensing devices for fitness. In Proceedings of the SIGCHI Conference on Human Factors in Computing Systems, Toronto, ON, Canada, 26 April–1 May 2014; ACM: New York, NY, USA, 2014; pp. 487–496.

27. Frie, K.; Hartmann-Boyce, J.; Jebb, S.; Albury, C.; Nourse, R.; Aveyard, P. Insights from Google Play Store user reviews for the development of weight loss apps: Mixed-method analysis. *JMIR MHealth UHealth* **2017**, *5*. [CrossRef] [PubMed]

28. Peng, W.; Kanthawala, S.; Yuan, S.; Hussain, S.A. A qualitative study of user perceptions of mobile health apps. *BMC Public Health* **2016**, *16*. [CrossRef] [PubMed]

Article

Comparing Interviewer-Administered and Web-Based Food Frequency Questionnaires to Predict Energy Requirements in Adults

Didier Brassard [1,2], Simone Lemieux [1,2], Amélie Charest [1,2], Annie Lapointe [1,2], Patrick Couture [1], Marie-Ève Labonté [1,2] and Benoît Lamarche [1,2,*]

[1]   Institute of Nutrition and Functional Foods (INAF), Laval University, Quebec City, QC G1V 0A6, Canada; didier.brassard.1@ulaval.ca (D.B.); simone.lemieux@fsaa.ulaval.ca (S.L.); Amelie.Charest@fsaa.ulaval.ca (A.C.); Annie.Lapointe@fsaa.ulaval.ca (A.L.); patrick.couture@crchudequebec.ulaval.ca (P.C.); Marie-Eve.Labonte@fsaa.ulaval.ca (M.-È.L.)
[2]   School of Nutrition, Laval University, Quebec City, QC G1V 0A6, Canada
*   Correspondence: Benoit.Lamarche@fsaa.ulaval.ca; Tel.: +1-418-656-2131 (ext. 4355)

Received: 6 August 2018; Accepted: 10 September 2018; Published: 12 September 2018

Abstract: Traditional food frequency questionnaires (FFQs) are influenced by systematic error, but web-based FFQ (WEB-FFQs) may mitigate this source of error. The objective of this study was to compare the accuracy of interview-based and web-based FFQs to assess energy requirements (mERs). The mER was measured in a series of controlled feeding trials in which participants daily received all foods and caloric drinks to maintain stable body weight over 4 to 6 weeks. FFQs assessing dietary intakes and hence mean energy intake were either interviewer-administered by a registered dietitian (IA-FFQ, $n = 127$; control method) or self-administered using a web-based platform (WEB-FFQ, $n = 200$; test method), on a single occasion. Comparison between self-reported energy intake and mER revealed significant under-reporting with the IA-FFQ (-9.5%; 95% CI, -12.7 to -6.1) and with the WEB-FFQ (-11.0%; 95% CI, -15.4 to -6.4), but to a similar extent between FFQs ($p = 0.62$). However, a greater proportion of individuals were considered as accurate reporters of energy intake using the IA-FFQ compared with the WEB-FFQ (67.7% vs. 48.0%, respectively), while the prevalence of over-reporting was lower with the IA-FFQ than with the WEB-FFQ (6.3% vs. 17.5%, respectively). These results suggest less accurate prediction of true energy intake by a self-administered WEB-FFQ than with an IA-FFQ.

Keywords: food frequency questionnaire; dietary assessment; web; under-reporting; over-reporting; energy intake

---

1. Introduction

Dietary assessment is central to nutritional epidemiology, which forms the basis of dietary guidelines [1,2]. Twenty-four-hour recalls (24HRs) and food frequency questionnaires (FFQs) are common instruments to collect self-reported dietary intakes [3]. However, the validity of self-reported data obtained via such memory-based dietary assessment methods, and hence the whole value of nutrition epidemiology, is being challenged based on their purported inability to correctly reflect true food and nutrient consumption [4–7]. However, others have argued that despite recognized limitations, relying on self-reported dietary intake data in epidemiological studies has been instrumental in developing impactful dietary guidelines and recommendations over the years [1,2,7–10]. One of the fundamental issues in this heated debate relates to whether 24HRs and FFQs can measure true energy intake, due among other factors to significant random and systematic errors [1,11–14].

New methods of dietary assessment using recent technologies are being developed and examined [15–17] and there is growing interest in the ability of web-based alternatives to improve the efficiency of data collection. Web-based tools increase the efficiency of the data processing; they can be completed at any time or location, they offer unique advantage regarding portion size presentation and food recognition, and they are cost-effective [16,18]. However, the extent to which web-based delivery methods may mitigate some of the errors seen with more traditional interview-administered (IA) methods such as FFQs remains uncertain. Previous data have suggested that web-based 24HRs may be less prone to social desirability bias compared with IA tools [18–20]. The use of digital pictures in a web-based 24HR has also been proposed to facilitate portion size estimation compared with an IA-24HR [21]. A recent review of Canadian epidemiological studies reported that web-based dietary assessment instruments have not yet been used [22]. Thus, the value of web tools needs to be examined carefully for robustness, validity and reproducibility before their use can be expanded in large epidemiological studies.

To the best of our knowledge, no study has yet compared the accuracy of an IA-FFQ and self-administered web-based FFQ (WEB-FFQ) to predict an objective measure of energy requirements (mERs). The primary objective of this study was therefore to compare the accuracy of an IA-FFQ and a WEB-FFQ to assess the mER. Our hypothesis was that the WEB-FFQ is more accurate in assessing the mER than the IA-FFQ.

2. Materials and Methods

2.1. Study Design and Population

As a secondary analysis, subjects included in this study were participants from a series of nine randomized and fully controlled feeding trials (six published to date) conducted at the Institute of Nutrition and Functional Foods in Quebec City and at the Richardson Centre For Functional Foods and Nutraceuticals in Winnipeg from 2008 to 2017. All trials were devised to test the impact of different diets and nutrients on cardiometabolic risk factors [23–28]. Briefly, participants in these trials were between 18 and 65 years of age, were non-smokers, and had no history of cardiovascular disease, type 1 or type 2 diabetes, monogenic dyslipidemia, or uncontrolled endocrine disorder. Participants had to have maintained a stable body weight (within 2.5 kg) for at least 3 months before the onset of the interventions. All trials considered in the present study were conducted in weight-stable participants. All participants gave their informed consent for inclusion before they participated in the trials included in the present study, which were approved by local ethic boards.

2.2. Anthropometric Assessment

Body weight, waist and hip circumference were measured according to standardized procedures after a 12-hour overnight fast before and after each intervention period [29]. In addition, body weight was measured continuously throughout all feeding phases, three to five times per week [23–28].

2.3. Reported Energy Intake (rEI)

The IA- and WEB-FFQs were previously validated for use in French-speaking adults and details have been published elsewhere [15,30]. Briefly, the IA-FFQ is a face-to-face interviewer-administered FFQ designed to reflect dietary intakes of the past 30 days. The questionnaire is based on typical food items available in the province of Quebec with a special focus on components of the Mediterranean diet in a North-American context, which was required for the trials conducted at the time. The IA-FFQ has 91 items and food models were used in the interviews to facilitate portion size estimation. Administration of the IA-FFQ by a registered dietitian took approximately 30–45 min using standardized language across all participants.

The WEB-FFQ is a self-administered web-based questionnaire also designed to reflect dietary intakes over the past 30 days. Participants completed the WEB-FFQ on-site or at home, using Internet.

The questionnaire has 136 questions which were based on the Willet FFQ and the previously validated IA-FFQ [30]. Several serving sizes based on the *Supplementation en Vitamines et Mineraux Antioxydants* (SU.VI.MAX) Food atlas [31] were digitally photographed using standardized dinnerware. Participants completed either the IA-FFQ or the WEB-FFQ, once during the run-in period (i.e., 0 to 4 weeks) preceding the first phase of each controlled feeding trial.

2.4. Measured Energy Requirement (mER)

Energy expenditure for each participant was first estimated with validated equations [32] and from the results of the IA- or WEB-FFQ prior to undertaking the intervention phases of the trials. During all phases, participants were asked to come to the laboratory of participating centers at least three times a week in order to pick up meals and snacks and for body weight measurement. Participants were instructed to consume all and only the foods and caloric drinks provided. Dietetic technicians prepared all meals and snacks in the metabolic kitchen of participating centers to the nearest 0.1 g. Participants received all foods and caloric drinks on a daily basis under isoenergetic conditions to maintain body weight constant over feeding phases of 4 to 6 weeks. Food provision was adjusted when body weight fluctuated by more than 2 kg over one week or with any major change in reported hunger or fullness. Participants were instructed to maintain their usual physical activity habits.

mER is considered as a valid estimate of true energy expenditure because energy intake during the feeding trials was adjusted constantly to achieve body weight stability [23–28]. Furthermore, controlled feeding studies conducted at the Institute of Nutrition and Functional Foods have been previously used to assess the validity of another web-based instrument [33]. Only the first phase of each trial was considered in the present study due to temporal proximity with the completion of either FFQ. The mER was the mean daily total energy provided to each participant during the fourth week of all feeding phases. Compliance with the dietary intervention was assessed using various approaches. Self-reported compliance assessed using checklists was high across all interventions (>98%) with a large proportion of the prescribed diets (between 30–40%) consumed on-site under direct supervision of the research staff [23–28]. Subjects included in the analyses were also in weight stable conditions throughout the various isoenergetic protocols. Changes in main cardiometabolic outcomes (mostly plasma lipids) in the trials were consistent with expected changes from other studies in the literature [34–36]. Finally, changes in plasma fatty acid profiles were also consistent with the dietary intervention [26]. Post- vs. pre-intervention differences in body weight were examined to further confirm body weight stability and hence isoenergetic feeding conditions. Based on the post- vs. pre-intervention body weight difference of all participants, an arbitrary cut-off of ±1.5 SD (0 ± 1.85 kg) change in body weight was chosen to exclude subjects with a large body weight variation after the intervention. A change within ±1.85 kg most likely reflects normal day-to-day variation in body weight, of which most is due to body water fluctuation [37].

2.5. Statistical Analyses

The statistical software package SAS® Studio (v3.6, Cary, NC, USA) was used for all analyses. Extreme values of rEI were excluded on the basis of the Outlier Labeling Rule [38]. Outliers are individual values above $Q3 + 2 \times (Q3 - Q1)$ or below $Q1 - 2 \times (Q3 - Q1)$ where Q1 and Q3 represent the 25th and 75th percentiles of the rEI distribution, respectively. Baseline characteristics of the participants were compared using two-sided Student *t* tests and chi-square tests, where appropriate.

Mean rEI and mER were compared using MIXED models with self-report flag (indicator variable for rEI or mER), age, sex, body mass index (BMI), ethnicity, trial and post vs. pre-intervention body weight difference as fixed effects, and subject as a random effect. Potential statistical differences between the IA- and WEB-FFQ were assessed with addition of the interaction term FFQ method (IA- or WEB-FFQ) × self-report flag to the MIXED models. Spearman correlations (r_s) were used to examine the association between rEI and mER with adjustment for age, sex, BMI, ethnicity, trial and post vs. pre-intervention body weight.

Participants were also classified as under-reporters, accurate reporters, or over-reporters on the basis of their ratio of rEI to mER (i.e., a ratio of 1.00 would indicate exact correspondence between both measures). Confidence limits (CL) were calculated around the rEI:mER ratio based on the coefficient of variation (CV) for rEI and mER to account for measurement errors and normal variation in energy expenditure:

$$95\% \; CL = \pm 2 \times \sqrt{(\frac{CV_{rEI}^{2}}{d} + CV_{mER}^{2})}. \tag{1}$$

The CV_{rEI} (29.3%) is the within-individual CV in rEI obtained from the WEB-FFQ [15]. Repeated measurement data for the IA-FFQ were unavailable and the same CV_{rEI} was used for both FFQs. The CV_{rEI} was subsequently divided by the number of days (*d*) recalled by the FFQs (i.e., 30 days). The CV_{mER} is obtained from regression equations of doubly labelled water studies and corresponds to measurement error and variation in energy expenditure (i.e., 9.1%) over a time span of 8 weeks [39]. This specific time span was chosen to account for the length of both the dietary intervention and the run-in period of all trials in the present study, as rEI and mER were not measured concurrently. A multiplicative factor of 2 was applied to the combined CV to obtain 95% confidence limits. Thus, individuals were classified as under-reporters or over-reporters if their rEI:mER ratio was below 0.79 or above 1.21, respectively.

Log-binomial regression models were used to assess the association between BMI and sex and the likelihood of under-reporting. Covariates included in the adjusted models, where appropriate, were sex, BMI, ethnicity, trial and post vs. pre-intervention body weight difference. A two-sided alpha level of less than 0.05 was used to assess statistical significance.

3. Results

3.1. Participants

Data from a total of 448 men and women were considered for this study. Twenty-four were excluded because they did not complete the first phase of the feeding trials, one participant was excluded because pre-intervention body weight was missing, 12 participants were considered outliers on the basis of their rEI (*n* = 5 for the WEB-FFQ and *n* = 7 for the IA-FFQ) and 84 participants were excluded because of a post- vs. pre-intervention body weight difference greater than ± 1.85 kg (*n* = 54 for the WEB-FFQ and *n* = 30 for the IA-FFQ; Figure 1).

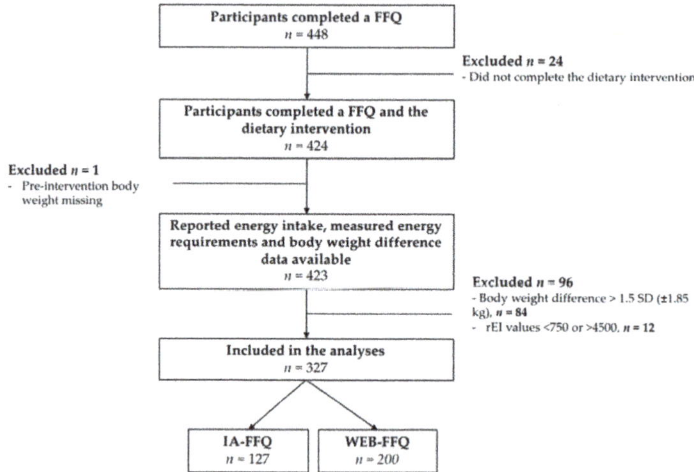

Figure 1. Flow chart of participants. FFQ: food frequency questionnaire; IA: interviewer-administered; rEI: reported energy intake.

Characteristics of the participants included in the analyses are presented for the IA-FFQ ($n = 127$) and WEB-FFQ ($n = 200$) in Table 1. Participants in the IA-FFQ group were slightly younger, had a lower body weight, waist circumference, and BMI, and included more women than participants in the WEB-FFQ group (all p values ≤ 0.02). The median (interquartile range) time for completion of the WEB-FFQ was 42.9 (34.0–59.3) min. Mean post- vs. pre-intervention body weight difference was -0.4 kg (95%CI, -0.6 to -0.3) in men and -0.6 kg (95%CI, -0.8 to -0.5) in women (both p values < 0.0001), which is within expected range (Table S1).

Table 1. Characteristics of the 327 men and women included in the analyses [1].

	IA-FFQ	WEB-FFQ	p [2]
	$N = 127$	$N = 200$	
Sex, n (%)			0.01
Men	51 (40.2)	108 (54.0)	
Women	76 (59.8)	92 (46.0)	
Ethnicity, n (%)			0.01
Caucasian	121 (95.3)	172 (86.0)	
Other	6 (4.7)	28 (14.0)	
Age [3], mean (SD) years	40.9 (16.8)	44.5 (15.4)	0.04
19–34, n (%)	58 (45.7)	67 (33.5)	
35–49, n (%)	14 (11.0)	44 (22.0)	
50–70, n (%)	55 (43.3)	89 (44.5)	
Time to completion, minutes	-	42.9 (34.0 to 59.3)	
Body weight, mean (SD) kg	72.8 (16.6)	84.6 (15.8)	<0.0001
Body mass index, mean (SD) kg/m^2	25.8 (5.4)	29.7 (4.4)	<0.0001
Normal, n (%)	69 (54.3)	27 (13.5)	
Overweight, n (%)	31 (24.4)	81 (40.5)	
Obese, n (%)	27 (21.3)	92 (46.0)	
Waist circumference, mean (SD) cm	88.4 (14.8)	100.6 (11.8)	<0.0001

[1] Values are means (SD) for continuous variables except for time to completion which is median (interquartile range). FFQ: food frequency questionnaire; IA: interviewer-administered. [2] p values indicate differences between the IA-FFQ and the WEB-FFQ, determined by Student's t test or Chi-squared test. [3] Analyses were performed on log-transformed data.

3.2. Reported Energy Intake Compared with Measured Energy Requirements

Mean differences between rEI and mER and rEI:mER ratios are presented by FFQ method and subgroups in Table 2. Results were similar either expressed as the absolute (in kcal) or relative (in %) difference between rEI and mER for both FFQs in all subgroups. Among all participants, the rEI derived from the IA-FFQ was significantly lower than the mER, by -229 kcal (95% CI, -324 to -133; $p < 0.0001$). The rEI derived from the WEB-FFQ was also significantly lower than mER (-166 kcal; 95% CI, -292 to -39; $p < 0.0001$). The mean differences between rEI and mER were similar between FFQs ($p = 0.62$). The IA-FFQ underestimated mean mER in men and women, as well as in non-obese and obese participants. The WEB-FFQ underestimated mean mER only in men and in obese individuals. Analyses stratified by sex and body weight classification revealed similar rEI to mER differences between the IA- and the WEB-FFQ (all p values > 0.30).

Spearman correlations between rEI and mER are presented in Table 2. Among all participants, the correlation was stronger with the IA-FFQ ($r_s = 0.50$; $p < 0.0001$) than with the WEB-FFQ ($r_s = 0.34$, $p < 0.0001$). In men, the correlation between rEI and mER was significant with the WEB-FFQ ($r_s = 0.40$; $p = 0.0001$), but not the IA-FFQ ($r_s = 0.23$; $p = 0.12$). Inversely, in women, the correlation between the rEI and mER was significant with the IA-FFQ ($r_s = 0.63$; $p < 0.0001$), but not with the WEB-FFQ ($r_s = 0.20$; $p = 0.06$).

Table 2. Comparison of reported energy intake (rEI) with measured energy requirement (mER) for maintenance of body weight during a controlled feeding phase of 4 to 6 weeks [1].

	n	rEI, kcal	mER, kcal	Δ rEI-mER, kcal	Δ rEI-mER, % *	Spearman CC
IA-FFQ						
All	127	2413 ± 602	2642 ± 558	−229 (−324 to −133) †	−9.5 (−12.7 to −6.1)	0.50 ‡
Sex						
Men	51	2744 ± 605	3161 ± 467	−417 (−600 to −234) †	−14.3 (−19.6 to −8.6)	0.23
Women	76	2191 ± 491	2294 ± 265	−102 (−197 to −8) †	−6.1 (−9.9 to −2.0)	0.63 ‡
BMI						
Non-obese	100	2415 ± 610	2594 ± 546	−179 (−282 to −76) †	−7.8 (−11.5 to −4.0)	0.51 ‡
Obese	27	2409 ± 582	2822 ± 573	−413 (−649 to −176) †	−15.3 (−21.9 to −8.1)	0.13
WEB-FFQ						
All	200	2519 ± 962	2684 ± 536	−166 (−292 to −39) †	−11.0 (−15.4 to −6.4)	0.34 ‡
Sex						
Men	108	2764 ± 991	3056 ± 414	−292 (−469 to −116) †	−14.9 (−20.5 to −8.9)	0.40 ‡
Women	92	2231 ± 845	2248 ± 265	−17 (−198 to 163)	−6.3 (−13.1 to 1.1)	0.20
BMI						
Non-obese	108	2583 ± 1021	2543 ± 520	40 (−132 to 212)	−4.1 (−10.4 to 2.7)	0.39 ‡
Obese	92	2443 ± 888	2850 ± 508	−407 (−585 to −230) †	−18.5 (−24.3 to −12.3)	0.27 ‡

[1] Values are means (SD) or means (95% CI). BMI: body mass index; CC: correlation coefficient; FFQ: food frequency questionnaire; IA: interviewer-administered; mER: measured energy requirement; rEI: reported energy intake; Δ: delta. * Mean percentage differences between rEI and mER were calculated as 100× exponential (mean of log rEI − mean log mER value) − 100; † Indicates a significant difference with mean rEI as determined by mixed models, $p < 0.05$. Analyses were performed on log-transformed data. ‡ Indicates a significant correlation, $p < 0.05$.

3.3. Under-Reporting and Over-Reporting

Prevalence and likelihood of under-reporting and over-reporting are shown in Table 3 and Figure 2 respectively. Among all participants, under-reporting was more prevalent with the WEB-FFQ than with the IA-FFQ (34.5% vs. 26.0%) but the difference did not reach statistical significance. The prevalence of under-reporting among obese participants was similar with the WEB-FFQ and the IA-FFQ (46.7% vs. 33.3%; $p = 0.24$) and also among non-obese participants (24.1% vs. 24.0% respectively). Obese individuals were more likely to under-report rEI than non-obese individuals with the WEB-FFQ (prevalence ratio, 1.97; 95% CI, 1.32 to 2.95), but not with the IA-FFQ (prevalence ratio, 0.75; 95% CI, 0.34 to 1.66; Figure 2). The prevalence of under-reporting was similar between the WEB-FFQ and the IA-FFQ among women (30.4% vs. 21.1%, respectively) and men (38.0% vs. 33.3%, respectively). Data presented in Figure 2 suggest that women were similarly likely to under-report rEI compared with men with both FFQs (IA-FFQ: prevalence ratio, 0.66; 95% CI, 0.29 to 1.50; WEB-FFQ: prevalence ratio, 0.95; 95% CI, 0.62 to 1.45). Finally, over-reporting was more prevalent with the WEB-FFQ than the IA-FFQ among all participants ($p = 0.0005$), while subgroup differences were statistically significant only in non-obese participants (Table 3).

Table 3. Prevalence of under- and over-reporting of energy intake according to agreement with measured energy requirements [1].

	FFQ Method	n	Under-Reporters	Accurate Reporters	Over-Reporters	p *
All	IA	127	26.0 (18.6 to 34.5)	67.7 (58.9 to 75.7)	6.3 (2.8 to 12.0)	0.0005
	WEB	200	34.5 (27.9 to 41.5)	48.0 (40.9 to 55.2)	17.5 (12.5 to 23.5)	
Sex						
Men	IA	51	33.3 (20.8 to 47.9)	60.8 (46.1 to 74.2)	5.9 (1.2 to 16.2)	0.12
	WEB	108	38.0 (28.8 to 47.8)	46.3 (36.7 to 56.2)	15.7 (9.5 to 24.0)	
Women	IA	76	21.1 (12.5 to 31.9)	72.4 (60.9 to 82.0)	6.6 (2.2 to 14.7)	0.0063
	WEB	92	30.4 (21.3 to 40.9)	50.0 (39.4 to 60.6)	19.6 (12.0 to 29.2)	

Table 3. *Cont.*

	FFQ Method	n	Under-Reporters	Accurate Reporters	Over-Reporters	p *
BMI						
Non-obese	IA	100	24.0 (16.0 to 33.6)	69.0 (59.0 to 77.9)	7.0 (2.9 to 13.9)	0.0019
	WEB	108	24.1 (16.4 to 33.3)	51.9 (42.0 to 61.6)	24.1 (16.4 to 33.3)	
Obese	IA	27	33.3 (16.5 to 54.0)	63.0 (42.4 to 80.6)	3.7 (0.1 to 19.0)	0.24
	WEB	92	46.7 (36.3 to 57.4)	43.5 (33.2 to 54.2)	9.8 (4.6 to 17.8)	

[1] Values are percentages (95% CI). Accurate reporters are individuals of which their corresponding rEI:mER ratio are within the 95% confidence limits of an agreement ratio of 1.00. Under-reporters and over-reporters had a ratio below 0.79 and above 1.21, respectively. BMI: body mass index; CI: confidence intervals; FFQ: food frequency questionnaire; IA: interviewer-administered; mER: measured energy requirements; rEI: self-reported energy intake. * p values indicate at least one significant difference between the IA-FFQ and the WEB-FFQ as determined by the Chi-squared test.

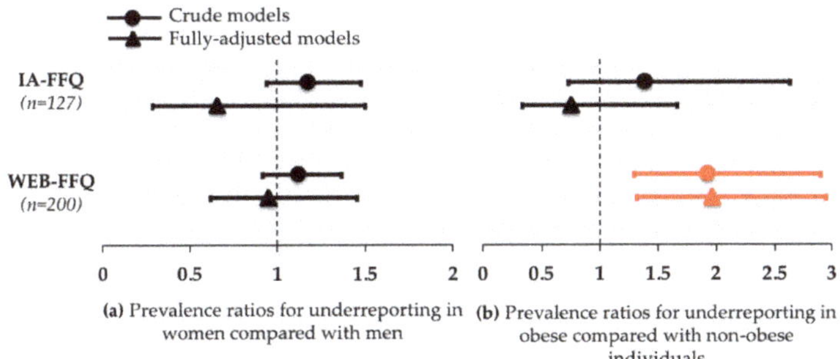

(a) Prevalence ratios for underreporting in women compared with men

(b) Prevalence ratios for underreporting in obese compared with non-obese individuals

Figure 2. Log-binomial regression analysis showing prevalence ratios with 95% CI for under-reporting in: (a) women vs. men and; (b) in obese vs. non-obese individuals. Crude models are shown with circles and adjusted models are shown with triangles. Under-reporters are individuals of which their corresponding rEI:mER ratio is below 0.79. CI: confidence interval; FFQ: food frequency questionnaire; IA: interviewer-administered.

4. Discussion

The aim of this study was to compare the accuracy of IA- and WEB-FFQ to assess an objective measure of energy requirements. Consistent with previous investigations [11], we found that both FFQs resulted in significant under-reporting of mER by −11.0% (WEB-FFQ) and −9.5% (IA-FFQ). In general, and contrary to our hypothesis, results indicated that the IA-FFQ performs slightly better than the WEB-FFQ in attenuating the prevalence of under-reporting and over-reporting in most subgroups based on sex and body weight classification.

Web-based tools such as the WEB-FFQ are being increasingly used in research for several reasons, including greater efficiency in administration process and facilitated data management [15–17]. However, studies that have compared the accuracy of traditional IA- and WEB-FFQs in predicting energy and nutrient intake are scarce to date. Park et al. [40] have recently conducted a large study comparing self-reported intakes using self-administered web-based instruments, including a FFQ, against recovery biomarkers. Energy was significantly under-reported by −29 to −34% on average compared with the doubly labeled water technique, the gold standard reference. The degree of under-reporting in the study by Park et al. was greater than in the current study (−11% for the WEB-FFQ), possibly due to different methodologies. Nonetheless, the results by Park et al. are consistent with results in the present study revealing systematic error when estimating energy intakes with a WEB-FFQ.

Kato et al. [41] compared two self-administered FFQs that differed only by their format (paper- vs. web-based) in their ability to accurately predict energy intake, using weighted 12-day food records as reference for "true" energy intake. Energy intake derived from the WEB-FFQ correlated weakly with energy intake derived from the food records, but the correlation was slightly higher among men than among women (Spearman's deattenuated correlation coefficients = 0.42 and 0.18, respectively). This observation is somewhat consistent with our results as the correlation between rEI and mER for the WEB-FFQ was significant in men, but not in women. Of note, the use of an objective reference method to assess ER in our study may have yielded weaker correlation between rEI and mER than those observed in this study by Kato et al. [41]. These authors also found that energy intakes derived from the WEB-FFQ were within acceptable limits of agreement in men (Bland-Altman method, 54–178%), and were slightly overestimated in women (Bland-Altman method, 55–220%) compared with the energy intakes derived from food records. Results from the present study showed that women had a similar likelihood to under-report energy intake compared with men when using the WEB-FFQ. The weaker correlation between rEI and mER in women when using the WEB-FFQ may be due to over-reporting being more prevalent in women than in men, which is consistent with the overestimation observed by Kato et al. [41]. Nonetheless, these observations contradict previous IA-24HR data that showed greater under-reporting in women [6]. Future studies should provide additional insight on potential sex-based differences on the accuracy of web-based tools in predicting energy requirements.

Another recent study used the doubly labeled water technique to examine the accuracy with which a WEB-FFQ (i.e., MiniMeal-Q) and a web-based 4 days food record (i.e., Riksmaten method) predict energy intake [42]. Pearson's correlations between rEI and mER were 0.28 (non-significant) for the WEB-FFQ and 0.40 ($p < 0.05$) for the food records. The WEB-FFQ resulted in a higher prevalence of under-reporters compared with the foods records (57.5% vs. 40%, respectively) and also a higher prevalence of over-reporters (15% vs. 5%, respectively). Although this study compared two web-based dietary assessment tools, the results support that current WEB-FFQ may not be better than other dietary assessment tool to estimate true energy intake. This observation is consistent with results from our study in that the WEB-FFQ produced weaker correlation between rEI and mER and a greater prevalence of under- and over-reporters compared with the IA-FFQ.

The rather large number of participants in this study compared with previous studies along with measured energy requirements are important strengths. Potential limitations also need to be considered. Firstly, the WEB-FFQ and the IA-FFQ have notable differences including the number of food items (greater with the WEB-FFQ) as well as different approaches to present serving sizes (food models vs. digital images), which could explain, at least partly, differences in the accuracy of mER prediction. Secondly, different study participants completed the IA- or WEB-FFQ. Therefore, differences observed in this study may not solely be due to the administration technique *per se* (i.e., IA vs. WEB) but could also reflect differences among study participants, although analyses were adjusted for these differences (i.e., age, sex, BMI, ethnicity). Thirdly, the controlled feeding phases were conducted in free-living conditions and some of the foods and beverages provided may not have been entirely consumed. However, the high self-reported compliance combined with the fact that a large proportion of the foods provided were consumed in the presence of study coordinators, the consistency of the cardiometabolic changes induced by the interventions and analysis of plasma biomarkers suggest that the risk of noncompliance in these studies is low. Finally, the significantly lower post- vs. pre-intervention body weight may suggest insufficient food provision (mER) in the feeding phases, but the weight difference was small and also added as a covariate in the analyses.

5. Conclusions

In conclusion, results from this study suggest that an IA-FFQ slightly attenuates the prevalence of under- and over-reporting of mER compared with a WEB-FFQ. Accordingly, the use of the WEB-FFQ resulted in accurate reporting of energy intake in 48% of all participants compared with 68% with the IA-FFQ. Considering the efficiency of web-based questionnaires and the importance of dietary assessment for population-based nutrition studies, our results support the urge to increase the quality of web-based dietary assessment tools and to further develop objective and innovative assessment techniques. Future studies should also examine if specific foods or nutrients are more likely to be under- or over-reported in web-based compared with traditional tools. The use of metabolomics and passive measure of one's food intake through digital imaging and video also have the potential to improve our ability to assess dietary intake [17,43].

Supplementary Materials: The following are available online at http://www.mdpi.com/2072-6643/10/9/1292/s1. Table S1: Anthropometric characteristics before and after controlled feeding phases of 4 to 6 weeks in men and women.

Author Contributions: Conceptualization, D.B. and B.L.; Data curation, D.B.; Formal analysis, D.B.; Funding acquisition, P.C., B.L., S.L.; Project administration, A.C. and P.C.; Software, M.-È.L., B.L.; Supervision, B.L.; Writing—original draft, D.B.; Writing—review and editing, S.L., A.C., A.L., P.C., M.-È.L. and B.L.

Funding: This research received no external funding.

Acknowledgments: We express our gratitude to the study participants, without whom the trials would not have been possible.

Conflicts of Interest: The authors declare no conflict of interest.

References

1. Hébert, J.R.; Hurley, T.G.; Steck, S.E.; Miller, D.R.; Tabung, F.K.; Peterson, K.E.; Kushi, L.H.; Frongillo, E.A. Considering the value of dietary assessment data in informing nutrition-related health policy. *Adv. Nutr.* **2014**, *5*, 447–455. [CrossRef] [PubMed]

2. Labonté, M.; Kirkpatrick, S.I.; Bell, R.C.; Boucher, B.A.; Csizmadi, I.; Koushik, A.; L'Abbé, M.R.; Massarelli, I.; Robson, P.J.; Rondeau, I.; et al. Dietary assessment is a critical element of health research-Perspective from the Partnership for Advancing Nutritional and Dietary Assessment in Canada. *Appl. Physiol. Nutr. Metab.* **2016**, 1–4. [CrossRef] [PubMed]

3. Thompson, F.E.; Kirkpatrick, S.I.; Subar, A.F.; Reedy, J.; Schap, T.E.; Wilson, M.M.; Krebs-Smith, S.M. The National Cancer Institute's Dietary Assessment Primer: A Resource for Diet Research. *J. Acad. Nutr. Diet.* **2015**, *115*, 1986–1995. [CrossRef] [PubMed]

4. Archer, E.; Marlow, M.L.; Lavie, C.J. Controversy and debate: Memory based methods paper 1: The fatal flaws of food frequency questionnaires and other memory-based dietary assessment methods. *J. Clin. Epidemiol.* **2018**. [CrossRef] [PubMed]

5. Archer, E.; Marlow, M.L.; Lavie, C.J. Controversy and debate: Memory-based dietary assessment methods paper #3. *J. Clin. Epidemiol.* **2018**. [CrossRef]

6. Archer, E.; Hand, G.A.; Blair, S.N. Validity of U.S. nutritional surveillance: National health and nutrition examination survey caloric energy intake data, 1971–2010. *PLoS One* **2013**, *8*. [CrossRef] [PubMed]

7. Subar, A.F.; Freedman, L.S.; Tooze, J.A.; Kirkpatrick, S.I.; Boushey, C.; Neuhouser, M.L.; Thompson, F.E.; Potischman, N.; Guenther, P.M.; Tarasuk, V.; et al. Addressing current criticism regarding the value of self-report dietary data. *J. Nutr.* **2015**, *145*, 2639–2645. [CrossRef] [PubMed]

8. Willett, W.C.; Rimm, E.B.; Hu, F.B. Reply to E Archer. *Am. J. Clin. Nutr.* **2017**, *106*, 950–951. [PubMed]

9. Martín-Calvo, N.; Martínez-González, M. Controversy and Debate: Memory based methods paper 2. *J. Clin. Epidemiol.* **2018**. [CrossRef]

10. Martín-Calvo, N.; Martínez-González, M. Controversy and Debate: Memory based methods paper 4: Please, no more idle talk on memory-based methods in science. *J. Clin. Epidemiol.* **2018**. [CrossRef] [PubMed]

11. Freedman, L.S.; Commins, J.M.; Moler, J.E.; Arab, L.; Baer, D.J.; Kipnis, V.; Midthune, D.; Moshfegh, A.J.; Neuhouser, M.L.; Prentice, R.L.; et al. Pooled results from 5 validation studies of dietary self-report instruments using recovery biomarkers for energy and protein intake. *Am. J. Epidemiol.* **2014**, *180*, 172–188. [CrossRef] [PubMed]

12. Livingstone, M.B.; Black, A.E. Markers of the validity of reported energy intake. *J. Nutr.* **2003**, *133* (Suppl. 3), 895S–920S. [CrossRef]

13. Subar, A.F.; Kipnis, V.; Troiano, R.P.; Midthune, D.; Schoeller, D.A.; Bingham, S.; Sharbaugh, C.O.; Trabulsi, J.; Runswick, S.; Ballard-Barbash, R.; et al. Using intake biomarkers to evaluate the extent of dietary misreporting in a large sample of adults: the OPEN study. *Am. J. Epidemiol.* **2003**, *158*, 1–13. [CrossRef] [PubMed]

14. Freedman, L.S.; Commins, J.M.; Moler, J.E.; Willett, W.; Tinker, L.F.; Subar, A.F.; Spiegelman, D.; Rhodes, D.; Potischman, N.; Neuhouser, M.L.; et al. Pooled results from 5 validation studies of dietary self-report instruments using recovery biomarkers for potassium and sodium intake. *Am. J. Epidemiol.* **2015**, *181*, 473–487. [CrossRef] [PubMed]

15. Labonté, M.; Cyr, A.; Baril-Gravel, L.; Royer, M.M.; Lamarche, B. Validity and reproducibility of a web-based, self-administered food frequency questionnaire. *Eur J. Clin. Nutr.* **2012**, *66*, 166–173. [CrossRef] [PubMed]

16. Illner, A.K.; Freisling, H.; Boeing, H.; Huybrechts, I.; Crispim, S.P.; Slimani, N. Review and evaluation of innovative technologies for measuring diet in nutritional epidemiology. *Int. J. Epidemiol.* **2012**, *41*, 1187–1203. [CrossRef] [PubMed]

17. Stumbo, P.J. New technology in dietary assessment: A review of digital methods in improving food record accuracy. *Proc. Nutr. Soc.* **2013**, *72*, 70–76. [CrossRef] [PubMed]

18. Touvier, M.; Kesse-Guyot, E.; Méjean, C.; Pollet, C.; Malon, A.; Castetbon, K.; Hercberg, S. Comparison between an interactive web-based self-administered 24 h dietary record and an interview by a dietitian for large-scale epidemiological studies. *Br. J. Nutr.* **2011**, *105*, 1055–1064. [CrossRef] [PubMed]

19. Probst, Y.C.; Faraji, S.; Batterham, M.; Steel, D.G.; Tapsell, L.C. Computerized dietary assessments compare well with interviewer administered diet histories for patients with type 2 diabetes mellitus in the primary healthcare setting. *Patient Educ. Couns.* **2008**, *72*, 49–55. [CrossRef] [PubMed]

20. Kesse-Guyot, E.; Assmann, K.; Andreeva, V.; Castetbon, K.; Méjean, C.; Touvier, M.; Salanave, B.; Deschamps, V.; Péneau, S.; Fezeu, L.; et al. Lessons Learned From Methodological Validation Research in E-Epidemiology. *JMIR Public Health Surveill.* **2016**, *2*. [CrossRef] [PubMed]

21. Kirkpatrick, S.I.; Potischman, N.; Dodd, K.W.; Douglass, D.; Zimmerman, T.P.; Kahle, L.L.; Thompson, F.E.; George, S.M.; Subar, A.F. The use of digital images in 24-hour recalls may lead to less misestimation of portion size compared with traditional interviewer-administered recalls. *J. Nutr.* **2016**, *146*, 2567–2573. [CrossRef] [PubMed]

22. Kirkpatrick, S.I.; Vanderlee, L.; Raffoul, A.; Stapleton, J.; Csizmadi, I.; Boucher, B.A.; Massarelli, I.; Rondeau, I.; Robson, P.J. Self-report dietary assessment tools used in canadian research: A scoping review. *Adv. Nutr.* **2017**, *8*, 276–289. [CrossRef] [PubMed]

23. Motard-Bélanger, A.; Charest, A.; Grenier, G.; Paquin, P.; Chouinard, Y.; Lemieux, S.; Couture, P.; Lamarche, B. Study of the effect of trans fatty acids from ruminants on blood lipids and other risk factors for cardiovascular disease. *Am. J. Clin. Nutr.* **2008**, *87*, 593–599. [CrossRef] [PubMed]

24. Richard, C.; Couture, P.; Desroches, S.; Charest, A.; Lamarche, B. Effect of the mediterranean diet with and without weight loss on cardiovascular risk factors in men with the metabolic syndrome. *Nutr. Metab. Cardiovasc. Dis.* **2011**, *21*, 628–635. [CrossRef] [PubMed]

25. Lacroix, E.; Charest, A.; Cyr, A.; Baril-Gravel, L.; Lebeuf, Y.; Paquin, P.; Chouinard, P.Y.; Couture, P.; Lamarche, B. Randomized controlled study of the effect of a butter naturally enriched in trans fatty acids on blood lipids in healthy women. *Am. J. Clin. Nutr.* **2012**, *95*, 318–325. [CrossRef] [PubMed]

26. Senanayake, V.K.; Pu, S.; Jenkins, D.A.; Lamarche, B.; Kris-Etherton, P.M.; West, S.G.; Fleming, J.A.; Liu, X.; McCrea, C.E.; Jones, P.J. Plasma fatty acid changes following consumption of dietary oils containing n-3, n-6, and n-9 fatty acids at different proportions: preliminary findings of the Canola Oil Multicenter Intervention Trial (COMIT). *Trials* **2014**, *15*. [CrossRef] [PubMed]

27. Drouin-Chartier, J.P.; Gagnon, J.; Labonté, M.; Desroches, S.; Charest, A.; Grenier, G.; Dodin, S.; Lemieux, S.; Couture, P.; Lamarche, B. Impact of milk consumption on cardiometabolic risk in postmenopausal women with abdominal obesity. *Nutr. J.* **2015**, *14*. [CrossRef] [PubMed]

28. Brassard, D.; Tessier-Grenier, M.; Allaire, J.; Rajendiran, E.; She, Y.; Ramprasath, V.; Gigleux, I.; Talbot, D.; Levy, E.; Tremblay, A.; et al. Comparison of the impact of SFAs from cheese and butter on cardiometabolic risk factors: A randomized controlled trial. *Am. J. Clin. Nutr.* **2017**, *105*, 800–809. [CrossRef] [PubMed]

29. Lohman, T.; Roche, A.; Martorell, R. *The Airlie (VA) Consensus Conference. Anthropometric Standardization Reference Manual*; Human Kinetics: Champaign, IL, USA, 1988; pp. 39–80.

30. Goulet, J.; Nadeau, G.; Lapointe, A.; Lamarche, B.; Lemieux, S. Validity and reproducibility of an interviewer-administered food frequency questionnaire for healthy French-Canadian men and women. *Nutr. J.* **2004**, *3*. [CrossRef] [PubMed]

31. Le Moullec, N.; Deheeger, M.; Preziosi, P.; Monteiro, P.; Valeix, P.; Rolland-Cachera, M.-F.; Potier De Courcy, G.; Christides, J.-P.; Chevrouvrier, F.; Galan, P.; et al. Validation of photographic document used to estimate the amounts of foods eaten by subjects in the SU.VI.MAX study. *Cah. Nutr. Diet.* **1996**, *31*, 158–164.

32. Roza, A.M.; Shizgal, H.M. The Harris Benedict equation reevaluated: Resting energy requirements and the body cell mass. *Am. J. Clin. Nutr.* **1984**, *40*, 168–182. [CrossRef] [PubMed]

33. Lafrenière, J.; Lamarche, B.; Laramée, C.; Robitaille, J.; Lemieux, S. Validation of a newly automated web-based 24-hour dietary recall using fully controlled feeding studies. *BMC Nutr.* **2017**, *3*. [CrossRef]

34. Mensink, R.P.; Zock, P.L.; Kester, A.D.; Katan, M.B. Effects of dietary fatty acids and carbohydrates on the ratio of serum total to HDL cholesterol and on serum lipids and apolipoproteins: A meta-analysis of 60 controlled trials. *Am. J. Clin. Nutr.* **2003**, *77*, 1146–1155. [CrossRef] [PubMed]

35. Siri-Tarino, P.W.; Chiu, S.; Bergeron, N.; Krauss, R.M. Saturated fats versus polyunsaturated fats versus carbohydrates for cardiovascular disease prevention and treatment. *Annu. Rev. Nutr.* **2015**, *35*, 517–543. [CrossRef] [PubMed]

36. Shen, J.; Wilmot, K.A.; Ghasemzadeh, N.; Molloy, D.L.; Burkman, G.; Mekonnen, G.; Gongora, M.C.; Quyyumi, A.A.; Sperling, L.S. Mediterranean Dietary Patterns and Cardiovascular Health. *Annu. Rev. Nutr.* **2015**, *35*, 425–449. [CrossRef] [PubMed]

37. Bhutani, S.; Kahn, E.; Tasali, E.; Schoeller, D.A. Composition of two-week change in body weight under unrestricted free-living conditions. *Physiol. Rep.* **2017**, *5*. [CrossRef] [PubMed]

38. Hoaglin, D.C.; Iglewicz, B. Fine-tuning some resistant rules for outlier labeling. *J. Am. Stat. Assoc.* **1987**, *82*, 1147–1149. [CrossRef]

39. Black, A.E.; Cole, T.J. Within- and between-subject variation in energy expenditure measured by the doubly-labelled water technique: implications for validating reported dietary energy intake. *Eur. J. Clin. Nutr.* **2000**, *54*, 386–394. [CrossRef] [PubMed]

40. Park, Y.; Dodd, K.W.; Kipnis, V.; Thompson, F.E.; Potischman, N.; Schoeller, D.A.; Baer, D.J.; Midthune, D.; Troiano, R.P.; Bowles, H.; et al. Comparison of self-reported dietary intakes from the Automated Self-Administered 24-h recall, 4-d food records, and food-frequency questionnaires against recovery biomarkers. *Am. J. Clin. Nutr.* **2018**, *107*, 80–93. [CrossRef] [PubMed]

41. Kato, E.; Takachi, R.; Ishihara, J.; Ishii, Y.; Sasazuki, S.; Sawada, N.; Iwasaki, M.; Shinozawa, Y.; Umezawa, J.; Tanaka, J.; et al. Online version of the self-administered food frequency questionnaire for the Japan Public Health Center-based Prospective Study for the Next Generation (JPHC-NEXT) protocol: Relative validity, usability, and comparison with a printed questionnaire. *J. Epidemiol.* **2017**. [CrossRef] [PubMed]

42. Nybacka, S.; Bertéus Forslund, H.; Wirfält, E.; Larsson, I.; Ericson, U.; Warensjö Lemming, E.; Bergström, G.; Hedblad, B.; Winkvist, A.; Lindroos, A.K. Comparison of a web-based food record tool and a food-frequency questionnaire and objective validation using the doubly labelled water technique in a Swedish middle-aged population. *J. Nutr. Sci.* **2016**, *5*. [CrossRef] [PubMed]

43. Guasch-Ferré, M.; Bhupathiraju, S.N.; Hu, F.B. Use of metabolomics in improving assessment of dietary intake. *Clin. Chem.* **2017**. [CrossRef] [PubMed]

 *nutrients*

Article

Comparison of Nutrient Estimates Based on Food Volume versus Weight: Implications for Dietary Assessment Methods

Emma K. Partridge [1], Marian L. Neuhouser [1,2], Kara Breymeyer [2] and Jeannette M. Schenk [2,*]

[1]   Nutritional Sciences Program, University of Washington, Seattle, WA 98109-1024, USA;
      emma.k.partridge@uwalumni.com (E.K.P.); mneuhous@fredhutch.org (M.L.N.)
[2]   Cancer Prevention, Fred Hutch Cancer Research Center, Seattle, WA 98109, USA; kbreymey@fredhutch.org
*   Correspondence: jschenk@fredhutch.org; Tel.: +1-206-667-6860

Received: 14 May 2018; Accepted: 17 July 2018; Published: 27 July 2018

Abstract: Novel technology-based dietary assessment methods use volume estimates of foods to assess dietary intake. However, the nutrient content of standard databases is based on food weight. The goal of this study is to evaluate the accuracy of the United States Department of Agriculture National Nutrient Database for Standard Reference (USDA-SR) estimates of volume and the corresponding macronutrient content of the foods. The weights of 35 individual food volumes were measured (on trial) and compared to the USDA-SR-determined weight for the food volume. Macronutrient content corresponding to the trial weight and the USDA-SR weight for the food volume (USDA) were determined using the USDA-SR, and the differences were calculated. There were statistically significant differences between the USDA and trial weights for 80% of foods measured. Calorie estimates by USDA weight were significantly lower than that of trial weight for 54% of foods but were significantly greater for 26% of foods. Differences in macronutrient estimates by trial and USDA weight varied by food type. These findings suggest that nutrient databases based on food weight may not provide accurate estimates of dietary intake when assessed using food volumes. Further development of image-assisted dietary assessment methods which measure food volumes will necessitate evaluation of the accuracy of the processes used to convert weight to volume in nutrient databases.

Keywords: nutrition; food measurement; nutrient database; dietary assessment

1. Introduction

Traditional dietary assessment tools, such as multiple-day food records and interviewer-assisted 24-h recalls, rely on self-assessment of the amounts of foods eaten. However, it is a well-documented fact that people cannot accurately recall or estimate the amount of food they consume [1–3]. Emerging technology-based dietary assessment methods that use images to assess the types and amounts of foods people consume have the potential to provide more objective estimates of dietary intake. As part of the development and validation of these new methods, it will be important to consider the accuracy of standard nutrient databases to estimate nutrient content information from food volumes as opposed to weights.

The United States Department of Agriculture National Nutrient Database for Standard Reference (USDA-SR) [4] is the primary source of food composition data in the US and serves as the foundation for most public and private food and nutrient databases, such as the Nutrition Data System for Research (NDS-R) [5]. USDA-SR is compiled of data from published and unpublished sources, including the Food and Nutrient Database for Dietary Studies (FNDDS), studies conducted by the USDA and contractors, lab analyses, algorithms, factors, or recipes [6]. The nutrient content information in these databases is largely based on food weights, not volumes, as food weight is considered the gold standard of measurement [6,7]. Although density factors have been developed for many of the foods to enable their conversion into household (volume) measures, the algorithms or processes used within USDA-SR are not fully known.

In this study, we evaluate the extent to which estimates of food portion sizes measured by volume differ from those measured by weight, and assess the subsequent differences in estimated macronutrient content of these food portion sizes when based on volume and weight.

2. Materials and Methods

2.1. Sample Size and Food Selection

Trials were performed on a total of 35 individual foods. Foods from each of the six USDA MyPlate [8] food categories (fruits, vegetables, grains, dairy, protein foods, and fats/oils) were selected to reflect the foods most commonly consumed by Americans, while allowing for variation in water content and shape [9,10]. Combination or mixed foods, such as soups and casseroles, were excluded for these experiments. A single preparation method was selected for most foods, and for a small number of foods (n = 4), multiple preparation methods were applied in order to assess differences in weight and nutrient content for one food prepared in multiple ways. Independent trials were completed on ten percent of foods chosen at random (potato (½ cup and 10 fries), chicken breast (whole and chopped), ice cream, regular salad dressing) for quality control.

2.2. Trial Volumes

Selected trial volumes for most individual food trials were based on MyPlate portion sizes [8]. Fruits and vegetables were measured as ½ cup-equivalents, grains as two ounce-equivalents, and dairy as one-cup equivalents, except ice cream, which was measured according to the serving size portion listed on the Nutrition Facts panel (½ cup). Protein foods were measured as individual portion sizes (patty, breast, large egg) between one or three ounce-equivalents, except bacon (three slices), which was measured according to the serving size portion listed on the Nutrition Facts panel. Fats/oils were measured between one and two tablespoons, depending on the individual food [8].

2.3. Preparation Methods

All foods were prepared in a commercial-grade metabolic research kitchen at the Fred Hutch Cancer Research Center (FHCRC) Human Nutrition Lab (HNL) by a single trained dietetic technician (EKP). Foods needing no preparation, such as raspberries, were measured 'as purchased'. If a food needed to be manipulated, alterations included first removing inedible portions, then size being manipulated (for example, sliced, chopped, or diced). Foods not commonly consumed raw were cooked according to protocols used by the FHCRC HNL, or by packaging instructions. The preparation method for each food was chosen based on available options in the USDA-SR 28 [4]. The LanguaL Thesaurus [11] was consulted to define standard size manipulations, and parchment paper with cut size markings was used for guidance. Details of selected foods, volumes, and preparation methods are available in Table 1.

Table 1. Descriptions of selected foods and preparation methods used for trials.

	Measured Portion Sizes	Preparation Method	As Purchased or Edible Portion	USDA Food Description	Purchased Food Description (If Applicable)	USDA Number
Fruits						
Strawberries	½ cup	Halves	Edible Portion	Strawberries, raw		09316
Cantaloupe	½ cup	Cubed [2]	Edible Portion	Melons, cantaloupe, raw		09181
Peaches	½ cup	Sliced [2]	Drained of Liquids	Peaches, canned, heavy syrup, drained		09370
Oranges	½ cup	Sectioned, with and without membranes	Edible Portion	Oranges, raw, all commercial varieties	Navel Oranges	09200
Raspberries	½ cup	Whole	As Purchased	Raspberries, raw		09302
Apples	½ cup	Quartered, sliced, and chopped [2]	Edible Portion	Apples, raw, with skin	Gala Apples	09003
Seedless Grapes	½ cup	Whole	Edible Portion	Grapes, red or green (European type, such as Thompson seedless), raw	Red Grapes	09132
Bananas	½ cup	Sliced [2]	Edible Portion	Bananas, raw		09040
Avocados	½ cup	Cubed [2]	Edible Portion	Avocados, raw, all commercial varieties		09037
Raisins	¼ cup [1]	Packed	As Purchased	Raisins, seedless	Generic brand (Kroger), seedless from green grapes	09298
Vegetables						
Iceberg Lettuce	½ cup	Chopped (loosely packed)	Edible Portion	Lettuce, iceberg (includes crisphead types), raw		11252
Tomatoes	½ cup	Chopped [2]	Edible Portion	Tomatoes, red, ripe, raw, year round average		11529
Potatoes, French fries [3]	½ cup & 10 fries [1]	Oven-heated	As purchased, heated	Potatoes, French fried, crinkle or regular cut, salt added in processing, frozen, oven-heated	Ore Ida frozen French-fried potatoes, regular cut	11360
Onions	½ cup	Sliced and chopped [2]	Edible Portion	Onions, raw	Yellow onions	11282
Sweet Corn	½ cup	Drained of liquids	As Prepared	Corn, sweet, yellow, canned, vacuum pack, regular pack	Santiam golden sweet whole kernel corn, canned	11176

Table 1. *Cont.*

	Measured Portion Sizes	Preparation Method	As Purchased or Edible Portion	USDA Food Description	Purchased Food Description (If Applicable)	USDA Number
Grains						
Bread	2 slices	Pre-Sliced	As Purchased	Bread, wheat	Generic brand (Kroger), wheat bread	18064
Pasta	1 cup	Boiled, not packed	As Purchased	Pasta, cooked, enriched, without added salt	Generic brand (Kroger), enriched spaghetti	20121
Rice	1 cup	Boiled	As Purchased	Rice, white, long-grain, regular, enriched, cooked	Generic brand (Kroger), long grain	20045
Dairy						
Cheese	$^1/_3$ cup	Shredded	As Purchased	Cheese, cheddar	Generic brand (Kroger), all natural medium cheddar—purchased in a block and grated	01009
Ice Cream [3]	½ cup	None	As Purchased	Ice creams, vanilla	Generic brand (Kroger), Vanilla	19095
Yogurt	1 cup	None	As Purchased	Yogurt, plain, skim milk, 13 grams protein per 8 ounce	Generic brand (Fred Meyer), non-fat, plain	01118
Protein						
Beef, ground 7% fat	4″ diameter patty	Pan-broiled	As Purchased	Beef, ground, 93% lean meat/7% fat, patty, cooked, pan-broiled	Generic brand (Kroger), 4″ diameter using 114 g (¼ pound) of raw meat (patties were formed raw then cooked per HNL protocols)	23474
30% fat	4″ diameter patty	Pan-broiled	As Purchased	Beef, ground, 70% lean meat/30% fat, patty cooked, pan-broiled	Fresh from butcher (Marketime foods)	13496
Chicken Breast, without skin [3]	Roasted	1 breast	Roasted, whole and chopped [2]	As Purchased	Chicken, broilers or fryers, breast, meat only, cooked, roasted	Foster Farms, breast fillets	05064

Table 1. *Cont.*

	Measured Portion Sizes	Preparation Method	As Purchased or Edible Portion	USDA Food Description	Purchased Food Description (If Applicable)	USDA Number
Protein						
Egg, large	Scrambled	1 large Egg, Scrambled	Edible Portion	Egg, whole, cooked, scrambled	Generic brand (Fred Meyer) large eggs, grade AA [3]	01132
Bacon	Pan-fried	3 slices, Pan-fried	As Purchased	Pork, cured, bacon, pre-sliced, cooked, pan-fried	Generic brand (Fred Meyer), traditional cut, sugar cured	10862
Cashews	⅛ cup	Raw, whole	As Purchased	Nuts, cashew nuts, raw	Raw, whole, purchased in bulk	12087
Almonds	⅛ cup	Raw, whole	As Purchased	Nuts, almonds	Blue Diamond, raw, whole	12061
Pecans	⅛ cup	Raw, halves	As Purchased	Nuts, pecans	Diamond of California, raw	12142
Peanut Butter	2 tablespoons	None	As Purchased	Peanut butter, smooth style, with salt	Generic brand (Fred Meyer)	16098
Fats, Oils						
Margarine	1 tablespoon	None	As Purchased	Margarine, regular, hard, soybean (hydrogenated)	Generic brand (Fred Meyer)	04073
Canola Oil	2 tablespoons	None	As Purchased	Oil, canola	Generic brand (Kroger)	04582
Mayonnaise	2 tablespoons	None	As Purchased	Salad dressing, mayonnaise, regular	Generic brand (Fred Meyer)	04025
Dressing, Italian [3]	2 tablespoons	None	As Purchased	Salad dressing, Italian dressing, commercial, regular	Generic brand (Fred Meyer), Zesty Italian	04114
Salsa	2 tablespoons	None	As Purchased	Sauce, salsa, ready-to-serve	Pace brand, Picante Medium	06164
Ketchup	2 tablespoons	None	As Purchased	Catsup	Generic brand (Fred Meyer)	11935

[1] MyPlate volumetric equivalents. ½ cup fruit equivalent = ½ cup hydrated fruit = ¼ cup dried fruit. ½ cup vegetable equivalent = 10 French fries. [2] Standard preparation methods and their parameters are as follows: Sliced: between 0.5 cm and 1.5 cm [11]. For consistency, the investigator measured slices to 1 cm. Cubed: >1.5 cm. For consistency, the investigator attempted to cut each cube side to 2 cm. Chopped: Item divided into pieces with a thickness <0.3 cm [11]. For consistency, the investigator chopped items as close to 0.25 cm as possible. [3] Foods that were re-measured for quality control tests. [4] Grade AA eggs have thick, firm whites and high, round yolks. USDA: United States Department of Agriculture National Nutrient Database; HNL: Human Nutrition Lab.

2.4. Data Collection

For each trial, a prepared food was weighed at the test volume to determine its weight in grams, herein referred to as trial weight. Ten replicates per preparation method were completed for each food. Foods were measured at the test volume using standard food measuring tools; the same instrument was used throughout each trial and cleaned (washed and dried) between replicates. For each food trial, the USDA weight and macronutrient content for the test volume was determined by entering the test volume directly into USDA-SR to yield the corresponding gram weight and macronutrient content. For a small number of foods (*n* = 3), a volume option was not available in the USDA-SR; thus, the Nutrition Data System for Research (NDSR) database (University of Minnesota, Minneapolis, MN, USA) [5] was used to determine the corresponding gram weight and macronutrient content.

2.5. Statistical Analysis

For each food trial, the means and standard deviations of the 10 trial weight replicates were calculated. The USDA weight was set for each test volume at the value obtained from the USDA-SR database. Percentage differences between the trial and USDA weights, defined as the difference of trial weight subtracted from USDA weight divided by USDA weight, were determined for each replicate, and overall mean percentage differences between the trial and USDA weight were calculated. Similar methods were used to determine absolute differences in macronutrient content between trial and USDA weight for the selected trial volume of an individual food. For each food, one-sample *t*-tests were used to evaluate whether the mean differences between trial (average of 10 replicates), USDA weights, and nutrient contents were significantly different from zero. Statistical analyses were conducted using Statistics and Data (STATA) software (Release 14, College Station, TX, USA).

3. Results

Table 2 summarizes the mean trial weight, USDA weight, and mean percentage difference in trial and USDA weight for test volumes of individual foods. For 80% of food trials, there were statistically significant relative differences between the USDA and trial weights of the selected trial volume, ranging from -103.4% for sliced onions to $+38.7\%$ for shredded cheddar cheese. Within individual food groups, relative differences between USDA and trial weights were statistically significant for 65% of fruit and vegetable, 67% of grain, 100% of dairy, 77% of protein, and 100% of fat/oil foods, though there were no discernable patterns in either the direction or magnitude of relative weight differences across food categories.

Table 3 provides estimates of calorie and macronutrient content corresponding to the USDA and trial weights, and their differences, for selected volumes of individual foods. Absolute differences between USDA and mean trial weight-derived calorie estimates for selected food volumes ranged from 0 to 60 kcal, and largely mirrored those reported for weights (Table 2). For 52% of food trials, calories determined by USDA weight were significantly lower than by trial weight, and for 26% of foods, calories determined by USDA were significantly greater than by trial weight, although the absolute value of these differences was small for many foods. The largest calorie differences between USDA and trial weight were found for dairy foods; calories determined by trial weight for ice cream were 60 ± 3 kcal less than by USDA weight ($p < 0.0001$). Conversely, for shredded cheddar cheese, calories determined by trial weight were 59 ± 2 kcal greater than by USDA weight ($p < 0.0001$).

Table 2. Comparison of mean measured (trial) weights of individual foods to the USDA-SR (United States Department of Agriculture National Nutrient Database for Standard Reference) 28 database weight for selected food volumes.

		Selected Test Volume	Trial Weight (g) Mean (sd) [1]	USDA Weight (g) Mean [2]	% Difference [3] Mean (se) [4]
Fruits					
Strawberries, halved		½ cup	75.8 (5.5)	76.0	0.3 (2.3)
Cantaloupe, cubed		½ cup	85.7 (8)	80.0	−7.2 (3.2) [6]
Peaches, sliced		½ cup	110.6 (4.4)	111.0	0.4 (1.3)
Orange sections	Membrane	½ cup	98.9 (5.1)	90.0	−9.9 (2.0) [8]
	No Membrane	½ cup	98.9 (3.9)	90.0	−9.9 (1.4) [8]
Raspberries, whole		½ cup	61.8 (3.8)	61.5	−0.5 (2.0)
	Quartered	½ cup	76.7 (5.1)	62.5	−22.6 (2.6) [8]
Apples	Sliced	½ cup	87.3 (5.2)	54.5	−60.2 (3.0) [8]
	Chopped	½ cup	67.31 (2.5)	62.5	−7.7 (1.3) [7]
Grapes, seedless, whole		½ cup	91.8 (2.8)	75.5	−21.6 (1.2) [8]
Bananas, sliced		½ cup	76.7 (4.5)	75.0	−2.3 (1.9)
Avocados, cubed		½ cup	90.0 (7.3)	75.0	−20.0 (3.1) [8]
Raisins, packed		½ cup	44.2 (2.2)	41.25	−7.1 (1.7) [7]
Vegetables					
Iceberg Lettuce, chopped		½ cup	30.0 (2.9)	28.5	−5.3 (3.2)
Tomatoes, chopped		½ cup	107.1 (4.1)	90.0	−19.0 (1.4) [8]
Potatoes, French fries		10 fries	47.9 (6.8)	69.0	30.6 (2.2) [8]
		½ cup [5]	51.6 (3.6)	52.0	0.9 (1.6)
Onions, raw	Sliced	½ cup	116.9 (15.5)	57.5	−103.4 (8.5) [8]
	Chopped	½ cup	81.3 (3.9)	80.0	−1.6 (1.5)
Sweet Corn, canned, drained		½ cup	87.5 (5.7)	105.0	16.6 (1.7) [8]
Grains					
Bread, wheat, sliced		2 slices	58.4 (2.7)	58.0	−0.6 (1.4)
Pasta, enriched, spaghetti		1 cup	119.0 (3.4)	124.0	4.0 (0.9) [7]
Rice, white, long grain, enriched		1 cup	182.7 (17.7)	158.0	−15.6 (3.5) [7]
Dairy					
Cheddar cheese, shredded		$^1/_3$ cup	23.1 (1.5)	37.6	38.7 (1.2) [8]
Ice Cream, vanilla		½ cup	95.3 (6.5)	66.0	−44.4 (1.5) [8]
Yogurt, skim		1 cup	260.8 (1.2)	245.0	−6.4 (0.2) [8]
Protein					
Beef, patty	7% fat	4″ patty [5]	84.2 (3.2)	81.6	−3.1 (1.2) [6]
	30% fat	4″ patty	71.2 (3.4)	77.0	7.5 (1.4) [8]
Chicken Breast,	Whole	1 breast	206.2 (43.8)	172.0	−19.9 (5.7) [7]
without skin, roasted	Chopped	1 cup	139.1 (3.9)	140.0	0.7 (0.7)
Egg, large, scrambled		1 large egg	43.9 (2.5)	61.0	28.0 (1.3) [8]
Bacon, regular cut, pan-fried		3 slices	26.9 (5.3)	34.5	22.1 (4.8) [7]
Cashews, raw, whole		⅛ cup [5]	24.3 (1.9)	16.1	−50.7 (3.8) [8]
Almonds, raw, whole		⅛ cup	24.8 (1.9)	17.9	−38.7 (3.3) [8]
Pecans, raw, halved		⅛ cup	20.2 (1.3)	12.4	−63.2 (3.3) [8]
Peanut Butter, smooth		2 T	29.7 (0.9)	32.0	7.2 (0.9) [8]
Fats and Oils					
Margarine		1 T	14.3 (0.2)	14.1	−1.6 (0.5) [6]
Canola Oil		2 T	25.5 (0.3)	28.0	8.8 (0.4) [8]
Mayonnaise		2 T	29.0 (0.4)	27.6	−5.1 (0.4) [8]
Dressing, Italian, regular		2 T	30.6 (0.9)	29.4	−4.0 (0.7) [8]
Salsa		2 T	32.9 (1.5)	36.0	8.8 (1.3) [8]
Ketchup		2 T	34.3 (0.3)	34.0	−1.0 (0.2) [7]

[1] Mean and standard deviation determined by 10 experimental weight replicates. [2] Estimate of variance not available from USDA-SR 28. [3] (USDA Weight—Trial Weight)/USDA weight. [4] Standard error from one-sample *t*-test. [5] Nutrient content unavailable in USDA-SR 28, value obtained from NDSR (Nutrition Data System for Research). [6] $p < 0.05$, [7] $p < 0.01$, [8] $p < 0.001$.

Table 3. Mean macronutrient values obtained from USDA-SR 28 based on trial volume and USDA volume and their differences (USDA Volume–Trial Volume).

	Calories (kcal)			Total Fat (g)			Protein (g)			Carbohydrate (g)		
	Weight		Diff: Mean (se)	Weight		Diff: Mean (se)	Weight		Diff: Mean (se)	Weight		Diff: Mean (se)
	Trial	USDA		Trial	USDA		Trial	USDA		Trial	USDA	
Fruits												
Strawberries, halved	24	24	−0.3 (0.56)	0.2	0.2	0 (0.01)	0.5	0.5	0 (0.01)	5.8	5.8	0.02 (0.13)
Cantaloupe, cubed	29	27	−2.3 (0.9) *	0.2	0.2	−0.01 (0) *	0.7	0.7	−0.05 (0.02) *	7	6.5	−0.47 (0.21) *
Peaches, sliced	80	80	0.4 (0.99)	0.2	0.2	0 (0)	0.6	0.6	0 (0.01)	20	20	0.08 (0.26)
Orange sections Membrane	46	42	−4.4 (0.76) ***	0.1	0.1	−0.01 (0) **	0.9	0.9	−0.08 (0.02) ***	12	11	−1.04 (0.19) ***
Orange sections No Membrane	47	42	−4.5 (0.56) ***	0.1	0.1	−0.01 (0) ***	0.9	0.9	−0.08 (0.01) ***	12	11	−1.04 (0.15) ***
Raspberries, whole	32	32	0 (0.67)	0.4	0.4	0 (0)	0.7	0.7	0 (0.01)	7.4	7.3	−0.04 (0.14)
Apples Quartered	40	32	−7.9 (0.81) ***	0.1	0.1	−0.02 (0) ***	0.2	0.2	−0.04 (0) ***	11	8.6	−1.96 (0.22) ***
Apples Sliced	45	28	−17.4 (0.87) ***	0.2	0.1	−0.06 (0) ***	0.3	0.1	−0.09 (0) ***	12	7.5	−4.53 (0.23) ***
Apples Chopped	35	32	−3 (0.37) ***	0.1	0.1	−0.01 (0) *	0.2	0.2	−0.01 (0) ***	9.3	8.6	−0.67 (0.11) ***
Grapes, seedless, whole	63	69	5.7 (0.58) ***	0.2	0.2	0.01 (0) ***	0.7	0.7	0.06 (0.01) ***	17	18	1.48 (0.16) ***
Bananas, sliced	68	67	−1.3 (1.26)	0.3	0.3	0 (0)	0.9	0.8	−0.02 (0.02)	18	17	−0.39 (0.32)
Avocados, cubed	144	120	−24.1 (3.73) ***	13	0.1	−2.19 (0.34) ***	1.8	1.5	−0.3 (0.05) ***	7.7	6.4	−1.28 (0.2) ***
Raisins, packed	132	123	−9.1 (2.09) **	0.2	0.2	−0.02 (0) **	1.4	1.3	−0.09 (0.02) **	35	33	−2.3 (0.56) **
Vegetables												
Iceberg Lettuce, chopped	4	4	−0.1 (0.1)	0	0	0 (0)	0.3	0.3	−0.01 (0.01)	0.9	0.9	−0.04 (0.03)
Tomatoes, chopped	19	16	−3.3 (0.26) ***	0.2	0.2	−0.03 (0) ***	0.9	0.8	−0.15 (0.01) ***	4.2	3.5	−0.67 (0.05) ***
Potatoes, French fries 10 fries	79	115	35.6 (2.5) ***	2.5	3.5	1.08 (0.08) ***	1.2	1.7	0.53 (0.04) ***	13	19	5.82 (0.42) ***
Potatoes, French fries ½ cup	86	85	−0.55 (1.35)	2.7	2.7	0.07 (0.04)	1.3	1.2	−0.06 (0.02) **	14	14	0.25 (0.22)
Onions, raw Sliced	47	23	−23.7 (1.94) ***	0.1	0.1	−0.06 (0) ***	1.3	0.6	−0.66 (0.05) ***	11	5.4	−5.55 (0.46) ***
Onions, raw Chopped	32	32	−0.4 (0.52)	0.1	0.1	0 (0)	0.9	0.9	−0.01 (0.01)	7.6	7.5	−0.12 (0.12)
Sweet Corn, canned, drained	69	83	13.9 (1.39) ***	0.4	0.5	0.08 (0.01) ***	2.1	2.5	0.42 (0.04) ***	17	20	3.39 (0.35) ***
Grains												
Bread, wheat, sliced	156	155	−0.7 (2.19)	1.9	1.9	−0.01 (0.03)	6.3	6.2	−0.04 (0.09)	28	28	−0.18 (0.41)
Pasta, enriched, spaghetti	188	196	7.8 (1.7) *	1.1	1.2	0.04 (0.01) **	6.9	7.2	0.29 (0.06) **	37	38	1.53 (0.33) **
Rice, white, long grain, enriched	246	205	−40.5 (2.24) ***	0.5	0.4	−0.09 (0) ***	5.1	4.3	−0.83 (0.05) ***	53	45	−8.7 (0.48) ***
Dairy												
Cheddar cheese, shredded	93	152	58.8 (1.9) ***	7.7	13	4.85 (0.16) ***	5.3	8.6	3.34 (0.11) ***	0.7	1.2	0.45 (0.01) ***
Ice Cream, vanilla	197	137	−60.25 (3.02) ***	10	7.3	−3.23 (0.16) ***	3.3	2.3	−1.03 (0.05) ***	23	16	−6.92 (0.35) ***
Yogurt, skim	146	137	−9.2 (0.2) ***	0.5	0.4	−0.03	15	14	−0.9 (0.02) ***	20	19	−1.21 (0.03) ***

Table 3. *Cont.*

		Calories (kcal)			Total Fat (g)			Protein (g)			Carbohydrate (g)		
		Weight		Diff: Mean (se)	Weight		Diff: Mean (se)	Weight		Diff: Mean (se)	Weight		Diff: Mean (se)
		Trial	USDA		Trial	USDA		Trial	USDA		Trial	USDA	
					Protein								
Beef, patty [a]	7% fat	153	155	1.8 (1.83)	6.7	6.8	0.01 (0.08)	22	22	0.56 (0.26)	0.1	0	−0.05 (0)
	30% fat	170	183	13.5 (2.53) ***	11	12	0.9 (0.17) ***	16	18	1.32 (0.24) ***	NA	NA	NA
Chicken Breast,	Whole	340	284	−56.3 (16.14) **	7.4	6.1	−1.22 (0.35) **	64	53	−10.61 (3.03) **	NA	NA	NA
without skin, roasted	Chopped	230	231	1.3 (1.45)	5	5	0.03 (0.03)	43	43	0.28 (0.27)	NA	NA	NA
Egg, large, scrambled		65	91	25.8 (1.15) ***	4.8	6.7	1.88 (0.09) ***	4.4	6.1	1.7 (0.08) ***	0.7	1	0.27 (0.01) ***
Bacon, regular cut, pan-fried		126	161	35.2 (7.83) **	9.4	12	2.68 (0.58) **	9.1	12	2.58 (0.56) **	0.5	0.6	0.13 (0.03) **
Cashews, raw, whole		134	89	−45.4 (3.35) ***	11	7.1	−3.58 (0.27) ***	4.4	2.9	−1.49 (0.11) ***	7.3	4.9	−2.47 (0.19) ***
Almonds, raw, whole		144	103	−40.5 (3.39) ***	12	8.9	−3.46 (0.29) ***	5.2	3.8	−1.47 (0.12) ***	5.4	3.9	−1.5 (0.13) ***
Pecans, raw, halved		140	86	−53.7 (2.8) ***	15	8.9	−5.64 (0.3) ***	1.9	1.1	−0.72 (0.04) ***	2.8	1.7	−1.08 (0.06) ***
Peanut Butter, smooth		178	191	13.3 (1.8) ***	15	16	1.19 (0.15) ***	6.6	7.1	0.52 (0.07) ***	6.6	7.1	0.52 (0.07) ***
					Fats and Oils								
Margarine		103	101	−2.1 (0.5) **	12	11	−0.18 (0.06) *	0.1	0.1	0 (0.00)	0.1	0.1	0 (0.00)
Canola Oil		226	248	22.1 (0.85) ***	26	28	2.46 (0.10) ***	0	0	0 (0.00)	NA	NA	NA
Mayonnaise		197	188	−9.4 (0.79) ***	22	21	−1.05 (0.08) ***	0.3	0.3	−0.02 (0) ***	0.2	0.2	−0.01 (0) *
Dressing, Italian, Regular		73	71	−2.4 (0.5) ***	6.5	6.2	−0.25 (0.04) ***	0.1	0.1	−0.01 (0) ***	3.7	3.6	−0.15 (0.02) ***
Salsa		10	10	0.5 (0.17) *	0.1	0.1	0.003 (0)	0.5	0.6	0.05 (0.01) ***	2.2	2.4	0.21 (0.03) ***
Ketchup		35	34	−0.7 (0.15) **	0	0	0 (0.00)	0.4	0.4	−0.01 (0) **	9.4	9.3	−0.09 (0.02) **

NA: Not Applicable; [a] Nutrient content unavailable in USDA-SR 28, value obtained from NDSR. * $p < 0.05$, ** $p < 0.01$, *** $p < 0.001$.

Differences in estimated macronutrient content between USDA and trial weight were dependent on food type (Table 3). Higher-fat foods, like shredded cheddar cheese and nuts, tended to have the largest absolute differences in estimates of fat content, although the direction of differences was inconsistent. In general, the absolute differences in estimated protein content were small for fruit and vegetable foods, which have lower protein content, but were quite large for protein and dairy foods. Similarly, absolute differences in estimated carbohydrate content between USDA and trial weight were largest for grains, dairy, and vegetables, but were relatively small for protein foods and fats/oils (Table 3).

4. Discussion

In this study we compared weights for selected food volumes measured in a research kitchen with those derived from the USDA-SR database. Overall, we found statistically significant differences between the USDA-derived and trial weights for 76% of the foods tested. In addition, there were significant differences in corresponding calorie estimates derived from the USDA and trial weights for 78% of foods. These findings suggest that the processes used to convert weight into volume in the USDA-SR may not provide accurate estimates of volume for many foods and may subsequently lead to inaccurate estimates of caloric and nutrient intake.

Efforts to develop improved methods of dietary assessment that employ more objective measures of intake have recently gained attention [12–14], with many innovative technologies focusing on the use of images to estimate food volumes [15–20]. Using images to calculate volumes, these methods hold promise to provide more accurate estimates of the amounts of foods which people eat. The potential of these novel approaches, however, may be limited by the fact that nutrient content information in available databases is currently based on food weights, and estimates of food density, or weight for unit food volume, are required to convert volume into weight. For many foods in these databases, food density has been generated; however, little information is available about the algorithms and processes used to convert weight to volume, and the accuracy of these data are uncertain [6,21,22].

For most of the foods evaluated in this study, USDA weights for the selected trial volume tended to be greater than the measured weights. For some foods—primarily, fruits, vegetables, and fats/oils—the absolute differences between the USDA-derived and trial weights for food volume were modest, indicating that the algorithms or processes used to convert weight to volume for these foods were relatively accurate. For other foods, such as dairy, high-protein, and some manipulated or prepared foods, there were substantial differences between the measured and USDA weights for the trial food volume. Differences between the USDA and measured weight for a given volume may, in part, have been due to the protocols followed for manipulating or preparing foods. For example, many foods that required cooking preparations, such as potato, chicken, egg, bacon, and rice, had statistically significant differences between the USDA and trial weights (all $p < 0.05$). Cooking time, heat intensity, and water retention/release can vary through the cooking process and between protocols used (trial vs. USDA), which may impact cooking yield, thereby contributing to the observed differences in USDA and trial weight for the selected test volume. For foods that were manipulated, variations in packing and differences in protocols used for manipulation may account for some differences between the USDA and measured weight for a given food volume. In our trials, manipulation methods were defined by the LanguaL Thesaurus and were standardized across trials. For some foods in the USDA-SR, different manipulation methods of the same food are grouped together, such as quartered and chopped apples. As a result, the weight and nutrient content for the same volume of each form of the food is identical, even though the size, shape, and air space upon packing differs greatly. It is important to note that image-assessed food volumes will inherently include air space, due to food packing; therefore, nutrient database conversions for weight to volume will need to be equivalently determined.

Because weight is the standard by which USDA-SR determines nutrient content, differences between the USDA-derived and trial weight for a given food volume yielded corresponding differences in calorie and macronutrient content estimates. Over 70% of foods, regardless of food group, that had significant differences between USDA-derived and trial weight for a given food volume also had corresponding significant differences in calorie estimates, though we found no apparent pattern within or across food groups in the magnitude or direction of these differences. However, the corresponding differences between USDA and trial weight-derived macronutrient content was dependent on the nutrient composition of the individual food; foods dense in a specific macronutrient tended to have greater differences in that macronutrient. For example, for high-fat foods such as pecans, even small absolute differences between the USDA-derived and trial weights yielded substantial differences in calorie and fat content estimates (difference between USDA–trial weight-derived weight, calories and fat: -7.8 g, -53.7 kcal; -5.6 g fat (both $p < -0.0001$)). The overall impact of differences between weight and volume-based measures of dietary intake will depend heavily on the individual foods people eat. In order to further evaluate this potential impact, the extent to which estimates of food portion sizes measured by volume differ from those measured by weight would need to be measured for an extensive list of foods.

To our knowledge, this is the first study to report differences in nutrient database information by volume and weight. Foods were systematically selected based on popularity in the US diet, and were measured and prepared via standardized methods defined by the LanguaL Thesaurus. In addition, multiple replicates were measured for each food and food preparation method, to align with the sampling methods used for USDA-SR. However, this study is not without limitations. Data from individual replicates were not publicly available for USDA-SR; thus, our estimates of mean differences between the trial and USDA relied only on a single value of the USDA mean, which may be anti-conservative. For some foods, the number of replicates used in this study ($n = 10$) may be less than that assessed by the USDA, which would reduce the accuracy of our measurements compared to those made by the USDA. In addition, while most preparation methods were available via the LanguaL Thesaurus, cooking heat and time were unavailable for most cooked foods. Instead, protocols defined by the Human Nutrition Laboratory at the Fred Hutchinson Cancer Research Center were followed, and may differ from those used by the USDA. For our trials, we purposely selected generic options; thus for some foods, the differences reported here may also reflect differences in the exact foods measured. Lastly, this study was limited to single and non-mixed foods. Although the foods chosen represent foods commonly eaten by Americans, they may be less representative of foods prevalent in everyday diets.

5. Conclusions

This study demonstrates that for selected food volumes, substantial differences existed between the corresponding USDA-derived and trial weights measured in a research kitchen. The differences between the USDA-derived and trial weights for a selected food volume also resulted in parallel differences between estimated macronutrient content. As the primary source of food composition data in the US, researchers rely heavily on the USDA-SR database, either directly or indirectly, to estimate dietary intake of nutrients.

Given the development of new image-assisted dietary assessment methods that provide objective measures food volume, it is important to assess the accuracy of nutrient databases to estimate nutrient content based on food volumes. The findings reported here suggest that the estimation of dietary intake using food volumes may not provide accurate estimates of nutrient intake. Most nutrient databases commonly used in the US are based on USDA-SR nutrient data, and thus would be affected by the same inaccuracies. Whether the same issues are apparent in other food and nutrient databases around the world is unknown. In order to better understand the impact of these discrepancies on assessment of dietary and micronutrient intake, further evaluation of the accuracy of processes used to convert weight to volume in the USDA-SR is warranted.

Author Contributions: E.K.P. and K.B. collected the data. E.K.P. and J.M.S. wrote the first draft with contributions from M.L.N. All authors reviewed and commented on subsequent drafts of the manuscript.

Funding: Please add: This research was funded in part by National Cancer Institute grants numbered U01-CA135133 and P30-CA15704.

Acknowledgments: This work was supported by P30 CA15704 and U01-CA135133.

Conflicts of Interest: The authors declare no conflict of interest.

References

1. Acheson, K.; Campbell, I.; Edholm, O.; Miller, D.; Stock, M. The measurement of food and energy intake in man—An evaluation of some techniques. *Am. J. Clin. Nutr.* **1980**, *33*, 1147–1154. [CrossRef] [PubMed]
2. Young, L.R.; Nestle, M. Portion sizes in dietary assessment: Issues and policy implications. *Nutr. Rev.* **1995**, *53*, 149–158. [CrossRef] [PubMed]
3. Westerterp, K.R.; Goris, A.H. Validity of the assessment of dietary intake: Problems of misreporting. *Curr. Opin. Clin. Nutr. Metab. Care* **2002**, *5*, 489–493. [CrossRef] [PubMed]
4. United States Department of Agriculture (USDA). Composition of Foods: Raw, Processed, Prepared. USDA National Nutrient Database for Standard Reference, Release 28. Available online: https://www.ars.usda.gov/northeast-area/beltsville-md-bhnrc/beltsville-human-nutrition-research-center/nutrient-data-laboratory/docs/usda-national-nutrient-database-for-standard-reference/ (accessed on 15 September 2015).
5. Schakel, S.; Sievert, Y.; Buzzard, I. Sources of data for developing and maintaining a nutrient database. *J. Am. Diet. Assoc.* **1988**, *88*, 1268–1271. [PubMed]
6. United States Department of Agriculture ARS; Food Surveys Research Group. *The USDA Food and Nutrient Database for Dietary Studies, 4.1–Documentation and User Guide*; Department of Agriculture ARS: Beltsville, MD, USA, 2010.
7. Raper, N.; Perloff, B.; Ingwersen, L.; Steinfeldt, L.; Anand, J. An overview of USDA's dietary intake data system. *J. Food Compos. Anal.* **2004**, *17*, 545–555. [CrossRef]
8. United States Department of Agriculture (USDA). Dietary Guidelines for Americans. Available online: https://www.choosemyplate.gov/ (accessed on 10 August 2015).
9. United States Department of Agriculture (USDA) Beltsville Human Nutrition Research Center; Food Surveys Research Group. *Data tables: Results from USDA's 1994–1996 Continuing Survey of Food Intakes by Individual and 1994–1996 Diet and Health Knowledge Survey*; Food Surveys Research Group: Beltsville, MD, USA, 1998.
10. Hoy, M.; Goldman, J.; Bowman, S.; Moshfegh, A. Foods contributing to fruit and vegetable intakes of US adults: What We Eat in America, NHANES 2009–2010 (369.3). *FASEB J.* **2014**, *28*, 363.
11. Møller, A.; Ireland, J. Langual™ 2017. The Langual™ Thesaurus. Available online: http://www.langual.org/langual_Thesaurus.asp (accessed on 18 July 2018).
12. Stumbo, P.J. New technology in dietary assessment: A review of digital methods in improving food record accuracy. *Proc. Nutr. Soc.* **2013**, *72*, 70–76. [CrossRef] [PubMed]
13. Gemming, L.; Utter, J.; Ni Mhurchu, C. Image-assisted dietary assessment: A systematic review of the evidence. *J. Acad. Nutr. Diet.* **2015**, *115*, 64–77. [CrossRef] [PubMed]
14. Thompson, F.E.; Subar, A.F.; Loria, C.M.; Reedy, J.L.; Baranowski, T. Need for technological innovation in dietary assessment. *J. Am. Diet. Assoc.* **2010**, *110*, 48–51. [CrossRef] [PubMed]
15. Wang, D.-H.; Kogashiwa, M.; Kira, S. Development of a new instrument for evaluating individuals' dietary intakes. *J. Am. Diet. Assoc.* **2006**, *106*, 1588–1593. [CrossRef] [PubMed]
16. Zhu, F.; Bosch, M.; Woo, I.; Kim, S.; Boushey, C.J.; Ebert, D.S.; Delp, E.J. The use of mobile devices in aiding dietary assessment and evaluation. *IEEE J. Sel. Top. Signal Process.* **2010**, *4*, 756–766. [PubMed]
17. Shang, J.; Pepin, E.; Johnson, E.; Hazel, D.; Teredesai, A.; Kristal, A.; Mamishev, A. Dietary intake assessment using integrated sensors and software. In *Multimedia on Mobile Devices 2012; and Multimedia Content Access: Algorithms and Systems VI*; International Society for Optics and Photonics: Bellingham, DC, USA, 2012.
18. Jia, W.; Chen, H.-C.; Yue, Y.; Li, Z.; Fernstrom, J.; Bai, Y.; Li, C.; Sun, M. Accuracy of food portion size estimation from digital pictures acquired by a chest-worn camera. *Public Health Nutr.* **2014**, *17*, 1671–1681. [CrossRef] [PubMed]
19. Weiss, R.; Stumbo, P.J. Divakaran A: Automatic food documentation and volume computation using digital imaging and electronic transmission. *J. Acad. Nutr. Diet.* **2010**, *110*, 42–44.

20. Martin, C.K.; Han, H.; Coulon, S.M.; Allen, H.R.; Champagne, C.M.; Anton, S.D. A novel method to remotely measure food intake of free-living individuals in real time: The remote food photography method. *Br. J. Nutr.* **2009**, *101*, 446–456. [CrossRef] [PubMed]
21. Schakel, S.F.; Buzzard, I.M.; Gebhardt, S.E. Procedures for estimating nutrient values for food composition databases. *J. Food Compos. Anal.* **1997**, *10*, 102–114. [CrossRef]
22. Powers, P.; Hoover, L. Calculating the nutrient composition of recipes with computers. *J. Am. Diet. Assoc.* **1989**, *89*, 224–232. [PubMed]

Article

Accuracy of Automatic Carbohydrate, Protein, Fat and Calorie Counting Based on Voice Descriptions of Meals in People with Type 1 Diabetes

Piotr Ladyzynski [1],*, Janusz Krzymien [2], Piotr Foltynski [1], Monika Rachuta [2] and Barbara Bonalska [2]

1   Nalecz Institute of Biocybernetics and Biomedical Engineering of the Polish Academy of Sciences, 4 Trojdena Street, 02-109 Warsaw, Poland; piotr.foltynski@ibib.waw.pl
2   Department of Diabetology and Internal Medicine, Medical University of Warsaw, 1A Banacha Street, 02-097 Warsaw, Poland; janusz.krzymien@hotmail.com (J.K.); m.rachuta@op.pl (M.R.); klindiab@wum.edu.pl (B.B.)
*   Correspondence: piotr.ladyzynski@ibib.waw.pl or pladyzynski@ibib.waw.pl; Tel.: +48-22-592-59-42

Received: 26 February 2018; Accepted: 19 April 2018; Published: 21 April 2018

Abstract: The aim of this work was to assess the accuracy of automatic macronutrient and calorie counting based on voice descriptions of meals provided by people with unstable type 1 diabetes using the developed expert system (VoiceDiab) in comparison with reference counting made by a dietitian, and to evaluate the impact of insulin doses recommended by a physician on glycemic control in the study's participants. We also compared insulin doses calculated using the algorithm implemented in the VoiceDiab system. Meal descriptions were provided by 30 hospitalized patients (mean hemoglobin A1c of 8.4%, i.e., 68 mmol/mol). In 16 subjects, the physician determined insulin boluses based on the data provided by the system, and in 14 subjects, by data provided by the dietitian. On one hand, differences introduced by patients who subjectively described their meals compared to those introduced by the system that used the average characteristics of food products, although statistically significant, were low enough not to have a significant impact on insulin doses automatically calculated by the system. On the other hand, the glycemic control of patients was comparable regardless of whether the physician was using the system-estimated or the reference content of meals to determine insulin doses.

Keywords: carbohydrate counting; protein and fat counting; calorie counting; automatic bolus calculator; voice description of meals; insulin dosage; glycemic control; diabetes mellitus

---

1. Introduction

Technical innovations create many possibilities in supporting the treatment of people with diabetes. According to the Statista Inc. report, smartphone user penetration as a percentage of the total global population exceeded 25% in 2015 [1]. This percentage is forecast to reach 37% by the year 2020. Advances in information and communication technologies (ICT) bring a significant opportunity to develop the integrated healthcare system, which is so difficult to achieve with the current traditional model of healthcare delivery. Telemedicine has the potential to become a key element of future integrated care—an important component of the new healthcare model according to the World Health Organization (WHO) [2]. In 2013, only about 6000 medical applications, i.e., medical "apps", were available in Google Play for Android-based smartphones [3], and this number increased rapidly in the following years to exceed 45,000 in 2017 [4]. The primary goal of majority of these medical apps is to help and coordinate continuous healthcare at home [5,6]. There is an ongoing discussion as to whether those mobile health (m-health) applications facilitate the gain of clinical benefits, to what extent they can be integrated with the current healthcare system and, finally, whether they are safe and do not create potential health risks for the patient. The US Food and Drug Administration (FDA) classifies the mobile application as a medical device if it is used to prevent, diagnose, care for or cure the disease. Such an app requires the approval of the Agency before it appears on the market [7]. The app should only be recommended to patients by health care professionals if its effectiveness has been scientifically confirmed.

Diabetes is one of the chronic diseases that requires a lot of attention from both the patient and the healthcare team. Regardless of the type of diabetes, patients require full information about the disease through continuous education and promotion of health-seeking behaviors as well as regular glucose monitoring, individual treatment plans, and an early diagnosis to prevent the health threats associated with complications of diabetes. Telemedicine provides a number of tools that could be helpful in choosing the right treatment plan, supporting actions to change a patient's lifestyle, strengthening motivation regarding health-related activities, facilitating a patient's ability to self-monitor and control their condition, and achieving the intended therapeutic goal.

Proper dietary treatment is one of the most important components of diabetes therapy, because it significantly affects glycemic control. The growing health awareness of patients has increased the interest in the use of new technologies that can help with dietary intervention and provide nutritional advice. Compared to traditional methods of diet planning and nutrition assessment, new technologies have many advantages, including the ability to quickly provide personalized advice. Several studies have shown that the use of new technologies that provide information and advice on diet can lead to positive changes in the dietary regimen of a patient, affecting the intake of selected nutrients [8]. Although there is still debate regarding the effectiveness of using new technologies in promoting a healthy diet, patients prefer applications that are quickly available and easy to use, increase the awareness of the type of food consumed and facilitating body weight control. In nine randomized controlled trials on the use of smartphone apps for promoting a healthy diet and nutrition, the use of such apps led to the selection of foods recommended by nutritionists, i.e., foods of higher quality, with lower calorific value and low-fat content, as well as participation in significantly more intense physical activity. These changes in the lifestyle resulted in significantly greater weight loss in comparison with people who did not use mobile apps [9]. However, to provide the personalized dietary advice, an appropriate method for measuring and evaluating food intake is required.

The digital revolution has made it possible to develop new instruments for the quantitative assessment of consumed food products [10]. Currently, new solutions supporting the estimation of food consumption use the Internet, mobile technology or both. They are preferred in comparison with traditional methods by both young people and adults. A global consistent increase in Internet access over the last few decades has resulted in the emergence of a number of websites that allow estimation of the consumption of products for both research and commercial purposes [11]. They can be easily accessed using desktop computers, but also from mobile devices, such as tablets or smartphones. In contrast to on-paper nutrition assessment methods, online systems have a few advantages—they can be pre-programmed and digital images of food items can be used to increase food recognition accuracy and facilitate estimation of portion size. There are four basic methods of food coding: the electronic food diary, the photo assisted tutorial, analysis of food photography by trained dieticians and the automatic analysis of digital food images [12]. Smartphones have enormous potential; apps enable cheap interventions among large populations [13], they make it possible to record data in real time, they are convenient to use, and they can provide continuous monitoring of consumed foods because users usually carry smartphones with them [14].

Automatic or semi-automatic food image analysis systems for dietary assessment are under continuous development. They achieve recognition accuracy below 90% when tested on databases consisting of up to a few hundred images of meals/dishes [15]. In recent years, image transducers have been developed that take serial photos, documenting the consecutive stages of a meal intake and enabling the estimation of the quantity of a leftover, uneaten meal [16]. Some of these lifelogging devices, such as the Microsoft SenseCam camera, along with the data obtained from a conventional food diary, allow for the improvement of the accuracy of calorie intake calculations [17]. Alternative approaches, which are based on the voice description of meals [18,19] or the monitoring of activities related to the meal consumption, e.g., chewing or swallowing [20,21] have been also reported.

Accurate assessment of meals consumed, which includes the correct calculation of carbohydrate exchange units (CU) (and in some applications, also protein and fat) and energy content in a meal, is one of the key elements of type 1 diabetes treatment. It is a challenge for many people with diabetes to estimate the appropriate insulin dose that correctly reflects the size and content of the meal, the pre-prandial glucose level and the expected level of physical activity. This may be one of the reasons why many people with type 1 diabetes do not achieve their therapeutic goals, which is expressed by an elevated level of glycated hemoglobin A1c (HbA1c) of 7.5% (58–64 mmol/mol) or more [22,23]. The prolonged lack of adequate glycemic control in this group of patients results in increased rates of complications and mortality [24]. Difficulties that exist in adjusting the prandial insulin dose are, in many cases, the major cause of both postprandial hypoglycemia and, even more often, hyperglycemia. Therefore, many research works have focused on the evaluation of applications aimed at the improvement of metabolic control, the reduction of the risk of hypoglycemic episodes, body weight reduction and improvement of quality of life as well as decision support regarding prandial insulin dose adjustment based on carbohydrate (CHO) counting. Recently, Tascini et al. pointed out that new insight concerning the effect of dietary macronutrients on postprandial glycemic control confirm that prandial insulin doses should combine CHO counting with protein and fat counting [25]. However, these authors also claimed that a successful application of protein and fat counting requires suitable and usable algorithms to be developed. Therefore, only a few reports so far have calculated prandial insulin doses based on integrated CHO, protein and fat counting with simultaneous evaluation of the accuracy or benefits of these calculations [26–28]. None of these reports presented data related to the accuracy of prandial insulin dose calculation based on automatic meal content estimation using a voice description of the meal. Foltynski et al. evaluated the efficacy of such a system (VoiceDiab) in controlling postprandial blood glucose concentrations in persons with type 1 diabetes treated with a continuous, subcutaneous insulin infusion under ambulatory conditions [18]. One of the limitations of that study was the lack of data regarding the meals eaten because patients were treated under ambulatory conditions. Other limitations were the fact that almost 75% of the study group were young patients (<18 years of age) and the mean HbA1c at baseline was 7% (53 mmol/mol), which means that the majority of participants were achieving the recommended target metabolic control according to the American Diabetes Association (ADA). These limitations indicate that it was not possible to determine the difference between the actual meal content and the meal content, which was estimated based on the voice description of the meals provided by the patient. Even if it were possible to calculate such differences, they might have been biased by the fact that participants of that study had, on average, good metabolic control. The accurate estimation of the meal content by the system based on the voice description provided by the person with diabetes is one of the necessary conditions to effectively help such a person calculate the proper insulin dose to compensate for the meal. Such a help is much more desirable in patients who have problems in achieving adequate metabolic control. However, the question arises of whether such patients are able to describe meals verbally in a way that makes it possible to automatically estimate the meal content with an accuracy suitable to calculate insulin doses compensating for these meals. The present study tries to answer this question. This question is also very important from the point of view of the possible application of systems using the voice description of meals in people who do not require exogenous insulin, such as the majority of patients with type 2 diabetes, some women with gestational diabetes and people without diabetes who want to monitor their diet, for example to control their body weight.

The previously published report by Foltynski et al. [18] aimed to evaluate only the effect of the use of the VoiceDiab system on postprandial blood glucose concentrations in ambulatory-treated children and young people with type 1 diabetes. In contrast to that study, the objective of the current work is to assess and demonstrate, for the first time, the accuracy of automatic CHO, protein, fat and calorie counting based on voice descriptions of meals provided by hospitalized adult persons with unstable type 1 diabetes using the VoiceDiab system, in comparison with reference counting results made by the dietitian, and to evaluate the effectiveness of the diabetes treatment, depending on whether the physician determined the insulin dosage based on the composition of meals calculated by the dietitian or whether it was automatically estimated by the VoiceDiab system. We also evaluated the effect of differences between macronutrient counting provided by the system and the dietitian regarding the discrepancy between insulin boluses calculated using the bolus calculator implemented in the voice expert system. Hence, both studies are significantly different not only because of different objectives but also due to their different study groups, the design of each study and the analyzed parameters.

2. Materials and Methods

2.1. Study Group

During short-term hospitalization because of unstable diabetes, 30 patients with type 1 diabetes treated with continuous, subcutaneous insulin infusion were familiarized with the VoiceDiab expert system and used it to verbally describe meals they intended to eat in order to automatically estimate the amount of CHO, protein, fat and calories in these meals. The inclusion criteria were as follows: 18 to 50 years of age, duration of diabetes $\geq$1 year and ability to comply with dietary recommendations and hospital procedures. Exclusion criteria included metabolic acidosis, dehydration and electrolytic disorders, other diagnosed endocrine diseases, chronic kidney disease (serum creatinine >1.5 mg/dL), proliferative retinopathy and concomitant infections. The study group consisted of 23 women and 7 men, aged 23.8 $\pm$ 4.6 (mean $\pm$ SD) years (from 19 to 38 years), with a duration of diabetes of 12.2 $\pm$ 6.5 years (from 3 to 26 years) and a varied level of metabolic control, expressed by the mean HbA1c, of 8.4 $\pm$ 1.5% (68 $\pm$ 16 mmol/mol) (from 6.1 to 12.6%, i.e., from 43 to 114 mmol/mol).

All subjects gave their informed consent for inclusion before they participated in the study. The study was conducted in accordance with the Declaration of Helsinki, and the protocol was approved by the Ethics Committee of the Medical University of Warsaw (KB/16/2014).

For each participant, a unified medical history was collected, concerning diet and eating habits, physical activity, insulin therapy (with particular emphasis on insulin boluses and basal infusion), the number of daily blood glucose tests, the frequency and severity of hypoglycemic episodes, and information about other diagnosed diseases, meditation used, smoking habits and alcohol and drug abuse. The physician analyzed this data to identify factors that could affect the glycemic control of the study participants.

2.2. The Voice System Design

The system consisted of an Android-controlled smartphone with the client application communicating wirelessly with servers to perform the following tasks: (1) automatic speech recognition (ASR) and transformation of the voice description of the meal into text; (2) analysis of the textual description to determine the composition of the meal; (3) calculation of the insulin dose compensating the meal according to the algorithm, taking into account either only the CHO content, or the CHO, protein and fat contents in the meal. A detailed description of the system can be found elsewhere [29]. It is noteworthy that the database of the system contains characteristics of 900 unique food products and 5000 terms, facilitating effective speech-to-text conversion, including foods that were present in the hospital menu. However, neither the number of calories, the quantities of CHO, protein and fat that characterize each product, nor any other data stored in the system database, were adapted to the characteristics of the hospital menu.

Safety of the patient is a priority in developing technical systems to support the treatment of people with diabetes. The VoiceDiab system contains a few levels of data input validation and control to ensure that the content of meal is calculated based on real data, i.e., that the system correctly "understands" the verbal description of the meal that the patient intends to deliver: (1) the text that results from the speech-to-text conversion is displayed in full to let the user validate the correctness of the automatic speech recognition; (2) each meal segment is associated with three icons indicating whether the system was able to extract from the voice description of the meal its full characteristics, i.e., the name of the food product, portion size or unit of measure and the number indicating the amount of food (a green icon color indicates that the trait associated with this icon has been recognized on the basis of the verbal description, yellow means that the feature has been recognized using contextual and grammatical analysis of the verbal description and red indicates that the system has not recognized the food segment); (3) for each recognized product, the full characteristics (i.e., name of the product, portion size or unit of measure and the number) are shown together with the estimated total mass and the content of CHO, protein, fat and energy to make it possible for the user to verify the correctness of the data; (4) the user must confirm that the meal was recognized in accordance with the verbal description to activate the bolus calculator; (5) a bolus exceeding the individually-configured threshold triggers the display of a warning message.

2.3. Voice System Usage and Built-In Bolus Calculator

Each study participant used the system in the following way. Before starting a meal, the participant verbally described its composition, giving the name and size (either in units such as grams, ounces or liters, or in customary units of measure, such as spoons, cups or portions) of each food product present in the meal. The description was transmitted to the server, and after speech-to-text conversion, each food product was identified and displayed on the smartphone screen for verification by the participant. If the identification failed, a warning message showed that the recognition had been unsuccessful due to an ASR failure or a lack of necessary information in the meal description, e.g., when the patient had specified a food product that was not present in the database of the system. In case of ASR failure, the patient repeated the description of the food product that had not been properly identified. For each recognized product, the system calculated the calorie content and CHO, protein and fat contents in grams.

Upon activation, the bolus calculator summarized the total caloric value, the carbohydrate exchange units (CU) and protein–fat exchange units (PFU) in the whole meal, and finally, the insulin dose required to compensate for the meal. The PFU was calculated using the following equation:

$$PFU = (4 \times Protein\ [kcal] + 9 \times Fat\ [kcal])/100. \tag{1}$$

If the PFU is greater than 1.0 a dual-wave bolus is recommended consisting of a simple bolus and a square-wave bolus lasting for 4 to 8 h depending on the value of PFU. The total prandial insulin dose was determined based on the following equation [30]:

$$IB\ [U] = CU \times ICR + PFU \times ICR, where\ ICR\ is\ the\ insulin\ to\ CHO\ ratio. \tag{2}$$

The first part of the sum in Equation (2) denotes the amount of insulin administered in the simple bolus and the second part denotes the insulin administered in the square-wave infusion of the variable duration. If the PFU is less than 1.0, then the system reduces it to zero and, consequently, recommends an insulin dose in the form of a simple bolus [30].

Regardless of the data received through the VoiceDiab system, the dietitian carefully calculated the content of each meal based on the weight and exact composition of each food product (provided by a supplier on the product label) present in the meal. The dietitian estimated the caloric value of the meal, CU and PFU, which were treated as the ground true or reference values. Based on the reference values of CU and PFU, the reference insulin doses were calculated manually using the same algorithm that was implemented in the VoiceDiab system. Additionally, a simple bolus (i.e., the first part of the sum in Equation (2) was calculated as if the patient was using a pen insulin injector. For the paired comparison of insulin boluses, the same insulin to CHO ratio (ICR) values were used in the insulin bolus calculations for each patient, which were equal to 1.5 and 1.0 for breakfast and the other meals, respectively. The reference counts of calories, CU, PFU and prandial insulin boluses were used to assess the accuracy and safety of the estimates provided by the VoiceDiab system. We used two ICR values to calculate insulin doses in all patients to account for the most important circadian changes in this parameter, but also to clearly show how differences in the meal contents calculated using both methods were reflected in differences in insulin doses. The VoiceDiab system makes it possible to program values of ICR for 8 time periods with flexible time limits during the day. Thanks to this feature, the circadian rhythm of ICR fluctuates, and changes related to illness or menstruation can also be taken into consideration. The VoiceDiab system cannot automatically estimate ICR. However, there has not been any other automatic bolus calculator reported that could do that. The values of ICR have to be programmed by the physician and they can be altered by the educated patient to adjust for changes in life conditions, e.g., illness.

2.4. Impact of the Method of Macronutrient Counting on Glycemic Control

To assess whether the automatic estimation of meal content based on the voice description of a meal can be used to control glycemia, the study group was randomly divided into two subgroups. In the first one, consisting of 14 subjects, aged 23.5 ± 3.8 years with HbA1c equal to 8.6 ± 1.8% (70 ± 20 mmol/mol), the insulin boluses were decided by a physician based on the reference meal content data. In the second one, involving subjects aged 23.7 ± 5.4 years with HbA1c equal to 8.5 ± 1.3% (69 ± 14 mmol/mol), the physician only had access to the data provided by the system when determining the insulin dosage. Each study participant was monitored using the continuous glucose monitoring system.

The following parameters were compared between the subgroups: the mean plasma glucose concentration (PG), the percentage of time when glucose concentration was normoglycemic, i.e., higher than 3.9 mmol/L (70 mg/dL) and lower than 10.0 mmol/L (180 mg/dL) (PNPG), the mean maximum increase in PG after the main meals and the number and duration of hypoglycemic episodes (i.e., glucose concentration equal or lower than 3.9 mmol/L or 70 mg/dL).

2.5. Statistical Analysis

The discrepancy in the distribution of the assessed variables from normality was assessed using the Shapiro–Wilk W test. The results indicated that the distribution of the variables differed from the normal distribution. Thus, the non-parametric Wilcoxon signed-rank test was used to analyze the significance of differences between the reference values of caloric content, CU and PFU calculated by the dietitian and the values of these parameters estimated by the VoiceDiab system. The same test was used to analyze differences between prandial boluses calculated according to the above-mentioned algorithm. The statistical analysis was carried out using Statistica version 10 (StatSoft, Inc., Tulsa, OK, USA). All data are presented as means ± SDs and their ranges, i.e., minimum and maximum values. Differences were considered to be statistically significant when $p < 0.05$.

3. Results

3.1. Accuracy of Macronutrient and Calorie Counting Based on Voice Descriptions of Meals

During their stay in hospital, patients received five meals a day, including three main meals and two snacks. All of the study participants used the VoiceDiab system to verbally describe 535 meals consisting of 1644 food products. The routine hospital diet that was served to participants during the study contained 85 unique food products in different combinations and of different sizes/weights. Plain bread, butter, potatoes, cottage cheese, tomatoes, apples, ham and boiled eggs were repeated in the hospital menu most often. Individual meals consisted of 1 to 6 unique food products. The average breakfast consisted of 4.1 ± 0.5 products, the morning snack, 2.2 ± 0.5 products, lunch, 4.0 ± 0.6 products, the afternoon snack, 1.2 ± 0.6 products, and dinner, 4.1 ± 0.4 products.

Table 1 presents the results of calorie counting done by the dietitian based on accurate, carefully collected data regarding the weights and compositions of meals in comparison with the VoiceDiab system estimates based on approximate information provided by the study participants.

The average calorie content in both snacks estimated by the system did not differ from those calculated by the dietitian. In the case of the main meals, the differences were statistically significant. Overall, the system tended to underestimate the calorie count, but the mean differences were relatively small and equal to −7.2 ± 24.4 kcal (−1.7 ± 6.2%), −55.6 ± 54.8 kcal (−10.8 ± 10.4%) and −6.5 ± 26.0 kcal (−1.2 ± 5.4%) for breakfast, lunch and dinner, respectively.

Table 1. Calorie content estimated by the dietician and the VoiceDiab system.

Meal	N	Calorie Content (kcal)		p
		Dietician Mean ± SD [1] Min–Max	System Mean ± SD Min–Max	
Breakfast	110	388 ± 85 166–602	381 ± 84 159–586	<0.0001
Morning snack	80	92 ± 38 23–324	90 ± 28 25–238	0.43
Lunch	130	507 ± 98 338–806	451 ± 97 283–766	<0.0001
Afternoon snack	101	58 ± 31 46–163	58 ± 28 34–168	0.46
Supper	114	410 ± 86 221–661	403 ± 79 217–661	<0.0001

[1] SD, the standard deviation.

In the case of each meal, except for lunch, the system estimated values of CU which were higher than those calculated by the dietitian (Table 2). The mean differences were equal to 0.3 ± 0.3 CU (8.8 ± 6.4%), 0.0 ± 0.6 CU (0.6 ± 12.4%) and 0.3 ± 0.2 CU (9.2 ± 5.9%), for the consecutive main meals starting with breakfast. In total, for the three main meals, the difference between the CHO content estimated by the system and by the dietitian was lower or equal to ±1 CU (i.e., ±10 g of CHO) in 96.3% of cases.

Table 2. Carbohydrate exchange unit (CU) and the protein–fat exchange unit (PFU) counting by the dietician and the VoiceDiab system.

Meal	N	Carbohydrate Exchange Units (CU)			Protein-Fat Exchange Units (PFU)		
		Dietician Mean ± SD Min–Max	System Mean ± SD Min–Max	p	Dietician Mean ± SD Min–Max	System Mean ± SD Min–Max	p
Breakfast	110	3.8 ± 0.8 2.2–6.0	4.1 ± 0.9 2.4–6.5	<0.0001	2.3 ± 0.6 0.7–3.7	2.2 ± 0.6 0.6–3.4	<0.0001
Morning snack	81	1.1 ± 0.4 0.5–3.1	1.1 ± 0.5 0.5–3.2	0.01	0.4 ± 0.3 0.0–1.9	0.4 ± 0.2 0.0–1.1	0.01
Lunch	130	5.5 ± 1.0 2.3–9.1	5.5 ± 1.0 2.9–8.0	0.78	2.8 ± 0.8 1.2–5.1	2.3 ± 0.8 0.7–5.2	<0.0001
Afternoon snack	100	1.2 ± 0.4 1.0–2.6	1.3 ± 0.4 0.9–2.5	<0.0001	0.05 ± 0.14 0.0–0.5	0.05 ± 0.13 0.0–0.6	0.28
Supper	114	3.9 ± 0.9 2.3–6.7	4.2 ± 0.9 2.7–6.5	<0.0001	2.5 ± 0.6 0.8–4.4	2.3 ± 0.5 0.7–4.2	<0.0001

The remaining 3.7% estimates differed by not more than ±2 CU. The percentage of the results within the range of ±1 CU was equal to 99.1% for breakfast, 90.8% for lunch and 100% for dinner.

The protein and fat contents were underestimated by the system for the main meals with mean differences of -0.1 ± 0.3 PFU ($-3.8 \pm 12.5\%$), -0.5 ± 0.5 PFU ($-17.4 \pm 17.6\%$) and -0.1 ± 0.3 PFU ($-4.5 \pm 9.4\%$), respectively. The results were not different for the afternoon snack, which only sporadically contained protein or fat. For the morning snack, the mean difference was positive with a large variability between estimates provided by the system and the dietitian ($13.3 \pm 32.6\%$).

3.2. Effect of Differences in Macronutrient Counting on Insulin Doses Estimated Using the Built-In Bolus Calculator of the VoiceDiab System

Figure 1a shows a comparison between the insulin boluses calculated based on meal composition provided by the dietitian versus the system, whereas Figure 1b illustrates the absolute differences between these insulin doses.

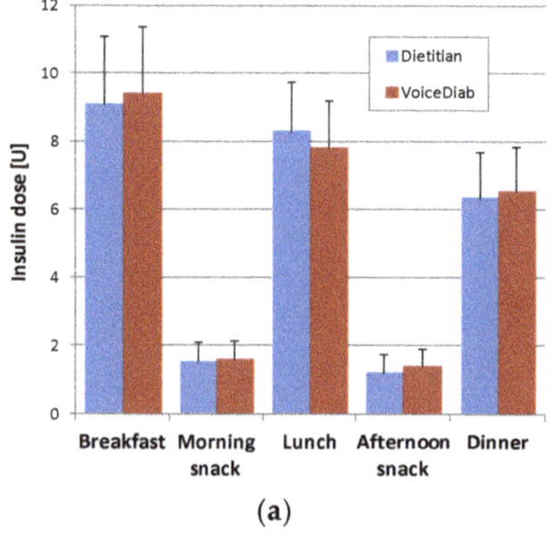

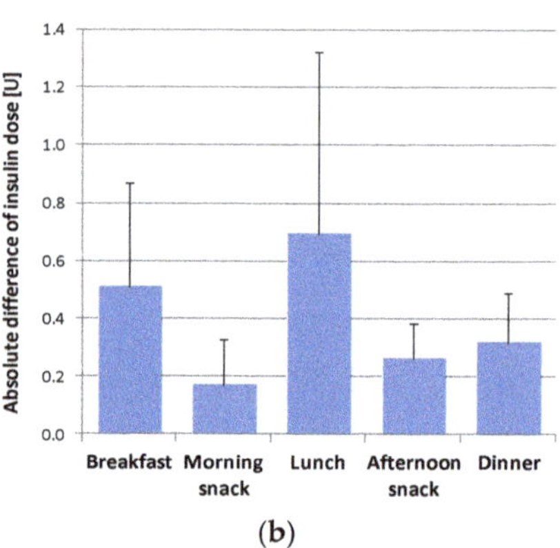

Figure 1. (**a**) Comparison of insulin doses, and (**b**) absolute differences of insulin doses calculated based on carbohydrate (CHO), protein and fat contents provided by the dietitian and the VoiceDiab system.

In the case of all meals, except for the morning snacks, the differences between insulin boluses were statistically significant ($p < 0.001$). However, the mean absolute difference did not exceed 0.70 U for any meal, and it was below 0.32 U for both snacks and dinner. The mean daily prandial insulin dose in all full days of hospitalization was equal to 25.6 ± 4.6 U when calculated based on the meal content estimated by the dietitian and 25.8 ± 4.4 U when the estimates provided by the VoiceDiab system were used, meaning that the average difference was equal to just 0.2 ± 0.8 U ($p = 0.059$).

Figure 2a shows, for each meal and for all meals together, the percentage of the prandial insulin doses calculated based on the meal estimates made by the system which were equal to their reference values, those that were in the range of 0.0–0.5 U, 0.5–1.0 U, 1.0–2.0 U and those that differed by more than 2 U from the reference values. The majority of the insulin doses (78.7%) differed by ± 0.5 U at most from the reference values and only 1.3% went beyond the ±2 U range. When we used the values of CU to calculate simple insulin boluses (Figure 2b), neglecting the protein and fat contents in meals, the results were similar, i.e., 81.7% of boluses were different from their reference values by 0.5 U or less, and only 1.1% differed by more than 2 U (of which 0.9% concerned insulin doses compensating breakfast).

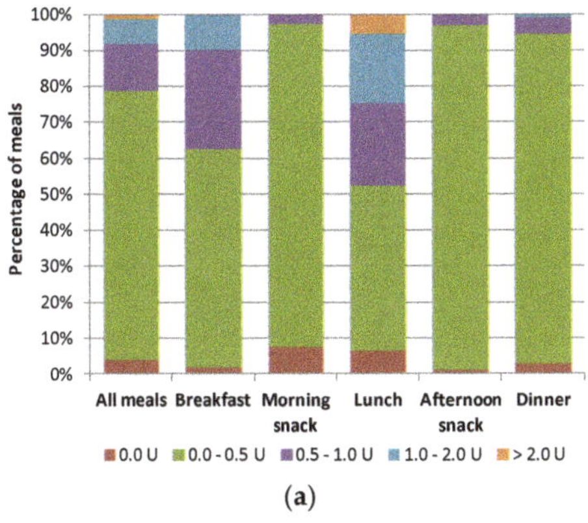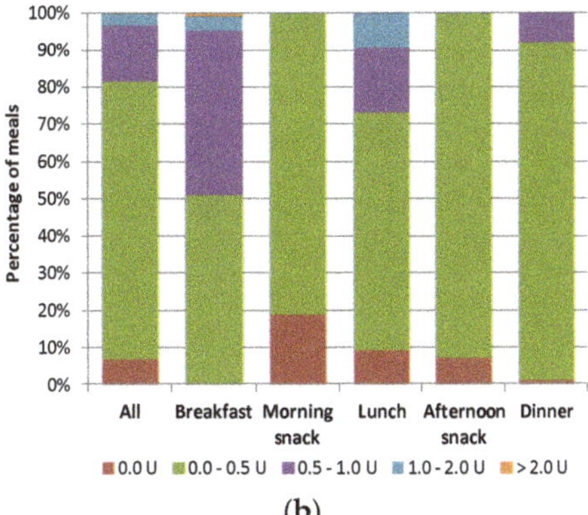

(a)   (b)

Figure 2. Mean relative absolute differences of insulin doses calculated according to Equation (2) based on estimates of carbohydrate exchange units (CU) and protein–fat exchange units (PFU) provided by the dietitian and the VoiceDiab system for (**a**) dual-wave boluses; (**b**) simple boluses. The insulin to carbohydrate ratio (ICR) was assumed to be 1.5 U/CU for breakfast and 1.0 U/CU for all other meals.

3.3. Impact of the Method of Macronutrient Counting on Glycemic Control

The average PG and PNPG were similar for both subgroups, i.e., 7.3 $\pm$ 0.8 mmol/L (131 $\pm$ 15 mg/dL) vs. 7.5 $\pm$ 0.9 mmol/L (135 $\pm$ 16 mg/dL), and 76 $\pm$ 7% (p = 0.42) vs. 75 $\pm$ 7% (p = 0.79), respectively. The maximum increase in PG was equal to 4.3 $\pm$ 1.4 mmol/L (77 $\pm$ 25 mg/dL) vs. 4.7 $\pm$ 1.8 mmol/L (85 $\pm$ 33 mg/dL) (p = 0.37) after breakfast, 3.7 $\pm$ 1.7 mmol/L (67 $\pm$ 30 mg/dL) vs. 4.0 $\pm$ 1.7 mmol/L (72 $\pm$ 30 mg/dL) (p = 0.55) after lunch and 3.9 $\pm$ 1.3 mmol/L (71 $\pm$ 24 mg/dL) vs. 4.3 $\pm$ 0.9 mmol/L (77 $\pm$ 17 mg/dL) (p = 0.11) after dinner, in the first and the second subgroups, respectively. In the first subgroup, hypoglycemia episodes occurred 2.1 $\pm$ 0.8 times per day, whereas in the second subgroup, they occurred 2.0 $\pm$ 1.3 per day (p = 0.77). The daily duration of hypoglycemic episodes was equal to 120 $\pm$ 70 min in the first subgroup and 95 $\pm$ 74 min in the second subgroup (p = 0.35).

4. Discussion and Conclusions

Since the mid 1990s, ICT technology has been used to support the treatment of people with type 1 diabetes, based primarily on the telemonitoring of patients' metabolic states and courses of treatment as well as teleconsultations. Telehome care systems have been used to support type 1 diabetes treatment in a few clinical trials, demonstrating a few benefits of this type of the care over routine periodical check-ups of patients' states in the physician's office [31–35]. Rapid development and a widespread use of smartphones created the basis for the development and clinical validation of m-health solutions, making it possible to monitor or support the treatment of people with diabetes in real time [36,37]. Currently, using smartphones, it is possible to transfer a lot of data to the treatment team, such as the results of glucose monitoring, meal size and composition or information on physical activity. However, the more data that is transferred, the more time and effort the therapeutic team needs to analyze these data and effectively support patients. Automatic bolus calculators can reduce the burden on the treatment team, helping patients to adjust their insulin doses to the size and composition of meals [38]. As indicated in the study by Franc et al., frequent support of treatment with a smartphone coupled to a website and the use of automatic bolus calculators may lead to a significant reduction in HbA1c in people with poorly-controlled type 1 diabetes. With less frequent use of smartphone apps, it is beneficial to frequently use teleconsultation services [39]. However, determination of the appropriate dose of insulin administered before the meal depends on the relatively accurate assessment of a meal by calculating, primarily, its CHO content.

The ability of the patient to accurately estimate the size of a meal depends on many factors, including quality of education, frequency of recurrent training and daily practice of such calculations by the patient. According to the survey carried out in people with type 1 diabetes and poor glycemic control, the average error in CHO counting in meals consumed during the day (three main meals and two snacks) was equal to 4.2 CU, and it tended to increase in people with long-term diabetes subjected to systematic education [40]. Brazeau et al., analyzed the differences between CHO estimates made by patients with long-term type 1 diabetes and calculations carried out by dieticians using a computer analysis program and found that the average absolute difference was equal to 15.4 ± 7.8 g (20.9 ± 9.7%) of CHO per meal, which, on average, contained 72.4 ± 34.7 g of CHO [41]. In this study, the CHO content in 63% out of the total of 448 meals was underestimated. Bishop et al. showed, by analyzing the most frequently eaten foods, that, in a group of teenagers, only 23% were able to estimate CHO content with an error not exceeding ±10 g in their daily diet despite the selection of common meals [42]. In up to 52% of these teenage patients, the difference between their calculations and those made by dieticians was within the range of ±30 g. Similar mistakes in CHO counting are made by adults with type 1 diabetes [43]. Currently, new systems are emerging which are aimed at supporting people with type 1 diabetes in CHO counting. The GoCARB system uses computer vision technology for this purpose. The user places a reference card next to the meal and takes two images using a smartphone camera. The system was developed based on the following minimum assumptions: the image contains only one dish/plate, which must be round, and various food products are not mixed on the plate. After taking photos, the images are transmitted to a dedicated server via a WiFi network, where a series of computer image processing operations are performed. All computer vision modules operate on the server, while the mobile phone is used only to acquire images, calculate CU and visualize the results. A comparison was made between the calculations performed by people with type 1 diabetes without system support and with the use of the system. An error below ±20 g per meal from the total of 114 meals was noted in 58.8% and 80.7% of participants, respectively, showing the advantage of using the system [44]. The caloric, CHO, protein and fat contents in food products can be calculated using numerous available publications as well as computer programs and mobile apps that have built-in calculators or databases to facilitate the capacity to obtain information on foods [45]. Often the quality of calculations is affected by the size of the portions.

Patients with type 1 diabetes often face a difficult choice of whether they need only to account for the CHO content of the meal or whether they should also include the contents of protein and fat to determine the insulin dose. Bell et al. attempted to determine differences between postprandial glucose concentrations after eating a high-fat, high-protein meal compared to a low-fat and low-protein meal with the same CHO content, and to determine differences in the insulin doses that should be applied following each of these meals to achieve the best postprandial glycemic control. The authors showed that in the case of a meal containing 40 g of fat and 27 g of protein in addition to 50 g of CHO, the insulin dose should be increased by 65 ± 10% using a dual-wave bolus carried out for 2.4 h to achieve adequate post-meal glycemic control [46]. Wolpert et al. assessed that the consumption of a high-fat lunch caused an average increase in insulin demand of about 42% with significant differences in individual patients [47]. A few studies have shown that the consumption of high-protein and high-fat meals results in an increase and delay in the postprandial glucose rise [48,49], indicating the need to include these nutrients in the determination of insulin doses. The VoiceDiab system make it possible to compensate for such meals using the dual-wave insulin bolus. Nevertheless, it is also possible to pre-program the system to calculate insulin doses based solely on CHO content in meals to be administered in a form of a simple bolus or an injection with a pen injector.

The question remains as to what extent errors in the calculation of CU affect glycemic control. Based on studies in groups of children and adolescents with diabetes, it has been demonstrated that an inaccuracy of ±10 g does not impair postprandial glycemia [49], but a discrepancy of ±20 g or more significantly influences glycemic control after the meal [50]. In spite of the fact that in our study, differences between CU counts estimated by the VoiceDiab system and the dietitian were significant from a statistical point of view, they were lower than ± 10 g of CHO in more than 96% of the analyzed main meals and in 100% of snack meals. They were lower than ± 20 g in the remaining 4% of the main meals. This result shows that inaccuracies that are introduced both by patients during subjective estimation of the size of a meal, and by the system which uses average characteristics of food products, should not have a significant negative impact on the setting of insulin doses and consequently, on postprandial glycemic control. Such a statement is further confirmed by (1) the results of the application of the VoiceDiab system in a group of ambulatory patients with type 1 diabetes characterized by good metabolic control and treated using insulin pumps, in whom the system proved to be effective in increasing the percentage of 2-hour postprandial glucose in the normoglycemic range by applying the insulin bolus algorithm implemented in the VoiceDiab system [18]; and (2) the results of the glycemic control of the participants of this study, where an expert determined the insulin dosages.

Through analyzing and summarizing the impact of differences in CU and PFU calculations on the resultant insulin doses estimated according to the algorithm implemented in the VoiceDiab system, we demonstrated that in over 91% of meals, the absolute difference in insulin doses were smaller than 1 U. The biggest discrepancies were noted in the calculation of pre-lunch insulin doses, where 26% of the differences were larger than 1 U.

Confirmation of the ability of the VoiceDiab system to provide meal content estimates that are similar to those calculated by a dietitian is an important step on the way to the further utilization of modern technology to support people with type 1 diabetes as well as other groups of patients who need to estimate the macronutrient contents of their meals. However, from the type 1 diabetes perspective, it should be emphasized that the automatic bolus calculator is just one element of the whole infrastructure that should be present to make it possible to adequately train and follow-up people with diabetes and ensure safe and efficient usage of new technology. This infrastructure is necessary for the patients but also for clinicians who have expertise in insulin management and are willing to accept the responsibility for ensuring that each patient receives adequate training and follow-up [51]. To reduce the risk of patients using inappropriate parameters in their daily regimens, patients should fully understand how to use bolus calculators. Otherwise, they may be exposed to the avoidable risk of potentially dangerous changes in glycemic control. Hirsch and Parkin, in their report on the safety and efficacy of smartphone bolus calculator apps, listed four key components of automated bolus calculator training: (1) determine the patient's competency in utilizing insulin therapy and self-management skills; (2) assess the appropriateness of the patient's basal dose and key insulin parameters, including the insulin sensitivity factor and ICR, blood glucose targets, and prescribed dosage adjustments for exercise and changes in health status; (3) utilize structured self-monitoring of blood glucose with patients; (4) monitor patient therapy consistently [51]. Generally, these should be key components of the training of any person with type 1 diabetes regardless of whether she or he is going to use an automatic bolus calculator. In fact, all these components were a part of the treatment plan or the education program of the patients participating in the current study. However, due to the study's main purpose and the fact that after hospitalization, the study participants had not been using the VoiceDiab system, it was not determined whether it would be possible to reproduce the results that were obtained in younger patients with good metabolic control in the previously reported study in this study group of individuals under ambulatory conditions [18]. Nevertheless, during hospitalization, the current study participants achieved lower mean PG values than those participating in the other study.

The meal content data estimated by the system can also be effectively used by the physician in a less formally-defined expert algorithm to effectively manage the intensive insulin treatment; this was confirmed by the comparable glycemic control of study participants whose insulin boluses were determined by the physician based on the reference meal data and those for whom meal content data calculated by the system was used. The differences were not significant regardless whether we analyzed indices characterizing the average daily glycemia, the glucose concentration rise after meals or the frequency and duration of hypoglycemic episodes. However, the lack of differences in glycemic control between both subgroups during this study should be interpreted with due caution, because, despite the fact that efforts were made to exclude patients with diagnosed endocrine disorders other than diabetes, it cannot be ruled out that participants might have had other medical conditions affecting their glycemic control. It should be mentioned, however, that none of the patients reported any coexistent disease or use of medications that may affect glycemic control. Hence, based on the results of the medical history that was collected, it can be stated that in case of the study participants, the most probable causes of unstable metabolic control were related to diabetes and included inappropriate health behaviors, insufficient daily blood glucose tests, fear of hypoglycemia and difficulties related to the proper adjustment of the bolus to the meal content. Nevertheless, participants could have intentionally or accidentally not informed the physician about medical conditions affecting their glycemic control.

Summing up, people with type 1 diabetes, despite education, face several difficulties in adjusting their insulin dosage based on their own estimates of the CHO content of meals. These difficulties may be even more pronounced when a complex insulin bolus is determined to compensate for not only CHO, but also protein and fat content. The developed system, which uses an intuitive user interface, is simple to use and quickly provides information on meal composition that may be used to automatically calculate prandial insulin doses. The obtained results and the literature data indicate that the accuracy of CU and PFU estimates computed by the system is sufficient to calculate insulin doses, either automatically using the algorithm implemented in the VoiceDiab system, or manually using the algorithm based on the knowledge and experience of a physician; these doses were also shown to be close to those calculated based on the reference values of CU and PFU established by the dietitian.

Supplementary Materials: The following are available online at http://www.mdpi.com/2072-6643/10/4/518/s1, Table S1: Basic characteristics of the study group, Table S2: Carbohydrate exchange units (CU), protein–fat exchange units (PFU) and calorie contents of meals calculated by the dietitian and estimated by the VoiceDiab system.

Acknowledgments: This study, including the costs to publish in open access, was funded by the National Center for Research and Development (grant No. PBS1/B9/13/2012).

Author Contributions: P.L. and J.K. conceived and designed the study; J.K., M.R. and B.B. performed the experiments and collected the data; P.L. and P.F. developed the VoiceDiab system and analyzed the data; P.L., J.K. and P.F interpreted the data and wrote the paper. P.L. and J.K. contributed equally to this paper. All the authors approved the final version of the paper.

Conflicts of Interest: The authors declare no conflict of interest.

References

1. Statista, Inc.: Technology & Telecommunications-Telecommunications: Share of Global Population that Uses a Smartphone 2014–2020. Available online: https://www.statista.com/statistics/203734/global-smartphone-penetration-per-capita-since-2005/ (accessed on 12 January 2018).
2. Stroetmann, K.A.; Kubitschke, L.; Robinson, S.; Stroetmann, V.; Cullen, K.; McDaid, D. How can telehealth help in provision of integrated care? *WHO Health Syst. Policy Anal.* **2010**, *13*, 1–39.
3. Eng, D.S.; Lee, J.M. The promise and peril of mobile health applications for diabetes and endocrinology. *Pediatr. Diabetes* **2013**, *14*, 231–238. [CrossRef] [PubMed]
4. AppBrain: Most Popular Google Play Categories. Available online: http://www.appbrain.com/stats/android-market-app-categories (accessed on 12 January 2018).
5. Daaleman, T.P. The medical home: Locus of physician formation. *J. Am. Board Fam. Med.* **2008**, *21*, 451–457. [CrossRef] [PubMed]
6. Rosenthal, T.C. The medical home: Growing evidence to support a new approach to primary care. *J. Am. Board Fam. Med.* **2008**, *21*, 427–440. [CrossRef] [PubMed]

7. Mobile Medical Applications. Guidance for Industry and Food and Drug Administration Staff. Available online: https://www.fda.gov/downloads/MedicalDevices/DeviceRegulationandGuidance/GuidanceDocu-ments/UCM263366.pdf (accessed on 12 February 2018).

8. Forster, H.; Walsh, M.C.; Gibney, M.J.; Brennan, L.; Gibney, E.R. Personalised nutrition: The role of new dietary assessment methods. *Proc. Nutr. Soc.* **2016**, *75*, 96–105. [CrossRef] [PubMed]

9. Coughlin, S.S.; Whitehead, M.; Sheats, J.Q.; Mastromonico, J.; Hardy, D.; Smith, S.A. Smartphone applications for promoting healthy diet and nutrition: A literature review. *Jacobs J. Food Nutr.* **2015**, *2*, 1–14.

10. Stumbo, P.J. New technology in dietary assessment: A review of digital methods in improving food record accuracy. *Proc. Nutr. Soc.* **2013**, *72*, 70–76. [CrossRef] [PubMed]

11. Forster, H.; Fallaize, R.; Gallagher, C.; O'Donovan, C.B.; Woolhead, C.; Walsh, M.C.; Macready, A.L.; Lovegrove, J.A.; Mathers, J.C.; Gibney, M.J.; et al. Online dietary intake estimation: The Food4Me food frequency questionnaire. *J. Med. Internet Res.* **2014**, *16*, e150. [CrossRef] [PubMed]

12. Sharp, D.B.; Allman-Farinelli, M. Feasibility and validity of mobile phones to assess dietary intake. *Nutrition* **2014**, *30*, 1257–1266. [CrossRef] [PubMed]

13. Free, C.; Phillips, G.; Galli, L.; Watson, L.; Felix, L.; Edwards, P.; Patel, V.; Haines, A. The effectiveness of mobile-health technology-based health behaviour change or disease management interventions for health care consumers: A systematic review. *PLoS Med.* **2013**, *10*, e1001362. [CrossRef] [PubMed]

14. Carter, M.C.; Burley, V.J.; Nykjaer, C.; Cade, J.E. 'My Meal Mate' (MMM): Validation of the diet measures captured on a smartphone application to facilitate weight loss. *Br. J. Nutr.* **2013**, *109*, 539–546. [CrossRef] [PubMed]

15. Mezgec, S.; Koroušic Seljak, B. NutriNet: A deep learning food and drink image recognition system for dietary assessment. *Nutrients* **2017**, *9*, 657. [CrossRef] [PubMed]

16. Sun, M.; Burke, L.E.; Mao, Z.H.; Chen, Y.; Chen, H.C.; Bai, Y.; Li, Y.; Li, C.; Jia, W. eButton: A wearable computer for health monitoring and personal assistance. In Proceedings of the 2014 51st ACM/EDAC/IEEE Design Automation Conference (DAC), San Francisco, CA, USA, 1–5 June 2014.

17. O'Loughlin, G.; Cullen, S.J.; McGoldrick, A.; O'Connor, S.; Blain, R.; O'Malley, S.; Warrington, G.D. Using a wearable camera to increase the accuracy of dietary analysis. *Am. J. Prev. Med.* **2013**, *44*, 297–301. [CrossRef] [PubMed]

18. Foltynski, P.; Ladyzynski, P.; Pankowska, E.; Mazurczak, K. Efficacy of automatic bolus calculator with automatic speech recognition in patients with type 1 diabetes: A randomized crossover trial. *J. Diabetes* **2018**. [CrossRef] [PubMed]

19. Mazurczak, K.; Pankowska, E.; Ladyzynski, P.; Foltynski, P. The first use of bolus calculator with speech analyzer. *J. Diabetes Sci. Technol.* **2017**, *11*, 7–11. [CrossRef] [PubMed]

20. Pasler, S.; Wolff, M.; Fischer, W.J. Food intake monitoring: An acoustical approach to automated food intake activity detection and classification of consumed food. *Physiol. Meas.* **2012**, *33*, 1073–1093. [CrossRef] [PubMed]

21. Sazonov, E.; Schuckers, S.; Lopez-Meyer, P.; Makeyev, O.; Sazonova, N.; Melanson, E.L.; Neuman, M. Non-invasive monitoring of chewing and swallowing for objective quantification of ingestive behavior. *Physiol. Meas.* **2008**, *29*, 525–541. [CrossRef] [PubMed]

22. Calvert, M.; Shankar, A.; McManus, R.J.; Lester, H.; Freemantle, N. Effect of the quality and outcomes framework on diabetes care in the United Kingdom: Retrospective cohort study. *BMJ* **2009**, *338*, b1870. [CrossRef] [PubMed]

23. Charpentier, G.; Benhamou, P.Y.; Dardari, D.; Clergeot, A.; Franc, S.; Schaepelynck-Belicar, P.; Catargi, B.; Melki, V.; Chaillous, L.; et al. The Diabeo software enabling individualized insulin dose adjustments combined with telemedicine support improves HbA1c in poorly controlled type 1 diabetic patients: A 6-month, randomized, open-label, parallel-group, multicenter trial (TeleDiab 1 Study). *Diabetes Care* **2011**, *34*, 533–539. [CrossRef] [PubMed]

24. The Diabetes Control and Complications Trial (DCCT)/Epidemiology of Diabetes Interventions and Complications (EDIC) Study Research Group. Mortality in type 1 diabetes in the DCCT/EDIC versus the general population. *Diabetes Care* **2016**, *39*, 1378–1383. [CrossRef]

25. Tascini, G.; Berioli, M.G.; Cerquiglini, L.; Santi, E.; Mancini, G.; Rogari, F.; Toni, G.; Esposito, S. Carbohydrate counting in children and adolescents with type 1 diabetes. *Nutrients* **2018**, *10*, 109. [CrossRef] [PubMed]

26. Kordonouri, O.; Hartmann, R.; Remus, K.; Blasig, S.; Sadeghian, E.; Danne, T. Benefit of supplementary fat plus protein counting as compared with conventional carbohydrate counting for insulin bolus calculation in children with pump therapy. *Pediatr. Diabetes* **2012**, *13*, 540–544. [CrossRef] [PubMed]

27. Chase, H.P.; Saib, S.Z.; Mac Kenzie, T.; Hansen, M.M.; Garg, S.K. Post-prandial glucose excursions following four methods of bolus insulin administration in subjects with type 1 diabetes. *Diabet. Med.* **2002**, *19*, 317–321. [CrossRef] [PubMed]

28. Lee, S.W.; Cao, M.; Sajid, S.; Hayes, M.; Choi, L.; Rother, C.; de León, R. The dual-wave bolus feature in continuous subcutaneous insulin infusion pumpscontrols prolonged post-prandial hyperglycaemia better than standard bolus in type 1 diabetes. *Diabetes Nutr. Metab.* **2004**, *17*, 211–216. [PubMed]

29. Foltynski, P.; Ladyzynski, P.; Wojcicki, J.M.; Pankowska, E.; Migalska-Musial, K.; Krzymien, J. System for calculating of compensating insulin dose from voice description of meal. In Proceedings of the 6th European Conference of the International Federation for Medical and Biological Engineering (MBEC 2014), Dubrovnik, Croatia, 7–11 September 2014; Lacković, I.; Vasić, D., Eds. *IFMBE Proc.* **2015**, *45*, 728–731. [CrossRef]

30. Pankowska, E.; Szypowska, A.; Lipka, M.; Szpotanska, M.; Blazik, M.; Groele, L. Application of novel dual wave meal bolus and its impact on glycated hemoglobin A1c level in children with type 1 diabetes. *Pediatr. Diabetes.* **2009**, *10*, 298–303. [CrossRef] [PubMed]

31. Ladyzynski, P.; Wojcicki, J.M.; Krzymien, J.; Blachowicz, J.; Jóźwicka, E.; Czajkowski, K.; Janczewska, E.; Karnafel, W. Teletransmission system supporting intensive insulin treatment of out-clinic type 1 diabetic pregnant women. Technical assessment during 3 years' application. *Int. J. Artif. Organs* **2001**, *24*, 157–163. [CrossRef] [PubMed]

32. Albisser, A.M.; Harris, R.I.; Sakkal, S.; Parson, I.D.; Chao, S.C. Diabetes intervention in the information age. *Med. Inform.* **1996**, *21*, 297–316. [CrossRef]

33. Larizza, C.; Bellazzi, R.; Stefanelli, M.; Ferrari, P.; De Cata, P.; Gazzaruso, C.; Fratino, P.; D'Annunzio, G.; Hernando, E.; Gomez, E.J. The M^{2}DM project—The experience of two Italian clinical sites with clinical evaluation of a multi-access service for the management of diabetes mellitus patients. *Methods Inf. Med.* **2006**, *45*, 79–84. [PubMed]

34. Wojcicki, J.M.; Ladyzynski, P.; Krzymien, J.; Jozwicka, E.; Blachowicz, J.; Janczewska, E.; Czajkowski, K.; Karnafel, W. What we can really expect from telemedicine in intensive diabetes treatment: Results from 3-year study on type 1 pregnant diabetic women. *Diabetes Technol. Ther.* **2001**, *3*, 581–589. [CrossRef] [PubMed]

35. Montori, V.M.; Helgemoe, P.K.; Guyatt, G.H.; Dean, D.S.; Leung, T.W.; Smith, S.A.; Kudva, Y.C. Telecare for patients with type 1 diabetes and inadequate glycemic control: A randomized controlled trial and meta-analysis. *Diabetes Care* **2004**, *27*, 1088–1094. [CrossRef] [PubMed]

36. Farmer, A.; Gibson, O.; Hayton, P.; Bryden, K.; Dudley, C.; Neil, A.; Tarassenko, L. A real-time, mobile phone-based telemedicine system to support young adults with type 1 diabetes. *Inform. Prim. Care* **2005**, *13*, 171–177. [CrossRef] [PubMed]

37. Ladyzynski, P.; Wojcicki, J.M.; Krzymien, J.; Foltynski, P.; Migalska-Musial, K.; Tracz, M.; Karnafel, W. Mobile telecare system for intensive insulin treatment and patient education. First applications for newly diagnosed type 1 diabetic patients. *Int. J. Artif. Organs* **2006**, *29*, 1074–1081. [CrossRef] [PubMed]

38. Franc, S.; Charpentier, G. Devices to support treatment decisions in type 1 diabetes: The Diabeo system. In *Technological Advances in the Treatment of Type 1 Diabetes*; Bruttomesso, D., Grassi, G., Eds.; Front Diabetes: Basel, Karger, 2015; Volume 24, pp. 210–225. [CrossRef]

39. Franc, S.; Borot, S.; Ronsin, O.; Quesada, J.L.; Dardari, D.; Fagour, C.; Renard, E.; Leguerrier, A.M.; Vigeral, C.; Moreau, F.; et al. Telemedicine and type 1 diabetes: Is technology per se sufficient to improve glycaemic control? *Diabetes Metab.* **2014**, *40*, 61–66. [CrossRef] [PubMed]

40. Krzymien, J.; Rachuta, M.; Kozlowska, I.; Ladyzynski, P.; Foltynski, P. Treatment of patients with type 1 diabetes—Insulin pumps or multiple injections? *Biocyber. Biomed. Eng.* **2016**, *36*, 1–8. [CrossRef]

41. Brazeau, A.S.; Mircescu, H.; Desjardins, K.; Leroux, C.; Strychar, I.; Ekoé, J.M.; Rabasa-Lhoret, R. Carbohydrate counting accuracy and blood glucose variability in adults with type 1 diabetes. *Diabetes Res. Clin. Pract.* **2013**, *99*, 19–23. [CrossRef] [PubMed]

42. Bishop, F.K.; Maahs, D.M.; Spiegel, G.; Owen, D.; Klingensmith, G.J.; Bortsov, A.; Thomas, J.; Mayer-Davis, E.J. The carbohydrate counting in adolescents with type 1 diabetes (CCAT) study. *Diabetes Spectrum.* **2009**, *22*, 56–62. [CrossRef]

43. Meade, L.T.; Rushton, W.E. Accuracy of carbohydrate counting in adults. *Clin. Diabetes* **2016**, *34*, 142–147. [CrossRef] [PubMed]

44. Rhyner, D.; Loher, H.; Dehais, J.; Anthimopoulos, M.; Shevchik, S.; Botwey, R.H.; Duke, D.; Stettler, C.; Diem, P.; Mougiakakou, S. Carbohydrate estimation by a mobile phone-based system versus self-estimations of individuals with type 1 diabetes mellitus: A comparative study. *J. Med. Internet Res.* **2016**, *18*, e101. [CrossRef] [PubMed]

45. Davis, N. The role of food diaries in diabetes self care. *Today's Dietician* **2014**, *16*, 14.

46. Bell, K.J.; Toschi, E.; Steil, G.M.; Wolpert, H.A. Optimized mealtime insulin dosing for fat and protein in type 1 diabetes: Application of a model-based approach to derive insulin doses for open-loop diabetes management. *Diabetes Care* **2016**, *39*, 1631–1634. [CrossRef] [PubMed]

47. Wolpert, H.A.; Atakov-Castillo, A.; Smith, S.A.; Steil, G.M. Dietary fat acutely increases glucose concentrations and insulin requirements in patients with type 1 diabetes: Implications for carbohydrate-based bolus dose calculation and intensive diabetes management. *Diabetes Care* **2013**, *36*, 810–816. [CrossRef] [PubMed]

48. Pankowska, E.; Blazik, M.; Groele, L. Does the fat-protein meal increase postprandial glucose level in type 1 diabetes patients on insulin pump: The conclusion of a randomized study. *Diabetes Technol. Ther.* **2012**, *14*, 16–22. [CrossRef] [PubMed]

49. Smart, C.E.M.; Evans, M.; O'Connell, S.M.; McElduff, P.; Lopez, P.E.; Jones, T.W.; Davis, E.A.; King, B.R. Both dietary protein and fat increase postprandial glucose excursions in children with type 1 diabetes, and the effect is additive. *Diabetes Care* **2013**, *36*, 3897–3902. [CrossRef] [PubMed]

50. Smart, C.E.; King, B.R.; McElduff, P.; Collins, C.E. In children using intensive insulin therapy, a 20-g variation in carbohydrate amount significantly impacts on postprandial glycaemia. *Diabet. Med.* **2012**, *29*, e21–e24. [CrossRef] [PubMed]

51. Hirsch, I.B.; Parkin, C.G. Unknown safety and efficacy of smartphone bolus calculator apps puts patients at risk for severe adverse outcomes. *J. Diabetes Sci. Technol.* **2016**, *10*, 977–980. [CrossRef] [PubMed]

 *nutrients*

Article

Pilot Testing a Photo-Based Food Diary in Nine- to Twelve-Year Old Children from Dunedin, New Zealand

Brittany K. Davison [1], Robin Quigg [2] and Paula M. L. Skidmore [1,*

[1]   Department of Human Nutrition, University of Otago, Dunedin 9054, New Zealand;
      brittany_davison@hotmail.com

[2]   Cancer Society Social and Behavioural Research Unit, Department of Preventive and Social Medicine,
      Dunedin School of Medicine, University of Otago, Dunedin 9054, New Zealand; robin.quigg@otago.ac.nz

*   Correspondence: paula.skidmore@otago.ac.nz; Tel.: +64-03-479-8374

Received: 17 December 2017; Accepted: 9 February 2018; Published: 20 February 2018

Abstract: The purpose of the study was to investigate if an Evernote app-based electronic food diary is an acceptable method to measure nutrient intake in children aged 9–12 years. A convenience sample of 16 nine- to twelve-year-olds from Dunedin, New Zealand, completed a paper-based food dairy on four days, followed by four more days using a photo-based diary on an iPod. This photo-based diary used a combination of photographs and short written descriptions of foods consumed. The photo-based diaries produced similar results to written diaries for all macronutrients and major micronutrients (e.g., calcium, fibre, vitamin C). Spearman correlation coefficients between the two methods for all nutrients, except sugars, were above 0.3. However, burden on researchers and participants was reduced for the photo-based diary, primarily due to the additional information obtained from photographs. Participating children needed less help from parents with completing the electronic diaries and preferred them to the paper version. This electronic diary is likely to be suitable, after additional formal validity testing, for use in measuring nutrient intake in children.

Keywords: children; dietary assessment; nutrients

1. Introduction

A healthy diet is essential in childhood because it is associated with current and future health [1,2]. Therefore, dietary assessment methods appropriate for children are vital. Several traditional methods are used to measure energy and nutrient intake, including 24-h recalls, food frequency questionnaires (FFQs) and food records. The 24-h recall method is used in large-scale surveys in New Zealand, such as the New Zealand Adult Nutrition Survey [3]. An important limitation of this method is recall bias, where people cannot accurately remember everything they consumed [4]. Using 24-h recalls is especially difficult in children as several people may need to be interviewed to ensure all the food consumed is accurately reported [5]. For example, parents, teachers and friends' parents may need to be consulted depending on where the child was on the specified day [5]. Children may not always be able to accurately describe what they have eaten. For similar reasons, while comprehensive self-completed FFQs are used commonly in large-scale studies of adults to determine long-term dietary intake [4,5], they are not ideal for comprehensive dietary assessment in children.

Weighed food diaries are the gold standard of dietary assessment, but are not suitable for children as this method is time consuming and places a high burden on participants and caregivers. Estimated food diaries are an alternative method, but accurate portion size estimation is an issue for children, requiring additional parental help [5]. A more recent method of collecting dietary data is using food photography. This involves participants taking photos of the food and drink they are going to consume, then another photo when they are finished [6]. A description of the photo can be added to provide extra information. The image can be used to determine what the person is eating and how much of it he/she ate. Benefits of this method are that portion size does not need to be estimated and there is a low participant burden [6]. Research suggests that younger people are more compliant with electronic nutrient data collection methods compared to paper-based methods (75% compared to 50% compliance) [7]. An increasing number of young people have smartphones, or other smart devices, therefore developing an electronic application (app) for these may be a cost-effective, low burden method of data collection. Previous U.S. research results suggest food photography is a valid and practical way to measure adult nutrient intake [8] and children's food intake [9] when compared to food records. Therefore, the aim of this pilot study was to determine the acceptability of an Evernote app-based food diary (photo-based diary) on an iPod for measuring dietary intake in children, in comparison to traditional written food records, and to assess its usability in this population.

2. Materials and Methods

2.1. Subjects and Study Design

We aimed to recruit a convenience sample of 16 children aged nine to twelve years recruited from Dunedin, New Zealand via word of mouth. Investigators asked people they knew with children of eligible age or those who worked with children of eligible age to contact us for information if they wanted to take part. They in turn also spoke to parents of eligible children about the study.

Only children who (a) were literate and therefore able to complete the diaries, either on their own, or with help from parents, (b) were available throughout the study period (i.e., not going away from Dunedin during the study period) and (c) gave permission for audio-recording of the group interview (where applicable) were eligible to participate.

Parents and children were required to provide written informed consent before entering the study. Ethical approval for the study was obtained from the University of Otago Human Ethics Committee (Ref 13/265, 20 September 2013).

2.2. Data Collection

We used the sequential explanatory mixed method for this study. Participants and their caregiver met with a researcher and were first given a written food diary, as the reference method, to record all food and drinks consumed over 4 days. Child/parent pairs were given verbal instructions on how to complete the diary, and written instructions were contained within the diary. Both of these were tailored to be understood by children of eligible age. Child participants were asked to complete the diary on 4 non-consecutive days, including a weekend day, with help from their primary caregiver if necessary. Participants were asked to include brand names of food and drinks to improve the accuracy of the final results. Where, possible the researcher would meet with the participant and their parent after the first day of recording to ensure all the necessary information was recorded.

A few days after completing the written food diary, each participant was given an iPod with the Evernote app on it. The Evernote app contained a basic food dairy, set up in the same way as the paper diary. This photo-based diary contained defined sections to record each eating occasion with space to write a short summary of the food and drink consumed. It also contained a designated space to add photographs of the meal before and after consumption, to estimate the proportion of food consumed. Participants were shown how to use Evernote, to photograph effectively and record details of each entry underneath each photo. As with the written diaries, they were provided with tailored instructions and examples; these focused on the electronic aspects of recording intake. Participants were asked to complete the photo-based diary on 4 non-consecutive days, including a weekend day. After all diet records were completed, all child and parent participants were invited to group interviews to gain feedback on the photo-based diary, particularly with respect to its ease of use compared to the paper diary. Child participants were invited to 1 interview and parents to a separate interview afterwards, so that any additional topics of interest that arose from the child interview could be covered in the parent interview.

2.3. Food Record Coding and Statistical Analyses

A trained researcher entered data from all food diaries into Kai-culator, a bespoke dietary assessment software application developed by the Department of Human Nutrition, which uses the 2014 version of the New Zealand food composition database 'NZ FOODfiles' (Version 1.08d, Department of Human Nutrition, University of Otago, Dunedin, New Zealand).

The photographs obtained from the photo-based diary were used to augment written information provided by participants, including pictures of additional helpings, if present. If foods in the diaries were not in the database, a similar product was substituted. For example, one participant consumed a German-made chocolate biscuit. The nutrient data for this product were searched for on Google, and the closest matched New Zealand biscuit was used.

When insufficient data were available to match food exactly, standardized substitutions were assumed. For example, if a 'handful' or 'scoopful' of hot chips was recorded in the food diary, the quantity was estimated if there was a photo, using standard portion photos developed for use in New Zealand national nutrition surveys, or it was assumed to be equal to 144 g, a typical portion size for this age group in New Zealand, using data from the most recent national survey. Nutrient information was obtained for all participants from all diaries, and simple descriptive statistics (mean and SD) were undertaken. Spearman's correlation coefficients (SCC) were calculated to assess agreement between the nutrient information obtained from the electronic and written food diaries. As suggested by experts in the field of dietary assessment methodology [10,11], SCC of 0.3 and above were considered acceptable.

3. Results

All participants completed at least three days for both the paper and photo-based diaries. A total of 64 days of entries from a possible 64 was included in the final analysis of the electronic diaries and 58 days for the written diaries. The results from Table 1 show that nutrient intakes generated from the photo-based diary were similar to those from the written food diary for all participants together and for boys and girls separately. SCCs for all participants for all nutrients, with the exception of sugars, were above 0.3. SCCs conducted for boys and girls separately showed similar results with the exception of sugars, where the SCC for girls was 0.3. Intakes from the written and photo-based diaries were broadly comparable to, but lower than those intakes from children who participated in the most recent Children's Nutrition Survey in New Zealand (CNS) in 2002 [12]. When data from boys and girls were combined, carbohydrate intake was 30 g higher from the written diary (around 5% of a child's energy intake) compared to the photo-based diary. Boys had a higher energy intake than girls, as expected.

Table 1. Summary of key nutrients from written and photo-based diaries ($n = 16$) compared to data from the Children's Nutrition Survey in New Zealand (CNS) plus SCC for written and photo-based diaries.

	Written Food Diary [1]			Photo-Based Diary [1]			CNS Data 7–10-Year-Olds [2]		SCC [3]
	All Participants	Males ($n = 8$)	Females ($n = 8$)	All Participants	Males ($n = 8$)	Females ($n = 8$)	Males	Females	All Participants
Energy (kJ/d)	6825 ± 853	7502 ± 1034	6444 ± 684	6615 ± 1006	7309 ± 162	6273 ± 1079	9015	7844	0.73 **
Protein (g/d)	62.8 ± 12.0	63.0 ± 15.9	58.9 ± 8.18	66.1 ± 10.5	71.7 ± 8.82	67.1 ± 14.2	73	64	0.32
Total fat (g/d)	64.1 ± 13.6	67.4 ± 19.0	59.0 ± 7.75	63.3 ± 11.1	70.3 ± 2.47	61.9 ± 13.1	81	70	0.61 *
Carbohydrate (g/d)	203 ± 27.4	237 ± 8.35	196 ± 31.1	189 ± 40.0	210 ± 11.9	171 ± 45.5	286	251	0.59 *
Fibre (g/d)	19.0 ± 4.60	24.7 ± 3.80	17.0 ± 3.86	18.4 ± 4.53	19.0 ± 3.07	17.9 ± 5.17	19.1	16.8	0.63 *
Saturated fat (g/d)	28.3 ± 5.94	29.6 ± 7.60	25.4 ± 3.53	25.9 ± 5.90	31.0 ± 3.80	26.5 ± 6.52	35.8	30.5	0.44 *
Sugar (g/d)	87.3 ± 18.2	96.9 ± 18.4	94.8 ± 21.0	80.2 ± 18.4	80.8 ± 5.54	68.5 ± 16.0	130	115	−0.03
Calcium (mg/d)	737 ± 280	592 ± 217	704 ± 153	727 ± 222	608 ± 47.1	796 ± 195	806	653	0.83 **
Vitamin C (mg/d)	77.4 ± 32.1	107 ± 18.0	56.2 ± 15.1	71.1 ± 39.3	68.9 ± 30.8	46.1 ± 22.2	123	103	0.35

[1] Mean ± SD, [2] New Zealand Children's Nutrition Survey data 2002, $n = 3275$, mean, [3] SCC = Spearman's correlation coefficients for written and photo-based food diaries, * $p < 0.05$, ** $p < 0.0001$.

The information provided in the photo-based diary made data entry easier and more straightforward than the written diary for several reasons. Firstly, typed information was easier to read than the children's handwriting. Photos in the photo-based diary provided additional information compared to what children included in some of the written diaries. Examples of this were a lack of detail in the written entries on foods that do not make up a main component of a meal, e.g., not documenting tomato sauce when consuming chips, grated cheese added to the top of pies, custard or cream added to a cake, or the exact composition of sandwiches. When reported in both diaries, some cheese sandwich photographs from the electronic food diaries showed additional food information not commonly reported in the written food diaries, such as tomatoes, vegetables or salad and bread type. Similarly, the amounts of butter added to toast or nuts contained in a handful were able to be more accurately estimated from photographs than from the paper diaries. Investigation of the photographs showed that other foods such as gravy or sauces tended not to be reported in the written diary or in the photo-based diary, even though these could be seen in the photographs. Drinks were not always reported by all participants, even when a can or carton of a drink was photographed as part of a meal.

Five children and three parents attended the group interviews, and the other child and parent participants provided written feedback on the photo-based diary. Individual feedback from all participants and their parents showed that the photo-based diary was more acceptable than the written diary, as it reduced the burden of writing everything down. If the children were unable to record everything in writing at the time of consumption, e.g., at a birthday party or a family gathering, they appreciated being able to photograph the food and add additional information when they returned home. There were several other examples of this, particularly relating to busy time periods. One was where participants made their school lunch the night before. They took a photo then, and as they did not have time to complete a full diary entry at lunchtime, they took a picture of the leftovers at the time and provided the written text that evening.

Parents and children reported that the iPod had the advantage of novelty over the written diary, which led to the children being more motivated to fill in the photo-based diary, compared to the written diary. Specific comments from the children were that the iPod is 'a lot more portable than the paper diary', 'writing everything down is boring' and 'the autocorrect for spelling in the iPod diary also make recording everything easy'. Two children reported that the electronic diaries were quicker to complete than the paper diaries, but the other three said that the two methods took similar amounts of time. Two participants needed help from parents to fill in the paper diary as 'It's neater when my mum does it', but were able to complete the photo-based diary on their own. Similarly, all participants had some help with estimating portion sizes for the paper diaries from parents. This was corroborated by comments from parents such as 'I did quite a lot of the writing' and 'My daughter was more concerned with how she was spelling things and whether people would be able to read, so I did most of the writing on the paper diaries'.

Generally, although participants preferred the photo-based diary, one participant found the keyboard on the iPod too small, and 'struggled to type' as it is 'really hard because you might type the wrong letter but when you have the book you can write it correctly'. When this was discussed further, all participants thought it would be easy to complete on a larger smartphone. One participant was not allowed to use the iPod at school, but they took photos of all the food they took to school and kept leftovers and packaging of food bought at school to photograph at the end of the school day. Reponses from parents indicated that they preferred the photo-based diary as less input was needed from them than for the paper diary. When parents helped with the photo-based diary, it was with the food descriptions, not photographs, e.g., how a particular food was cooked or information on some of the ingredients in composite dishes. One participant noted that their 'mum doesn't like technology', but that they were willing for them to use it for 'important' purposes, such as school or research; the other participants agreed with this comment.

4. Discussion

Electronic food diaries produced similar results to written food diaries in children aged 9–12 years. Nearly all nutrient values were comparable between the food diaries. The only substantial difference was a higher carbohydrate intake from the iPod diaries. As we found acceptable correlations of 0.3 or above [10,11] for all nutrients except sugar, the results indicate that the photo-based diary shows promise as a valid dietary assessment tool for this age group. However, we must interpret these results with caution, due to the small sample size, and a further, larger validity study is needed to confirm these results. However, these results strengthen current research that suggests that technology may be an appropriate tool to measure nutrient intake [8,9].

In terms of comparing our study results with those from the only available nationally-representative dataset, the results are broadly comparable, although some small differences are seen. This is likely to be due to the different time periods in which data were collected as the CNS data are from 2002. Data from the two most recent adult nutrition surveys suggest that major changes in food consumption have occurred in New Zealand over the 10-year period between these surveys [13], which may account for some of the differences seen. The CNS used one 24-h recall from a parent to collect dietary information, which may have contributed to the observed differences.

Photo-based collection methods are becoming more feasible due the wide-spread use of smartphones by people of all ages, including older children. Current smartphones with high quality cameras are now available and inexpensive [13], meaning that this is technology accessible to most people, and smartphones, or other devices, can be provided by researchers to those without. iPods were provided for children as they do not generally have access to smart phones. The photo-based diary is suitable for and has been tested on a variety of Android-based mobile phones, as well as iPhones and iPads.

Electronic food diaries have the benefit of providing extra details not always included in written food diaries. Items such as mayonnaise on chips are often omitted from food diaries, possibly resulting in underreporting of energy intake [7–9]. As such foods can be seen in photos, they can be accounted for by the researcher, possibly leading to more accurate results [7–9], and this may explain the additional carbohydrate intakes seen in the iPod data. An important advantage of electronic food diaries is that it is that researchers can more easily gain enough information to enter food records into the database for more accurate nutrient intake estimates. When using the paper food diary, parents often had to be asked extra questions about their child's food intake because not enough detail was provided, e.g., how many slices of bread a sandwich had or how many potatoes were eaten with dinner. The photo-based diary overcomes this because it can be seen clearly in the photos how much food was eaten. This reduces researcher and participant burden. A further advantage is that leftovers can be photographed. Children often do not eat all that is provided; recording these results accurately is important.

Another important benefit of taking photos of food and drinks is that the burden of estimating portion size is transferred from the participant to the researcher. Children find it difficult to accurately estimate the amount of food they are consuming, resulting in inaccurate nutrient intake results [9]. Researchers and dietitians who have training in portion size estimation have been shown to produce estimates that highly correlate with weighed food portions [14]. Electronic food diaries are therefore likely to have a high accuracy level.

Study limitations included the small sample size and representation of New Zealand's population; some results and feedback may not be generalizable to all New Zealand children. During this study, furthermore, children sometimes forgot to take the iPod with them and were forced to rely on memory to write down what they had eaten at a later time. This may have reduced the accuracy of the results as no photo was provided and information was not recorded straight away. However, the same problem occurred with the written food diary, showing that the photo-based diary should be no less accurate than a written diary. An interesting observation was some parents and one school seemed to resist technology use, such as portable devices, but these devices were permitted to be used for academic or research purposes. This challenge would require further thought in a larger validation study.

A further possible limitation is that by asking all participants to complete the paper diary first, followed almost immediately (usually 4–7 days later) by the photo-based diary could lead to over-estimation of the agreement between methods. Indeed, experts in the validation of dietary methods recommend that a sufficient time period elapse between completion of the dietary assessment methods used to minimise learning effects. However, most of these recommendations come from the field of FFQ validation [11], where the learning effects are primarily related to test-retest reliability. As the principal of the two methods tested in this study was essentially the same, the information provided around writing down information in both diaries was the same and the only additional instruction for the photo-based dairy was based on the photos, learning effects should be minimal between the two methods. However, we acknowledge that (a) randomising participants as to which diary they complete first and (b) using a greater washout period may have led to lower agreement between methods. It is important to note that this study is not a formal validity study. It was designed to initially assess the feasibility of the diary as a first step before a larger, formal validity study with a larger sample size and appropriate statistical methods such as Bland–Altman.

In conclusion, this study supports the use of electronic food diaries in children, pending further formal validation. Electronic food diaries produce comparable results to written food diaries, have the advantage of being more fun for participants to fill in and provide more information to facilitate data entry for researchers. Using iPods reduces the burden on participants by replacing the need to write down comprehensive descriptions of food consumed with information from photos and may have an important role in child research in the future.

5. Conclusions

This novel method of dietary data collection reduces burden for participants and researchers, and allows for the more accurate coding of diet records, as it requires less estimation around portion sizes from participants [4]. The detail available from the photographs makes coding decisions more straightforward than from traditional diaries. As participants reported enjoying completing the electronic diaries, greater compliance may be seen in larger studies compared to paper diaries.

Acknowledgments: Brittany K. Davison was funded by a Summer Scholarship from the Foodstuffs Community Trust, New Zealand.

Author Contributions: Brittany K. Davison, Robin Quigg and Paula M. L. Skidmore conceived of and designed the project. Brittany K. Davison collected and analysed the data and wrote the first draft of the manuscript under the supervision of Robin Quigg and Paula M. L. Skidmore. Robin Quigg and Paula M. L. Skidmore provided critical reviews of the manuscript.

Conflicts of Interest: The authors declare no conflict of interest.

References

1. Craigie, A.M.; Lake, A.A.; Kelly, S.A.; Adamson, A.J.; Mathers, J.C. Tracking of obesity-related behaviours from childhood to adulthood: A systematic review. *Maturitas* **2011**, *70*, 266–284. [CrossRef] [PubMed]
2. Kaikkonen, J.E.; Mikkilä, V.; Raitakari, O.T. Role of childhood food patterns on adult cardiovascular disease risk. *Curr. Atheroscler. Rep.* **2014**, *16*, 443. [CrossRef] [PubMed]
3. University of Otago; Ministry of Health. *A Focus on Nutrition: Key Findings of the 2008/09 New Zealand Adult Nutrition Survey*; Ministry of Health: Wellington, New Zealand, 2011.
4. Lee, R.D.; Nieman, D.C. *Nutritional Assessment*, 6th ed.; McGraw-Hill Higher Education: New York, NY, USA, 2012.
5. Livingstone, M.B.E.; Robson, P.J.; Wallace, J.M.W. Issues in dietary intake assessment of children and adolescents. *Br. J. Nutr.* **2004**, *92*, S213–S222. [CrossRef] [PubMed]
6. Martin, C.K.; Han, H.; Coulon, S.M.; Allen, H.R.; Champagne, C.M.; Anton, S.D. A novel method to remotely measure food intake of free-living individuals in real time: The remote food photography method. *Br. J. Nutr.* **2009**, *101*, 446–456. [CrossRef] [PubMed]

7. Stumbo, P.J. New technology in dietary assessment: A review of digital methods in improving food record accuracy. *Proc. Nutr. Soc.* **2013**, *72*, 70–76. [CrossRef] [PubMed]

8. Williamson, D.A.; Allen, H.R.; Davis, M.P.; Alfonso, A.J.; Gerald, B.; Hunt, A. Comparison of digital photography to weighed and visual estimation of portion sizes. *J. Am. Diet. Assoc.* **2003**, *103*, 1139–1145. [CrossRef]

9. Martin, C.K.; Newton, R.L., Jr.; Anton, S.D.; Allen, R.; Alfonso, A.; Han, H.; Stewart, T.; Sothern, M.; Williamson, D.A. Measurement of children's food intake with digital photography and the effects of second servings upon food intake. *Eat. Behav.* **2007**, *8*, 148–156. [CrossRef] [PubMed]

10. Willett, W.C. *Nutritional Epidemiology*, 3rd ed.; Oxford University Press: New York, NY, USA, 2012.

11. Cade, J.E.; Burley, V.J.; Warm, D.L.; Thompson, R.L.; Margetts, B.M. Development, validation and utilisation of food-frequency questionnaires—A review. *Public Health Nutr.* **2002**, *5*, 567–587. [CrossRef] [PubMed]

12. Parnell, W.; Scragg, R.; Wilson, N.; Schaaf, D.; Fitzgerald, E. *NZ Food, NZ Children: Key Results of the 2002 National Children's Nutrition Survey*; Ministry of Health: Wellington, New Zealand, 2013. Available online: https://www.health.govt.nz/system/files/documents/publications/nzfoodnzchildren.pdf (accessed on 12 January 2017).

13. Smith, C.; Gray, A.R.; Mainvil, L.A.; Fleming, E.A.; Parnell, W.R. Secular changes in intakes of foods among New Zealand adults from 1997 to 2008/09. *Public Health Nutr.* **2015**, *18*, 3249–3259. [CrossRef] [PubMed]

14. Kikunaga, S.; Tin, T.; Ishibashi, G.; Wang, D.H.; Kira, S. The application of a handheld personal digital assistant with camera and mobile phone card (Wellnavi) to the general population in a dietary survey. *J. Nutr. Sci.* **2007**, *53*, 109–116. [CrossRef]

 *nutrients*

Review

Assessing Eating Behaviour Using Upper Limb Mounted Motion Sensors: A Systematic Review

Hamid Heydarian [1], **Marc Adam** [1,2], **Tracy Burrows** [2,3], **Clare Collins** [2,3] and **Megan E. Rollo** [2,3,*]

1   School of Electrical Engineering and Computing, Faculty of Engineering and Built Environment, The University of Newcastle, Callaghan, NSW 2308, Australia; hamid.heydarian@uon.edu.au (H.H.); marc.adam@newcastle.edu.au (M.A.)
2   Priority Research Centre for Physical Activity and Nutrition, The University of Newcastle, Callaghan, NSW 2308, Australia; tracy.burrows@newcastle.edu.au (T.B.); clare.collins@newcastle.edu.au (C.C.)
3   School of Health Sciences, Faculty of Health and Medicine, The University of Newcastle, Callaghan, NSW 2308, Australia
*   Correspondence: megan.rollo@newcastle.edu.au

Received: 28 February 2019; Accepted: 22 May 2019; Published: 24 May 2019

Abstract: Wearable motion tracking sensors are now widely used to monitor physical activity, and have recently gained more attention in dietary monitoring research. The aim of this review is to synthesise research to date that utilises upper limb motion tracking sensors, either individually or in combination with other technologies (e.g., cameras, microphones), to objectively assess eating behaviour. Eleven electronic databases were searched in January 2019, and 653 distinct records were obtained. Including 10 studies found in backward and forward searches, a total of 69 studies met the inclusion criteria, with 28 published since 2017. Fifty studies were conducted exclusively in laboratory settings, 13 exclusively in free-living settings, and three in both settings. The most commonly used motion sensor was an accelerometer (64) worn on the wrist (60) or lower arm (5), while in most studies (45), accelerometers were used in combination with gyroscopes. Twenty-six studies used commercial-grade smartwatches or fitness bands, 11 used professional grade devices, and 32 used standalone sensor chipsets. The most used machine learning approaches were Support Vector Machine (SVM, $n = 21$), Random Forest ($n = 19$), Decision Tree ($n = 16$), Hidden Markov Model (HMM, $n = 10$) algorithms, and from 2017 Deep Learning ($n = 5$). While comparisons of the detection models are not valid due to the use of different datasets, the models that consider the sequential context of data across time, such as HMM and Deep Learning, show promising results for eating activity detection. We discuss opportunities for future research and emerging applications in the context of dietary assessment and monitoring.

Keywords: eating activity detection; hand-to-mouth movement; wrist-mounted motion tracking sensor; accelerometer; gyroscope

---

1. Introduction

Recent advances in the accuracy and accessibility of wearable sensing technology (e.g., commercial inertial sensors, fitness bands, and smart watches) has allowed researchers and practitioners to utilise motion sensors mounted on the upper limbs (i.e., lower arm/wrist, upper arm) to assess dietary intake and eating behaviour in both laboratory and free-living conditions. Inertial sensors such as accelerometers (e.g., [1,2]) and gyroscopes (e.g., [3,4]), as well as proximity sensors (e.g., radio-frequency identification (RFID) [5,6]), can be used to detect and quantify characteristic hand-to-mouth gestures associated with food and beverage consumption. As such, compared to other types and/or positioning of sensors (e.g., mounted to a user's neck or head), this technology offers advantages in terms of detecting the timing and amounts of eating behaviour in an unobtrusive, accessible, and affordable way that yields high levels of technology acceptance [7–9]. Disadvantages, including the limited ability to detect brief snacks and the type and amounts of food being consumed [10], can be addressed by combining these sensors with other active (e.g., self-reporting with a food record or recall) and passive capture methods (e.g., microphone, video). In this vein, one can use the data gained from upper limb motion sensors to (1) improve and complement traditional dietary assessment methods [11] (e.g., by triggering reminders to actively take a photo when an eating occasion is detected), and (2) to support the delivery of dietary behaviour change interventions, for instance by capturing characteristic hand-to-mouth movements (e.g., [1,12]).

Overall, the field of wrist-mounted motion tracking sensors for the measurement of eating behaviour has evolved rapidly over the past decade. In 2012, Dong and colleagues [13] used a relatively expensive device (about US $2000 per device) called the InertiaCube3 (tri-axial accelerometer, tri-axial gyroscope, and tri-axial magnetometer) that was wired to a separate reader device. Within only a few years, the price of these sensors dropped to less than US $200. At the same time, sensors have now substantially reduced in size, operate wirelessly, and are powered by rechargeable batteries (e.g., [9,14]). High-quality motion tracking sensors are available within off-the-shelf, commercial smart watches that can be purchased at a fraction of the price of earlier devices and by the general population, not just researchers. Fuelled by these technological advances, and with wearable motion tracking devices experiencing rapid growth in areas such as fitness [15], the field of monitoring and assessing eating behaviour and dietary intake using these technologies is evolving. Another factor contributing to the rapid proliferation of motion tracking devices is their high level of technology acceptance; such devices have become increasingly culturally acceptable and unobtrusive to wear [7–9,16]. Combined with machine learning methods, the collection of movement data from wrist-mounted motion tracking sensors can be used to extract meaningful information about a person's daily activities (e.g., eating behaviour and physical activity) in a continuous, scalable, and discreet way [2,17].

Several recent reviews have explored how wearable sensors have been applied to the field of nutrition for assessment or monitoring of eating behaviour and/or dietary intake [10,15,16,18–20]. However, previous reviews have focused on a wider range of wearable sensors (e.g., camera, microphone) located on various parts of the body (e.g., ear, temple, torso) and included a range of eating-related activities [10,16,18,19] (e.g., chewing, swallowing), or focused on smoking [15]. Hassannejad and colleagues [18] reviewed two main approaches used to try and automate dietary monitoring. The first approach was to automatically extract information on food content based on image analysis and the second approach was to extract the information sourcing data from wearable sensors to detect eating behaviour. Kalantarian and colleagues [10] provided a general overview of dietary monitoring technology (e.g., acoustic, image, inertial, and manual food diaries). Prioleau and colleagues [16] focused specifically on wearable sensors such as cameras, microphones, and motion sensors placed on different body locations (e.g., ear, mouth, neck, upper limb). Vu and colleagues [19] provided an overview of data analytic and sensing platforms for wearable food intake monitoring, covering a wide range of systems including acoustic, inertial, muscle activity, proximity, and visual sensors, rather than focusing on a specific type of wearable sensors. Parate and colleagues [15] reviewed approaches designed for detecting eating and smoking behaviours where hand gestures are involved. Doulah and colleagues [20] conducted a systematic review of studies focusing on estimating dietary energy intake (i.e., amount of consumption, energy density, and food recognition), covering a range of image-based approaches (e.g., depth, smartphone, or wearable cameras) and wearable sensors (e.g., chewing sound, jaw motion, or wrist motion sensors). However, to the best of our knowledge, there is currently no systematic review of the current state of research available that specifically focuses on the use of upper limb-mounted motion tracking sensors for assessing eating behaviour.

This review differs from existing reviews in the area in that it includes a systematic search and thereby follows a rigorous process with two reviewers to provide an overview of the current state of research on upper limb-mounted motion sensors for assessing eating behavior across the 69 identified studies. Given the wide availability and affordability of upper limb-mounted motion sensors, an understanding of the study settings, sensor configurations, detection approaches, and eating behaviour assessment in the extant literature is important in order to progress research in this area and inform the application of these approaches in practice. Hence, the aim of the current review is to summarise the current evidence on use of upper limb-mounted motion sensors for assessing eating behaviour.

2. Materials and Methods

2.1. Definition of Common Terms

Table 1 provides an overview of the terms and definitions employed in the current review. Throughout this review, we use the term *motion sensor* to refer to wearable motion tracking or wearable motion detection sensors, unless specified otherwise. Wearable motion sensors are usually integrated into a tracking device mounted on the wrist or other parts of the upper limbs (e.g., activity tracker, fitness tracker, smart watch). The tracking device commonly consists of several different motion sensors such as inertial sensors and proximity sensors. A *proximity sensor* can detect the presence of nearby objects and therefore requires a separate sensing device. An *inertial sensor* can detect changes in linear or angular momentum. The two most widely-used inertial sensors are three-dimensional micromachined microelectromechanical systems (MEMS) accelerometers and gyroscopes. While the tri-axial *accelerometer* measures magnitude and direction of acceleration on X, Y and Z axes, the tri-axial *gyroscope* measures the rate of rotation on yaw, pitch, and roll axes. The studies across the field have used a variety of different terms to refer to the same concept. Action classes are the desired types of events to be detected through the artificial intelligence models. The action classes vary depending on the machine learning approach taken and the behaviour assessment outcomes expected. These classes need to be predefined with labels (i.e., tagged), and used in the process of data annotation to mark the events (e.g., using video cameras or self-report push buttons). The events are marked with the start time, end time, and a label (action class) that described what the event is about. An event may be marked with multiple labels (e.g., drinking, left hand).

Table 1. Terms used in this review with synonyms and definitions.

Term	Synonyms Used in the Literature	Definition
Action classes	Event/activity classifications/categories	Different categories that the classifiers (detection models) are trained and tested to classify
Artificial intelligence model	Artificial intelligence approach, machine learning algorithm	The approach used for automatic eating behaviour detection
Backward search		Search through references of included studies to find other relevant studies that are not found through database search
Eating activity		Eating and drinking activity
Eating behaviour assessment	Food intake detection, eating detection, ingestion monitoring	Assessing whether the participant is eating (including drinking) and what their eating characteristics are
Forward search		Search for relevant studies that cited included studies
F-score	F1 score, F-measure	F-score is a measure of accuracy. While accuracy is the total number of correctly classified items divided by all classified items, F-score is harmonic average of the precision and recall.
Hand-to-mouth gesture	Hand-to-mouth movement	The movement of hand carrying food with or without utensils to the mouth
Motion sensors	Motion tracking sensors, motion detection sensors, activity tracker	Sensors used to detect movements. Wearable motion sensors focused on in the current review include upper limb-mounted motion sensors.
Participant	Subject	An individual who has successfully participated in a study (i.e., not counting individuals who were invited but did not participate or individuals with failed measurements)
Upper limb	Arm	Region of body that includes shoulder, upper arm, lower arm, wrist, and hand

2.2. Search Strategy

For the current review, we included studies that (1) used at least one wearable motion sensor, (2) that was mounted to the wrist, lower arm, or upper arm (referred to as the upper limb in this review), (3) for eating behaviour assessment or human activity detection, where one of the classified activities is eating or drinking. We explicitly also included studies that additionally employed other sensors on other parts of the body (e.g., cameras, microphones, scales). In order to identify studies that meet these criteria, we constructed the search string to include three parts (motion sensor, mounted to upper limb, eating behaviour assessment). The search string was then iteratively developed from a set of key studies that were identified in an initial search as well as from MeSH headings and consultation with a medical librarian. Using multiple combinations of search terms shown in Tables A1 and A2 a comprehensive search was conducted to interrogate electronic archives across medical and health sciences as well as computing disciplines for studies published in English. In computing the ACM digital library, AIS electronic library (AISeL), IEEE Xplore, ScienceDirect, SpringerLink archives and in health sciences the CINAHL, MEDLINE, EMBASE, Ovid, Web of Science and Scopus archives, eleven in total were searched. In order to account for the breadth of publications in health and computing-focused outlets, the search covered peer-reviewed studies published in book chapters, journals, and full conference proceedings (excluding abstract-only/extended-abstract papers). Particularly in computing, studies are often published as full conference papers. The search terms combination was adapted to each electronic archive due to their limitation on search input. The search was conducted in January 2019 and backward and forward search was done after the included studies were identified.

The review protocol was registered with Prospero system (the CRD42018089493). The primary outcomes assess upper limb-mounted motion sensors and devices used to detect hand-to-mouth gestures associated with eating. This is to identify what types of sensors were used, how the sensors were combined or used together, and where on the upper limb they are mounted. The secondary outcomes assess the algorithms and techniques utilised to analyse the output of the sensors used on body for motion tracking associated with eating occasions, the environmental conditions under which the experiments were been conducted (e.g., setting, food items, serving vessels and eating utensils), and the characteristics of eating behaviour that were assessed (e.g., bite count, duration of eating, quantification of amounts, and type of the food eaten).

2.3. Selection Process

The results of the database search were imported into a web-based tool (Covidence [21]), duplicate items were identified and removed, and the rest of the studies were title- and abstract-screened by two of four independent reviewers (H.H., M.A., T.B., M.E.R.) to identify studies that potentially meet the inclusion criteria. The full text articles were then retrieved and assessed for eligibility by two of the four independent reviewers, with discrepancies resolved by discussion with a third independent reviewer who was not involved in assessing that particular study.

Following the selection of studies, two reviewers independently extracted relevant information using a custom-made data collection form; any discrepancies regarding this data were resolved by discussion with a third reviewer. Data from the selected studies were captured and summarised in Table 2 which was constructed for the purpose of this review. These were initially pilot tested with seven studies to ensure all data was extracted and appropriate. Due to the nature of this review evaluating the performance of technology, a risk of bias assessment was not deemed to be necessary/appropriate by the research team. Countries of data collection were categorised by economies according to a UN report [22].

Table 2. Included studies ($n = 69$).

Author (Year), Country [Ref]	Type	Participants (Environment)	Upper Limb Sensor (Frequency) [Position]	Comparator	Algorithm (Detection)
Amft et al. (2005), Switzerland [23]	Conference	2 (lab)	Acc/Gyro (100 Hz) [wrist, both]	NR	FSS, HMM (GD)
Amft et al. (2007), Switzerland [24]	Conference	1 (lab)	Acc/Gyro (NR) [lower arm, both]	NR	PCFG (AD)
Amft et al. (2008), Switzerland [25]	Journal	4 (lab)	Acc/Gyro (100 Hz) [lower arm, both]	NR	FSS (GD)
Junker et al. (2008), Switzerland [26]	Journal	4 (Lab)	Acc/Gyro (100 Hz) [lower arm, both]	NR	HMM (GD)
Laerhoven et al. (2008), Germany [27]	Conference	2 (free-living)	Acc (NR) [wrist, dominant]	Other self-report	KNN (AD)
Pirkl et al. (2008), Germany [28]	Conference	1 (lab)	Acc/Gyro (50 Hz), Prox [lower arm, dominant]	NR	DT (GD)
Tolstikov et al. (2008), Singapore [29]	Conference	NR (lab)	Acc (50 Hz), Prox [wrist, dominant]	Camera	DBN (AD)
Zhang et al. (2008), Singapore [30]	Conference	NR (lab)	Acc (NR) [wrist, both]	NR	HTM (GD)
Dong et al. (2009), USA [13]	Conference	10 (lab)	Acc/Gyro (60 Hz) [wrist, dominant]	Camera	C/RB (GD)
Teixeira et al. (2009), USA [31]	Conference	1 (free-living)	Acc/Gyro (NR) [wrist, dominant]	NR	FSM (AD)
Amft et al. (2010), Netherlands [32]	Conference	9 (lab)	Acc/Gyro (87 Hz), Prox [wrist, dominant]	Camera	DT, FSS (GD)
Dong et al. (2011), USA [53]	Conference	4 (free-living)	Acc/Gyro (60 Hz) [wrist, dominant]	Diary	C/RB (AD)
Dong et al. (2012), USA [3]	Journal	102 (both)	Acc/Gyro (NR) [wrist, dominant]	Camera, Diary	C/RB (GD)
Grosse-Puppendahl et al. (2012), Germany [34]	Chapter	7 (lab)	Acc (NR), Prox (NR) [wrist, dominant]	NR	SVM (AD)
Kim et al. (2012), South Korea [35]	Conference	13 (lab)	Acc (30 Hz) [wrist, dominant]	Camera	DT (CD, GD)
Mousavi Hondori et al. (2012), USA [36]	Conference	1 (lab)	Acc (NR) [utensil, both]	NR	NR (NR)
Varkey et al. (2012), USA [8]	Journal	1 (lab)	Acc/Gyro (20 Hz) [wrist, dominant]	NR	SVM (AD)
Farooq et al. (2013), USA [5]	Conference	13 (free-living)	Prox (NR) [wrist, dominant]	Diary, Push Button	ANN, SVM (AD)
Fontana et al. (2013), USA [37]	Conference	12 (free-living)	Prox (10 Hz) [wrist, dominant]	Push Button	RF (AD)
Kim & Choi (2013), South Korea [38]	Conference	8 (lab)	Acc (30 Hz) [wrist, dominant]	Camera	DT, NB (AD, CD, GD)
Ramos-Garcia & Hoover (2013), USA [39]	Conference	273 (lab)	Acc/Gyro (15 Hz) [wrist, dominant]	Camera	HMM, KNN (GD)
Desendorf et al. (2014), USA [40]	Journal	15 (lab)	Acc (80 Hz) [wrist, dominant]	NR	C/RB (GD)
Dong et al. (2014), USA [41]	Journal	43 (free-living)	Acc/Gyro (NR) [lower arm, dominant]	Mobile App	C/RB, NB (GD, AD)
Fontana et al. (2014), USA [6]	Journal	12 (free-living)	Prox (NR) [wrist, dominant]	Mobile App, Push Button	ANN (AD)
Ramos-Garcia et al. (2015), USA [12]	Journal	25 (lab)	Acc/Gyro (15 Hz) [wrist, dominant]	Camera	HMM, KNN (GD)
Sen et al. (2015), Singapore [42]	Conference	6 (lab)	Acc/Gyro (100 Hz) [wrist, dominant]	Wearable Camera	C/RB (GD)
Thomaz et al. (2015), USA [2]	Conference	28 (both)	Acc (25 Hz) [wrist, dominant]	Camera, Wearable Camera	DBSCAN, KNN, RF, SVM (AD, GD)
Ye et al. (2015), USA [43]	Conference	10 (lab)	Acc (50 Hz) [wrist, dominant]	Camera	DT, NB, SVM (GD)
Zhou et al. (2015), Japan [9]	Journal	5 (lab)	Acc (NR) [finger, dominant]	NR	DT, KNN (GAD)
Fan et al. (2016), USA [44]	Journal	1 (lab)	Acc/Gyro (50.1 Hz) [wrist, dominant]	NR	ANN, DT, KNN, NB, Reg, RF, SVM (AD)
Farooq & Sazonov (2016), USA [45]	Conference	12 (free-living)	Prox. (NR) [wrist, dominant]	Mobile App, Push Button	DT, Reg (AD)
Fortuna et al. (2016), USA [46]	Conference	3 (free-living)	Acc/Gyro (10 Hz) [wrist, dominant]	Camera, Wearable Camera	NB (GD)
Kim et al. (2016), South Korea [4]	Conference	15 (lab)	Acc/Gyro (NR) [wrist, dominant]	Camera	C/RB (GD)
Maramis et al. (2016), Greece [47]	Conference	8 (lab)	Acc/Gyro (NR) [wrist, dominant]	Camera	SVM (GD)
Mirtchouk M. et al. (2016), USA [48]	Conference	6 (lab)	Acc/Gyro (15 Hz) [wrist, both]	Camera	RF (CD)
Parra-Sanchez et al. (2016), Mexico [49]	Conference	7 (lab)	Electro-hydraulic (NR) [arm, both]	Camera	DT, HMM (GAD)

Table 2. *Cont.*

Author (Year), Country [Ref]	Type	Participants (Environment)	Upper Limb Sensor (Frequency) [Position]	Comparator	Algorithm (Detection)
Rahman et al. (2016), Canada [50]	Conference	8 (free-living)	Acc/Gyro (15 Hz) [wrist, dominant]	Mobile App	DT, Reg, RF, SVM (AD, CD)
Sharma et al. (2016), USA [51]	Conference	94 (free-living)	Acc/Gyro (15 Hz) [wrist, dominant]	NR	C/RB, NB (AD, GD)
Shen et al. (2016), USA [52]	Conference	215 (lab)	Acc/Gyro (15 Hz) [wrist, dominant]	Camera	HMM (GD)
Shoaib et al. (2016), Netherlands [53]	Conference	11 (lab)	Acc/Gyro (50 Hz) [wrist, dominant]	NR	DT, RF, SVM (AD)
Ye et al. (2016), USA [54]	Conference	7 (free-living)	Acc (50 Hz) [wrist, dominant]	Mobile App	SVM (GD)
Zhang et al. (2016), USA [55]	Conference	15 (lab)	Acc/Gyro (31 Hz) [wrist, both]	Camera	DBSCAN, DT, NB, Reg, RF, SVM (GD, AD)
Alexander et al. (2017), USA [56]	Journal	4 (lab)	Acc/Gyro (20 Hz) [wrist, dominant]	Time Sync	NB (AD)
Bi et al. (2017), USA [57]	Journal	37 (free-living)	Acc (80 Hz) [wrist, dominant]	Other self-report	HMM, SVM (AD)
Dong & Biswas (2017), USA [14]	Journal	14 (lab)	Acc (100 Hz) [wrist, dominant]	Camera, Push Button	C/RB, HMM, SVM (GD, AD)
Egilmez et al. (2017), USA [58]	Journal	9 (lab)	Acc/Gyro (5 Hz) [wrist, dominant]	NR	NB, Reg, RF, SVM (GAD)
Garcia-Ceja et al. (2017), Mexico [59]	Journal	3 (lab)	Acc/Gyro (31 Hz) [wrist, dominant]	NR	RF (GAD)
Kyritsis et al. (2017), Greece [60]	Conference	8 (lab)	Acc/Gyro (62 Hz) [wrist, dominant]	Camera	HMM, SVM (AD, GD)
Kyritsis et al. (2017), Greece [61]	Conference	10 (lab)	Acc/Gyro (62 Hz) [wrist, dominant]	Camera	DL, SVM (AD, GD)
Loke & Abkenar (2017), Australia [62]	Journal	4 (lab)	Acc (NR) [wrist, dominant]	NR	DT (GAD)
Moschetti et al. (2017), Italy [63]	Conference	12 (lab)	Acc/Gyro (50 Hz) [wrist, dominant]	NR	GMM, KM, RF, SVM (GD)
Sen et al. (2017), Singapore [7]	Journal	28 (both)	Acc/Gyro (NR) [wrist, dominant]	Camera, Diary	DT, RF, SVM (AD, GD)
Shen et al. (2017), USA [1]	Journal	271 (lab)	Acc/Gyro (15 Hz) [wrist, dominant]	Camera	C/RB (GD)
Soubam et al. (2017), India [64]	Conference	11 (lab)	Acc (186 Hz) [wrist, dominant]	Time Sync	DT, NB, RF, SVM (CD, GD)
Thomaz et al. (2017), USA [65]	Conference	14 (lab)	Acc/Gyro (30 Hz) [wrist, both]	Camera	RF (AD)
Yoneda & Weiss (2017), USA [66]	Conference	51 (lab)	Acc/Gyro (20 Hz) [wrist, dominant]	NR	DT, KNN, RF (GAD)
Zhang et al. (2017), USA [67]	Conference	8 (free-living)	Acc/Gyro (31 Hz) [wrist, both]	Wearable Camera	RF (GD)
Anderez et al. (2018), UK [68]	Conference	NR (lab)	Acc/Gyro (100 Hz) [wrist, dominant]	NR	DL, RF (GD)
Anderez et al. (2018), UK [69]	Conference	NR (lab)	Acc (100 Hz) [wrist, dominant]	NR	KNN (GD)
Balaji et al. (2018), India [70]	Conference	NR (NR)	Acc/Gyro (100 Hz) [wrist, NR]	NR	C/RB (GAD)
Cho & Choi (2018), South Korea [71]	Conference	8 (lab)	Acc (50 Hz) [wrist, dominant]	Camera	DL (CD, GD)
Clapés et al. (2018), Spain [72]	Journal	14 (lab)	Acc/Gyro (25 Hz) [wrist, dominant]	Camera	ANN, Opt (GAD, GD)
Kyritsis et al. (2018), Greece [73]	Conference	10 (lab)	Acc/Gyro (100 Hz) [wrist, dominant]	Camera	DL (AD)
Manzi et al. (2018), Italy [74]	Journal	20 (lab)	Acc/Gyro (NR) [wrist, dominant]	NR	RF (GAD)
Papadopoulos et al. (2018), Greece [75]	Conference	10 (lab)	Acc/Gyro (62 Hz) [wrist, dominant]	Camera	semi-supervised DL (AD)
Schibon & Amft (2018), Germany [76]	Conference	6 (free-living)	Acc/Gyro (NR) [wrist, both]	Diary	SVM (GD)
Shen et al. (2018), USA [77]	arXiv	269 (lab)	Acc/Gyro (15 Hz) [wrist, dominant]	Camera	HMM (GD)
Zambrana et al. (2018), Spain [78]	Journal	21 (lab)	Acc (20 Hz) [wrist, both]	Camera	KNN, RF, SVM (GAD)
Zhang et al. (2018), USA [17]	Journal	10 (lab)	Acc/Gyro (NR) [wrist, dominant]	Camera	RF (GD)

<u>Note:</u> The table only includes performance comparisons where studies have directly compared different algorithms using the same dataset. Acc = Accelerometer, AD = Eating Activity Detection, ANN = Artificial Neural Network, CD = Eating Characteristics Detection, Chap = Book Chapter, Conf = Conference, C/RB = Custom Rule-Based, DBN = Dynamic Bayesian Network, DBSCAN = Density-Based Spatial Clustering of Applications with Noise, DL = Deep Learning, DT = Decision Tree, FL = Free-Living, FSM = Finite State Machine, FSS = Feature Similarity Detection, GAD = General Activity Detection, GD = Eating Gesture Detection, GMM = Gaussian Mixture Model, Gyro = Gyroscope, HMM = Hidden Markov Model, HTM = Hierarchical Temporal Memory, Jour = Journal, KM = K-Means, KNN = K-Nearest Neighbours, NB = Naive Bayes, NR = Not Reported, Opt = Monte Carlo Optimization method, PCFG = Probabilistic Context-Free Grammar, Prox = Proximity, Reg = Regression, RF = Random Forest, SVM = Support Vector Machine.

3. Results

In total, 792 studies were identified through the search strategy, after removing 139 duplicates, 653 studies were screened on title and abstract. Of these, 111 were full-text reviewed independently by two authors, with a third author coming in if consensus was needed. With six studies found through backward search and four studies found through forward search, 69 studies were included in the review (Figure 1).

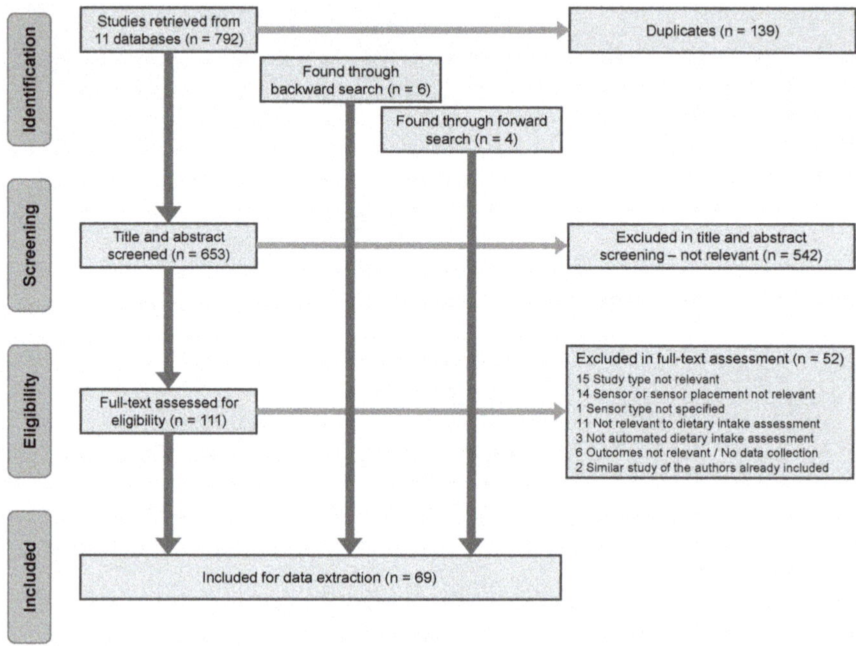

Figure 1. Flow diagram of article selection process in the systematic review.

This review provides a narrative synthesis of the findings from the included studies and uses these finding to structure a conceptual framework (Figure 2). In particular, we reviewed the selected studies to identify common components and implicit design choices that are involved in carrying out research in this area. We then synthesised this knowledge into a conceptual overview. The framework depicts an overview of the process of assessing eating behaviour using upper limb-mounted motion sensors and the different components involved in the process. Thereby, *study design* pertains to the environmental conditions that the participants experience as well as the requirements, instruments, and instructions for data collection process. In contrast, *sensor configuration* summarises the specific type, sampling frequency, and position of the employed motion sensor(s). These are the main components required to build a model to detect eating behaviour shown under *detection approach*. This process leads to identifying and assessing dietary behaviour which is depicted under *behaviour assessment*. The framework provides a structure for the synthesis and presentation of results in this review. Please note that some subcategories are not shown in Figure 2 because no studies were identified for them. For instance, none of the reviewed studies used sensor frequencies between 21 and 24 Hz.

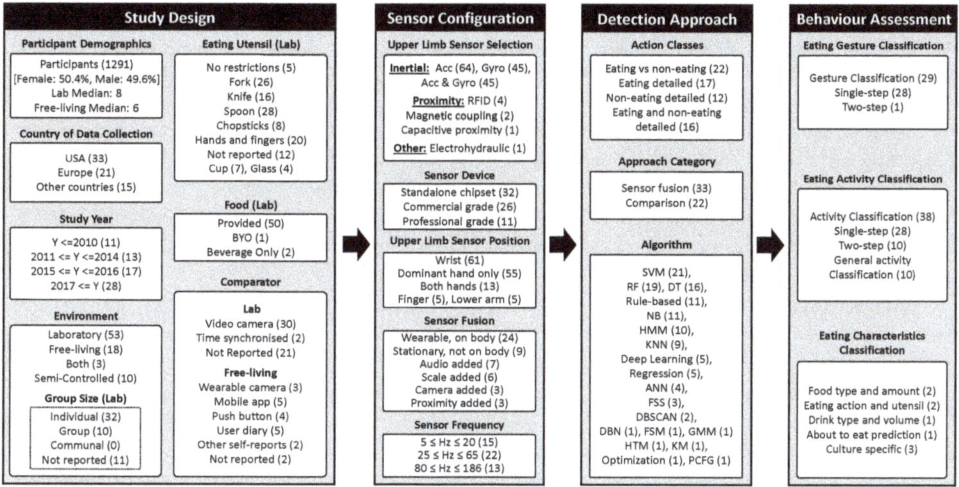

Figure 2. Conceptual framework of components for assessing eating behaviour with upper limb-mounted motion sensors.

3.1. Study Design

3.1.1. Participant Demographics

The number of participants ranges from one (i.e., [8,24,28,31,36,44]) to 276 [52] (median: 8 in lab setting, 6 in free-living setting). The total number of participants who successfully participated in the experiments was 1291. Of the included studies that reported participant gender ($n = 36$, 52.2%), 50.4% of participants were female and 49.6% were male. According to the demographic data where social class was reported, the participants were commonly university students.

3.1.2. Country of Data Collection

In all studies, all data collection was done in the country of the first author's affiliation. Most studies were conducted in the US ($n = 33$, 47.8%), followed by Europe ($n = 21$, 30.4%). Singapore, South Korea, India, and Mexico had two studies each. Australia, Canada, and Japan hosted one study each. As it can be seen the data is mostly collected in high-income countries (94.2%). Only two studies were conducted in a lower-middle-income country (India). No study collected data in a low-income country.

3.1.3. Study Year

Up to and including 2010, only 11 studies (15.9%) were published in this field, with 13 Studies (18.8%) published between years 2011 and 2014. Seventeen studies (24.6%) were published between years 2015 and 2016. Interestingly, 40.6% of the studies (28 studies) were published in 2017 or later.

3.1.4. Environment

The majority of the studies were conducted exclusively in controlled laboratory settings ($n = 50$, 72.5%; e.g., [55,65]), followed by exclusively free-living settings ($n = 15$, 21.7%; e.g., [5,51]), with fewer being conducted in both settings ($n = 3$, 4.3%; [2,3,7]). One study (1.4%) did not report the environment setting. The three studies that considered laboratory as well as field data used the data collected in the laboratory for training the machine learning models and the data from the free-living environment for evaluating the model's performance (e.g., [2,3,46]). The laboratory environment may affect the participant's natural behaviour in the progress of experiment. Therefore, 10 studies (14.5%) conducted semi-controlled experiments in laboratory environments (i.e., [2,14,65]) or in a cafeteria, restaurant, or dining hall (i.e., [1,12,39,52,60,61,77]).

The laboratory environment commonly involved participants sitting individually (e.g., [25]) at a table or in a group (e.g., four people [1,39]) around a table recorded with video camera(s) to capture the eating session. In a study by Amft and Tröster [25], participants were instructed to perform non-eating tasks such as reading a newspaper (including turning pages), scratching their head, and answering a simulated mobile phone call. The leftover food from the participant's meal could either be weighed throughout the experiment to keep track of food consumed [48] or at the end of the session to estimate the total amount of the food consumed [17,55]. However, few studies measured leftover food [17,39,48,52,55].

Studies in free-living environments commonly allowed participants to perform their daily activities during the day while wearing the sensor(s). The longer duration experiments involved more non-eating associated activities (e.g., driving, watching TV and working on a computer) than eating activities. Thomaz and colleagues [2] conducted an experiment in both settings. For the laboratory setting, the average duration of the data collection was 31 min which included 48% eating activities. In contrast, of the two experiments conducted in free-living conditions, one had an average duration of 6 h and included 6.7% eating activities while the other one was carried out over 31 days and included only 3.7% eating activities. Several studies indicate challenges associated with field data collection. In a free-living study by Dong and colleagues [41], data from ten out of a subsample of 30 individuals were discarded due to poor compliance with keeping manual records of activities (e.g., misinterpreting the instructions and starting/stopping recording for meals only). In a study by Sharma and colleagues [51], data collected from 10% of the 104 individuals were discarded because they failed to wait ten minutes between wearing the device and the first meal.

Among the 53 studies conducted in the laboratory, 32 studies (60.4%) asked participants to eat individually from a discrete plate (e.g., [26,65]), 10 studies (18.9%) were carried out in groups comprising between two to four people (e.g., [1,12]), and 11 studies (20.8%) did not report the group size. In group settings, participants were still provided with discrete plates of food and/or asked to self-serve on to their own individual plate. No experiment was reported asking the participants to share food from one or more plates (communal eating).

3.1.5. Eating Utensils

The utensils most commonly used in laboratory experiments were spoons ($n = 28$, 52.8%), followed by forks ($n = 26$, 49.1%), knifes ($n = 16$, 30.2%), and chopsticks ($n = 8$, 15.1%). Five studies (9.4%) conducted in the laboratory applied no restriction on the type of eating utensils. Eating with hands or fingers was reported in 20 studies (37.7%) conducted in the laboratory. Twelve studies (22.6%) conducted in the laboratory did not report what utensils were used. The studies that reported drinking vessels used cups ($n = 7$, 13.2%) or glasses ($n = 4$, 7.5%). However, participants were served with yogurt in a mug in one study [24] and the use of a straw to drink beverages was reported in four studies [3,17,40,48]. Zhang and colleagues [17] reported that drinking from a straw for longer than 30 s produced unusual motion sensor data which was disregarded as a single gesture.

3.1.6. Food

In the experiments conducted in free-living conditions the participants consumed their own food ($n = 18$, 26%). In contrast, food in the laboratory settings was commonly provided for participants by researchers. Only in one laboratory study [3], participants were asked to bring their own foods and drinks. One of the most reported foods eaten with a fork and knife in the experiments was lasagne (i.e., [2,23–26,65], $n = 6$). Rice (e.g., [6,35,42]) and soup (e.g., [35,40,53]) were commonly eaten with a spoon, whereas pizza (e.g., [12,40,48]) and bread (e.g., [12,43,48]) were commonly eaten with hands. Kim and colleagues [35] collected data from participants eating rice with both chopsticks and spoon.

Some of the other food items reported in laboratory settings were chips/fries, burger/sandwich, fruit, meat/steak, pasta, salad, vegetables, yoghurt, and snack foods (e.g., cake, candy, chocolate, ice-cream, popcorn). In addition, various beverages (e.g., coffee, juice, smoothie, soda, tea, and water) were provided for participants to drink while consuming food. Some studies (e.g., [1,12,48]) provided the participants with multiple food options so they could self-select amounts and types of food. These studies were usually conducted in a semi-controlled environment. In comparison, two laboratory studies (i.e., [32,64]) exclusively examined drinking behaviour. In one of these studies, Amft and colleagues [32] used nine different drink containers to investigate the recognition of container types and the volume of fluid consumed from the container.

The duration of an uninterrupted eating episode in a controlled environment depends on the number of hand-to-mouth gestures and chewing time, which is directly related to the food type. Sen and colleagues [7] observed that eating episodes ranged from 51 s for fruit to 19 min for rice.

3.1.7. Comparator

To facilitate sensor data analysis, collected data must be annotated with differing labels to represent actions and events that occurred. The annotated data is then used to train the machine learning models and evaluate their performance. One approach for data annotation is to let participants self-report the investigated activities in real-time using a mobile app and/or a push-button technique (i.e., [5,6,14,37,45,54]). Further, some studies in free-living environments combined a push-button approach with a pen and paper diary (e.g., [5]) or an electronic food diary on a smartphone (e.g., [54]) completed by the participant. However, these commonly employed comparator techniques rely on participants to provide an accurate and complete record of activities. Hence, it is not possible to unambiguously establish ground truth. By contrast, for experiments conducted in laboratory settings, ground truth can be established by using objective observation instruments. This is commonly achieved through video cameras.

Of the 53 laboratory studies, 32 (60.4%) reported the comparator. Thirty (56.6%) used video recordings to establish ground truth (mostly surveillance video with one study using a wearable camera [42]), while the other two studies (3.8%) used different time synchronisation mechanisms (timestamps for predetermined tasks [56] or alarms to instigate drinking [64]). Of the 18 free-living studies, only two (11.1%) did not report the comparator. Five studies (27.8%) used a diary, five (27.8%) used a self-report mobile app, and four (22.2%) used a button on the wearable sensor device. Interestingly three studies (16.7%) used wearable camera to establish ground truth on the free-living environment and two studies (11.1%) used other self-report/self-recall approaches.

3.2. Sensor Configuration

3.2.1. Sensor Selection on the Upper Limbs

The most commonly used motion sensors that were mounted on the upper limbs are (tri-axial) accelerometers ($n = 64$, 92.8%) and (tri-axial) gyroscopes ($n = 45$, 65.2%). Interestingly, all 45 studies that used a gyroscope also used an accelerometer. Seven studies (10.1%) used proximity sensors on the upper limbs. This includes RFID sensors (four studies; [5,6,37,45]), magnetic coupling sensors (two studies; [28,32]), and capacitive proximity sensor (one study, combined with accelerometer; [34]). One study [49] used electrohydraulic sensors. Additional proximity sensors mounted to the drinking vessel [29] or the eating utensils (fork, knife and cup) [36] were also reported. Amft and colleagues [32] used a magnetic coupling sensor where the field emitting sensor was attached to the shoulder while the receiver unit was attached to the wrist.

3.2.2. Sensor Device

The majority of studies directly used standalone sensor chipsets rather than an integrated recording device ($n = 32$, 46.4%). Twenty-six studies (37.7%) used off-the-shelf, commercial-grade smartwatches or fitness bands, such as Microsoft Band and Pebble watch. Eleven studies (15.9%) used professional grade devices with embedded sensors such as Shimmer and XSens. In recent years, more studies have tended to use off-the-shelf, commercial-grade smartwatches or fitness bands and less studies employed standalone sensor chipsets. One study [41] used the accelerometer and gyroscope embedded in a smartphone (iPhone 4) mounted on the forearm (wrist). However, a smartphone was used in another study [2] to conduct a pilot formative experiment before collecting data using accelerometer and gyroscope sensor modules. One study [28] used a professional grade device (Xsens) as well as a standalone sensor chipset.

3.2.3. Sensor Position on Upper Limbs

Sixty-one studies (88.4%) used at least one motion sensor on the wrist and five studies (7.2%) reported at least one motion sensor mounted to the lower arm. Four studies [44,58,63,74] used an inertial sensor on a finger in addition to the wrist, while another study [9] only used an accelerometer worn on an index finger. Five studies (7.2%; [23–26,28]) used motion sensors on the upper arm as well as wrist or lower arm. One study [36] used motion sensors only on utensils (fork, knife and cup), and another study [49] used electro-hydraulic sensors on both hands. Fifty-five studies (79.7%) used the motion sensors only on the dominant eating hand, while thirteen studies (18.8%) used the motion sensors on both hands.

3.2.4. Sensor Fusion

Thirty-three studies (47.8%) combined upper limb-mounted motion sensors with other sensors on different parts of the body or in the environment. Twenty-four of these studies (34.8%) used different types of sensors on or attached to the participants' body (i.e., torso, chest, upper back, head, jaw, throat, ear, foot) or in participants' pocket in addition to their upper limbs. The other studies (n=9, 13.0%) used sensors placed in the participants' environment (e.g., camera, scale, and proximity). For example, Amft and Tröster [25] used (inertial) motion sensors including accelerometer, gyroscope, and compass on lower arm, upper arm, and upper back, all attached onto a jacket to detect movement activities. Further, they used an ear microphone (electret miniature condenser microphone) to detect chewing activities as well as a stethoscope microphone mounted to the hyoid and an electromyogram (EMG) mounted to the infra-hyoid throat to detect swallowing activities. Six studies (8.7%) used scales to measure the weight of food consumed throughout the experiment (i.e., [17,39,48,52,55,64]). Further, several studies combined motion sensor data with audio (*n* = 7, 10.1%; [24,25,48,50,57,59,62]) or video camera recordings (*n* = 3, 4.1%; [31,72,74]) to detect eating behaviour. For instance, Mirtchouk and colleagues [48] combined accelerometer data from each participant's both wrists and head with audio data recorded from a pocket audio recorder. Garcia-Ceja and colleagues [59] combined accelerometer data with audio data collected from a smartphone placed on a table in the same room as the participant to record environmental sound.

3.2.5. Sensor Sampling Frequency

Sensor sample rate (frequency) is the number of data items the sensor collects per second. Forty-nine studies (71%) reported the sample rate, with frequencies for the wrist-mounted motion sensors, ranging from 5 Hz [58] to 186 Hz [64]. Among these, 15 (21.7%) used a frequency of lower or equal to 20 Hz, 22 (31.9%) used a frequency between 25 Hz and 65 Hz, and 13 (18.8%) used a frequency of 80 Hz or more. The median sampling frequency was 50 Hz. Five studies [1,12,39,51,52] used both an accelerometer and a gyroscope with a 15 Hz sample rate frequency, whereas three studies [23–25] also used both an accelerometer and a gyroscope but with a higher rate of 100Hz.

3.3. Detection Approach

This section discusses the categories that eating detection approaches fall into, algorithms used to build detection approaches, and types of gestures and activities defined for prediction, referred to in this review as action classes. Detection approaches commonly involved three consecutive stages: pre-processing, feature extraction, and building an eating action detection model.

3.3.1. Action Classes

The action classes at the simplest level (binary) were eating and non-eating actions ($n = 22$, 31.9%). Thereby, we can distinguish between *gesture detection* (characteristic low-level actions) and *activity detection* (high-level actions). In 17 studies (24.6%) only eating associated actions were detailed to subcategories. In 12 studies (17.4%) only non-eating associated actions were detailed to subcategories. In 16 studies (23.2%) both eating and non-eating associated actions were subcategorised. Kim and colleagues [35], defined the classes to detect the utensil type in addition to the eating action. Amft and colleagues [32] defined nine different drinking vessels as the action classes for the purpose of container type and fluid level recognition.

3.3.2. Approach Category

We can identify two approaches for eating behaviour assessment: eating gesture detection and eating activity detection. At the lower level, in *eating gesture detection* ($n = 29$, 42%), the aim is to detect characteristic eating gestures that are the building blocks of eating occasions while in *eating activity detection* ($n = 38$, 55.1%), the aim is to detect the occasions when the participant was eating. For instance, a period of time can be categorised as an eating occasion when at least a certain number of eating gestures occur in a row. There are mainly two different approaches to implement an eating activity detection solution single-step and two-step. In the *single-step* approach ($n = 28$, 40.6%; e.g., [6,42,47]), the eating detection model is trained on pre-processed motion data with the aim of detecting the pre-defined activities (e.g., eating events versus non-eating events). In the *two-step* approach, two different models are consecutively employed where typically the first model is responsible to detect the desired hand gestures using pre-processed data as input. The model at the second step uses the output of the first step as its input to detect the desired activities ($n = 10$, 14.5%; e.g., [7,51,60]).

Further, sensor fusion methods may also be utilised in the two above-mentioned approaches. In the *fusion* approach (e.g., [6,59,72]), researchers collect data using multiple sensors on different body parts or combine wearable and stationary sensors, as opposed to collecting data from sensor(s) mounted on one position on body. In this approach typically multiple classifiers are used where the outputs of the classifiers will be aggregated to detect desired activities based on action classes.

3.3.3. Algorithm

Table 3 provides an overview of the machine learning algorithms and detection approaches used in the reviewed studies. It also demonstrates the experiments conducted to compare the performance of the machine learning algorithms. Thereby, in order to avoid repetitions, each comparison study is only listed once, namely for the algorithm where the comparison yielded the best performance. Twenty-two studies (31.9%) compared the performance of different algorithms. Naive Bayes was used for benchmarking where multiple algorithms were compared.

Table 3. Machine learning algorithms used in the included studies as well as and detection approaches and performance comparisons conducted in the studies.

Algorithm (# and % Studies)—Approach	Best Performing Algorithm (vs. Comparison Algorithm/s)	Performance Comparison Results
SVM (21, 30.4%) AD: [5,8,34,44,53,57], AD/CD: [50], CD/GD: [64], GAD: [58,78], GD: [43,47,54,63,76], GD/AD: [2,7,14,55,60,61]	SVM (vs. DT, NB) [43] (GD)	Using only wrist or head motion data SVM accuracy for eating detection was between 0.895 to 0.951 (combined wrist/head: 0.970).
	SVM (vs. KNN, RF) [78] (GAD)	Best accuracy of SVM using time-domain features was 0.957 (F = 0.957, two-second window). Best accuracy of RF model using frequency-domain features was 0.939 (F = 0.940, four-second window).
	SVM (vs. DT, RF) [53] (AD)	F-scores for detecting eating, drinking, and smoking with SVM were 0.910, 0.780, and 0.830, compared to RF with 0.860, 0.780, and 0.840, and DT with 0.820, 0.690, and 0.780.
RF (19, 27.5%) AD: [37,44,53,65], AD/CD: [50], CD: [48], CD/GD: [64], GAD: [58,59,66,74,78], GD: [17,63,67,68], GD/AD: [2,7,55]	RF (vs. SVM, KNN) [2] (GD)	FR outperformed SVM and 3-NN in detecting eating gestures in two free-living settings (seven participants in one day, F = 0.761; one participant in 31 days (F = 0.713).
	RF (vs. DT, NB, Reg, SVM) [55] (GD)	Eating gesture detection with RF yielded F = 0.753 compared to F = 0.708 (SVM), F = 0.694 (Reg), F = 0.647 (DT), and F = 0.634 (NB).
	RF (vs. DT, SVM) [7] (GD)	The accuracies achieved by RF, DT, and SVM were 0.982, 0.966, and 0.857, respectively.
	RF (vs. SVM) [63] (GD)	Using leave one person out cross-validation method the accuracies achieved by RF and SVM were 0.943 (F = 0.949) and 0.882 (F = 0.895).
	RF (vs. NB, Reg, SVM) [58] (GAD)	F-scores for general activity detection with FR, Reg, SVM, and NB were 0.788, 0.661, 0.613, and 0.268, respectively (eating was among action classes).
	RF (vs. DT, NB, SVM) [64] (GD)	The accuracies of RF, SVM, DT, and NB in person independent drinking versus eating detection were 0.924, 0.905, 0.881 and 0.871, respectively.
	RF (vs. DT, Reg, SVM) [50] (AD)	Person independent "about-to-eat" detection achieved F = 0.690 (RF) compared to 0.660 (SVM), 0.660 (REG), and 0.640 (DT).
	RF (vs. DT, KNN) [66] (GAD)	Accuracies of RF, DT, and KNN were 0.997, 0.980, and 0.888, respectively (using accelerometer data).
DT (16; 23.2%) AD: [44,45,53], AD/CD: [50], CD/GD: [35,64], CD/GD/AD: [38], GAD: [9,49,62,66], GD: [28,32,43], GD/AD: [7,55]	DT (vs. NB) [38] (AD/CD)	Best F-scores of DT and NB were 0.930 and 0.900 (eating activity detection) and 0.780 and 0.700 (eating type detection), respectively.
	DT (vs. NB) [35] (CD/GD)	Best F-scores of DT and NB were 0.750 and 0.650 (eating utensil detection) and 0.280 and 0.190 (eating action detection), respectively.
HMM (10, 14.5%) AD: [57], GAD: [49], GD: [12,23,26,39,52,77], GD/AD: [14,60]	HMM (vs. SVM) [57] (AD)	HMM outperformed SVM by 6.82% on recall in family meal detection, while the average precision and recall were 0.807 and 0.895.
	HMM (vs KNN) [39] (GD)	Accuracies of HMM, and KNN were 0.843 and 0.717, respectively.
	HMM-1 (vs. HMM-S, HMM-N, N:2-6) [77] (GD)	Accuracies of HMM-S and gesture-to-gesture HMM-1 were 0.852 and 0.895 respectively. According to the figure provided in the study the accuracy for HMM-2 to 4 stays similar and decreases for HMM-5 and 6.
	HMM-6 (vs. HMM-N, N:1-5, KNN, S-HMM) [12] (GD)	The accuracies of HMM-6, HMM-5, HMM-4, HMM-3, HMM-2, HMM-1, sub-gesture HMM and KNN were 0.965, 0.946, 0.922, 0.896, 0.880, 0.877, 0.843 and 0.758, respectively.

Table 3. *Cont.*

Algorithm (# and % Studies)—Approach	Best Performing Algorithm (vs. Comparison Algorithm/s)	Performance Comparison Results
KNN (9, 13%) AD: [27,44], GAD: [9,66,78], GD: [12,39,69], GD/AD: [2]	KNN (vs. ANN, DT, NB, Reg, RF, SVM) [44] (AD)	The accuracies of KNN, RF, Reg, SVM, ANN, NB and DT were 0.936, 0.933, 0.923, 0.920, 0.913, 0.906 and 0.893, respectively.
	KNN (vs. DT, NB) [9] (GAD)	The precision (and recall) values of KNN, DT and NB were 0.710 (0.719), 0.670 (0.686) and 0.657 (0.635), respectively.
DL (5, 7.2%) AD: [73,75], CD/GD: [71], GD: [68], GD/AD: [61]	RNN (vs. HMM) [61,73] (AD)	Replacing HMM with RNN in a SVM-HMM model improved F-score from 0.814 to 0.892 [61]. In a subsequent study a single-step end-to-end RNN reached almost the same performance (F = 0.884) [73].
ANN (4, 5.8%) AD: [5,6,44], GD/GAD: [72]	ANN (vs. SVM) [5] (AD)	ANN achieved accuracy of 0.869 (±0.065) compared to SVM (0.819, ±0.092) for eating activity detection (12 participants). ANN achieved accuracy of 0.727 compared to SVM (0.636, ±0.092) for number of meals detection (1 participant).
KM (1, 1.4%) GD: [63]	KM (vs. GMM) [63] (GD)	In an inter-person comparison, the accuracies of unsupervised approaches KM and GMM were 0.917 (F = 0.920) and 0.796 (F = 0.805), respectively.

Other: C/RB (11, 15.9%, AD [33], GAD [70], GD [1,3,4,13,40,42], GD/AD [14,41,51]), NB (11, 15.9%, AD [44,56], CD/GD [35,64], CD/GD/AD [38], GAD [58], GD [43,46], GD/AD [41,51,55]), Reg (5, 7.2%, AD [44,45], AD/CD [50], GAD [58], GD/AD [55]), FSS (3, 4.3%, GD: [23,25,32]), DBSCAN (2, 2.9%, GD/AD [2,55]), DBN (1, 1.4%, AD [29]), FSM (1, 1.4%, AD [31]), HTM (1, 1.4%, GD [30]), Opt (1, 1.4%, GD/GAD [72]), PCFG (1, 1.4%, AD [24])

Note: AD = Eating Activity Detection, ANN = Artificial Neural Network, CD = Eating Characteristics Detection, C/RB = Custom Rule-Based, DBN = Dynamic Bayesian Network, DBSCAN = Density-Based Spatial Clustering of Applications with Noise, DL = Deep Learning, DT = Decision Tree, F = F-score, FSM = Finite State Machine, FSS = Feature Similarity Detection, GAD = General Activity Detection, GD = Eating Gesture Detection, GMM = Gaussian Mixture Model, HMM = Hidden Markov Model, HMM-S = single-gesture HMM, HTM = Hierarchical Temporal Memory, KM = K-Means, KNN = K-Nearest Neighbours, NB = Naive Bayes, Opt = Monte Carlo Optimization method, PCFG = Probabilistic Context-Free Grammar, Reg = Regression, RF = Random Forest, RNN = Recurrent Neural Network, SVM = Support Vector Machine.

3.4. Eating Behaviour Assessment

3.4.1. Eating Gesture Classification

The aim of eating gesture classification is to detect characteristic gestures involved in ingestive behaviours (e.g., hand-to-mouth gestures). Such gestures are produced when an individual picks up food and moves it towards his/her mouth (hand-to-mouth movements, with or without utensils). Twenty-nine studies (42%) targeted only different aspects of eating gesture classification. Detecting eating gestures is often achieved with a single-step classification technique. However, researchers in [67] used two steps for eating gesture classification. They used a sliding window technique to first detect stationary periods, where the participants were more likely to eat, and the model then detected eating-associated gestures in the next step.

3.4.2. Eating Activity Classification

Twenty-eight studies (40.6%) used a direct detection approach for eating activity classification, i.e., detecting eating activities without detecting eating gestures first (e.g., [6,33,65]). Ten studies (14.5%) built eating gesture detection models as the first step to then detect eating activities in the second step (e.g., [2,51,60]). In other words, these studies employed a two-step detection approach, where the eating gestures detected in the first step are used to build a model in the second step to differentiate eating and non-eating activities (e.g., brushing teeth, combing hair, talking on the phone, walking, watching TV, and writing). Ten studies (14.5%) conducted general activity detection where eating activities were included in the data collection process along with a range of other activities and then classified in the activity detection approach (e.g., ambient assisted living).

3.4.3. Eating Characteristics Classification

In addition to detecting eating gestures and eating activities, six studies (8.7%) aimed to detect further characteristics of eating behaviour, i.e., food type and amount detection ($n = 2$, 2.9%; [38,48]), eating action and utensil detection ($n = 2$, 2.9%; [35,71]), drink type and volume detection ($n = 1$, 1.4%; [64]), and also about-to-eat and time until the next eating event prediction ($n = 1$, 1.4%; [50]). Mirtchouk and colleagues [48] investigated food type detection and amount consumed. Kim and colleagues [35] detected different types utensils (i.e., chopsticks, hand, spoon) as well as eating and non-eating gestures such as stirring, picking up rice, and using tissue. Rahman and colleagues [50] designed a system to predict the next eating occasion. Soubam and colleagues [64] detected drink type and volume in addition to eating and drinking gesture detection. Three studies (4.3%, [35,38,71]) specifically explored the Asian eating style. Cho and Choi [71] focused on eating action and utensil detection specifically for Asian-style food intake pattern estimation (chopsticks vs. spoon).

4. Discussion

The current review set out to synthesise existing research that describes the use of the upper limb-mounted motion sensors for assessing eating behaviour. Based on the 69 studies identified in our search, we are able to document the current body of research in the detection of *eating activities* (e.g., drinking, eating) and individual *eating gestures* (e.g., specific hand-to-mouth movements). To this date, most studies were carried out in laboratory conditions with university student (young healthy adults), with limited application in free-living settings or in diverse publication groups. Devices used were predominantly accelerometers in combination with gyroscopes worn on the wrist of the dominant hand, and the focus so far lied on distinguishing eating from non-eating activities.

4.1. Research Environments and Ground Truth

The conditions and restrictions of the research environments have implications for different aspects of the eating detection approach; these are important considerations, given that the majority of the included studies were conducted in a laboratory setting. As a result, the accuracy achieved in testing models with data collected from the free-living settings may be lower compared to models trained and tested on the laboratory data. However, few studies collected data from free-living environments for evaluation purposes. Using data collected from free-living environment for training purposes will likely help improve the performance of detection models in less controlled settings. Future studies may overcome this issue by combining laboratory and free-living data approaches in a multi-stage approach to study design. Few studies have combined lab and free-living data (e.g., [2,3,46]) to date. For instance, Ye and colleagues [43] first trained a model in a laboratory study. In a follow-up study [54], they then used buttons on a smart watch (Pebble) and an app (Evernote) to confirm or reject detected eating occasions when testing the model in free-living setting.

To implement a machine learning model to automatically identify eating gestures, accurate data containing the target activities or the "ground truth" is required. The machine learning model then learns from this data and can later be used for automated eating activity detection. Objective ground truth tools (e.g., video cameras) are more practical in laboratory settings. Such controlled settings are imperative to increase the accuracy of data annotation which is crucial for building and evaluating classifiers. Only a few studies in free-living settings have used passive capture of video as the measure of ground truth (e.g., [31]). In contrast, most studies in free-living settings rely on participants self-reporting the target activities by using tools such as diaries or push buttons on a device [5]. However, even for data for which a video recording exists, the annotation of the exact start and end times of eating gestures can be ambiguous, which in turn may affect a model's accuracy. Difficulties could include the assessment of the exact moment when the hand-to-mouth movement starts and when the hand returns to an idle state, synchronisation across multiple devices or sensors (e.g., wrist sensor for gesture capture with video of eating activity; [48]), obstruction of ground truth measurement due to unrelated movements, people, or objects in certain settings such as communal eating.

4.2. Eating Context and Population Groups

The characteristics of eating movements, and the volume of food consumed, may change in different contexts (e.g., when the participant is stressed, walking, or working). However, the impact of context on the accuracy of automatically detecting eating gestures is yet to be explored. Snacking or in-between meal eating has widely been disregarded in the surveyed studies, possibly because it is difficult to detect sporadic eating-associated with hand-to-mouth movements in a free-living setting and it could easily be confused with other movements. Eating behaviour assessment is often based on a two-step approach that links individual eating gestures to timeframes of eating activities. Further, the majority of lab studies provided food to participants, often with a limited variety in type, which is in contrast to the wide variety of food available in free-living settings. Further, the majority of studies in laboratories were carried out with university students, therefore the movement data may not be representative for other population segments (e.g., elderly, young children, clinical populations). Another important contextual factor is eating culture. For instance, Cho and Choi [71] and Kim and colleagues [35,38] specifically explored the Asian eating style and found that hand movements associated with eating with a spoon are characteristically different from those associated with eating with chopsticks. Different cultural aspects of eating behaviour have been overlooked in the literature. For instance, at this stage there are no studies that consider data from communal and shared plate eating (e.g., with servings from a shared dish [79]). Abkenar and colleagues [62] investigated a context where two participants shared a meal together, yet this did not involve a shared dish. Communal eating is an important form of eating in many cultures (e.g., [79–81]). Further, there has been no study that has considered using upper limb motion sensors for detecting eating behaviour of individuals from low and lower-middle income countries. All of the settings mentioned will likely include additional challenges due to characteristic hand movements associated with serving food from communal dishes to individual serving vessels.

4.3. Advanced Models and Deep Learning

Machine learning algorithms employed to detect eating behaviour are distinguished by whether and how they consider the sequential context. Classifiers such as K-nearest neighbours (KNN) or support vector machine (SVM) do not explicitly utilise the sequential aspect of data. By contrast, classifiers such as Hidden Markov Model (HMM) and Recurrent Neural Networks (RNN) take into account the sequential context, using previous states of data to predict the current state of data. The latter types have gained more attention recently ([61,68,71,73,75]). In the current review most studies used approaches that do not model the sequential context of data across time (e.g., 21 SVM, 19 Fandom Forest, 16 Decision Tree, 9 KNN) while recently more studies have considered the sequential context (10 HMM, 4 RNN). These recent models have shown promising results. For instance, Ramos-Garcia & Hoover [39] found that HMM outperforms KNN by approximately 13% when distinguishing between four activities (rest, bite, drink, using utensils). Further, they found that taking into account inter-gesture sequential dependencies further improves model performance (up to 96.5% accuracy). Kyritsis and colleagues [61] showed that replacing HMM with RNN improves the performance of the model even more. Taken together, these results hint at the importance of utilising the sequential context.

Notably, up to 2017, there was no study that utilised deep learning to detect eating behaviour in this context. Driven by the growing computing power, and specifically the availability of GPU-based high-performance computing, researchers increasingly explore the application of deep networks such as CNN and RNN (specifically Long Short-Term Memory networks, LSTM) to various classification problems (e.g., since 2010 in human affect recognition [82]). Since 2017, five studies have investigated the application of deep learning for assessing eating behaviour based on movement sensors ([61,68, 71,73,75]). Results show that in an end-to-end deep learning solution a combination of CNN and RNN performs significantly better than a CNN-only solution while the models have no knowledge of micro-movements, also known as sub-gestures [73]. This will also simplify the annotation process since less detailed labelling regime will be required. As another example, Papadopoulos and colleagues [75] showed how an eating detection dataset can be used to (pre)train a LSTM and then fine-tune it on unlabelled data to adapt to a new participant using semi-supervised approved, allowing for a more personalised approach. Another application of deep learning is sensor fusion.

4.4. Public Database Development

Deep learning may not have been applied earlier in the eating behaviour context due to the inherent need for large datasets to train deep networks. Notably, compared to other domains such as object and human affect (e.g., face) recognition, there are few publicly available eating behaviour datasets with the total number of observations being relatively small (e.g., compared to affective computing where public datasets with millions of records exist; [83]). A related problem is that in order to accurately compare the performance of different classifiers, the models need to be evaluated using the same data. Hence, collecting and publishing reusable datasets can help researchers to compare the accuracy of models implemented based on different detection approaches. In recent years a few databases have been made public. In 2015, Thomaz and colleagues [2] published a lab and two free-living datasets (20 lab participants, seven free-living participants, one longitudinal free-living participant; http://www.ethomaz.com). In 2016, Mirtchouk and colleagues [48] published a wrist motion and audio sensors dataset (six participants; http://www.skleinberg.org/data.html). In 2017, Kyritsis and colleagues [61] published a food intake cycle dataset (10 participants; https://mug.ee.auth.gr/intake-cycle-detection). Finally, in 2018, Shen and colleagues [77] published a dataset that consists of 51,614 manually labelled gestures from 169 participants that was developed over the course of several studies (http://cecas.clemson.edu/$\sim$ahoover/cafeteria). This highlights the considerable amount of time and effort to prepare such a dataset. The growing availability of such datasets will help advance training classifiers in this area. In particular, publicly available datasets can provide the opportunity to pre-train models that can then be enhanced and improved on for specific hand gestures, or for a specific participant [75]. Further, this will allow better comparison and reconciliation of different ways of annotating eating gestures, which in turn facilitates enhanced comparison of the accuracy achieved across different types of sensors and algorithms.

4.5. Granularity of Eating Behaviour Detection and Sensor Fusion

In the context of dataset availability, it is noteworthy that the majority of studies, and especially those published in earlier calendar years, exclusively focus on a binary detection in terms of eating versus non-eating; both in terms of detecting overall eating occasions as well as individual hand gestures. While this binary classification provides a range of interesting insights (e.g., in terms of identifying the time, duration, and speed of eating), it does not consider other important aspects of eating such as the type (e.g., rice vs noodle [38]; distinguishing different drinks [64]) and amount of food being consumed (e.g., drink volume [64]), the category of eating utensil and serving vessel used (e.g., distinguishing chopsticks, hand, and spoon [35]), or related hand gestures (e.g., using cutlery to prepare food items for intake, using spoon to transfer food into serving vessel). Over time, the binary detection of eating occasions and individual hand-to-mouth movements has improved substantially. However, improving the detection of eating utensils and the amount of food that is being consumed will require more sophisticated models, larger reference datasets, and synthesis with established dietary assessment tools. Image-based food records [84] are well suited to complement data capture of hand-to-mouth movement data, due to the collection of type and amount of food, in addition to timing, and are preferred to traditional methods such as weighed food records [85]. Leveraging the potential of automating model configuration and employing end-to-end models that require less detailed annotations could be important steps in this direction.

In terms of sensor fusion, studies combined (1) different kinds of motion sensors (e.g., accelerometer, gyroscope, magnetic coupling and RFID sensors), (2) upper limb-mounted motion sensors with motion sensors mounted to other body parts (e.g., torso, jaw; [37]), and (3) motion sensors with other different types of sensors (e.g., camera, microphone, scales). Particularly when non-motion sensors are used, the goal is usually to narrow down the location (e.g., which room in smart homes, [31]) or activity of the user and, hence, reduce or remove confounding gesture types in free-living settings. Further, in earlier studies, some primarily focused on accelerometers because at that time gyroscopes required considerable amounts of energy. However, with the recent advances in gyroscope and battery technologies, these obstacles have been overcome for most settings. Further, in an effort to save energy, some studies used a hybrid approach where the gyroscope was only activated when the accelerometer detected a series of eating associated gestures [70]. A similar approach was used to start recordings with wearable cameras [7]. Shibon and Amft [76] applied a controller to the sensing and processing system to increase the sample and processing rate once a rotational hand gesture is detected. Hence, despite the progress in technology, these approaches might still be useful in scenarios where access to power is limited (e.g., in low and lower-middle income country settings) or where motion data is to be complemented with energy or storage intensive video recordings. However, concerns on privacy of wearable cameras need to be acknowledged, and the impact on behaviours relating to eating has not been determined. Alternatively, active image capture methods, such as image-based food records collected via mobile devices [84], allow for collection of data on food type and amount, meal composition and temporal eating patterns which could be combined with wrist motion sensor data in new ways such as to verify intake data from such self-reported tools.

4.6. Applicability in Dietary Assessment and Eating Behaviour Interventions

While initially, studies relied on specialised research equipment or dedicated hardware prototypes, recent advances in accuracy and affordability of wearable sensing technology have made commercial-grade sensors widely accessible. Increasingly, studies rely on off-the-shelf devices such as smart watches, demonstrating that such devices are considered reliable and accurate for detecting eating behaviour (e.g., [61,62,67]). This has important implications for the real-world feasibility of using this technology for dietary assessment and monitoring [86]. In particular, because watches have been worn on the wrist for more than a century, using wearable sensors on the wrist is an unobtrusive solution for collecting movement data. Hence, readily available smartwatches could provide the infrastructure to implement end-user applications that allow to track eating behaviour (e.g., [86,87]). However, the software infrastructure is yet to be developed to collect, store, and analyse personal data. For instance, the computing power of smart watches could be used for an online detection of eating behaviour and the delivery of context-sensitive behavioural recommendations. Further, by establishing a data exchange with health practitioners and others, such systems could provide targeted recommendations that promote positive health outcomes [88]. In the case of disease management, for instance, this data could be used by health practitioners to keep track of a patient's dietary intake behaviour and characteristics and provide them with useful dietary advice.

4.7. Strengths and Limitations of the Current Review

The current review has strengths and limitations that should be considered in the interpretation of its findings. A strength is that it is the first systematic review on the automatic detection of eating behaviour based on upper limb-mounted motion sensors following a rigorous review approach. Based on this, this review provides the first comprehensive overview of study settings, sensor configurations, action classes, performance comparisons, and detection approaches for assessing eating behaviour from upper limb motion sensors. The developed framework conceptualises the components and implicit design choices that researchers and practitioners need to consider when carrying out studies and may hence facilitate further research in this area. Further, by searching across 11 different databases, we cover health and dietary assessment journals as well as computing-focused ones. Nevertheless, it needs to be acknowledged that only considering studies published in English language may constitute a limitation. Further, due to the limitation of number of search terms, our search string only covers plural forms for word combinations. This is based on the advice of a medical librarian we consulted with that search databases will automatically detect plural forms for single terms (e.g., "smartphone" will cover "smartphones") but not for word combinations (e.g., "arm movement" will not cover "arm movements"). Finally, focusing only on upper limb-mounted wrist sensors does not take into account other sensor positions (e.g., head, neck) and associated sensor fusion approaches (e.g., microphone).

5. Conclusions

To date, 69 studies have investigated upper limb-mounted motion sensors for automatic eating behaviour recognition. These studies were predominantly laboratory based and were undertaken by university students, employed shallow machine learning architectures, and focused on distinguishing eating from non-eating activities. At this stage, five studies have successfully employed deep learning architectures in this context. The availability of large public databases will be paramount to progressing the development of more fine-grained eating behaviour assessment approaches. This will allow future research to directly compare the accuracy of different classifiers, consider multiple contextual factors inherent to eating (e.g., communal eating, culture), and to transfer those models from controlled laboratory conditions to practical free-living settings in different countries (e.g., low and lower-middle income) and eating contexts (e.g., home vs work environment, social gatherings).

Supplementary Materials: The following are available online at http://www.mdpi.com/2072-6643/11/5/1168/s1, Table S1: List of search strategy strings.

Author Contributions: The review protocol was developed by H.H., M.A., T.B., and M.E.R. Article retrieval and screening articles for inclusion was undertaken by H.H., M.A., T.B., and M.E.R. All authors (H.H., M.A., C.C., T.B., M.E.R.) provided content and were involved in the preparation of the manuscript. The final manuscript was approved by all authors.

Funding: Hamid Heydarian is supported by an Australian Government Research Training Program (RTP) Scholarship. Clare Collins is supported by an NHMRC Senior Research Fellowship and a Gladys M Brawn Senior Research Fellowship from the Faculty of Health and Medicine, the University of Newcastle.

Acknowledgments: The authors thank Clare Cummings, Kerith Duncanson, and Janelle Skinner for their help with extracting data from the reviewed studies.

Conflicts of Interest: The authors declare no conflict of interest.

Appendix A

Table A1. Search strategy string.

Search String
(accelerometer OR gyroscope OR smartwatch OR "inertial sensor" OR "inertial sensors" OR "inertial sensing" OR smartphone OR "cell phone" OR wristband) AND ("dietary intake" OR "dietary assessment" OR "food intake" OR "nutrition assessment" OR "eating activity" OR "eating activities" OR "eating behavior" OR "eating behaviour" OR "energy intake" OR "detecting eating" OR "detect eating" OR "eating episodes" OR "eating period") AND ("bite counting" OR "counting bites" OR "hand gesture" OR "hand gestures" OR "arm gesture" OR "arm gestures" OR "wrist gesture" OR "wrist gestures" OR "hand motion" OR "hand motions" OR "arm motion" OR "arm motions" OR "wrist motion" OR "wrist motions" OR "hand movement" OR "hand movements" OR "arm movement" OR "arm movements" OR "wrist movement" OR "wrist movements" OR "hand to mouth" OR "hand-to-mouth" OR "wrist-worn" OR "wrist-mounted")

Table A2. Search strategy databases (English only results).

Database name	Result
ACM	10
AIS Electronic Library	55
CINAHL	16
EMBASE [1]	
IEEE [2]	140
MEDLINE [1]	
Ovid databases [3]	54
ScienceDirect [4]	133
Scopus	161
SpringerLink	205
Web of Science	18
Total results before removing duplicates	792
Total results after removing duplicates	653

[1] Results from Ovid databases include results from databases EMBASE and MEDLINE. [2] Due to limitation on the length of the search string that IEEE database accepted, the search string was broken up to smaller parts (see Supplementary Material). [3] Ovid databases include Books@Ovid, Embase, Emcare, MEDLINE. The following option were chosen to be included in the results: AMED (Allied and Complementary Medicine) 1985 to September 2017, Books@Ovid September 25, 2017, Embase 1947 to present, Emcare (Nursing and Allied Health) 1995–present, International Pharmaceutical Abstracts 1970 to September 2017, Medline 1946–present, University of Newcastle Journals. [4] ScienceDirect database web search did not accept the search terms "hand-to-mouth" and "hand to mouth". Therefore, these two search terms were excluded from the search string submitted to ScienceDirect. Also due to limitation on the length of the search string that ScienceDirect database accepted, the search string was broken up to smaller parts (see Supplementary Material).

References

1. Shen, Y.; Salley, J.; Muth, E.; Hoover, A. Assessing the accuracy of a wrist motion tracking method for counting bites across demographic and food variables. *IEEE J. Biomed. Heal. Inform.* **2017**, *21*, 599–606. [CrossRef] [PubMed]
2. Thomaz, E.; Essa, I.; Abowd, G.D. A Practical Approach for Recognizing Eating Moments with Wrist-Mounted Inertial Sensing. In Proceedings of the ACM International Joint Conference on Pervasive and Ubiquitous Computing, Osaka, Japan, 7–11 September 2015; pp. 1029–1040.

3.	Dong, Y.; Hoover, A.; Scisco, J.; Muth, E. A new method for measuring meal intake in humans via automated wrist motion tracking. *Appl. Psychophysiol. Biofeedback* **2012**, *37*, 205–215. [CrossRef] [PubMed]

4.	Kim, J.; Lee, M.; Lee, K.-J.; Lee, T.; Bae, B.-C.; Cho, J.-D. An Eating Speed Guide System Using a Wristband and Tabletop Unit. In Proceedings of the ACM International Joint Conference on Pervasive and Ubiquitous Computing Adjunct, Heidelberg, Germany, 12–16 September 2016; pp. 121–124.

5.	Farooq, M.; Fontana, J.M.; Boateng, A.F.; Mccrory, M.A.; Sazonov, E. A Comparative Study of Food Intake Detection Using Artificial Neural Network and Support Vector Machine. In Proceedings of the 12th International Conference on Machine Learning and Applications, Miami, FL, USA, 4–7 December 2013; pp. 153–157.

6.	Fontana, J.M.; Farooq, M.; Sazonov, E. Automatic ingestion monitor: A novel wearable device for monitoring of ingestive behavior. *IEEE Trans. Biomed. Eng.* **2014**, *61*, 1772–1779. [CrossRef] [PubMed]

7.	Sen, S.; Subbaraju, V.; Misra, A.; Balan, R.K.; Lee, Y. Experiences in Building a Real-World Eating Recogniser. In Proceedings of the IEEE International on Workshop on Physical Analytics, New York, NY, USA, 19 June 2017; pp. 7–12.

8.	Varkey, J.P.; Pompili, D.; Walls, T.A. Human motion recognition using a wireless sensor-based wearable system. *Pers. Ubiquitous Comput.* **2012**, *16*, 897–910. [CrossRef]

9.	Zhou, Y.; Cheng, Z.; Jing, L.; Tatsuya, H. Towards unobtrusive detection and realistic attribute analysis of daily activity sequences using a finger-worn device. *Appl. Intell.* **2015**, *43*, 386–396. [CrossRef]

10.	Kalantarian, H.; Alshurafa, N.; Sarrafzadeh, M. A survey of diet monitoring technology. *IEEE Pervasive Comput.* **2017**, *16*, 57–65. [CrossRef]

11.	Rouast, P.; Adam, M.; Burrows, T.; Chiong, R. Using Deep Learning and 360 Video to Detect Eating Behavior for User Assistance Systems. In Proceedings of the European Conference on Information Systems, Portsmouth, UK, 23–28 June 2018; pp. 1–11.

12.	Ramos-Garcia, R.I.; Muth, E.R.; Gowdy, J.N.; Hoover, A.W. Improving the recognition of eating gestures using intergesture sequential dependencies. *IEEE J. Biomed. Heal. Inform.* **2015**, *19*, 825–831. [CrossRef] [PubMed]

13.	Dong, Y.; Hoover, A.; Muth, E. A Device for Detecting and Counting Bites of Food Taken by a Person during Eating. In Proceedings of the IEEE International Conference on Bioinformatics and Biomedicine, Washington, DC, USA, 1–4 November 2009; pp. 265–268.

14.	Dong, B.; Biswas, S. Meal-time and duration monitoring using wearable sensors. *Biomed. Signal Process. Control* **2017**, *32*, 97–109. [CrossRef]

15.	Parate, A.; Ganesan, D. Detecting Eating and Smoking Behaviors Using Smartwatches. In *Mobile Health*; Rehg, J.M., Murphy, S.A., Kumar, S., Eds.; Springer: Cham, Switzerland, 2017; pp. 175–201. ISBN 978-3-319-51393-5.

16.	Prioleau, T.; Moore II, E.; Ghovanloo, M. Unobtrusive and wearable systems for automatic dietary monitoring. *IEEE Trans. Biomed. Eng.* **2017**, *64*, 2075–2089. [CrossRef] [PubMed]

17.	Zhang, S.; Stogin, W.; Alshurafa, N. I sense overeating: Motif-based machine learning framework to detect overeating using wrist-worn sensing. *Inf. Fusion* **2018**, *41*, 37–47. [CrossRef]

18.	Hassannejad, H.; Matrella, G.; Ciampolini, P.; De Munari, I.; Mordonini, M.; Cagnoni, S. Automatic diet monitoring: a review of computer vision and wearable sensor-based methods. *Int. J. Food Sci. Nutr.* **2017**, *68*, 656–670. [CrossRef] [PubMed]

19.	Vu, T.; Lin, F.; Alshurafa, N.; Xu, W. Wearable food intake monitoring technologies: A comprehensive review. *Computers* **2017**, *6*, 4. [CrossRef]

20.	Doulah, A.; Mccrory, M.A.; Higgins, J.A.; Sazonov, E. A Systematic Review of Technology-Driven Methodologies for Estimation of Energy Intake. *IEEE Access* **2019**, *7*, 49653–49668. [CrossRef]

21.	Covidence: Better Systematic Review Management. Available online: https://www.covidence.org (accessed on 21 Feburary 2019).

22.	UN World Economic Situation and Prospects: Statistical Annex. Available online: https://www.un.org/development/desa/dpad/wp-content/uploads/sites/45/WESP2018_Annex.pdf (accessed on 21 Feburary 2019).

23.	Amft, O.; Junker, H.; Tröster, G. Detection of Eating and Drinking Arm Gestures Using Inertial Body-Worn Sensors. In Proceedings of the Ninth IEEE International Symposium on Wearable Computers, Osaka, Japan, 18–21 October 2005; pp. 1–4.

24. Amft, O.; Kusserow, M.; Tröster, G. Probabilistic Parsing of Dietary Activity Events. In Proceedings of the 4th International Workshop on Wearable and Implantable Body Sensor Networks, Aachen, Germany, 26–28 March 2007; pp. 242–247.

25. Amft, O.; Tröster, G. Recognition of dietary activity events using on-body sensors. *Artif. Intell. Med.* **2008**, *42*, 121–136. [CrossRef] [PubMed]

26. Junker, H.; Amft, O.; Lukowicz, P.; Tröster, G. Gesture spotting with body-worn inertial sensors to detect user activities. *Pattern Recognit.* **2008**, *41*, 2010–2024. [CrossRef]

27. Van Laerhoven, K.; Kilian, D.; Schiele, B. Using Rhythm Awareness in Long-Term Activity Recognition. In Proceedings of the IEEE International Symposium on Wearable Computers, Pittsburgh, PA, USA, 28 September–1 October 2008; pp. 63–66.

28. Pirkl, G.; Stockinger, K.; Kunze, K.; Lukowicz, P. Adapting Magnetic Resonant Coupling Based Relative Positioning Technology for Wearable Activitiy Recogniton. In Proceedings of the 12th IEEE International Symposium on Wearable Computers, Pittsburgh, PA, USA, 28 September–1 October 2008; pp. 47–54.

29. Tolstikov, A.; Biswas, J.; Tham, C.-K.; Yap, P. Eating Activity Primitives Detection—A Step Towards ADL Recognition. In Proceedings of the 10th International Conference on e-health Networking, Applications and Services, Singapore, 7–9 July 2008; pp. 35–41.

30. Zhang, S.; Ang, M.H.; Xiao, W.; Tham, C.K. Detection of Activities for Daily Life Surveillance: Eating and Drinking. In Proceedings of the IEEE Intl. Conf. on e-Health Networking, Applications and Service, Singapore, 7–9 July 2008; pp. 171–176.

31. Teixeira, T.; Jung, D.; Dublon, G.; Savvides, A. Recognizing Activities from Context and Arm Pose Using Finite State Machines. In Proceedings of the International Conference on Distributed Smart Cameras, Como, Italy, 30 August–2 September 2009; pp. 1–8.

32. Amft, O.; Bannach, D.; Pirkl, G.; Kreil, M.; Lukowicz, P. Towards Wearable Sensing-Based Assessment of Fluid Intake. In Proceedings of the 8th IEEE International Conference on Pervasive Computing and Communications Workshops, Mannheim, Germany, 29 March–2 April 2010; pp. 298–303.

33. Dong, Y.; Hoover, A.; Scisco, J.; Muth, E. Detecting Eating Using a Wrist Mounted Device During Normal Daily Activities. In Proceedings of the International Conference on Embedded Systems and Applications, Las Vegas, NV, USA, 18–21 July 2011; pp. 3–9.

34. Grosse-Puppendahl, T.; Berlin, E.; Borazio, M. Enhancing accelerometer-based activity recognition with capacitive proximity sensing. In *Ambient Intelligence*; Paternò, F., de Ruyter, B., Markopoulos, P., Santoro, C., van Loenen, E., Luyten, K., Eds.; Springer: Pisa, Italy, 2012; pp. 17–32.

35. Kim, H.J.; Kim, M.; Lee, S.J.; Choi, Y.S. An Analysis of Eating Activities for Automatic Food Type Recognition. In Proceedings of the Asia-Pacific Signal and Information Processing Association Annual Summit and Conference, Hollywood, CA, USA, 3–6 December 2012; pp. 1–5.

36. Mousavi Hondori, H.; Khademi, M.; Videira Lopes, C. Monitoring Intake Gestures Using Sensor Fusion (Microsoft Kinect and Inertial Sensors) for Smart Home Tele-Rehab Setting. In Proceedings of the 1st Annual Healthcare Innovation Conference of the IEEE EMBS, Houston, TX, USA, 7–9 November 2012; pp. 36–39.

37. Fontana, J.M.; Farooq, M.; Sazonov, E. Estimation of Feature Importance for Food Intake Detection Based on Random Forests Classification. In Proceedings of the Annual International Conference of the IEEE Engineering in Medicine and Biology Society, Osaka, Japan, 3–7 July 2013; pp. 6756–6759.

38. Kim, H.-J.; Choi, Y.S. Eating Activity Recognition for Health and Wellness: A Case Study on Asian Eating Style. In Proceedings of the IEEE International Conference on Consumer Electronics, Las Vegas, NV, USA, 11–14 January 2013; pp. 446–447.

39. Ramos-Garcia, R.I.; Hoover, A.W. A study of Temporal Action Sequencing during Consumption of a Meal. In Proceedings of the International Conference on Bioinformatics, Computational Biology and Biomedical Informatics, Wshington, DC, USA, 22–25 September 2013; pp. 68–75.

40. Desendorf, J.; Bassett, D.R.; Raynor, H.A.; Coe, D.P. Validity of the Bite Counter Device in a Controlled laboratory setting. *Eat. Behav.* **2014**, *15*, 502–504. [CrossRef] [PubMed]

41. Dong, Y.; Scisco, J.; Wilson, M.; Muth, E.; Hoover, A. Detecting periods of eating during free-living by tracking wrist motion. *IEEE J. Biomed. Heal. Inform.* **2014**, *18*, 1253–1260. [CrossRef]

42. Sen, S.; Subbaraju, V.; Misra, A.; Balan, R.K.; Lee, Y. The Case for Smartwatch-Based Diet Monitoring. In Proceedings of the IEEE International Conference on Pervasive Computing and Communication Workshops, St. Louis, MO, USA, 23–27 March 2015; pp. 585–590.

43. Ye, X.; Chen, G.; Cao, Y. Automatic Eating Detection Using Head-Mount and Wrist-Worn Accelerometers. In Proceedings of the 17th International Conference on E-Health Networking, Application and Services, Boston, MA, USA, 14–17 October 2015; pp. 578–581.

44. Fan, D.; Gong, J.; Lach, J. Eating Gestures Detection by Tracking Finger Motion. In Proceedings of the IEEE Wireless Health, Bethesda, MD, USA, 25–27 October 2016; pp. 1–6.

45. Farooq, M.; Sazonov, E. Detection of Chewing from Piezoelectric Film Sensor Signals Using Ensemble Classifiers. In Proceedings of the Annual International Conference of the IEEE Engineering in Medicine and Biology Society, Orlando, FL, USA, 16–20 August 2016; pp. 4929–4932.

46. Fortuna, C.; Giraud-Carrier, C.; West, J. Hand-to-Mouth Motion Tracking in Free-Living Conditions for Improved Weight Control. In Proceedings of the IEEE International Conference on Healthcare Informatics, Chicago, IL, USA, 4–7 October 2016; pp. 341–348.

47. Maramis, C.; Kilintzis, V.; Maglaveras, N. Real-Time Bite Detection from Smartwatch Orientation Sensor Data. In Proceedings of the ACM International Conference Proceeding Series, Thessaloniki, Greece, 18–20 May 2016; pp. 1–4.

48. Mirtchouk, M.; Merck, C.; Kleinberg, S. Automated Estimation of Food Type and Amount Consumed from Body-Worn Audio and Motion Sensors. In Proceedings of the ACM International Joint Conference on Pervasive and Ubiquitous Computing, Heidelberg, Germany, 12–16 September 2016; pp. 451–462.

49. Parra-Sánchez, S.; Gomez-González, J.M.; Ortega, I.A.Q.; Mendoza-Novelo, B.; Delgado-García, J.; Cruz, M.C.; Vega-González, A. Recognition of Activities of Daily Living Based on the Vertical Displacement of the Wrist. In Proceedings of the 14th Mexican Symposium on Medical Physics, México City, Mexico, 16–17 March 2016; pp. 1–6.

50. Rahman, T.; Czerwinski, M.; Gilad-Bachrach, R.; Johns, P. Predicting "about-to-eat" Moments for just-in-Time Eating Intervention. In Proceedings of the Digital Health Conference, Montreal, QC, Canada, 11–13 April 2016; pp. 141–150.

51. Sharma, S.; Jasper, P.; Muth, E.; Hoover, A. Automatic Detection of Periods of Eating Using Wrist Motion Tracking. In Proceedings of the IEEE International Conference on Connected Health, Washington, DC, USA, 27–29 June 2016.

52. Shen, Y.; Muth, E.; Hoover, A. Recognizing Eating Gestures Using Context Dependent Hidden Markov Models. In Proceedings of the IEEE 1st International Conference on Connected Health: Applications, Systems and Engineering Technologies, Washington, DC, USA, 27–29 June 2016; pp. 248–253.

53. Shoaib, M.; Scholten, H.; Havinga, P.J.M.; Incel, O.D. A Hierarchical Lazy Smoking Detection Algorithm Using Smartwatch Sensors. In Proceedings of the IEEE 18th International Conference on e-Health Networking, Applications and Services, Munich, Germany, 14–16 September 2016; pp. 1–6.

54. Ye, X.; Chen, G.; Gao, Y.; Wang, H.; Cao, Y. Assisting Food Journaling with Automatic Eating Detection. In Proceedings of the Conference on Human Factors in Computing Systems, San Jose, CA, USA, 7–12 May 2016; pp. 3255–3262.

55. Zhang, S.; Alharbi, R.; Stogin, W.; Pourhomayun, M.; Spring, B.; Alshurafa, N. Food Watch: Detecting and Characterizing Eating Episodes through Feeding Gestures. In Proceedings of the 11th EAI International Conference on Body Area Networks, Turin, Italy, 15–16 December 2016; pp. 91–96.

56. Alexander, B.; Karakas, K.; Kohout, C.; Sakarya, H.; Singh, N.; Stachtiaris, J.; Barnes, L.E.; Gerber, M.S. A Behavioral Sensing System that Promotes Positive Lifestyle Changes and Improves Metabolic Control among Adults with Type 2 Diabetes. In Proceedings of the IEEE Systems and Information Engineering Design Symposium, Charlottesville, VA, USA, 28 April 2017; pp. 283–288.

57. Bi, C.; Xing, G.; Hao, T.; Huh, J.; Peng, W.; Ma, M. FamilyLog: A Mobile System for Monitoring Family Mealtime Activities. In Proceedings of the IEEE International Conference on Pervasive Computing and Communications, Kona, HI, USA, 13–17 March 2017; pp. 1–10.

58. Egilmez, B.; Poyraz, E.; Zhou, W.; Memik, G.; Dinda, P.; Alshurafa, N. UStress: Understanding College Student Subjective Stress Using Wrist-Based Passive Sensing. In Proceedings of the IEEE WristSense Workshop on Sensing Systems and Applications using Wrist Worn Smart Devices, Kona, HI, USA, 13–17 March 2017; pp. 1–6.

59. Garcia-Ceja, E.; Galván-Tejada, C.E.; Brena, R. Multi-view stacking for activity recognition with sound and accelerometer data. *Inf. Fusion* **2017**, *40*, 45–56. [CrossRef]

60. Kyritsis, K.; Tatli, C.L.; Diou, C.; Delopoulos, A. Automated Analysis of in Meal Eating Behavior Using a Commercial Wristband IMU Sensor. In Proceedings of the International Conference of the IEEE Engineering in Medicine and Biology Society, Seogwipo, Korea, 11–15 July 2017; pp. 2843–2846.

61. Kyritsis, K.; Diou, C.; Delopoulos, A. Food Intake Detection from Inertial Sensors USING Lstm Networks. In Proceedings of the International Conference on Image Analysis and Processing, Catania, Italy, 11–15 September 2017; pp. 411–418.

62. Loke, S.W.; Abkenar, A.B. Assigning group activity semantics to multi-device mobile sensor data. *KI—Künstliche Intelligenz* **2017**, *31*, 349–355. [CrossRef]

63. Moschetti, A.; Fiorini, L.; Esposito, D.; Dario, P.; Cavallo, F. Daily Activity Recognition with Inertial Ring and Bracelet: An Unsupervised Approach. In Proceedings of the IEEE International Conference on Robotics and Automation, Singapore, 29 May–3 June 2017; pp. 3250–3255.

64. Soubam, S.; Agrawal, M.; Naik, V. Using an Arduino and a Smartwatch to Measure Liquid Consumed from any Container. In Proceedings of the 9th International Conference on Communication Systems and Networks, Bangalore, India, 4–8 January 2017; pp. 464–467.

65. Thomaz, E.; Bedri, A.; Prioleau, T.; Essa, I.; Abowd, G.D. Exploring Symmetric and Asymmetric Bimanual Eating Detection with Inertial Sensors on the Wrist. In Proceedings of the 1st Workshop on Digital Biomarkers, Niagara Falls, NY, USA, 23 June 2017; pp. 21–26.

66. Yoneda, K.; Weiss, G.M. Mobile Sensor-Based Biometrics Using Common Daily Activities. In Proceedings of the IEEE Annual Ubiquitous Computing, Electronics and Mobile Communication Conference, New York, NY, USA, 19–21 October 2017; pp. 584–590.

67. Zhang, S.; Alharbi, R.; Nicholson, M.; Alshurafa, N. When Generalized Eating Detection Machine Learning Models Fail in the Field. In Proceedings of the ACM International Joint Conference on Pervasive and Ubiquitous Computing and Proceedings of the ACM International Symposium on Wearable Computers, Maui, HI, USA, 11–15 September 2017; pp. 613–622.

68. Anderez, D.O.; Lotfi, A.; Langensiepen, C. A Hierarchical Approach in Food and Drink Intake Recognition Using Wearable Inertial Sensors. In Proceedings of the 11th PErvasive Technologies Related to Assistive Environments Conference on, Corfu, Greece, 26–29 June 2018; pp. 552–557.

69. Anderez, D.O.; Lotfi, A.; Langensiepen, C. A novel crossings-based segmentation approach for gesture recognition. In *Advances in Computational Intelligence Systems*; Springer: Cham, Switzerland, 2018; pp. 383–391.

70. Balaji, R.; Bhavsar, K.; Bhowmick, B.; Mithun, B.S.; Chakravarty, K.; Chatterjee, D.; Ghose, A.; Gupta, P.; Jaiswal, D.; Kimbahune, S.; et al. A Framework for Pervasive and Ubiquitous Geriatric Monitoring. In Proceedings of the Human Aspects of IT for the Aged Population. Applications in Health, Assistance, and Entertainment, Las Vegas, NV, USA, 15–20 July 2018; pp. 205–230.

71. Cho, J.; Choi, A. Asian-Style Food Intake Pattern Estimation Based on Convolutional Neural Network. In Proceedings of the IEEE International Conference on Consumer Electronics, Las Vegas, NV, USA, 12–14 January 2018; pp. 1–2.

72. Clapés, A.; Pardo, À.; Pujol Vila, O.; Escalera, S. Action Detection Fusing Multiple Kinects and a WIMU: An application to in-home assistive technology for the elderly. *Mach. Vis. Appl.* **2018**, *29*, 765–788. [CrossRef]

73. Kyritsis, K.; Diou, C.; Delopoulos, A. End-to-End Learning for Measuring in-Meal Eating Behavior From a Smartwatch. In Proceedings of the 40th Annual International Conference of the IEEE Engineering in Medicine and Biology Society, Honolulu, HI, USA, 18–21 July 2018; pp. 5511–5514.

74. Manzi, A.; Moschetti, A.; Limosani, R.; Fiorini, L.; Cavallo, F. Enhancing activity recognition of self-localized robot through depth camera and wearable sensors. *IEEE Sens. J.* **2018**, *18*, 9324–9331. [CrossRef]

75. Papadopoulos, A.; Kyritsis, K.; Sarafis, I.; Delopoulos, A. Personalised Meal Eating Behaviour Analysis via Semi-Supervised Learning. In Proceedings of the 40th Annual International Conference of the IEEE Engineering in Medicine and Biology Society, Honolulu, HI, USA, 18–21 July 2018; pp. 4768–4771.

76. Schibon, G.; Amft, O. Saving Energy on Wrist-Mounted Inertial Sensors by Motion-Adaptive Duty-Cycling in Free-Living. In Proceedings of the IEEE 15th International Conference on Wearable and Implantable Body Sensor Networks, Las Vegas, NV, USA, 4–7 March 2018; pp. 197–200.

77. Shen, Y.; Muth, E.; Hoover, A. The Impact of Quantity of Training Data on Recognition of Eating Gestures. Available online: https://arxiv.org/pdf/1812.04513.pdf (accessed on 7 February 2019).

78. Zambrana, C.; Idelsohn-Zielonka, S.; Claramunt-Molet, M.; Almenara-Masbernat, M.; Opisso, E.; Tormos, J.M.; Miralles, F.; Vargiu, E. Monitoring of upper-limb movements through inertial sensors—Preliminary results. *Smart Heal.* **2018**. [CrossRef]

79. Burrows, T.; Collins, C.; Adam, M.; Duncanson, K.; Rollo, M. Dietary assessment of shared plate eating: A missing link. *Nutrients* **2019**, *11*, 789. [CrossRef]

80. Savy, M.; Martin-Prével, Y.; Sawadogo, P.; Kameli, Y.; Delpeuch, F. Use of variety/diversity scores for diet quality measurement: relation with nutritional status of women in a rural area in Burkina Faso. *Eur. J. Clin. Nutr.* **2005**, *59*, 703–716. [CrossRef] [PubMed]

81. Savy, M.; Martin-Prével, Y.; Traissac, P.; Delpeuch, F. Measuring dietary diversity in rural Burkina Faso: comparison of a 1-day and a 3-day dietary recall. *Public Health Nutr.* **2007**, *10*, 71–78. [CrossRef] [PubMed]

82. Rouast, P.V.; Adam, M.T.P.; Chiong, R. Deep learning for human affect recognition: insights and new developments. *IEEE Trans. Affect. Comput.* **2019**, 1–20. [CrossRef]

83. Mollahosseini, A.; Hasani, B.; Mahoor, M.H. AffectNet: A database for facial expression, valence, and arousal computing in the wild. *IEEE Trans. Affect. Comput.* **2019**, *10*, 18–31. [CrossRef]

84. Boushey, C.J.; Spoden, M.; Zhu, F.M.; Delp, E.J.; Kerr, D.A. New mobile methods for dietary assessment: review of image-assisted and image-based dietary assessment methods. *Proc. Nutr. Soc.* **2017**, *76*, 283–294. [CrossRef] [PubMed]

85. Rollo, M.; Ash, S.; Lyons-Wall, P.; Russell, A.; Rollo, M.E.; Ash, S.; Lyons-Wall, P.; Russell, A.W. Evaluation of a mobile phone image-based dietary assessment method in adults with type 2 diabetes. *Nutrients* **2015**, *7*, 4897–4910. [CrossRef] [PubMed]

86. Turner-McGrievy, G.M.; Boutté, A.; Crimarco, A.; Wilcox, S.; Hutto, B.E.; Hoover, A.; Muth, E.R. Byte by bite: Use of a mobile Bite Counter and weekly behavioral challenges to promote weight loss. *Smart Health* **2017**, *3–4*, 20–26. [CrossRef] [PubMed]

87. Scisco, J.L.; Muth, E.R.; Dong, Y.; Hoover, A.W. Slowing bite-rate reduces energy intake: an application of the Bite Counter device. *J. Am. Diet. Assoc.* **2011**, *111*, 1231–1235. [CrossRef] [PubMed]

88. Noorbergen, T.J.; Adam, M.T.P.; Attia, J.R.; Cornforth, D.J.; Minichiello, M. Exploring the design of mHealth systems for health behavior change using mobile biosensors. *Commun. Assoc. Inf. Syst.* **2019**, 1–37. (in press).

 *nutrients*

Review

Dietary Assessment of Shared Plate Eating: A Missing Link

Tracy Burrows [1,2,*], Clare Collins [1,2], Marc Adam [2,3], Kerith Duncanson [1,2] and Megan Rollo [1,2]

1   School of Health Sciences, Faculty of Health and Medicine, University of Newcastle, Callaghan, NSW 2308, Australia; clare.collins@newcastle.edu.au (C.C.); kerith.duncanson@newcastle.edu.au (K.D.); Megan.rollo@newcastle.edu.au (M.R.)
2   Priority Research Centre for Physical Activity and Nutrition, University of Newcastle, Callaghan, NSW 2308, Australia; marc.adam@newcastle.edu.au
3   School of Electrical Engineering and Computing, Faculty of Engineering and Built Environment, University of Newcastle, Callaghan, NSW 2308, Australia
*   Correspondence: tracy.burrows@newcastle.edu.au; Tel.: +61-2-4921-5514

Received: 22 February 2019; Accepted: 2 April 2019; Published: 5 April 2019

Abstract: Shared plate eating is a defining feature of the way food is consumed in some countries and cultures. Food may be portioned to another serving vessel or directly consumed into the mouth from a centralised dish rather than served individually onto a discrete plate for each person. Shared plate eating is common in some low- and lower-middle income countries (LLMIC). The aim of this narrative review was to synthesise research that has reported on the assessment of dietary intake from shared plate eating, investigate specific aspects such as individual portion size or consumption from shared plates and use of technology in order to guide future development work in this area. Variations of shared plate eating that were identified in this review included foods consumed directly from a central dish or shared plate food, served onto additional plates shared by two or more people. In some settings, a hierarchical sharing structure was reported whereby different family members eat in turn from the shared plate. A range of dietary assessment methods have been used in studies assessing shared plate eating with the most common being 24-h recalls. The tools reported as being used to assist in the quantification of food intake from shared plate eating included food photographs, portion size images, line drawings, and the carrying capacity of bread, which is often used rather than utensils. Overall few studies were identified that have assessed and reported on methods to assess shared plate eating, highlighting the identified gap in an area of research that is important in improving understanding of, and redressing dietary inadequacies in LLMIC.

Keywords: shared plate eating; dietary assessment; lower middle income countries

---

1. Introduction

The need to make dietary data more widely available has been reported as one of 10 global research priorities [1]. Access to accurate dietary data relies on use and publication of validated dietary assessment methodologies in a range of settings. Current evidence relating to dietary assessment is focused at the individual level without considering energy and nutrients that may be consumed from shared plate eating. Shared plate eating is an important factor to consider in dietary assessment as it may contribute a substantial proportion of energy and nutrient intake, particularly in those parts of the world where this is how the majority of food is consumed.

Internationally the way foods and dishes are consumed and the factors that influence consumption vary from region to region [2]. In many high-income countries (HIC) food items are most commonly served on discrete plates for individuals to consume. In other regions dishes are often served centrally with individuals consuming directly from a shared central plate. Shared plate eating has been shown to be evident in many countries, most commonly in Asian countries and low- and lower-middle income countries (LLMIC) [3]. It is likely that the low representation of dietary assessment information relating to shared plate eating is the result of dietary assessment methodology originating in HIC, where shared plate eating is less common.

Assessing dietary intake in low- and lower-middle income countries (LLMIC) is necessary for dietary data to become more available and more applicable to nutrition priorities, but has unique challenges. Compared to high-income countries (HIC), there is substantially less information reported about the way food is prepared, served, and eaten in LLMIC. Details related to how foods are consumed, methods of food preparation, recipes and how nutrient composition data has been compiled and nutrients analysed from the collected intake data and additionally food composition databases are sparse [2,3].

Dietary intake assessment occurs less often in LLMIC and predominantly relied heavily on adaptation of methods used in HIC. For these reasons the data collection methods have been tailored to food consumption from individual plates or servings. Therefore, minimal research attention has focused on shared plate eating where food is directly consumed into the mouth from a centralised dish rather than served individually onto a discrete plate for each person. Shared plate eating is more prevalent and often a defining feature of the way food is often consumed in many LLMIC [4] when compared generally to HIC. When shared plate eating does occur in HIC such as a group sharing pizza or hot chips, it is usually associated with an abundant supply of food. Examples include celebrations or cafeteria, or family-style American meals, all of which differ in context and content from shared meals in LLMIC.

Consuming food in this manner typically occurs multiple times throughout a meal and also when parents are feeding their children. Shared plate eating (sometimes referred to as communal eating) is often overlooked in dietary assessment. Challenges in quantification of shared plate eating include accurate estimation of the number of spoonfuls or handfuls of each dish consumed, the amount eaten from each spoonful/handful, and the highly variable nutrient composition of dishes for which nutrient content have not been characterised or where the composition of each spoonful or handful may vary due to the contents of the dish (for example a meat and vegetable soup where one spoonful may be more liquid based and contain less meat and vegetables and next spoonful may contain more meat and vegetables and less liquid). Additionally, the associated literacy and numeracy skills required by an individual to self-report or for a trained observer to estimate intake from shared plates have not been well described or quantified.

In addition to the complexity of shared plate eating common in LLMIC, food and nutrient databases are less available compared to HIC [5]. Additional reasons for less dietary intake research being conducted in LLMIC compared to HIC include a lack of context-specific validated dietary assessment tools, low availability of trained personnel to collect and analyse intake data, and limited infrastructure and resources to co-ordinate population-based surveys [5].

A review by Ngo et al. 2005 [6] summarised studies that have adapted traditional dietary assessment measures for use in ethnic and/or minority groups, with a specific focus on those of European immigrant groups. The most common dietary assessment methods included in the previous review were interviewer administered food frequency questionnaires (FFQs), 24-h recalls (24HR), and the weighed food record (WFR) [6]. Adaptations to these traditional dietary assessment tools for ethnically diverse groups in LLMIC included identifying key dishes or foods, which may differ from the general population, and determining relevant portion sizes prior to data collection. In addition to the dietary tools, issues also exist with respect to food quantification, limited recipes or unclear recipe construction and lack of inclusion of traditional dishes within nutrient databases [7]. Critical information needed to process dietary data is often also limited or missing in LLMIC, such as country-specific food composition databases and tables of conversion to allow quantification of context-specific portion size [7]. Visual aids have been used in previous studies to assist in quantification of portion size, however it was more common to use standard serving sizes from other studies or countries to quantify intake not population specific [5]. Using pre-defined serving sizes to estimate portion size is likely to incur a bias and incorrect estimation of food intake, especially when the common portion size of dishes in LLMIC are unknown.

In LLMIC, the use of 24HR has been recommended over other methods such as WFR, due to the perception of being less time consuming and having a lower participant burden [5]. A review of existing dietary assessment in LLMIC identified that while 24HR was most commonly performed using pen and paper, there were substantial costs and burden associated with using this method, particularly the increased time for researchers to code and then analyse data [5]. Unique costs associated within LLMIC have been previously identified and included costs for externally-based researchers to provide training and supervision to upskill research assistants, as well as costs to expand the food composition database, and for logistics (e.g., transportation) and equipment (e.g., internet connections, laptops, mobile phones, phone cards).

Electronic data capture and use of technology has been suggested as potentially very useful in LLMIC, given that it is likely to be less expensive and more time effective than traditional pen and paper methods [5]. In this way, electronic data capture may overcome some of the identified costs and also potential language/ communication issues. Standardised and streamlined technologies can provide improvements in a range of areas previously acknowledged to improve the ease, time and cost of data collection and processing, and also ensure high-quality standardised data entry, analysis, consistency and comparability across dietary data [7,8].

While externally-based researchers have expertise related to dietary assessment methodologies and understanding of the food supply in their respective countries, it has been identified that identification of foods is more accurate if it involves local people with food expertise [5]. Prynne et al. [5] reported that agreement relating to estimated energy intake by internal and external coders and researchers is quite high, but that at a micronutrient levels understanding of the local food supply and eating habits is essential for more reliable nutrient estimates.

Research focused on shared plate eating in LLMIC has not been previously reviewed and synthesised, but is important in improving the accuracy of assessing dietary intake in these settings. The aim of the current paper is to provide a narrative synthesis of current research that has reported on the assessment intake from shared plate eating, investigated specific aspects such as individual portion size estimation tools from shared plate eating and use of technology to guide future development work in this area.

2. Overview of Research

The majority of studies that have assessed shared plate eating were undertaken in LLMIC including: Gambia [9] Burkino Faso [10–12] and Egypt [13], two in Nepal [14,15], two in India [16,17], Sri Lanka [18] and Zambia [19] with one study each identified as undertaken in Japan and Israel [20] representing higher income countries [21] (Table 1). Studies on shared plate eating were carried out with mothers and children [10,11], children only [14,15,18,19], or adults [9,13,16,17,20,21]. Sample sizes ranged from 17 to 3908. Shared plate eating was found to contribute between 30 and 88% of total daily energy intake [20]. More frequent shared plate eating was reported in rural locations when compared with urban areas [20]. There was no trend identified towards more studies being published in recent years with three studies published in the 1990s and only 3 studies published in 2010 or later.

Table 1. Studies investigating shared plate eating in low- and lower-middle income countries (LLMIC) and reported diet, food, or nutrient outcomes.

Author Year Country	Study Design and Setting	Study Population	Participant Number and Gender	Dietary Assessment Method	Validated/Standardised Method	Primary Outcomes
			Studies involving direct observation			
Hudson 1995 [9] Gambia	Repeat cross sectional Rural African community	Unclear	Phase 1: 208 'sinkiros' (cooking unit within family structure) Phase 2: 12 families Phase 3: 7 males	Phase 1: All ingredients identified and weighed before being cooked	Direct observation: Each bowl weighed (1) when empty (2) after staple food added (3) after sauce added. Age, sex, body weight and amount of food waste was recorded for each participant. Phase 3: DLW study	1. Detailed observation and measurement of meal preparation to calculate nutrient intake from each meal. 2. Average weights of staple foods/sauces consumed. 3. Energy intake estimations from two main meals
Shankar et al. 1998 [15] Nepal	Case Control Sarlahi district—rural Nepal. 3 village development communities	Children 1-6yrs at risk of Vit A deficiency	162 households (81 case/ 81 control) Gender NR	Direct observation by 10 local Nepalese males trained for 3 months	Direct observation of control and case participants	1. Classification of feeding episodes: no food sharing/shared plate eating/interplate sharing 2. Shared plate vs individual plate eating 3. Average portion sizes 4. Odds of consuming different food groups by feeding type
Shankar et al. 2001 [14] Nepal	Validation study as part of larger longitudinal study Sarlahi District. Rural region of Nepal	Children aged 1–10 years old.	11 (6 male, 5 female) 17 field tests (9 individual plate, 8 shared plate setting)	Direct observation by 8 observers who undertook 3 months of training	Food weighing used as reference to determine accuracy of observers' visual direct observation of food intake.	1. Accuracy of observations in individual plate eating and shared plate eating 2. Comparison of estimates between observers
			Studies using 24 h recall dietary assessment method			
Abu-Saad et al. 2009 [20] Israel	Cross sectional Semi nomadic population in Southern Israel	Healthy 19–82-year-old semi-nomadic adults visiting hospital patients or attending Maternal and Child Health Care clinics	n = 451 (149 male, 302 female) >1× 24HR recall. 40 completed 3 × 24HR recalls	Modified USDA 24HR recall conducted by trained interviewers and administered using the multi-pass method	EI calculated using American Food Information Analysis System. Compared EI from 24HR recall with BMR using the Schofield equation	1. Eating patterns 2. Nutrient intakes 3. EI using Schofield vs recall 4. Day to day variation in 3-day results for 40 respondents
Caswell et al. 2015 [19] Zambia	Cross- sectional	Children aged 4–8 not yet enrolled in school	938 (479 male, 459 female)	24 h recall conducted on tablet by local interviewers	Nutrient intakes were calculated using food composition tables developed for Zambia by HarvestPlus. USDA National Nutrient Database and other local food composition tables.	1. Demographic Characteristics 2. Common foods consumed 3. Nutrient intakes

Table 1. *Cont.*

Author Year Country	Study Design and Setting	Study Population	Participant Number and Gender	Dietary Assessment Method	Validated/Standardised Method	Primary Outcomes
Savy et al. 2005 [11] 2007 [12] Burkina Faso	Cross sectional	Women living in randomly selected compounds with at least 1 child under 5 years of age.	691 females	Three-day dietary intake 24 h recall conducted by 14 local fieldworkers. Food variety score (FVS) and Dietary Diversity Score (DDS) calculated	NR	1. Relationship FVS + DDS and socio-demographic and economic characteristics 2. Relationships between DDS + FVS and anthropometry 3. Relationship between DDS + FVS and nutritional status
			Studies using an interview or questionnaire method			
Daniel et al. 2014 [16] India	Cross sectional 3 regions of India; New Delhi, Mumbai and Trivandrum. Selected to capture cancer registries	Aged 35–69 years old, resided in study area for at least 1 year.	3908 (male and female) completed DHQ, 3862 included in analysis after data cleaning	Interviews conducted by trained staff at home using New Interactive Nutrition Assistant–Diet in India Study of Health (NINA-DISH): (1) DHQ, (2) questions on meal times; (3) food-preparer QA and (4) 24HR recall	NR	1. Number of food items from food groups reported in DHQ & 24HR recall 2. Number of total food items and time taken to complete DHQ & 24HR recall 3. Top food contributors to nutrient values
Ferrucci et al. 2010 [17] India	Cross sectional from national registry (cancer specific content) three regions (New Delhi, Mumbai and Trivandrum)	Aged 35–69 years old, resided in study area for at least one year. Recruited one male and one female/household	3625 (male and female) (New Delhi n = 835, Trivandrum n = 2,044, Mumbai n = 746)	Computer-based diet QA using NINA-DISH software administered by trained field personnel	NR	1. Global spice consumption and cancer incidence 2. Consumption of spices and seasonings in participants 3. Consumption of commonly used cooking oils 4. Socio-demographic characteristics
Iwaoka et al. 2001 [21] Japan	Cohort College	Dietetics students and their mothers	64 females (32 households)	Approximated proportion	Individual-based food weighing method	1. Mean difference energy and nutrient intakes between methods
		Studies using dietary assessment tools of interest to shared plate eating				
Jerome 1997 [13] Egypt and Grenada	Case Study NR	Egypt: Kalama village, periurban community. Grenada	NR	Egypt: Household and individual intake, Grenada: Dietary information reported from each individual in the household (not shared plate)	NR	To use both case studies to highlight the importance of matching the dietary assessment method with the culture of the population being studied.

Table 1. *Cont.*

Author Year Country	Study Design and Setting	Study Population	Participant Number and Gender	Dietary Assessment Method	Validated/Standardised Method	Primary Outcomes
Thoradeniya et al. 2012 [18] Sri Lanka	Cross sectional Laboratory	School children 10–16 years	80 (32 male, 48 female)	Portion size estimation aids of 16 food items: (1) small photographs ($n = 11$ foods, 876 estimations), (2) life-size photographs ($n = 7$ foods, 558 estimations), (3) 2D life-size diagrams ($n = 16$ foods, 1271 estimations) and (4) household utensils ($n = 6$ foods, 475 estimations)	Actual weight of food	1. Precision and accuracy or portion size estimations tools for Asian Countries

Abbreviations: BMR: Basal Metabolic Rate; D: Dimensional; DDS: Diet Diversity Score; DHQ: Diet History Questionnaire; DLW: Doubly-labelled Water; EI: Energy Intake; FVS: Food Variety Score; N: Number; NR: Not Reported; PSEA: Portion Size Estimation Aid; USDA: United States Department of Agriculture.

2.1. Variants of Shared Plate Eating

The review identified that different forms of shared plate eating exist, with multiple people eating from one central dish the most common [15]. Others forms of shared plate eating identified included interpolate (or post-serve) sharing defined as two or more people eating from the same plate after serving from a central dish [15]. Food sharing was reported to occur at both meals and snacks and for both adults and children [15].

Shared plate eating was reported to involve complex rules around food distribution based on family structure [22]. For example, an adult male family member may eat first and be offered the protein components of meal first, while women and children will eat from what remains after the men have eaten. This may lead to certain individuals receiving disproportionately less of the food or substantially different meal compositions, and hence varying nutrient intakes at the household level. Further, the feeding of young children differs substantially between households and for children of different ages, which may determine whether a child is self-feeding or being fed by another [23].

2.2. Methods of Assessing Dietary Intake

Dietary assessment methods used to assess intakes from shared plate eating (Table 1) were varied and included 24HR (four studies) [11,12,19,20], two studies that used direct observation [14,15] or food weighing [10,21], and one study using a dietary survey study [17]. Two studies utilised multiple dietary methods; one study used [16] interviews, diet history questionnaire and 24HR while another study used direct observation and ingredient weighing to capture dietary intake [9]. Only one used an objective biomarker, which was doubly-labelled water, to estimate total energy expenditure and to compare energy intake assessed by direct observation [9].

2.3. Direct Observation Methods

Direct observation was used in a variety of ways to assess shared plate eating in three studies; two in Nepal [14,15] and one in Gambia [9]. In the study of adult males in Gambia [9] the contribution of two cooked meals per day to energy and nutrient intake of adult males was determined using doubly-labelled water and algorithms based on observation of household food preparation and consumption. The process involved identification and weighing of each ingredient prior to being added to each cooking pot. A researcher observed the preparation process and documented the addition of each ingredient. When the meal was ready for consumption the weight of each empty eating bowl was determined, then weighed again after the addition of the staple (i.e., rice, grains) and again after the addition of each respective meal component. The body weight of each person and the food they consumed from each dish was recorded. The observer remained in the house to weigh any remaining/ leftover food [9].

In the same study the average weights of six common staple foods (rice, sorghum, sanyo, findo, maize, cassava) consumed at each meal and who consumed these foods was determined through direct observation. Estimated intakes for common additions (such as sauces, spices, herbs and condiments) were also determined. Through use of this technique, an algorithm was created to quantify the distribution between individuals of food from shared plate dishes.

Doubly-labelled water was used to verify total energy intake of adult males, with urine collected over a period of ten days. The results indicated that estimation from two cooked meals was equivalent to 80% of an individual's total energy expenditure, with the remainder likely to be contributed by snacks between meals which were not assessed in the study. As data collection occurred periodically throughout the year, distinct seasonal changes in the total energy intakes consumed and associated weight status were reported. Higher energy intakes and weight status were reported from October to April coinciding with and following the harvest season in Gambia, and showed a steady decline in middle months of the year.

The accuracy of visual estimations of children's food intake during shared plate eating compared to individual-plate eating scenarios was investigated by Shankar et al, 2001 [14] in a study involving male and female Nepali children. In this study, eight trained observers estimated food portions consumed by children enacting common eating scenarios. Test foods were selected from food groups regularly eaten in this region (grains, vegetables, pulses, fruits, meats, dairy, mixed dishes). Foods were weighed at the start of the meal as a reference measure to improve estimations by trained observers, and at the end of meal to quantify volumes of leftover food. Foods were categorised by food group and categorised as individual-plate or shared-plate. Observed food weight estimates were compared to actual weights of 69 food portions of children eating alone and 26 portions where children were eating from a shared plate. Analyses revealed that observer estimates of dark green leafy vegetables (141%) and fruits (139%) tended to be overestimated by the trained observers whereas grains and mixed foods (98% and 96%) were closer to weighed method. Overall, food weights under field conditions were highly correlated with actual weights for individual-plate ($r = 0.89$) and less accurately for shared plate eating ($r = 0.84$). Accuracy of estimations was influenced by food weight with greater error associated with food quantities of less than 70 grams. Mothers or primary caretakers were not always present during a child's meal and therefore may not have observed the portion eaten, which suggests that proxy report for children's intake is not always suitable in these settings [14].

Another direct observation study that involved Nepali children [15] was used to investigate dietary differences between children with Vitamin A deficiency and those who were Vitamin A sufficient. Household intake was recorded, however the observers focused predominantly on child intake. Food was visually estimated by trained observers as amount consumed and amount lost to spillage, with total estimations completed for everyone except the last person eating as these were ascertained by subtraction. Each food consumed was categorised into a group. A code was assigned to each member of the shared plate eating episode and other members who joined the meal but not the shared plate, with a second food specific code used to readily identify shared plate eating. A feeding episode was defined as all food consumed within a 30 min time frame. For a child, the mean number of feeding episodes was 3.9 and, on average 2.6 people, were at a shared eating occasion. A meal was defined as when three or more people were eating. Shared plate eating accounted for 26% of all feeding episodes compared with 14% for interpolate feeding and seven percent classified as post-serve sharing. Children who ate from shared plates ate larger portions, and were more than twice as likely to consume grains, carotenoid rich vegetables, pulses, fruit, dairy, and meat as children eating from an individual plate. Results from this study identified that children in a shared plate eating situation were more likely to eat Vitamin A-rich foods than children eating individually.

2.4. 24-Hour Recalls

Four studies used 24HR to assess dietary intake from shared plate eating [11,12,19,20], each using variations of standard 24HR protocols that were reported as appropriate for the setting and study design.

A study in West Africa [11] involved assessment of shared plate eating or collective/ communal dishes by a trained field worker. A qualitative recall of all foods consumed during the previous 24 h was administered to women with children aged under five years. Collective/communal dishes were initially identified by the women in the compound, with the woman in charge then providing a complete list of all the ingredients that were used. The number of different ingredients was counted but quantification (i.e., nutrients) of intake was not measured. A food variety score (FVS) and diet diversity score (DDS) were determined based on either the number of different items or food groups that were consumed the day before the survey [11]. The mean FVS was 8.3 ± 2.9 items (range 4 to 20), indicating a low number of different ingredients. The DDS was 5.1 ± 1.7 food groups (range 2 to 10), indicating very basic diets. Market days were taken into consideration relative to when recalls were conducted, as diet diversity scores were higher on market days due to women eating more vegetables, although not a greater food quantity.

A subsequent study also by Savy et al. [12] was conducted in Burkina Faso to compare dietary diversity scores measured over a 1-day and a 3-day period, and to assess their relationships with socio-economic characteristics and the nutritional status of rural African women who eat communally. A single recall interview for the three previous days was conducted, and included a spontaneous description, followed by prompting for forgotten foods. Verification of ingredients in dishes mentioned was then conducted with the woman responsible for food preparation [12]. Food consumed outside the compound was accounted for through prompted questions. A dietary diversity score (DDS), defined as the number of different food groups consumed by each woman over a given reference period, was calculated by researchers. Foods were grouped using a nine-item classification: cereals/roots/tubers; pulses/nuts; vitamin-A-rich fruits/vegetables; other vegetables; other fruits; meat/poultry/fish; eggs; milk/dairy products; oils/fats. Quantification and food frequency were not considered, with the scores used in analysis as discrete quantitative variables and after categorisation into tertiles. The mean DDS was 3.5 for a 1-day recall, and increased to 4.4 when calculated from a 3-day recall ($p < 0.0001$). The DDS calculated from a 1-day recall was higher when a market day occurred during the recall period. Both scores were linked to the sociodemographic and economic characteristics of the women. Women in the lowest DDS tertile calculated from the 1-day recall had a mean BMI of 20.5 and 17.7% of them were underweight, versus 21.6 and 3.5% for those in the highest tertile ($p < 0.0003$ and $p < 0.0007$, respectively). Authors concluded that the DDS calculated from a 1-day dietary recall was suitable for predicting the women's nutritional status, with market days requiring consideration.

In an Israeli (defined as a HIC) study [20], it was identified that individuals could provide information at the individual level for bread and food served onto an individual plate, but accuracy was not known for eating from a common plate of varying sizes or eating directly from a larger platter. The United States Department of Agriculture (USDA) 24HR recall multiple pass method was modified for trained interviewers to record three eating practices; (i) individual plate (ii) eating from a common plate (small, medium or large) with bread, and (iii) eating directly from a larger platter. As bread is often used as the utensil for eating from common dishes, the 'carrying capacity of bread' was quantified for 28 common dishes prior to the 24HR recalls. The average carrying capacity of bread was reported as 1.3 grams of solid/semi-solid food per gram of bread and 1.0 grams liquid dishes per gram of bread [20]. The modified 24HR recalls were completed using photographs as reporting aids for shared plate foods. The photos showed shared plates with different relative portions removed, and participant selected the photograph that was representative of their portion. Portion sizes for individual foods were reported using standard 24HR recall methods. Mean (SE) energy intake was 9648 (276) kilojoules (kJ)/day for men and 8230 (172) kJ/day for women, of which carbohydrates accounted for 63 to 64%. Energy intake to estimated energy requirement (EER) ratios ranged from 0.87 to 0.93 among non-dieters who ate the usual amount on the recall day. The authors concluded that the modified 24HR recall produced plausible estimates of energy and nutrient intakes, comparable to those obtained in other populations. The modified questionnaire was proposed as a model for modifying instruments to quantify individual dietary intake in other populations that practice shared plate eating.

2.5. Weighed and Estimated Record

In a study by Iwaoka [21], Japanese mothers ($n = 64$) who prepared meals for their daughters were asked to weigh and record all the ingredients used for cooking. The mothers reported the proportions of the shared dish and/or food eaten by each household member. Results obtained from data collection by mothers were compared to independently collected, self-reported shared dish consumption by daughters. Mothers were reported to underestimate intake of their daughters when compared to self-reported intake of the daughters for energy intake (kJ), macronutrient contribution and within food types, including rice and soup dishes [21]. Fifty percent of under-reporting by the mothers was attributable to rice, the staple food.

2.6. Dietary Survey

Ferrucci et al. [17] analysed data from 3625 participants in the Indian Health study. The overarching health study included questions specific to household/communal spice and oil intake, acknowledging the nutritional contributions these make to Indian dietary intake. The number of spices consumed was collected via a 'food preparer questionnaire'. The questionnaire included detailed information on 19 spices and oils, in order to quantify how much was purchased (g or kg/number in household) within a particular timeframe (week/month). The gram weight of spices purchased from markets was known to the population group and was linked to the data on the number and ages of people in the household. To account for the varying amount of food consumed by different age groups, individuals less than five years were counted as 0.7 individuals, 5–12 years as 0.9 of a person unit and individuals greater than age 12 years were counted as 1.0. The total weight per item per household was then divided by the total person units to calculate per capita consumption of the spice.

2.7. Use of Technology in Assessment of Shared Plate Eating

Four identified studies reported on how technology had been modified to account for shared plate eating, or to improve the quantification of shared plate eating [10,16,17,20].

A variety of forms of technology were used, three were predominantly for assisting in the collection of dietary intake information. Ferruci et al. [17] and Daniel et al. [16] used a computer based diet questionnaire using software called Interactive Nutrition Assistant- Diet in India Study (NINA-DISH), which was comprised of four components (i) defined questions on frequency and portion size (ii) an open ended section for each meal time (iii) food preparer questionnaire and (iv) 24 h recall. The system includes a user interface, business logic and the database, so that it can be imported to any database with minimal modifications. The inclusion of multiple methods to assess dietary intake, combined with versatile computer software make such methods generalisable to assessment of shared plate eating in other LLMIC.

Prynn et al. [10] used an electronic method for direct entry for coding diet diaries which included shared plate eating and was constructed around the hierarchal food menu structure that allowed easy adaptation to the Gambian food database. This hierarchal structure starts with rice: rice alone, boiled rice mixed with each of the basic five sauces, rice cooked with ground nuts and thin rice porridge. The third level offers each of the preceding rice levels with common additions such as fish or vegetables.

Abu Saad et al. [20] modified the Unities States Department of Agriculture USDA 24HR multiple-pass recall for the three eating practices (i) eating an item as an individual plate (ii) eating from a common plate with bread (iii) eating directly from a larger platter, this tool was initially piloted in 40 locals and results confirmed that individuals could estimate the amount of bread consumed.

All four of these studies provide evidence of the potential for technology used in dietary assessment in HIC to be adapted for use in assessing shared plate eating in LLMIC.

2.8. Tools to Assist in Portion Size Estimation from Shared Plates

A study by Thoradeniya [18] investigated different types of portion size estimation tools used to quantify Asian foods. Small photographs, life photographs, line drawings, and use of utensils as aids were trialed. All aids except utensils correlated with actual intakes of foods, with household utensils found to only be correlated for vegetables ($r = 0.69$, $p < 0.01$). Estimations using line diagrams were the most accurate with correlations of $r = 0.73$ for cereal-based food and $r = 0.86$ for vegetables ($p < 0.01$). Line diagrams also performed well overall, with 64% correct estimations, 18% overestimated and 18.1% underestimated, compared to household utensils with 0.6% correct estimations. Higher accuracy and precision were achieved with small photographs for amorphous foods and line diagrams for non-amorphous foods. The combination of small photographs (for vegetables) and line diagrams (for other foods) achieved a high correlation ($r = 0.959$, $p \leq 0.001$), percentage correct estimations (68.3%) and low under estimations (19.9%) and over estimations (11.8%) [18].

Jerome et al. [13] collected ethnographic data on food consumption patterns in Egypt where shared plate eating is common. This case study focused on the local cultural rules regarding food distribution and consumption, and associated rules regarding the order of eating and drinking (who eats or drinks first or last) and how food-consumption priorities are assigned. It was acknowledged that it may not be culturally appropriate to collect individual-level dietary intake data in settings where food is served communally to a household, family or extended family, and highlighted the challenges of determining whether everyone ate something from every dish and how much of each item was consumed by each person in shared plate eating. The importance of improving quantification was emphasised, given that shared plate eating is commonplace in the majority of the 'non-Western world'.

3. Discussion

This narrative review identified that studies assessing shared plate eating were predominantly carried out in LLMIC's in addition to two HIC's, Israel and Japan. There was a particular focus on mothers and children particularly for reporting of dietary intakes. Overall, there were few studies identified, highlighting the identified gap in research in this area. Considering the publication year of studies reviewed here there were only few studies included published in the last 10 years. The lack of research in this area may be partly attributed to previously identified challenges associated of conducting dietary intake assessment research in LLMIC [5]. Challenges for LLMIC include language, food composition database limitations, unknown nutritional compositions of traditional foods and spices, high biodiversity of staples [24], variable portion sizes, and low access to trained workers familiar with dietary assessment and eating behaviours.

Most dietary assessment studies to date have been done in HICs where it is more common to eat from discrete or individual plates. Discrete plate eating in comparison to shared plate eating is easier to capture and quantify as individuals are likely to be more aware of what foods, and the amount they are consuming. Shared plate eating is not as frequent in the home setting in HICs where it is more common to serve or be served discrete plates of food for each individual in the household and when eating out. However, with increasing globalisation, including migration, shared plate eating is becoming more widespread. All of these factors contribute to making shared plate eating of high interest in the dietary assessment field.

It was identified that shared plate eating occurred at both meals and snacks [15], although most studies focused on consumption at meal times only. The importance of assessing between-meal dietary intake or across a whole 24 h period was highlighted in a doubly labeled water biomarker study that indicated that snacks accounted for 20% of total energy expenditure [9]. As research into shared plate eating progresses, consideration will need to be given to capturing dietary intake data from snacks, particularly where the eating occasion structure and the form of shared plate eating may vary at different meal occasions.

A variety of modes of shared plate eating were found to exist including: eating directly from a central dish, placing portions on to discrete plates to be consumed by individuals, or post-plate sharing whereby food from the central dish is placed on a secondary plate that is shared by multiple people. Post-plate shared eating was reported for both adults and children [12]. Therefore, collection of preliminary ethnographic data collection to ascertain the cultural norms about shared plate eating, before embarking on dietary assessment studies is of high importance [2]. Qualitative data analysis will allow for an appropriate dietary method to be selected and modified to ensure the data collected reflects the usual consumption [13].

In this review, mothers or the female household members were usually responsible for reporting and quantifying dietary intake data from shared eating episodes [13]. This is likely to be attributed to the mother's role in in the procurement and preparation of food, the cognitively challenging tasks of estimating foods consumed [13], and the age of children in the included studies, with many being young children under six years old [10,11,14]. However, in situations where the mother is not always at home for eating occasions [14], or when an individual within the commune is responsible for food preparation [9] the mother may not be the most appropriate dietary intake reporter. This could be pre-empted by collection of ethnographic data.

Food from shared plate eating contributed the majority of total daily energy intake in the two included studies that reported energy intake [9,20]. Despite the 24HR method being used in three other studies [6,7,10], the dietary intake data was used for purposes other than calculation of energy and nutrient intake such as food and diet variety. There is considerable potential for shared plate eating data collection to improve in order for more accurate and comparable dietary intake to be obtained and reported. Accurate assessment of shared plate eating is currently limited by difficulty in quantification, particularly when shared dishes vary in nutrient and fluid proportions [20]. Even if a single dish is served the nutrient composition of each portion is likely to be variable, demonstrating the complexity of this area of dietary assessment.

In all studies, observers or interviewers were reported to have undertaken training from researchers, however the components of the training were not well reported. Training is likely required including how many dishes are served, who is eating from each plate, how many people ate from a particular dish, the serving vessel (hands/ utensils/ breads) and nutrient compositions of each mouthful. A previous review of technology-based dietary assessment tools found that technologies exhibiting substantial practical constraints and a lack of demonstrated feasibility for use in LLMICs [8]. It has been previously recommended that to increase collection of dietary data in LLMICs, development of contextually adaptable, interviewer-administered dietary assessment platform areas would be of benefit. In the studies reviewed in the current paper that utilised technology, it was identified that the purpose was primarily assist in standardizing the collection of dietary information.

Recommendations apparent from this review for the progression of research to refine the dietary assessment methodology of shared plate eating include:

(1) Consideration of seasonality which influences the availability and dietary contribution of different foods at different times of year, and harvest season may result in a period of more plentiful food supply for several months, usually once per year [9].

(2) For optimal accuracy in dietary intake estimation, consideration should be given to weighing the staple food, as discrepancies in estimation of the staple are likely to account for the majority of overall daily energy discrepancy [21].

(3) Modified 24HR with photographs of shared plate with portions removed may serve as a model for shared plate eating assessment [20], as these may be easier for participants to estimate than photographs of individual portions.

(4) Clearly defined aims are required in order to adequately capture relevant dietary intake data. For example, if calcium consumption is of interest then increased attention to consumption of edible bones is important, or if micronutrients such as sodium are being assessed, condiment and sauce consumption requires more detailed assessment as these can be significant contributors [17].

(5) Combination approaches to portion size estimation are recommended, rather than one tool in isolation from other methods [18].

(6) Consideration if culturally appropriate to evaluate individual dietary intakes and maybe household intake i.e., group level might be in some regions more acceptable. A clearly defined preparation and planning phase with ethnographic data is essential [7].

(7) The use of consistent terminology to describe shared plate eating in published research would be valuable to and further the field of research for regions/areas where shared plate eating is the cultural norm, the method of quantification of shared plate eating should be reported so data can be consolidated across studies where possible. Alternately, if shared plate eating has not been taken into consideration in assessment where it is known to occur, this should be acknowledged a limitation of research.

(8)   Less intrusive methods of assessing shared plate eating, compared to direct observation, need to be developed to ensure dietary undertake assessment is undertaken as objectively as possible. Direct observation studies can influence the way people eat, can be prohibitively expensive and can be inaccurate compared to weighed intake [14].

The use of technology as a means of assessing dietary intake has increased in parallel to the development of image-based methods, wearable devices, and online methods of administering dietary assessment tools [25]. As evidenced in this review, the application of such approaches to shared plate eating remain relatively untested with very few studies reviewed in this last 10 years. However, Caswell et al. [19] have reported efficient collection of 24HR data using tailored software on a tablet platform in a rural district in central Zambia. The tool was considered easy to use by trained interviewers without prior nutrition training or computing experience to administer a 24HR to caregivers on dietary intakes of children participating in an efficacy trial. If technology approaches can be to individual dietary-level dietary assessment in similar demographic groups to that reported by Caswell et al. [19], the extension of this into shared plate eating warrants substantial research investment, particularly considering the need for improved dietary intake and nutritional status of populations who engage in shared plate eating [8]. For camera devices there is a need to investigate the acceptability of this approach, as it yet to be established and tested in a range of population groups and different ethnicities.

4. Conclusions

Shared plate eating is a very common food consumption modality, particularly in LLMIC, but is under-represented in dietary assessment literature. Key factors identified as contributing to improved assessment of shared plate eating were accurate assessment of staple food intake and the need for combined approaches to portion size estimation. It is recommended that dietary assessment methods match the cultural context in which data is being collected, and that technology methods be considered to replace direct observation. Progress in the dietary assessment of shared plate eating depends on use of consistent terminology and documentation of the methods used to quantify shared plate eating, so data can be consolidated across studies where possible.

Author Contributions: T.B. was responsible for data collection and drafting of the manuscript. K.D. checked data extraction, M.R., C.C. and M.A. were all involved in manuscript preparation and approved the final version of the manuscript.

Funding: T.B. is funded by a UON Brawn research fellowship, C.C. is a NHMRC SRF Research fellowship. This work was supported by the Bill & Melinda Gates Foundation [OPP1171389]

Acknowledgments: The author wish to acknowledge Janelle Skinner for her assistance in locating articles suitable for inclusion in this review.

Conflicts of Interest: The authors declare no conflict of interest.

References

1.   Haddad, L.; Hawkes, C.; Webb, P.; Thomas, S.; Beddington, J.; Waage, J.; Flynn, D. A new global research agenda for food. *Nature* **2016**, *540*, 30–32. [CrossRef] [PubMed]
2.   Tumilowicz, A.; Neufeld, L.; Pelto, G. Using ethnography in implementation research to improve nutrition interventions in populations. *Matern. Child Nutr.* **2015**, *11*, 55–72.
3.   Woolley, K.; Fishbach, A. Shared Plates, Shared Minds: Consuming from a Shared Plate Promotes Cooperation. *Psychol. Sci.* **2019**. [CrossRef] [PubMed]
4.   Kodish, S.; Aburto, N.; Nseluke Hambayi, M.; Dibari, F.; Gittelsohn, J. Patterns and determinants of small-quantity LNS utilization in rural Malawi and Mozambique: Considerations for interventions with specialized nutritious foods. *Matern. Child. Nutr.* **2017**, *13*. [CrossRef] [PubMed]
5.   Coates, J.C.; Colaiezzi, B.A.; Bell, W.; Charrondiere, U.R.; Leclercq, C. Overcoming Dietary Assessment Challenges in Low-Income Countries: Technological Solutions Proposed by the International Dietary Data Expansion (INDDEX) Project. *Nutrients* **2017**, *9*, 289.

6. Ngo, J.; Gurinovic, M.; Frost-Andersen, L.; Serra-Majem, L. How dietary intake methodology is adapted for use in European immigrant population groups—A review. *Br. J. Nutr.* **2009**, *101*, S86–S94. [CrossRef] [PubMed]

7. Micha, R.; Coates, J.; Leclercq, C.; Charrondiere, U.R.; Mozaffarian, D. Global Dietary Surveillance: Data Gaps and Challenges. *Food Nutr. Bull.* **2018**, *39*, 175–205. [CrossRef] [PubMed]

8. Bell, W.; Colaiezzi, B.A.; Prata, C.S.; Coates, J.C. Scaling up Dietary Data for Decision-Making in Low-Income Countries: New Technological Frontiers. *Adv. Nutr. Int. Rev. J.* **2017**, *8*, 916–932. [CrossRef] [PubMed]

9. Hudson, G.J. Food intake in a West African village. Estimation of food intake from a shared bowl. *Br. J. Nutr.* **1995**, *73*, 551–569. [CrossRef] [PubMed]

10. Prynne, C.; Paul, A.; Dibba, B.; Jarjou, L. Gambian Food Records: A New Framework for Computer Coding. *J. Food Compos. Anal.* **2002**, *15*, 349–357. [CrossRef]

11. Savy, M.; Martin-Prével, Y.; Sawadogo, P.; Kameli, Y.; Delpeuch, F. Use of variety/diversity scores for diet quality measurement: Relation with nutritional status of women in a rural area in Burkina Faso. *Eur. J. Clin. Nutr.* **2005**, *59*, 703–716. [CrossRef] [PubMed]

12. Savy, M.; Martin-Prével, Y.; Traissac, P.; Delpeuch, F. Measuring dietary diversity in rural Burkina Faso: Comparison of a 1-day and a 3-day dietary recall. *Public Health Nutr.* **2006**, *10*, 71–78. [CrossRef] [PubMed]

13. Jerome, N.W. Culture-specific strategies for capturing local dietary intake patterns. *Am. J. Clin. Nutr.* **1997**, *65*, 1166S–1167S. [CrossRef]

14. Shankar, A.V.; Gittelsohn, J.; Stallings, R.; West, K.P.; Gnywali, T.; Dhungel, C.; Dahal, B. Comparison of Visual Estimates of Children's Portion Sizes Under Both Shared-Plate and Individual-Plate Conditions. *J. Am. Diet. Assoc.* **2001**, *101*, 47–52. [CrossRef]

15. Shankar, A.V.; Gittelsohn, J.; West, K.P.; Stallings, R.; Gnywali, T.; Faruque, F. Eating from a Shared Plate Affects Food Consumption in Vitamin A–Deficient Nepali Children. *J. Nutr.* **1998**, *128*, 1127–1133. [CrossRef]

16. Daniel, C.R.; Kapur, K.; McAdams, M.J.; Dixit-Joshi, S.; Devasenapathy, N.; Shetty, H.; Hariharan, S.; George, P.S.; Mathew, A.; Sinha, R. Development of a field-friendly automated dietary assessment tool and nutrient database for India. *Br. J. Nutr.* **2014**, *111*, 160–171. [CrossRef] [PubMed]

17. Ferrucci, L.M.; Daniel, C.R.; Kapur, K.; Chadha, P.; Shetty, H.; Graubard, B.I.; George, P.S.; Osborne, W.; Yurgalevitch, S.; Devasenapathy, N.; et al. Measurement of spices and seasonings in India: Opportunities for cancer epidemiology and prevention. *Asian Pac. J. Cancer Prev.* **2010**, *11*, 1621–1629. [PubMed]

18. Thoradeniya, T.; De Silva, A.; Arambepola, C.; Atukorala, S.; Lanerolle, P. Portion size estimation aids for Asian foods. *J. Hum. Nutr. Diet.* **2012**, *25*, 497–504. [CrossRef]

19. Caswell, B.L.; Talegawkar, S.A.; Dyer, B.; Siamusantu, W.; Klemm, R.D.W.; Palmer, A.C. Assessing Child Nutrient Intakes Using a Tablet-Based 24-Hour Recall Tool in Rural Zambia. *Food Nutr. Bull.* **2015**, *36*, 467–480. [CrossRef] [PubMed]

20. Abu-Saad, K.; Shahar, D.R.; Abu-Shareb, H.; Vardi, H.; Bilenko, N.; Fraser, D. Assessing individual dietary intake from common-plate meals: A new tool for an enduring practice. *Public Health Nutr.* **2009**, *12*, 2464–2472. [CrossRef] [PubMed]

21. Iwaoka, H.; Yoshiike, N.; Date, C.; Shimada, T.; Tanaka, H. A Validation Study on a Method to Estimate Nutrient Intake by Family Members through a Household-based Food-weighing Survey. *J. Nutr. Sci. Vitaminol.* **2001**, *47*, 222–227. [CrossRef] [PubMed]

22. Engle, P.L.; Nieves, I. Intra-household food distribution among Guatemalan families in a supplementary feeding program: Behavior patterns. *Soc. Sci. Med.* **1993**, *36*, 1605–1612. [CrossRef]

23. Dettwyler, K.A. Styles of Infant Feeding: Parental/Caretaker Control of Food Consumption in Young Children. *Am. Anthropol.* **1989**, *91*, 696–703. [CrossRef]

24. Toledo, Á.; Burlingame, B. Biodiversity and nutrition: A common path toward global food security and sustainable development. *J. Food Compos. Anal.* **2006**, *19*, 477–483. [CrossRef]

25. Rollo, M.E.; Williams, R.L.; Burrows, T.; Kirkpatrick, S.I.; Bucher, T.; Collins, C.E. What Are They Really Eating? A Review on New Approaches to Dietary Intake Assessment and Validation. *Curr. Nutr. Rep.* **2016**, *5*, 307–314. [CrossRef]

 *nutrients*

Review

Potential Use of Mobile Phone Applications for Self-Monitoring and Increasing Daily Fruit and Vegetable Consumption: A Systematized Review

Floriana Mandracchia [1,†], Elisabet Llauradó [1,†], Lucia Tarro [1,2,*], Josep Maria del Bas [2], Rosa Maria Valls [1], Anna Pedret [1,2], Petia Radeva [3,4], Lluís Arola [2,5], Rosa Solà [1,2,6,*] and Noemi Boqué [2]

[1] Functional Nutrition, Oxidation, and Cardiovascular Diseases Group (NFOC-Salut), Facultat de Medicina i Ciències de la Salut, Universitat Rovira i Virgili, 43201 Reus, Spain; floriana.mandracchia1@urv.cat (F.M.); elisabet.llaurado@urv.cat (E.L.); rosamaria.valls@urv.cat (R.M.V.); anna.pedret@urv.cat (A.P.)
[2] Unitat de Nutrició i Salut, Centre Tecnològic de Catalunya, Eurecat, 43204 Reus, Spain; josep.delbas@eurecat.org (J.M.d.B.); lluis.arola@urv.cat (L.A.); noemi.boque@eurecat.org (N.B.)
[3] Facultat de Matemàtiques i Informàtica, Universitat de Barcelona, 08007 Barcelona, Spain; petia.ivanova@ub.edu
[4] Centre de Visió per Computador, 08193 Bellaterra (Barcelona), Spain
[5] Nutrigenomics Research Group, Department of Biochemistry and Biotechnology, Universitat Rovira i Virgili, 43007 Tarragona, Spain
[6] Hospital Universitari Sant Joan de Reus, 43204 Reus, Spain
*   Correspondence: lucia.tarro@urv.cat (L.T.); rosa.sola@urv.cat (R.S.); Tel.: +34-977-758-920
†   F.M. and E.L. contributed equally to this work.

Received: 25 February 2019; Accepted: 20 March 2019; Published: 22 March 2019

Abstract: A wide range of chronic diseases could be prevented through healthy lifestyle choices, such as consuming five portions of fruits and vegetables daily, although the majority of the adult population does not meet this recommendation. The use of mobile phone applications for health purposes has greatly increased; these applications guide users in real time through various phases of behavioural change. This review aimed to assess the potential of self-monitoring mobile phone health (mHealth) applications to increase fruit and vegetable intake. PubMed and Web of Science were used to conduct this systematized review, and the inclusion criteria were: randomized controlled trials evaluating mobile phone applications focused on increasing fruit and/or vegetable intake as a primary or secondary outcome performed from 2008 to 2018. Eight studies were included in the final assessment. The interventions described in six of these studies were effective in increasing fruit and/or vegetable intake. Targeting stratified populations and using long-lasting interventions were identified as key aspects that could influence the effectiveness of these interventions. In conclusion, evidence shows the effectiveness of mHealth application interventions to increase fruit and vegetable consumption. Further research is needed to design effective interventions and to determine their efficacy over the long term.

Keywords: mobile app; mHealth; fruits; vegetables; self-monitoring; healthy diet

---

1. Introduction

The health benefits of consuming fruits and vegetables have been extensively demonstrated. These beneficial effects are attributed to their high contents of fibre, vitamins, minerals and phytochemicals (mainly antioxidants) together with negligible amounts of fat. Increased consumption of fruits and vegetables has been associated with reduced risks of many chronic diseases, such as obesity, cardiovascular diseases, type II diabetes, osteoporosis and certain cancers, as well as all-cause mortality [1,2]. In fact, it was estimated that in 2013, 7.8 million premature deaths worldwide could be attributed to low fruit and vegetable intake [1]. Moreover, adequate intake of fruits and vegetables could avoid approximately 31% of ischaemic heart disease, 19% of stroke, 20% of oesophageal cancer and 19% of gastric cancer cases [3].

Health authorities, such as the World Health Organization (WHO), recommend a daily intake of at least 400 g of fruits and vegetables, which corresponds to 5 servings of 80 g per day [4]. For the purpose of encouraging fruit and vegetable consumption in all women, children and men so they meet the recommended intake, the WHO launched an international programme termed "5 a day", which has been adopted by most national governments, including Spain, France, Germany and the United Kingdom. Similarly, the Dietary Guidelines for Americans advise that one-half of the plate should be fruits and vegetables [5], and Canada's Food Guide recommends including plenty of vegetables and fruits in daily meals and snacks to prevent the risk of heart diseases [6].

International organizations and national governments have set increasing fruit and vegetable intake as a priority. Despite the numerous and diverse public health campaigns implemented in recent decades to promote increased consumption of fruits and vegetables in Western countries, the average intake remains far from these recommendations, reflecting the modest impact of these kinds of interventions. Data from a European Food Safety Authority (EFSA) analysis based on national dietary surveys revealed that only 4 of the European Union (EU) member states reported adequate consumption of fruits and vegetables [7]. The success of the last major campaigns conducted worldwide that intended to increase fruit and vegetable consumption has been reviewed by Rekhy et al. [8], who concluded that these interventions were quite effective in the short term but generally failed over the long term despite the enormous cost and effort they require. Importantly, it is inferred from the same work that the effectiveness of these health programmes is greater when factors such as behavioural changes, goal setting, clear messages and interactive approaches are included.

In recent years, strategies for promoting long-term adherence to different interventions have focused on multidisciplinary approaches. For example, management of weight loss depends on multiple factors, such as behaviour, a cognitive component, personality traits and even the patient-therapist interaction [9]. This multifaceted approach has been proven successful for weight loss maintenance over the long term (up to 42 months) by means of coaching strategies [10]. Integrative health coaching conducted by telephone calls has been used as a tool for enhancing treatment outcomes in type 2 diabetic patients, who are able to improve their adherence to medication and glycated haemoglobin, a marker of long-term blood glucose levels [11]. Johnson et al. showed that health coaching delivered by videoconference was an effective strategy for reducing weight and ameliorating insulin-resistance markers in obese individuals [12]. Overall, health coaching and behavioural changes have arisen as key elements for achieving substantial and long-term adherence to healthy habits. Therefore, such approaches represent a promising strategy for increasing the consumption of fruits and vegetables.

In this scenario, current advances in information and communication technologies (ICTs), also known as eHealth [13], might provide a wide array of supportive tools, allowing a wide deployment of coaching and behavioural change strategies to the general population. Importantly, it is inferred from the same work that the effectiveness of these health programmes is greater when factors such as self-monitoring, goal setting, clear messages and interactive approaches are included. MHealth applications, which are used on mobile phones and wireless devices, such as tablets, personal digital assistance (PDA) devices, and so on, could be a better method to improve people's lifestyles [13] than traditional face-to-face education methods [14]. The mHealth App Developer Economics study showed an increase of 25% year-to-year from 2015 to 2017 of the number of mHealth applications [15]. Moreover, from the mHealth Economics report, an increase from 2.1 billion smartphone users in 2016 to 2.5 billion in 2019 [16] is expected. Mobile technologies allow interactions with users in real-time and the delivery of health interventions at any time [17] and can act in different environmental and behavioural contexts [17]. Mobile technologies have been demonstrated to be a valid tool for dietary self-monitoring [18]. Toro-Ramos et al. showed that using a mobile phone application that provides nutritional and behavioural education together with coaching promoted clinically significant long-term weight loss, reduced blood glucose levels and improved different lipid markers in overweight and obese individuals [19]. There are several basic mobile and web journaling applications that allow users to set weight-loss goals, collect daily calorie target chart data to reflect trends over time, and record food consumption and exercise levels. The indicative paradigms of journaling applications are weight management applications, such as Weightbot© (2017 Meeco Labs, Linz, Austria), LoseIt© (2008–2019 FitNow, Inc, Saint Honoré, Paris), InsideTracker© (2009–2019 Segterra, Inc, Cambridge, MA, U.S.A.), FoodLog© (2013 foo.log, Inc, Tokyo, Japan), Cronometer© (2011–2019, Cronometer.com, Revelstoke, BC, Canada), MyFitnessPal© (2009–2019 MyFitnessPal, Inc, San Francisco, CA, U.S.A.), MyPlate© (2017 LIVESTRONG.COM, Santa Monica, CA, U.S.A.), EasyFit© (2016 Cellularline, Reggio Emilia, Italy), FatSecret© (2019, FatSecret, Victoria, Australia), MyNetDiary© (2018 MyNetDiary Inc, Marlton, NJ, U.S.A.), and so on, which enable the user to enter weight and body composition measurements, visualize curves, superimpose trends and track progress. Following standard paper-based analogues to obtain information on nutrition habits, mobile applications provide electronic forms and efficient interfaces to assist in logging food intake and beverages in terms of types, meal courses, total meal calories, recipes, photos, and so on. These applications calculate energy intake and balance, report additional parameters and visualize summaries [20]. Three systematic reviews demonstrated the efficacy of mHealth applications to prevent obesity in young people [21–23], but there is a lack of scientific evidence of the effects on fruit and vegetable consumption, while the majority of published studies focus on weight management and physical activity improvement [24]. Thus, new technologies represent a promising opportunity in fields such as nutrition and health monitoring [25–27].

Increasing and improving the consumption of fruits and vegetables in the general population represents a challenge for public health that has not yet been resolved. ICTs might represent an opportunity to achieve this objective. Therefore, we conducted a systematized review of the last 10 years to assess whether interventions based on mobile phone applications result in positive outcomes and to identify the main weaknesses of the different approaches used to date.

2. Materials and Methods

The present paper is a systematized review and has some characteristics of a narrative review and some of a systematic review [28].

2.1. Search Strategy

Article searches were limited to a recent time range of 10 years, considering that the use and availbility of mobile phone applications, which are the key tools evaluated in the present review, increased only a few years ago, starting from 2007, when they appeared on the market. In this sense, there is a lack of published trials on the use of mobile phone applications for health interventions before 2010 [29–31]. This systematized review was based on two electronic databases: PubMed and Web of Science. The search strategy involved peer-reviewed and English-language articles. For the search strategy, the following keywords were used separately or in combination: 'Self-monitoring' AND 'Fruit and Vegetables' OR 'Healthy meals', 'Fruit and vegetables' AND 'Mobile health applications' OR 'eHealth' OR 'mHealth' OR 'Mobile technology', and 'Mobile phone applications' AND 'Fruit and Vegetables'.

2.2. Selection Criteria and Data Collection

The PubMed and Web of Science databases were searched, resulting in a total of 1208 articles, as shown in the PRISMA (Preferred Reporting Items for Systematic Reviews and Meta-Analyses) flow diagram for systematic reviews and meta-analysis (Figure 1); of these articles, 228 were found in

MEDLINE (PubMed) and 980 in Web of Science. During the screening of possible articles, the reference lists of the full-text articles were assessed for eligibility and cross-checked; it was decided to evaluate articles from these lists when they focused on the same outcome, resulting in the identification of 14 articles for further screening. The titles and abstracts of the 1222 total articles were screened by two researchers (F.M. and E.L.) to determine if they fulfilled the eligibility criteria. The articles had to include clinical trials and other experimental studies designed to develop, test or validate a mobile phone application for dietary self-monitoring in which fruit and/or vegetable intake was one of the principal outcomes (primary and/or secondary). Studies including other health concerns (physical activity, weight control, sugar-sweetened beverage intake, takeout meals, dietary habits, etc.) were included, and no limitations were made in terms of the type of population (gender, age, race, health status). Studies using web-based self-monitoring technologies were excluded, which focused the search on mobile phone applications only. This selection process was performed by two reviewers (F.M. and E.L.). In cases of discrepancy, a third reviewer (L.T.) was consulted.

Following the screening, 1196 articles were excluded on the basis of their title or abstract. The remaining 26 articles were subjected to a detailed examination of the abstract to determine their eligibility on the basis of the inclusion criteria. Of these, 18 were excluded due to the type of technology tool used or lack of results, leaving 8 peer-reviewed papers included in the current review.

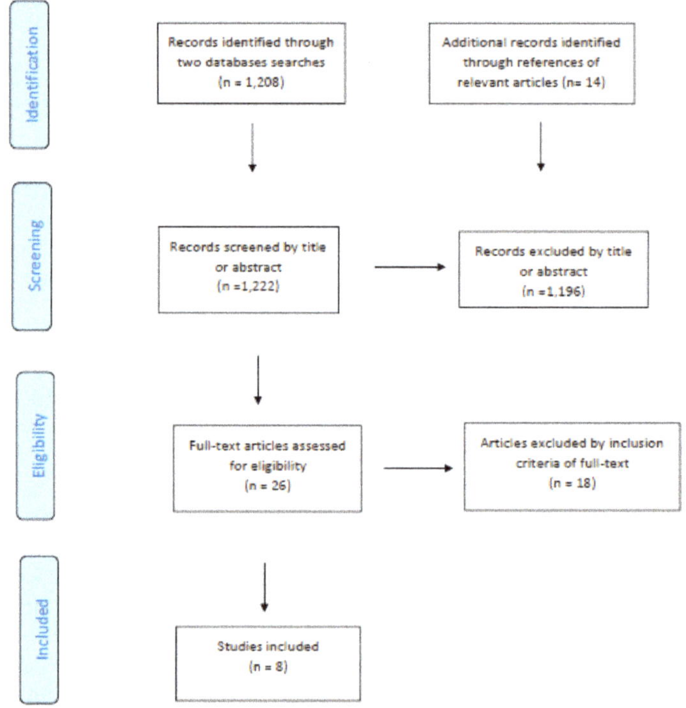

Figure 1. Preferred Reporting Items for Systematic Reviews and Meta-Analyses (PRISMA) 2009 flow diagram for the systematic review and meta-analysis of the article selection process.

2.3. Data Extraction

Data extraction from the included studies was performed by two reviewers working simultaneously (F.M. and E.L.) and revised by all others. The data extraction tables include the following study variables: name of the article, authors, year of publication, name of the intervention or mobile phone application, country, type of intervention, objective, intervention duration, number of participants, population, description of the mobile phone application (name and measurement of the outcome), brief description of the intervention, results of fruit and vegetable consumption variables, and conclusions of the study.

2.4. Quality of Studies Included

The quality of the included studies was assessed using the standardized framework of the Quality Assessment Tool for Quantitative Studies, developed by the Effective Public Health Practice Project. This tool consists of 8 items: selection bias, study design, confounders, blinding, data collection methods, withdrawals and dropouts, intervention integrity and analysis. This tool allows the categorization of each study's methodological quality as weak ($\geq$2 weak category ratings), moderate (1–3 strong category ratings and 1 weak category rating) or strong ($\geq$4 strong category ratings and no weak category ratings) (Table S1).

3. Results

The systematized search identified 1222 articles, of which eight studies were found to meet the inclusion criteria to be considered in the review (Table S2).

The most relevant characteristics of the data extracted from the included studies are presented in Table 1 and Table S2: Relevant information regarding the data extracted from the included studies.

The eight studies included were randomized controlled trials (RCTs) [32–39], one of which was a pilot study [32], and included a total of 1524 participants at baseline. Additionally, the participants covered a large age range, from 16 to 71 years, and one of the interventions was conducted through parents, although the target population was children. In the vast majority of cases, the participants did not present any disease. Most of the studies were conducted in the United States (n = 4), followed by Australia (n = 2), the Netherlands (n = 1) and Sweden (n = 1).

From the selected studies, two studies aimed to evaluate the effectiveness of mobile phone applications in stimulating both fruit and vegetable intake [34,38], two targeted only vegetable consumption [32,33], three tested whether a multicomponent intervention integrating a mHealth application could improve dietary habits (including the increase of fruit and/or vegetable intake) and physical activity [35,37,39], and only one assessed the effectiveness of a mobile phone application to achieve a healthy weight and healthy body fat percentage by changing daily servings of fruits and vegetables [36].

Moreover, different methodologies were used to improve fruit and/or vegetable intake: (a) three of the studies used personalized informative and motivational messages (text and/or audio) [34,38,39]; (b) seven added personal dietary feedback at a regular frequency [32–38]; (c) three sent push notifications to remind users about their goals [32,33,36]; (d) two provided rewards as incentives [35,37]; (e) three provided remote coaching support through mobile phone calls, emails and in-person meetings [35,37,39]; (f) one offered the possibility of receiving support from a dietitian or a psychologist [36]; and (g) two provided access to further informative material and information through a diet booklet [39] and a mobile phone application [36,39].

Fruit and/or vegetable intake was assessed by the Food Frequency Questionnaire (FFQ) in three RCTs [32–34], by a dietary record in one study [35], by self-monitoring through a mobile phone application in two studies [36,38], and by categorical questions in the other two studies [37,39]. Moreover, the results were expressed as servings, pieces/day or pieces/week of fruits and/or vegetables in six RCTs [32–35,37,38]; grams of fruits and/or vegetables in one RCT [36]; and percentage of participants who consumed $\geq$2 servings or pieces of fruits and/or vegetables per day in one RCT [39].

Six of the eight studies included were effective in increasing fruit and/or vegetable intake. Of these, five studies [32,33,35,37,39] demonstrated that the interventions were effective for increasing vegetable consumption, and interestingly, all of them included a self-monitoring component implemented by a mobile phone application. Some of the included studies used other methodologies apart from self-monitoring: four used dietary feedback [32,33,35,37] and three provided remote coaching support [35,37,39]. Furthermore, three studies [34,35,37] reported that the interventions were effective for increasing daily fruit consumption. All three mHealth interventions included a self-monitoring component by mobile phone application and personal dietary feedback, while two of them [35,37] provided a financial incentive as a reward and remote coaching support. The increase in intake ranged from +2.4 servings/day to +10.6 servings/day.

Moreover, from the effective interventions identified in the present review, two focused on overweight adults [32,33], three focused on adults with unhealthy lifestyles [34,35,37], and one focused on young adults characterized by unhealthy lifestyles [39].

Half of the included studies considered the improvement of fruit and/or vegetable intake [32–34,38] as the primary outcome, and the other half considered it to be a secondary outcome [35–37,39].

Of the four RCTs that designated increasing fruit and/or vegetable intake as the primary outcome [32–34,38], two were only effective in increasing vegetable intake [32,33], one showed was only effective in increasing fruit intake [34], and the last one presented no improvement [38]. These four RCTs lasted from 2 to 6 months and were population stratified by common characteristics.

Mummah et al. have iteratively developed a theory-driven mobile phone application called Vegethon to increase vegetable consumption through self-monitoring, goal setting, feedback, and social comparison [40]. Vegethon has been tested in two studies: A RCT pilot study [32] and a RCT [33]. The target population of the Vegethon pilot study comprised 17 overweight adults aged 18–50 years [31,32] who were randomized for the use of the Vegethon mobile phone application as the intervention group or to a wait-listed control condition. The intervention group was instructed to use the Vegethon application and encouraged to self-monitor and increase their vegetable intake. At 12 weeks, the results showed that vegetable intake was significantly increased in the intervention group by +7.5 servings/day (from 6.0 ± 2.7 to 13.5 ± 8.1) compared with the decrease in the control group of -3.1 servings/day (from 7.0 ± 5.9 to 3.9 ± 2.0), resulting in a significant difference of +10.6 servings/day between both groups ($p = 0.02$). Moreover, as mentioned, the effectiveness of Vegethon in increasing vegetable consumption was also verified in an RCT among 135 overweight adults aged 18–50 years [20]. The intervention was the same as in the pilot study. The intervention group reported an increase of +0.7 servings/day of vegetables (from 6.7 ± 5.2 to 7.4 ± 5.4), while the control group reported decreased vegetable consumption of -1.7 servings/day (from 8.1 ± 8.2 to 6.4 ± 4.3), resulting in a significant difference of +2.4 servings/day between both groups ($p = 0.04$). As a result, the Vegethon mobile phone application was effective in improving vegetable consumption.

Table 1. Description of the studies included in the present review.

Reference	Study Design	Duration	Type of Intervention and Target Population	Outcome	Effectiveness	Change in Fruit Consumption	Change in Vegetable Consumption
[32]	Randomized Controlled Pilot Study	3 months	Self-monitored by mobile-phone application. Directed to overweight adults aged 18–50. Total: 17 participants (intervention: $n = 8$; control: $n = 9$).	Primary	✔For vegetables	Fruit consumption not measured	**Intervention group:** +7.5 servings/day; **Control group:** −3.1 servings/day; **Difference between groups:** 10.6 servings/day
[33]	Randomized Controlled Trial	2 months	Self-monitored by mobile-phone application. Directed to overweight adults aged 18–50. Total: 135 participants (intervention: $n = 68$; control: $n = 67$).	Primary	✔For vegetables	Fruit consumption not measured	**Intervention group:** +0.7 servings/day; **Control group:** −1.7 servings/day; **Difference between groups:** 2.4 servings/day
[34]	Randomized Controlled Trial	6 months	Self-monitored by mobile-phone application, text-based or audio-based health messages. Directed to adults aged 16–71 who had not yet succeeded in consuming the daily recommended fruit and vegetable consumption. Total: 146 participants (intervention: $n = 88$; control: $n = 58$).	Primary	✔For fruits x For vegetables	**Textual feedback condition:** −0.6 pieces/week; **Auditory feedback condition:** +3.3 pieces/week; **Control group:** +0.4 pieces/week; **Differences between Textual- Control:** −0.2; **Differences between Auditory-Control:** 3.7	Non-significant difference
[35]	Randomized Controlled Trial (4-arm parallel groups)	6 months (3 weeks of treatment and 20 weeks of follow-up)	Self-monitored by a PDA, remote coach support and financial incentives. Directed to adults aged 21-60 who eat <5 servings of fruits and vegetables a day, consume ≥8% daily calories from saturated fat; do <150 min/week of Physical Activity and watch >120 min/week of leisure screen time. Total: 204 participants (group A: $n = 47$; group B: $n = 48$; group C: $n = 56$; group D: $n = 53$).	Secondary	✔For fruits and vegetables (only significant improvements in Behaviour C intervention)		**Baseline vs. 3 weeks** **Behaviour A (targeted saturated fat and physical activity):** +0.6 servings/day; **Behaviour B (targeted fruit/vegetable intake and physical activity):** +4.3 servings/day; **Behaviour C (targeted fruit/vegetable intake and sedentary behaviours):** +4.3 servings/day; **Behaviour D (targeted saturated fat and sedentary behaviour):** +0.5 servings/day

Table 1. *Cont.*

Reference	Study Design	Duration	Type of Intervention and Target Population	Outcome	Effectiveness	Change in Fruit Consumption	Change in Vegetable Consumption
[36]	Randomized Controlled Trial	6 months	Self-monitored, informed and supported by a mobile-phone application, weekly graphic feedback and facultative dietitian contact for further support. Directed to parents of children aged 4.5. Total: 315 participants (intervention: $n = 156$; control: $n = 159$).	Secondary	x For fruits x For vegetables	**Intervention group:** +2.9 ± 78.9 g/day; **Control group:** −12.1 ± 87.9 g/day	**Intervention group:** −6.7 ± 42.1 g/day; **Control group:** −3.6 ± 39.7 g/day
[37]	Randomized Controlled Trial (3-arm parallel groups)	9 months (6 months of treatment and 3 months of post-intervention follow-up)	Self-monitored by a mobile-phone application and an accelerometer, remote coaching by telephone and rewarding financial incentives. Directed to adults aged 18–65 characterized by those who eat <5 servings of fruits and vegetables a day, consume ≥8% daily calories from saturated fat; do <150 min/week of Physical Activity and watch >120 min/week of leisure screen time. Total: 212 participants (intervention A: $n = 84$; intervention B: $n = 84$; control: $n = 44$).	Secondary	✔For fruits and vegetables	**Baseline vs. 6 months** Group A (Simultaneous): +6.6 servings/day; Group B (Sequential): +7.4 servings/day; Control group: +0.5 servings/day.	
[38]	Randomized Controlled Trial (3-arm parallel groups)	6 months	Self-monitored by a mobile-phone application, dietary feedback messages and text messages via mobile-phone. Directed to young adults aged 18–30. Total: 212 participants (intervention A: $n = 84$; intervention B: $n = 84$; control: $n = 44$).	Primary	x For fruits x For vegetables	**Group A:** −0.2 ± 0.1 servings/day; **Group B:** −0.1 ± 0.1 servings/day; **Control Group:** −0.2 ± 0.1 servings/day; **Between group difference in Mean change ($p > 0.05$)**	**Group A:** +0.2 ± 0.1 servings/day; **Group B:** +0.4 ± 0.1 servings/day **Control Group:** +0.4 ± 0.1 servings/day; **Between group difference in Mean change ($p > 0.05$)**
[39]	Randomized Controlled Trial	3 months	Self-monitored and trained by a mobile-phone application, text messages, e-mails, coach calls, diet booklet and access to resources. Directed to young adults aged 18–35 characterized by fruit intake <2 servings/day and vegetable intake <5 servings/day. Total: 248 participants (intervention: $n = 123$; control: $n = 125$).	Secondary	✔For vegetables x For fruits	Non-significant difference	Intervention group: from 34.1% (baseline) to 64.3% (12 weeks) of people consuming ≥2 servings/day; **Control group:** from 36% (baseline) to 48% (12 weeks) of people consuming ≥2 servings/day

✔ Effective in increasing fruit or vegetable consumption (mentioned in the table). x No effective in increasing fruit or vegetable consumption (mentioned in the table).

3.1. Increasing Daily Fruit and/or Vegetable Consumption as the Primary Outcome

The study by Elbert et al. [34] provides evidence-based insight into the effects of a mobile health application in changing fruit and vegetable intake in a 6-month intervention. This study was a 3-arm RCT that included a population of 146 adults aged 16–71 years. The intervention groups (A and B) were exposed monthly to tailored health information and feedback in the form of either (A) an audio-based intervention or (B) a text-based intervention via mobile phone application over a 6-month period. Participants in the control group only completed the baseline and post-intervention measures. After 6 months, the average fruit intake, measured by a food frequency questionnaire, increased by +3.3 pieces/week (from 14.2 ± 10.6 to 17.5 ± 11.1) in intervention group A, whereas in intervention group B, the average fruit intake decreased by -0.6 pieces/week (from 14.8 ± 11.1 to 14.2 ± 6.9), and in the control group, the average fruit intake increased by +0.4 pieces/week (from 13.4 ± 10.4 to 13.8 ± 9.4). However, the intake of vegetables was not improved by these interventions.

In another 3-arm RCT, the Connecting Health and Technology study [38], Kerr et al. aimed to evaluate the effectiveness of tailored dietary feedback and weekly text messaging to improve the dietary intake of fruits and vegetables among other dietary improvements over a 6-month period in a population-based sample of men and women aged 18–30 years. Participants were randomized into three groups: (A) a group that received dietary feedback and weekly text messages, (B) a group that received dietary feedback only and (C) a control group (that received any intervention). Dietary intake was assessed using a mobile food record application in which participants captured images of the foods and beverages they consumed over 4 days at baseline and at 6 months post-intervention. After 6 months of intervention, participants in group B and the control group demonstrated a significantly increased daily intake of vegetable servings ($+0.4 \pm 0.1$, $p = 0.002$ and $+0.4 \pm 0.1$, $p = 0.02$, respectively), while group A demonstrated a significantly decreased daily intake of fruit servings (-0.2 ± 0.1; $p = 0.03$). However, no significant differences between groups in terms of fruit and vegetable intake were observed ($p < 0.05$).

3.2. Increasing Daily Fruit and/or Vegetable Consumption as a Secondary Outcome

Of the four RCTs that designated increasing fruit and/or vegetable intake as the secondary outcome [35–37,39], one revealed that the intervention was effective in both targets [37], two were partially effective [35,39], and the last one was not effective for either of the two targets [36]. These four RCTs lasted from 3 to 9 months, and all of the RCTs were population stratified.

The Make Better Choices (MBC) study [35] was a comparative 4-arm RCT designed to discern the optimal approach to simultaneously target diet and physical activity. The MBC study [35] consisted of a 6-month intervention (3-week intervention and 5-month follow-up) with 204 adults aged 21–60 years who were randomized into one of four behavioural change prescriptions. The MBC study compared four different behaviours: (1) 5 fruit/vegetable servings; (2) saturated fat consumption of less than 8% of total calories; (3) physical activity of at least 60 min/day; and (4) sedentary leisure of less than 90 min/day. The intervention consisted of present and remote coaches accessed by a mobile personal digital assistant (PDA) that tailored the behavioural strategies based on the baseline data of the participants. Moreover, participants received financial incentives when they reached the goals. The two groups targeted to increase fruit and vegetable intake (Group B and Group C) seemed more successful than the other two groups targeted to change other behaviours: Group B increased from 1.3 ± 1.1 servings/day at baseline to 5.6 ± 1.1 servings/day at the end of the intervention, and Group C increased from 1.2 ± 0.9 servings/day at baseline to 5.5 ± 1.0 servings/day at the end of the intervention. The two groups that were not targeted to improve fruit and vegetable intake (Group A and Group D) seemed less successful than the other two groups: Group A increased from 1.1 ± 0.9 servings/day at baseline to 1.7 ± 1.1 servings/day at the end of intervention; Group D increased from 1.4 ± 1.1 servings/day at baseline to 1.9 ± 1.6 at the end of intervention. However, the differences between baseline and the end of intervention regarding fruit and vegetable intake in each group and among groups were not reported by the researchers.

Another study related to the MBC study was the MBC 2 trial [37], a 3-arm RCT that tested whether a multicomponent intervention of 9 months (6-month intervention and 3-month follow-up) integrating a mHealth application, modest incentives and remote coaching could sustainably improve dietary habits and physical activity. Participants were randomly assigned to one of two interventions. Intervention group (A) targeted performing moderate to vigorous physical activity (MVPA) simultaneously with other diet and activity targets, and intervention group (B) targeted the same goals but sequentially. The control intervention group only addressed improving stress and sleep. After 6 months, fruit and vegetable intake increased by +6.6 servings/day in group A (simultaneous), by +7.4 servings/day in group B (sequential) and by +0.5 servings/day in the control group.

In the third study, Partridge et al. [39] performed a 2-arm RCT from a larger mHealth lifestyle program called "TXT2BFiT" to improve dietary and physical activity behaviours among 248 young adults aged 18–35 years who were at high risk for the development of obesity. The intervention group comprised 8 weekly motivational text messages, 5 personalized coaching calls, 1 weekly email, a diet booklet and a mobile phone application that provided education, self-monitoring, access to a community blog and support resource, over 3 months. Control group participants only received 4 text messages and dietary and physical activity guidelines. Intervention participants were more likely to consume greater quantities of vegetables after 3 months compared to control participants ($p = 0.009$). Additionally, at 3 months, the proportion of participants with a vegetable intake of ≥ 2 servings/day increased from 34.1% to 64.3% in the intervention group and from 36% to 48% in the control group.

The Mobile-Based Intervention to Stop Obesity in Pre-schoolers (MINISTOP) [36] aimed to help, through intervention by their parents, 315 children aged 4.5 years to improve their body status, nutritional habits and physical activity via a smartphone application during a 6-month intervention. Participants were randomly assigned to the intervention or control group: the intervention group received a 6-month mHealth application to register information about their child's food consumption and physical activity; and the control group received a pamphlet on healthy eating and physical activity. The differences between baseline and the follow-up for the intervention group resulted in an increase of 2.9 ± 78.9 g/day of fruits and -6.7 ± 42.1 g/day of vegetables consumed, while for the control group, decreases of -12.1 ± 87.9 g/day of fruits and -3.6 ± 39.7 g/day of vegetables were observed. However, no significant differences between groups were observed in fruit or vegetable intake.

3.3. Quality Appraisal and Risk of Bias in the Included Studies

Analysis of the quality of the included studies showed that all of the studies were of weak quality (≥ 2 weak category ratings). The best good quality items were the study design and dropouts, whereas the other items were of poor quality.

4. Discussion

The present review aimed to investigate the current literature on the potential use of mobile phone applications to self-monitor and increase the intake of fruits and/or vegetables. From a search of the literature, eight studies were included in the final screening for evaluation. The present review proposes that mobile phone applications that include a self-monitoring component have great potential in improving fruit and vegetable intake, supporting the important health benefits associated with technology-based interventions. Six of the eight studies included in the review were effective. These studies focused on overweight adults and adults or young adults with unhealthy lifestyles. Thus, it could be inferred from the current review that the interventions delivered through mobile phone applications that successfully improved the intake of fruits and/or vegetables had something in common: stratification of a specific population that has common interests and motivations. Focusing on the age of the participants, it was observed that 5 studies targeting the adult population (18–60 years) were effective, 2 studies targeting young adults (18–35 years) were not effective or partially effective, and 1 study targeting children through their parents was not effective. Age seems to be an irrelevant factor in determining the effectiveness of the intervention, while other factors, such as participants' common interests, play a key role in achieving an increase in fruit and vegetable intake. Indeed, population selection is one of the characteristics considered in social marketing principles to enable healthy choices [41]. Moreover, two [32,33] of the six effective studies focused on increasing fruit and/or vegetable intake in targeted overweight adults, suggesting that the effectiveness of these types of interventions could be influenced by specific motivations, such as overweight-associated health risks. Pre-existing health problems in the study population were related to increased effectiveness of the intervention compared with the effectiveness observed in the population without diseases, as demonstrated in a previous review [23].

Furthermore, it seems that both outcomes, increased fruit and vegetable daily intake, were better achieved when self-monitoring and dietary feedback were used in the intervention. However, two of the analysed RCTs that failed to increase fruit and vegetable consumption [36,38] also included these methodologies together with push notifications or motivational text messages. These contradictory results could be explained because in one of these studies, parents were responsible for improving the fruit and vegetable consumption of their children [36]; thus, it seems that monitoring the dietary intake of children via their parents is not effective in improving dietary habits.

On the other hand, increasing fruit and/or vegetable intake was not the primary outcome for all of the eight included studies. Four studies had decreasing body weight or body fat or improving diet and activity behaviours as their primary outcome, while increasing fruit and/or vegetable intake was set as a secondary outcome [35–37,39]. In these trials, the effects on fruit and/or vegetable intake evaluated as a secondary outcome were unclear. Thus, our results suggested that when the increase of fruit and/or vegetable intake was defined as the primary outcome, the intervention was more effective than when it was defined as a secondary outcome.

All of the eight studies included in this review implemented a self-monitoring and self-reporting component through a mobile phone application to set and control users' daily fruit and/or vegetable intake goals. Considering the other parts of the methodologies of the studies included in the review, all the studies were randomized controlled studies ranging from two to nine months of intervention and reported the need for further investigations to observe the effects over time. From the results presented in this systematized review, it was observed that an increase of fruit and/or vegetable intake could be observed from two to nine months, and an early rise in vegetable intake compared to that in fruit intake was found, which required more time to achieve an effective improvement.

The increased amount of fruit and/or vegetable intake is an important point to be discussed. Considering that one serving is equal to a minimum of 80 g [42], the minimum increase achieved in the eight studies included in the present review of +2.4 servings/day is an approximately 200 g increase in fruit and/or vegetable intake. Accordingly, an increase of 200 g/day of fruit and/or vegetable intake is associated with an 8%–16% reduction in the relative risk of coronary heart disease, 13%–18% reduction in the risk of stroke, 8%–13% reduction in the risk of cardiovascular diseases, 3%–4% reduction in the risk of cancer, and 10%–15% reduction in the risk of all-cause mortality [1].

The use of mHealth applications has increased, but the question of whether these applications are better than traditional methods is still open. Users have demonstrated general acceptability and adherence to mobile phone tools [43,44] in comparison with the traditional methods of dietary self-monitoring [45,46] because of their personal tailoring, low cost, and interactivity [47]. Furthermore, self-monitoring though mobile phone applications seems to provide easier and real-time dietary assessments [48] and is also associated with better quality dietary data compared with traditional methods, which could be affected by users' memory [44]. However, more evidence is needed for mHealth applications because the majority of these applications are developed with minimum feedback users and little support [49]. Although self-monitoring via mobile phone applications seems to have positive effects on fruit and/or vegetable intake, its relationship with an effective improvement of dietary habits has not yet been confirmed. Although mobile phone applications have been widely tested in weight-loss trials [24,50], their utilization for the improvement of specific target food group intake is still scarce.

Comparing different methodologies used to increase the consumption of fruits and/or vegetables, a systematic review from 2005 that considered 44 studies using diverse approaches, but not mobile phone applications, observed increases in fruit and vegetable intake from 0.1 to 1.4 servings/day in heathy adults [23]. Computer-tailored information and interventions using telephone contacts were found to represent an adequate alternative to face-to-face education and counselling-based interventions [23]. Notably, the improvements in fruit and/or vegetable intake observed in the studies included in the present review (range: +0.2 to +7.5 servings/day of fruit and vegetable), which used mobile phone applications, seem to be greater than those obtained from interventions employing more traditional methodologies.

Finally, although the use of different mobile phone applications in several studies [9,11,19,51] shows a positive outcome in increasing the awareness of the quality of food intake, improving dietary habits and educating individuals, it is clear that the implementation of mHealth applications for fruit and/or vegetable intake promotion can deeply affect the final outcome. Moreover, it seems that only monitoring fruit and/or vegetable intake may not be sufficiently engaging; thus, implementing smart techniques for individual engagement, such as expert feedback [9,10,12,19] or positive rewards, could affect the final success of the mobile phone application [12,19]. In other words, it is well known that the effectiveness of mHealth applications depends on the usability their interface, feedback, rewards, and so on. A complete comparison would require using different mobile phone applications and studying their usability and effectiveness in the same population group. Unfortunately, most of the mobile phone applications used are built in-house and thus are not publicly available for direct comparison. Moreover, implementation of an extensive usability study is beyond the scope of this paper.

Additionally, assessment of food intake through web and mobile app tools requires the collaboration of individuals and thus can be subjective, retrospectively biased, and suffer from low compliance. It is tedious for people to continuously annotate their food intake over long periods of time. In addition to the fact that having to annotate all meals is embarrassing and subjective, people generally do not remember all the food they have eaten. Another important drawback of manual annotation is food underreporting [20]. Moreover, many health applications are not created by nutritional professionals. Additionally, we cannot assume that food diaries based on personal annotations of a few days are representative of an individual's complete diet [20]. Recently, some web-based and mobile applications have included automatic food recognition that is based on smartphone pictures. Some applications have very recently claimed to introduce this option: LoseIt!© (2008–2019 FitNow, Inc., Saint Honoré, Paris), MyFitnessPal© (2009–2019 MyFitnessPal, Inc., San Francisco, CA, U.S.A.), CalorieMama© (2017, Azumio, Inc., Redwood city, CA, U.S.A) and FatSecret© (2019, FatSecret, Victoria, Australia). This ability makes the process of food intake reporting easier, faster and more pleasant, but it currently suffers from not being able to recognize a large amount of foods in the diet, demonstrating sub-optimal performance and limited recognition of different types of dishes.

In general, the majority of the articles included in the present systematized review discussed future interventions, such as larger-scale and longer trials, rather than technical improvements of mHealth applications. Regardless, some design elements could be taken into account for the future development of mHealth applications to improve vegetable and fruit intake, such as (a) inclusion of a validated tool to register food intake and to improve dietary assessments [38]; (b) weekly messages to reinforce the health recommendations about vegetable and fruit intake [39]; (c) remote connected coaching [37]; (d) remainders to buy fruits and vegetables when people are in the supermarket (e) interactive information between users and coaches [34,35]; and (f) tracking and sensor technologies as an interactive information system [34].

There are several limitations in this review. First, the majority of the trials included in this review presented the following limitations: the population sample was not representative of the community setting because of the small size, level of education, gender and origin [32–36,38,39] and low reliance of the data, which were self-reported by participants [32,33,35–39]. Second, the reviewed studies expressed results on the primary outcome (fruit and/or vegetable intake) using different units of measure, such as g/day, servings/day, pieces/week and percentage of people consuming ≥ 2 servings/day. Third, a description of the amount of grams considered to be a serving of fruit and/or vegetable intake was not provided. Fourth, the self-monitoring tool type was not the same for all the studies, which influenced the presentation of the final results (servings/day or week, g/day, percentage, etc.), and could be better expressed in future studies as servings/day or g/day. Fifth, only two databases were used to search for results: PubMed and Web of Science. Although these databases are the most commonly used, the inclusion of other databases could have increased and influenced the final results of the review. Sixth, the present paper is a systematized review, i.e., the review process is shorter than that of a systematic review, may or may not include comprehensive searching, may or may not include quality assessment, and describes the uncertainty around the findings and the limitations of the methodology [28]. Finally, although the study quality was not an inclusion criterion, the weakness of the majority of the included studies presents problems for the generalizability of the results of this systematized review.

5. Conclusions

The present review demonstrates that effective interventions to increase fruit and vegetable consumption using mobile phone applications last from two to nine months and are characterized by a stratified population that shares the same motivation to achieve better dietary habits. Furthermore, the inclusion of behavioural change techniques, such as dietary feedback together with self-monitoring and remote coaching support, has been identified as a key element that can definitively facilitate the adoption of new dietary habits. This issue strongly suggests that behavioural theory-based strategies must be considered when designing dietary mHealth application interventions. Further research on mHealth applications is needed to design more effective interventions and to determine their efficacy over the long term. Although evidence shows a promising future for mHealth applications to promote healthy nutrition, it is an open question as to how to ensure that the maturity and popularity of these applications is similar to those of other tools for the promotion of healthy habits, such as activity trackers.

Supplementary Materials: The following are available online at http://www.mdpi.com/2072-6643/11/3/686/s1, Table S1: Quality of the studies included in the systematized review; Table S2: Relevant information regarding the data extracted from the included studies.

Author Contributions: Each author has made substantial contributions to the conception or design of the work F.M., E.L., L.T., R.S., N.B.; the acquisition, analysis, or interpretation of data F.M., E.L., L.T., N.B., R.S.; the creation of new software used in the work or has drafted the work or substantively revised it F.M., E.L., L.T., J.M.d.B., R.M.V., A.P., P.R., L.A., R.S., N.B. Each author has approved the submitted version (and a version substantially edited by journal staff that involves the author's contribution to the study) and agrees to be personally accountable for the author's own contributions and for ensuring that questions related to the accuracy or integrity of any part of the work, even ones in which the author was not personally involved, are appropriately investigated, resolved, and documented in the literature.

Funding: Floriana Mandracchia has received funding from the European Union's Horizon 2020 research and innovation programme under the Marie Sklodowska-Curie grant agreement No. 713679 and from the Universitat Rovira i Virgili (URV) (Reference number: 2018 MFP-COFUND-24).

Acknowledgments: This publication is co-funded by the European Regional Development Fund (ERDF) of the European Union within the framework of the ERDF operative programme of Catalonia 2014–2020 aimed at an objective of investment in growth and employment. This publication is framed within the initiative of coordinated PECT TurisTIC en familia, Operation 12: "Healthy Meals". Anna Pedret has Torres Quevedo contract (Subprograma Estatal de Incorporación, Plan Estatal de Investigación Científica y Técnica y de Innovación). NFOC-Salut group is a consolidated research group of Generalitat de Catalunya, Spain (2017 SGR522).

Conflicts of Interest: The authors declare no conflict of interest.

References

1. Aune, D.; Giovannucci, E.; Boffetta, P.; Fadnes, L.T.; Keum, N.N.; Norat, T.; Greenwood, D.C.; Riboli, E.; Vatten, L.J.; Tonstad, S. Fruit and vegetable intake and the risk of cardiovascular disease, total cancer and all-cause mortality-A systematic review and dose-response meta-analysis of prospective studies. *Int. J. Epidemiol.* **2017**, *46*, 1029–1056. [CrossRef]
2. He, K.; Hu, F.B.; Colditz, G.A.; Manson, J.E.; Willett, W.C.; Liu, S. Changes in intake of fruits and vegetables in relation to risk of obesity and weight gain among middle-aged women. *Int. J. Obes.* **2004**, *28*, 1569–1574. [CrossRef] [PubMed]
3. Ezzati, M.; Lopez, A.D.; Rodgers, A.; Murray, C.J.L. Global and Regional Burden of Disease Attributable to Selected Major Risk Factors. *Comparat. Q. Health Risks Global Reg. Burd. Dis.* **2004**, 257–280. [CrossRef]
4. Nishida, C.; Uauy, R.; Kumanyika, S.; Shetty, P. The Joint WHO/FAO Expert Consultation on diet, nutrition and the prevention of chronic diseases: Process, product and policy implications. *Public Health Nutr.* **2004**, *7*, 245–250. [CrossRef]
5. United States Department of Health and Human Services & United States Department of Agriculture 2015–2020 Dietary Guidelines for Americans. 8th Edition. Available online: https://health.gov/dietaryguidelines/2015/ (accessed on 5 February 2019).
6. Her Majesty the Queen in Right of Canada, represented by the Minister of Health Canada. Eating Well with Canada's FoodGuide. 2011. Available online: https://www.canada.ca/content/dam/hc-sc/migration/hc-sc/fn-an/alt_formats/hpfb-dgpsa/pdf/food-guide-aliment/view_eatwell_vue_bienmang-eng.pdf (accessed on 15 February 2019).
7. Elmadfa, I.; Meyer, A.; Nowak, V.; Hasenegger, V.; Putz, P.; Verstraeten, R.; Remaut-DeWinter, A.M.; Kolsteren, P.; Dostálová, J.; Dlouhý, P.; et al. European Nutrition and Health Report 2009. *Forum Nutr.* **2009**, *62*, 1–405.
8. Rekhy, R.; McConchie, R. Promoting consumption of fruit and vegetables for better health. Have campaigns delivered on the goals? *Appetite* **2014**, *79*, 113–123. [CrossRef] [PubMed]
9. Marchesini, G.; Montesi, L.; El Ghoch, M.; Brodosi, L.; Calugi, S.; Dalle Grave, R. Long-term weight loss maintenance for obesity: A multidisciplinary approach. *Diabetes Metab. Syndr. Obes: Targets Ther.* **2016**, *9*, 37. [CrossRef]
10. Proeschold-Bell, R.J.; Steinberg, D.M.; Yao, J.; Eagle, D.E.; Smith, T.W.; Cai, G.Y.; Turner, E.L. Using a holistic health approach to achieve weight-loss maintenance: Results from the Spirited Life intervention. *Transl. Behav. Med.* **2018**. [CrossRef]
11. Wolever, R.Q.; Dreusicke, M.H. Integrative health coaching: A behavior skills approach that improves HbA1c and pharmacy claims-derived medication adherence. *BMJ Open Diabetes Res. Care* **2016**, *4*, e000201. [CrossRef]
12. Johnson, K.E.; Alencar, M.K.; Coakley, K.E.; Swift, D.L.; Cole, N.H.; Mermier, C.M.; Kravitz, L.; Amorim, F.T.; Gibson, A.L. Telemedicine-Based Health Coaching Is Effective for Inducing Weight Loss and Improving Metabolic Markers. *Telemed. e-Health* **2018**. [CrossRef] [PubMed]
13. World Health Organization (WHO) eHealth at WHO. Available online: https://www.who.int/ehealth/about/en/ (accessed on 5 February 2019).
14. Bakirci-Taylor, A.L.; Reed, D.B.; McCool, B.; Dawson, J.A. mHealth Improved Fruit and Vegetable Accessibility and Intake in Young Children. *J. Nutr. Educ. Behav.* **2019**. [CrossRef] [PubMed]
15. Pohl, M. 325,000 Mobile Health Apps Available in 2017–Android Now the Leading mHealth Platform. Available online: https://research2guidance.com/325000-mobile-health apps-available-in-2017/ (accessed on 21 February 2019).
16. EMarketer Number of Smartphone Users Worldwide from 2014 to 2020 (In Billions). In Statista–The Statistics Portal. Available online: https://www.statista.com/statistics/330695/number-of-smartphone-users-worldwide/ (accessed on 21 February 2019).
17. Patrick, K.; Griswold, W.G.; Raab, F.; Intille, S.S. Health and the Mobile Phone. *Am. J. Prev Med.* **2009**, *35*, 177–181. [CrossRef] [PubMed]
18. Goggin, K.J.; Ellerbeck, E.F. Mobile Technology for Obesity Prevention A Randomized Pilot Study in Racial and Ethnic Minority Girls. *Am. J. Prev Med.* **2015**, *46*, 404–408. [CrossRef]

19. Toro-Ramos, T.; Lee, D.-H.; Kim, Y.; Michaelides, A.; Oh, T.J.; Kim, K.M.; Jang, H.C.; Lim, S. Effectiveness of a Smartphone Application for the Management of Metabolic Syndrome Components Focusing on Weight Loss: A Preliminary Study. *Metab. Syndr. Relat. Disord.* **2017**, *15*, 465–473. [CrossRef] [PubMed]

20. McClung, H.L.; Ptomey, L.T.; Shook, R.P.; Aggarwal, A.; Gorczyca, A.M.; Sazonov, E.S.; Becofsky, K.; Weiss, R.; Das, S.K. Dietary Intake and Physical Activity Assessment: Current Tools, Techniques, and Technologies for Use in Adult Populations. *Am. J. Prev. Med.* **2018**, *55*, e93–e104. [CrossRef] [PubMed]

21. Hammersley, M.L.; Jones, R.A.; Okely, A.D.; Ave, N. Parent-Focused Childhood and Adolescent Overweight and Obesity eHealth Interventions: A Systematic Review and Meta-Analysis. *J. Med. Int. Res.* **2016**, *18*, e203. [CrossRef]

22. Nguyen, B.; Kornman, K.P.; Baur, L.A. A review of electronic interventions for prevention and treatment of overweight and obesity in young people. *Obes. Rev.* **2011**, *12*, e298–e314. [CrossRef]

23. Pomerleau, J.; Lock, K.; Knai, C.; Mckee, M. Interventions Designed to Increase Adult Fruit and Vegetable Intake Can Be Effective: A Systematic Review of the Literature. *Nutr. Epidemiol.* **2005**, 2486–2495. [CrossRef]

24. Mateo, G.F.; Granado-Font, E.; Ferré-Grau, C.; Montaña-Carreras, X. Mobile phone apps to promote weight loss and increase physical activity: A systematic review and meta-analysis. *J. Med. Int. Res.* **2015**.

25. Schaaf, M.; Chhabra, S.; Flores, W.; Feruglio, F.; Dasgupta, J.; Ruano, A.L. Does Information and Communication Technology Add Value to Citizen-Led Accountability Initiatives in Health? Experiences from India and Guatemala. *Health Human Rights* **2018**, *20*, 169–184.

26. Wildevuur, S.E.; Simonse, L.W. Information and Communication Technology–Enabled Person-Centered Care for the "Big Five" Chronic Conditions: Scoping Review. *J. Med. Int. Res.* **2015**, *17*, e77. [CrossRef] [PubMed]

27. Charlton, N.; Kingston, J.; Petridis, M.; Fletcher, B.C. Using Data Mining to Refine Digital Behaviour Change Interventions. In Proceedings of the 2017 International Conference on Digital Healt–DH '17; ACM Press: New York, NY, USA, 2017; pp. 90–98.

28. Duke University Systematic Reviews: The process: Types of Reviews. Available online: https://guides. mclibrary.duke.edu/sysreview (accessed on 22 February 2019).

29. Dennison, L.; Morrison, L.; Conway, G.; Yardley, L.; Dennison, L. Opportunities and Challenges for Smartphone Applications in Supporting Health Behavior Change: Qualitative Study. *J. Med. Int. Res.* **2013**, *15*, e86. [CrossRef] [PubMed]

30. Chatzipavlou, I.A.; Christoforidou, S.A.; Vlachopoulou, M. A recommended guideline for the development of mHealth Apps. *mHealth* **2016**, *2*, 1–7. [CrossRef] [PubMed]

31. Sandberg, R.; Rollins, M. *Business of Android Apps Development*, 2nd ed.; Apress, 2011; ISBN 978-1-4302-5008-1.

32. Mummah, S.A.; Mathur, M.; King, A.C.; Gardner, C.D.; Sutton, S. Mobile Technology for Vegetable Consumption: A Randomized Controlled Pilot Study in Overweight Adults. *JMIR mHealth uHealth* **2016**, *4*, e51. [CrossRef] [PubMed]

33. Mummah, S.; Robinson, T.N.; Mathur, M.; Farzinkhou, S.; Sutton, S.; Gardner, C.D. Effect of a mobile app intervention on vegetable consumption in overweight adults: A randomized controlled trial. *Int. J. Behav. Nutr. Phys. Act.* **2017**, *14*, 1–10. [CrossRef]

34. Elbert, S.P.; Dijkstra, A.; Oenema, A. A mobile phone app intervention targeting fruit and vegetable consumption:the efficacy of textual and auditory tailored health information tested in a randomized controlled trial. *J. Med. Int. Res.* **2016**, *18*. [CrossRef]

35. Spring, B.; Schneider, K.; McFadden, H.G.; Vaughn, J.; Kozak, A.T.; Smith, M.; Moller, A.C.; Epstein, L.H.; DeMott, A.; Hedeker, D.; et al. Multiple Behavior Changes in Diet and Activity. *Arch. Int. Med.* **2012**, *172*, 789–796. [CrossRef] [PubMed]

36. Nystrom, C.D.; Sandin, S.; Henriksson, P.; Henriksson, H.; Trolle-Lagerros, Y.; Larsson, C.; Maddison, R.; Ortega, F.B.; Pomeroy, J.; Ruiz, J.R.; et al. Mobile-based intervention intended to stop obesity in preschool-aged children: The MINISTOP randomized controlled trial. *Am. J. Clin. Nutr.* **2017**, *105*, 1327–1335. [CrossRef]

37. Spring, B.; Pellegrini, C.; McFadden, H.G.; Pfammatter, A.F.; Stump, T.K.; Siddique, J.; King, A.C.; Hedeker, D. Multicomponent mHealth intervention for large, sustained change in multiple diet and activity risk behaviors: The make better choices 2 randomized controlled trial. *J. Med. Int. Res.* **2018**, *20*, e10528. [CrossRef]

38. Kerr, D.A.; Harray, A.J.; Pollard, C.M.; Dhaliwal, S.S.; Delp, E.J.; Howat, P.A.; Pickering, M.R.; Ahmad, Z.; Meng, X.; Pratt, I.S.; et al. The connecting health and technology study: A 6-month randomized controlled trial to improve nutrition behaviours using a mobile food record and text messaging support in young adults. *Int. J. Behav. Nutr. Phys. Act.* **2016**, *13*, 1–14. [CrossRef]
39. Partridge, S.R.; McGeechan, K.; Hebden, L.; Balestracci, K.; Wong, A.T.; Denney-Wilson, E.; Harris, M.F.; Phongsavan, P.; Bauman, A.; Allman-Farinelli, M. Effectiveness of a mHealth Lifestyle Program With Telephone Support (TXT2BFiT) to Prevent Unhealthy Weight Gain in Young Adults: Randomized Controlled Trial. *JMIR mHealth uHealth* **2015**, *3*, e66. [CrossRef] [PubMed]
40. Mummah, S.A.; King, A.C.; Gardner, C.D.; Sutton, S. Iterative development of Vegethon: A theory-based mobile app intervention to increase vegetable consumption. *Int. J. Behav. Nutr. Phys. Act.* **2016**, *13*, 1–12. [CrossRef] [PubMed]
41. Andreasen, A.R. Social Marketing: Its Definition and Domain. *J. Public Policy Mark.* **1994**, *13*, 108–114. [CrossRef]
42. Food and Agriculture Organization of the United Nations. What Is a Serving? Available online: http://www.fao.org/giii/english/newsroom/focus/2003/fruitveg2.htm (accessed on 15 February 2019).
43. Payne, J.E.; Turk, M.T.; Kalarchian, M.A.; Pellegrini, C.A. Defining Adherence to Dietary Self-Monitoring Using a Mobile App: A Narrative Review. *J. Acad. Nutr. Diet.* **2018**, *118*, 2094–2119. [CrossRef]
44. Lieffers, J.R.L.; Hanning, R.M. Dietary assessment and self-monitoring: With nutrition applications for mobile devices. *Can. J. Diet. Pract. Res.* **2012**, *73*, 253–260. [CrossRef] [PubMed]
45. Burke, L.; Wang, J.; Sevick, M. Self-Monitoring in Weight Loss: A Systematic Review of the Literature. *J. Am. Diet. Assoc.* **2012**, *111*, 92–102. [CrossRef] [PubMed]
46. Burke, L.; Conroy, M.; Sereika, S.; Elci, O.; Styn, M.; Acharya, S.; Sevick, M.; Ewing, L.; Glanz, K. The Effect of Electronic Self-Monitoring on Weight Loss and Dietary Intake: A Randomized Behavioral Weight Loss Trial Lora. *Obesity (Silver Spring, Md.)* **2012**, *1*, 233–245. [CrossRef]
47. Tang, J.; Abraham, C.; Stamp, E.; Greaves, C. How can weight-loss app designers' best engage and support users? A qualitative investigation. *Br. J. Health Psychol.* **2015**, *20*, 151–171. [CrossRef] [PubMed]
48. Turner-McGrievy, G.M.; Beets, M.W.; Moore, J.B.; Kaczynski, A.T.; Barr-Anderson, D.J.; Tate, D.F. Comparison of traditional versus mobile app self-monitoring of physical activity and dietary intake among overweight adults participating in an mHealth weight loss program. *J. Am. Med. Inf. Assoc.* **2013**, *20*, 513–518. [CrossRef] [PubMed]
49. Roess, A. The Promise, Growth, and Reality of Mobile Health—Another Data-free Zone. *New Engl. J. Med.* **2017**, *377*, 2010–2011. [CrossRef] [PubMed]
50. Liu, F.; Kong, X.; Cao, J.; Chen, S.; Li, C.; Huang, J.; Gu, D.; Kelly, T.N. Mobile phone intervention and weight loss among overweight and obese adults: A meta-analysis of randomized controlled trials. *Am. J. Epidemiol.* **2015**, *181*, 337–348. [CrossRef] [PubMed]
51. Celis-Morales, C.; Livingstone, K.M.; Marsaux, C.F.M.; Forster, H.; O'Donovan, C.B.; Woolhead, C.; Macready, A.L.; Fallaize, R.; Navas-Carretero, S.; San-Cristobal, R.; et al. Design and baseline characteristics of the Food4Me study: A web-based randomised controlled trial of personalised nutrition in seven European countries. *Genes Nutr.* **2015**, *10*, 450. [CrossRef] [PubMed]

nutrients

Review

Evaluation of New Technology-Based Tools for Dietary Intake Assessment—An ILSI Europe Dietary Intake and Exposure Task Force Evaluation

Alison L. Eldridge [1],*, Carmen Piernas [2], Anne-Kathrin Illner [3], Michael J. Gibney [4], Mirjana A. Gurinović [5], Jeanne H.M. de Vries [6] and Janet E. Cade [7]

1. Nestlé Research, Vers-chez-les-Blanc, 1000 Lausanne 26, Switzerland
2. Nuffield Department of Primary Care Health Sciences, University of Oxford, Oxford OX2 6GG, UK; carmen.piernas-sanchez@phc.ox.ac.uk
3. College of Health Sciences, Polytechnic Institute UniLaSalle Beauvais, 60026 Beauvais, France; Anne-Kathrin.ILLNER@unilasalle.fr
4. Institute of Food and Health, University College Dublin, Dublin D04 V1W8, Ireland; mike.gibney@ucd.ie
5. Centre of Research Excellence in Nutrition and Metabolism, Institute for Medical Research, University of Belgrade, Belgrade 11000, Serbia; mirjana.gurinovic@gmail.com
6. Division of Human Nutrition and Health, Wageningen University, 6708WE Wageningen, The Netherlands; jeanne.devries@wur.nl
7. School of Food Science and Nutrition, University of Leeds, Leeds LS2 9JT, UK; J.E.Cade@leeds.ac.uk
* Correspondence: alison.eldridge@rdls.nestle.com; Tel.: +41-21-785-8365

Received: 27 November 2018; Accepted: 25 December 2018; Published: 28 December 2018

Abstract: Background: New technology-based dietary assessment tools, including Web-based programs, mobile applications, and wearable devices, may improve accuracy and reduce costs of dietary data collection and processing. The International Life Sciences Institute (ILSI) Europe Dietary Intake and Exposure Task Force launched this project to evaluate new tools in order to recommend general quality standards for future applications. Methods: A comprehensive literature search identified technology-based dietary assessment tools, including those published in English from 01/2011 to 09/2017, and providing details on tool features, functions and uses. Each of the 43 tools identified (33 for research and 10 designed for consumer use) was rated on 25 attributes. Results: Most of the tools identified (79%) relied on self-reported dietary intakes. Most (91%) used text entry and 33% used digital images to help identify foods. Only 65% had integrated databases for estimating energy or nutrients. Fewer than 50% contained any features of customization and about half generated automatic reports. Most tools reported on usability or reported validity compared with another assessment method (77%). A set of Best Practice Guidelines was developed for reporting dietary assessment tools using new technology. Conclusions: Dietary assessment methods that utilize technology offer many advantages for research and are often preferable to consumers over more traditional methods. In order to meet general quality standards, new technology tools require detailed publications describing tool development, food identification and quantification, customization, outputs, food composition tables used, and usability/validity testing.

Keywords: dietary assessment; mobile technologies; Web-based technologies

---

1. Introduction

The opportunities provided by the internet to link large scale food and nutrient databases with automated dietary recording has led to growth in the number of online dietary assessment tools [1]. New technologies for measuring diet can be categorized according to the type of technology being used, such as Web-based or online tools, mobile systems (apps), camera-based tools, and other developing technologies, such as consumer purchase data and wearable sensors. Traditional methods relied heavily on self-reporting of foods consumed either using food frequency questionnaires (FFQ) or with paper-based recalls or diaries. All of the traditional methods lacked accuracy as a result of problems including the ability to recall food consumed, difficulties with portion size estimations or limited food composition tables [2]. Considerable manual input and time was required for coding and converting foods recorded into nutrients. This meant that in large-scale cohort studies it was not generally possible to collect detailed food intake information, and studies relied on food frequency questionnaire data, which is subject to greater measurement error than other self-report measures [3,4]. Use of computerized tools facilitated data coding, and incorporation of the automated multiple-pass method (AMPM) standardized data collection for national surveys [5,6]. New methods have allowed for an expansion and potential improvement on the traditional methods. The use of the Internet makes larger-scale collection of food and nutrient information practical with lower costs and burden for both researchers and participants [7]. Study participants can be invited to take part in research electronically via email or text [8]. Users of new technology tools can more easily identify foods consumed through interactive searchable databases [9]. They can provide real-time results and feedback [1] and can include enhanced options for portion size description, such as using digital images [10], and more relevant lists of branded food items [9].

It is often not clear how relevant a particular dietary assessment tool is for research as a result of limited information provided on the development process and lack of validation. An evaluation of new technologies to assess diet may help understanding of their potential to replace, improve, or complement traditional methods. Due to the rapid development of new technologies, existing reviews of the area quickly become out of date, including obsolete technologies such as personal digital assistants or PDAs [11]. Highlighting features of new technologies, such as those found in Web-based recalls or apps, in comparison with tool elements reflecting traditional approaches may help to identify techniques that can enhance dietary measurement [12]. Recently, clear guidance in terms of dietary assessment tool choice and reporting has been published [2,13]. However, guidance on the development of new tools with quality criteria for their assessment is still lacking.

In 2016, the International Life Sciences Institute (ILSI) Europe Dietary Intake and Exposure Task Force (http://ilsi.eu/task-forces/food-safety/dietary-intake-and-exposure/) established an expert group on evaluation of new methods for dietary intake assessment. The aim of the group was to review new technologies for diet assessment in terms of features, sources and quality of data, and validity. The review presented here will help to understand the relative merits of particular new tools and applications currently available for dietary intake assessment. We have critically evaluated tools, including their sources of data, applicability for research, ease of use by different population groups, and ability to handle a wide range of foods and beverages. In a second step, we also suggest guidelines for quality standards to improve reporting of dietary intake assessment tools.

The objectives of this paper are to: (i) report on a comprehensive review of tools for dietary assessment using new technologies which are applicable for use in research, commercial, clinical and public health contexts; (ii) to develop guidelines for quality criteria required for a good quality tool; and (iii) to make recommendations for future reporting of dietary assessment tools using new technologies.

2. Materials and Methods

2.1. Inclusion Criteria and Search Strategy

Comprehensive literature searches were conducted to identify articles pertaining to new technologies for dietary intake assessment using key word searches with the following inclusion criteria: (1) publications were in English, (2) articles were published from January 2011 to September 2017, and (3) sufficient information was available to evaluate tool features, functions, and uses. Various search terms were used related to dietary or nutrition surveys, nutrition assessment, and the use of technologies, including mobile apps, Web-based tools, online or Internet tools, and software. PubMed, PLOS, BioMED, Science Direct and Ovid databases were used, each with slightly different search terms (Supplemental Table S1). The searches were limited to articles published after 1 January 2011 because the field of technology development for dietary intake assessment is advancing rapidly, and tools developed prior to 2011 have been previously evaluated [12]. Dietary assessment tools were identified, details of which were available in one or more publications.

2.2. Evaluation Criteria and Data Extraction

The Expert Group, comprised of the authors of this manuscript, identified 25 attributes related to data entry, identification and quantification of foods, customization, output, usability and validity, which were used to evaluate each dietary assessment tool (Supplemental Table S2). Under the heading of Data Entry, we assessed whether the tools relied on text entry, digital images and/or bar-code scanners, and whether they also collected information about health characteristics or physical activity. For the Identification and Quantification of Foods, we assessed whether the foods or beverages were automatically identified from an image or required manual identification, the source of food composition data used, and how the intake amounts were quantified, either by weights or household measures, or estimated from digital images. In the Customization section, we assessed whether the tool allowed the user to add missing foods, custom recipes or dietary supplements, and whether the program used machine learning to adapt the list of foods to user preferences. Under Output, we considered whether the tool provided data on energy, macro- and micro-nutrient intakes, food groups consumed, time of intake and meal name, and whether the tool generated automated reports. Finally, we assessed Usability and Validity by checking whether there were any reports of user feedback, time to complete the assessment, and whether any validation studies had been conducted.

The features of each dietary assessment tool were assessed independently by two members of the Expert Group from details provided in the publications, and any discrepancies were discussed at the Expert Group level. If the publications identified in the searches did not provide the sufficient detail to complete the assessment, additional literature, websites, contacts with authors, or tool use itself were used to attempt to fill gaps.

3. Results

3.1. Search Results

The PRISMA diagram showing the search flow and inclusion/exclusion of studies appears in Figure 1. A total of 4695 articles were initially identified. Duplicates were removed and the remaining articles screened (title and abstract) to eliminate those that were not relevant to meet the project objectives, yielding a total of 800 publications related to dietary intake databases, applications, and tools. The goal of this review was to identify unique technology-based tools for dietary intake assessment, including smartphone applications, those that captured digital images of foods and beverages for the purpose of dietary intake assessment, and dietary assessment tools available from the Web or that were accessed from a personal computer (PC). From the 800 articles that mentioned dietary assessment in the title or abstract, 151 were related to new technologies for dietary intake assessment, and of these, 66 were additional references for tools already identified. Papers describing the remaining 85 tools were reviewed in detail. A further 42 were excluded following the detailed review: 14 were deemed to be not relevant because they were editorials ($n = 1$), review papers ($n = 4$), or did not describe a new tool for dietary intake assessment ($n = 9$); 16 were missing sufficient detail to do our evaluation; seven of the tools were developed and reported on prior to 2011, thereby meeting our exclusion criteria; and five were eliminated because the publications referred to a tool that had been subsequently renamed. In the latter case, the updated tool name was retained for our evaluation. Consequently, we included 43 unique tools in our evaluation.

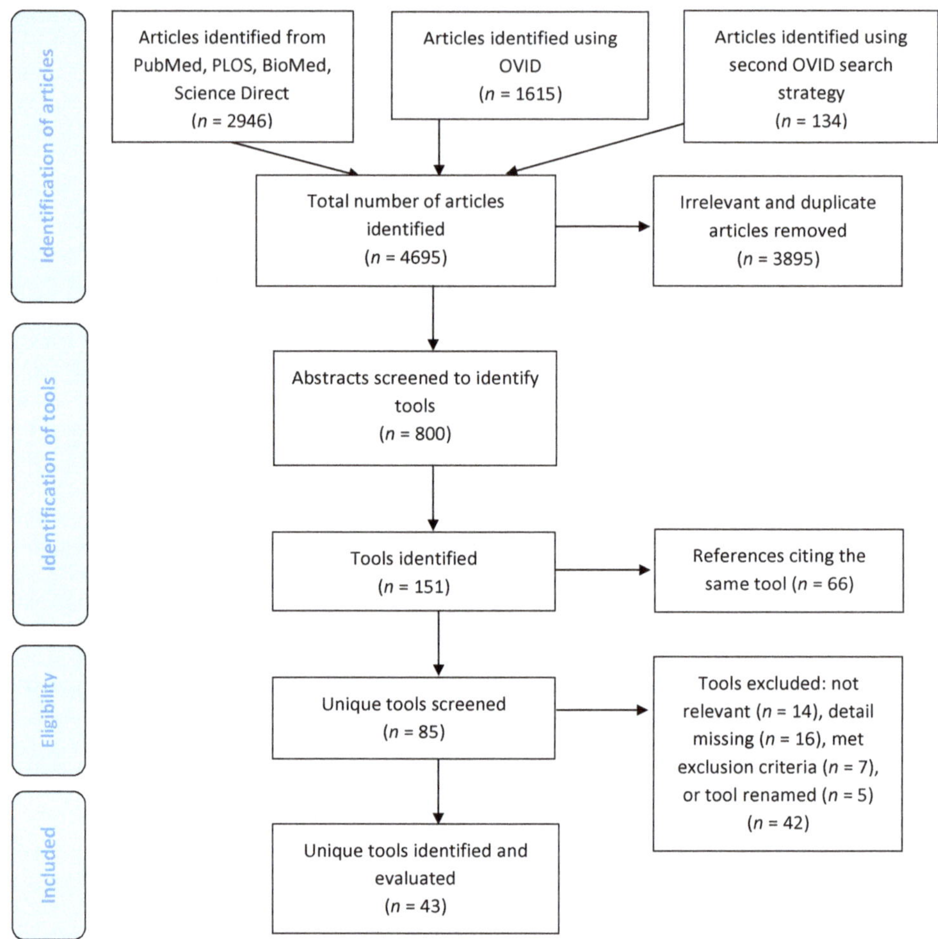

Figure 1. PRISMA diagram used to identify technology-based tools for dietary intake assessment.

3.2. Characteristics of Included Studies

In total, from the 43 tools identified, 33 tools were for use in research or surveillance and 10 tools intended for direct consumer use (Table 1), and since several of the attributes differed between the research/surveillance tools and those designed for consumers, we separated them. Of the 33 tools used for research or surveillance, *n* = 21 (64%) were Web-based to be used on a computer; *n* = 6 (18%) were optimized to be used on smartphones; *n* = 3 (9%) were for PC only (not Web-based); *n* = 2 (6%) used wearables for data collection and *n* = 1 (3%) was designed to be used on a tablet. Of the 10 tools identified for consumer use, *n* = 8 (80%) were optimized for smartphone use and *n* = 2 (20%) were Web-based to be used on a computer. Of the 33 tools designed to collect dietary data for research purposes, *n* = 16 (48%) were designed for adults exclusively, *n* = 11 (33%) were for all ages, and *n* = 6 (18%) were exclusively for children and/or adolescents. Of the 10 tools designed for consumer use, *n* = 7 (70%) were for adults exclusively, while *n* = 3 (30%) were designed for all ages. Among all the tools designed for research purposes, *n* = 17 (52%) collected dietary intake over the previous 24h using dietary recalls; *n* = 11 (33%) collected food records, while the rest collected intakes via food frequency questionnaires (*n* = 3; 9%) or imaging systems (*n* = 2; 6%). Of the 10 tools designed for consumer use, most of them collected food records ($n = 8$; 80%), while $n = 2$ (20%) collected food frequency questionnaires.

Although all of these tools used technology for dietary intake data collection, not all of the tools automatically coded the intake information to generate energy and nutrients (Table 1). Of the tools assessed here, 15 of the 43 (35%) were used for data capture only and required a dietitian or a coder to enter the items and portions in another tool later to estimate energy and nutrient intakes. These are identified as "not integrated into the tool" in Table 1. Another large difference in the tools was the source of food composition data and the number of items available. Tools designed to assess food consumption frequency (Evident II, Food4Me, GraFFS, IDQC, Oxford WebQ, and WebFFQ) included 135–200 individual line items (individual foods or aggregated food categories). Those designed for children varied, with SNAP and WebCaaFE including a limited list (49 and 32 foods and beverages, respectively), while WebFR and WebDASC included a more extensive list of 550 and 1300 items, respectively. Tools that relied on national food composition tables ranged from about 1000 items to more than 45,000 if branded foods were also included (e.g., myfood24), and were largely complete with respect to nutrients. The source of food composition was reported in all but one case, but the number of foods included in the database was missing for six of the tools. The daily time to complete each tool was reported in 18 of the 43 studies. The times ranged from an average low of 14 min to as much as 45–60 min, but most tools were completed within 15–35 min.

The use of images also differed considerably among tools. TADA, Snap-N-Eat, and DietCam automatically coded foods and beverages from digital images [14–16], and RFPM used semi-automatic coding of images to facilitate data entry. GoCARB automatically coded carbohydrate content of food categories identified from images. Chest-worn cameras, like eButton or Microsoft SenseCam, captured digital images throughout the day but required subsequent coding by nutritionists for nutrient intake estimates. Several tools, CHAT, FoodNow, NANA, NuDAM, and TECH, used digital images to enhance reporting of food intakes, along with text or voice recordings. FoodLog used images as a visual diary of food intakes for patients with diabetes, and Microsoft SenseCam used images as a memory aid for food records.

Table 1. Design characteristics of the technology-based tools used in dietary intake assessment.

Device Name	Country	Main Purpose of the Tool	Target Audience	Main Platform for Tool	Method of Data Collection and Entry	Food Composition Source	Approximate Number of Items	Time to Complete	References
					Tools for Use in Research or Surveillance (*n* = 33)				
ASA24 Automated Self-Administered 24 h Recall	USA, Canada, Australia	Dietary intake	Adults and children from 10 years	Web-based	24-h recall based on automated multiple-pass method (AMPM)	USDA's FNDDS [4], Canadian Food Composition and Australian food tables	10,000	Average of 24 min; most within 17–34 min	Baranowski et al. 2012 and 2014; Kirkpatrick et al. 2014; Thompson et al. 2015 [17–20]
CHAT Connecting Health and Technology; mobile food record	Australia	Food groups consumed	Adults and adolescents	Smartphone App	Food record based on images; dietitian identifies foods and food groups	Australia Guide to Healthy Eating, but not integrated into tool	2670	Not specified	Kerr et al. 2012 and 2016; Pollard et al. 2016 [21–23]
Compl-Eat	Netherlands	Dietary intake	Adults and adolescents from 16 years	Web-based	Interviewer-assisted or self-administered 24-h recall based on	Dutch Food Composition Database	2000	Close to 30 min	Meijboom et al. 2017 [24]
DAP Diet Assess and Plan	Serbia, Balkan region	Diet and physical activity	All ages	Web-based, PC	24-h recalls, food frequency questionnaires (FFQ), food propensity; dietitian enters data	Serbian and Balkan regional food composition databases	1450 Serbian and 1600 Balkan foods and recipes [9]	15–30 min	Gurinovic et al. 2016 and 2018; Zekovic et al. 2017 [25–27]
DES Diet Evaluation System	Korea	Dietary intake	All ages	Web-based	Interviewer-assisted 24-h recall	Korean food composition tables	8100	Average of 14 min	Jung et al. 2014 [28]
eButton	USA	Dietary intake, activity	All ages	Wearable	Imaging system with automated portion estimates; dietitian enters data	USDA's FNDDS, but not integrated into tool	8500	Not specified	Sun et al. 2010 and 2014; Jia et al. 2014 [29–31]
e-CA Electronic Carnet Alimentaire	Switzerland	Dietary intake	Adults	Web-based	Electronic food record; dietitian enters for coding	Prodi 6.3 software, but not integrated into the tool	900	Average of 19 min	Bucher Della Torre et al. 2017 [32]
eDIA Electronic Dietary Intake Assessment	Australia	Dietary intake	19–24 years old	Smartphone app	Food record	AUSNUT [5] 2007	4500	Not specified	Rangan et al. 2015 and 2016 [33,34]
EPIC-Soft [1]	European Union (EU)	Dietary intake	All ages	PC with Web-based management platform	Interviewer-assisted 24-h recall or dietitian enters data from food records	EPIC [6] software from all EU countries	10,000	Not specified	de Boer et al. 2011; Huybrechts et al. 2011; Freisling et al. 2014; Park et al. 2017 [35–38]

Table 1. *Cont.*

Device Name	Country	Main Purpose of the Tool	Target Audience	Main Platform for Tool	Method of Data Collection and Entry	Food Composition Source	Approximate Number of Items	Time to Complete	References
Food4Me	EU—7 European counties	Dietary Intake	Adults	Web-based	FFQ	WISP [7] software; based on McCance and Widdowson	157 items grouped into 11 categories	Not specified	Fallaize et al., 2014; Forster et al., 2014; Celis-Morales et al. 2016 [39–41]
FoodBook24	Ireland	Dietary intake	Adults	Web-based	Food record, FFQ, food choice	Irish National Adult Nutrition Survey food composition database	751	Average of 15 min [10]	Timon et al. 2017a and 2017b [42,43]
FoodNow	Australian	Diet and physical activity	Young adults	Smartphone; wearable armband for energy expenditure	Food record based on images, text or voice; dietitian enters data	2011–2013 Australian Food and Nutrient Database	5740	Not specified	Pendergast et al. 2017 [44]
GraFFS Graphical Food Frequency System	US	Dietary intake	Adults	Web-based	FFQ	NDSR and USDA's FNDDS	156	Not specified	Kristal et al. 2014 [45]
INTAKE24 Self-Completed Recall and Analysis of Nutrition [2]	UK	Dietary Intake	Adults and children from 11 years	Web-based	24-h recall based on AMPM	McCance and Widdowson	2800	Average of 13 min with flat interface	Foster et al. 2014; Bradley et al. 2016; Simpson et al. 2017 [46–48]
Microsoft SenseCam	Ireland, UK, Australia, others	Dietary intake, activity	Tested in athletes and different adult groups	Wearable	Imaging system to enhance 24-h recall interviews	WinDiets, but not integrated into tool	WinDiets has food databases from many countries	Not specified	O'Loughlin et al. 2013; Gemming et al. 2013 and 2015 [49–51]
myfood24	UK	Dietary Intake	Young Adults, Adults, Elderly	Web-based	24-h recall based on AMPM or food record	UK food composition database (branded and generic foods)	45,000	Average of 19 min (+/−7 min)	Carter et al. 2015; Albar et al. 2016 [52,53]
NINA-DISH New Interactive Nutrition Assistant	India: specifically New Delhi, Mumbai and Trivandrum	Dietary intake	Adults (35–69)	PC	Interviewer-assisted 24-h recalls, diet history, mealtime and food-preparer questionnaire	Indian FCT [8] augmented with data from UK, FNDDS, Singapore and Malaysia	1000	Not specified	Daniel et al. 2014 [54]
NANA Novel Assessment of Nutrition and Ageing	UK and USA	Dietary intake, activity, cognitive function	Elderly	Touch-screen computer with audio-recording	Food record based on images and voice; dietitian enters data	Windiets, but not integrated into tool	1200	Not specified	Astell et al. 2014; Timon et al. 2015 [55,56]

Table 1. *Cont.*

Device Name	Country	Main Purpose of the Tool	Target Audience	Main Platform for Tool	Method of Data Collection and Entry	Food Composition Source	Approximate Number of Items	Time to Complete	References
NuDAM Nutricam Dietary Assessment Method	Australia	Dietary intake	Adults	Smartphone/camera	Food record based on images and voice notes; dietitian enters data	FoodWorks 5.1, but not integrated into tool	13,000	Not specified	Rollo et al. 2011 and 2015 [57,58]
NutriNet Santé	France	Diet and physical activity	Adults	Web-based	24-h recall or food record based on AMPM	French food composition table	2600	Average of 31 ± 29 min; Median 25 min	Touvier et al. 2011 [59]
Oxford WebQ	UK	Diet and physical activity	Adults	Web-based, PC	24-h dietary checklist	McCance and Widdowson	200 items in 21 food groups	Average of 14 min; Median 12.5 min	Liu et al. 2011; Galante et al. 2016 [60,61]
R24W Self-Administered Web-based recall	French Canadian	Dietary intake	Adults and adolescents from 16 years	Web-based	24-h recalls based on AMPM	Canadian Nutrient file 2010 and Foods Commonly Consumed in Canada	4000	27.6% reported < 20 min, 31% 20–30 min, 24.1% 30–45 min, 7% 45–60 min	Jacques et al. 2016; Lafrenière et al. 2017 [62,63]
RFPM Remote Food Photography Method	USA	Dietary intake	All ages	Smartphone/ camera/ bar-code reader	Remote imaging system; semi-automated food identification	USDA's FNDDS, but not integrated into tool	8500	Not specified	Martin et al. 2012 and 2014; Nicklas et al. 2017 [64–66]
SNAP Synchronized Nutrition and Activity Program	UK	Diet and physical activity	Children	Web-based	Food records collected during eight time-points daily	UK food consumption database	49 (40 foods, nine beverages)	<25 min	Moore et al. 2013 [67]
SNAPA Synchronized Nutrition and Activity Program for Adults	UK	Diet and physical activity	Adults	Web-based	Food records collected during 4 time periods each day	UK food consumption database	120 (102 foods and 18 beverages)	Not specified	Hillier, et al. 2012 [68]
TADA Technology Assisted Dietary Assessment;mobile food record	USA	Dietary intake	Adults and children from 3 years	Smartphone App	Food record based on before and after images of foods and beverages; system calculates energy and nutrients	USDA's Food and Nutrient Database for Dietary Studies (FNDDS)	8500	Not specified	Daugherty et al. 2012; Ahmad et al. 2016; Boushey et al. 2015 and 2017 [69–72]

Table 1. *Cont.*

Device Name	Country	Main Purpose of the Tool	Target Audience	Main Platform for Tool	Method of Data Collection and Entry	Food Composition Source	Approximate Number of Items	Time to Complete	References
TECH Tool for Energy Balance in Children	Sweden	Diet and physical activity	2–5 years old	Smartphone App	Food record: Parents take images and provide short descriptions; dietitian enters data	Swedish Food Database, but this was not integrated into tool	Not reported	Not specified	Delisle et al. 2015; Henriksson et al. 2015; Delisle Nyström et al. 2016 [73–75]
VNP Virtual Nutri Plus	Brazil	Dietary intake	Patients undergoing gastric bypass surgery	PC	24-h recall or food record; dietitian enters data	Brazilian Food Chemical Composition Table	1711	Not specified	da Silva et al. 2014a and 2014b [76,77]
WebCAAFE Food Intake and Physical Activity of School-children	Brazil	Diet and physical activity	Children 6–12 years	Web-based	24-h recall	None; evaluates foods and beverages only	32 items in each of 6 eating events per day	Not specified	Davies et al., 2015; Kupek et al. 2016 [78,79]
WebDASC Web-Based Dietary Assessment Software for Children	Denmark	Dietary Intake	Children	Web-based	24-h recall	Danish National Survey of Diet and Physical Activity (DANSDA)	1300	Average of 15 min (after first day)	Biltoft-Jensen et al. 2012 and 2013; Andersen et al. 2015 [80–82]
Web-FFQ	Quebec, Canada	Dietary intake	Adults	Web-based	FFQ	Nutrition Data System for Research and the Canadian Nutrient File	136	45 min	Labonte et al. 2012 [83]
WebFR Web-based Food Record	Norway	Dietary Intake	Children	Web-based	24-h recall	Norwegian National Survey database (NORKOST)	550	Not specified	Medin et al. 2015, 2016, and 2017 [84–86]
Zambia Tablet-based 24h recall Tool	Zambia	Dietary intake	Children	Tablet	Interviewer-assisted 24-h recall	HarvestPlus and Zambia food comp tables	Not specified	Not specified	Caswell et al. 2015 [87]
Tools for Consumer Use ($n = 10$)									
Diabetics Diary, paired with Pebble smartwatch	Norway	Diabetes management Diet and physical activity	Adults	Android Smartphone plus Smart watch	Carbohydrate food log	None	Not reported	Not specified	Arsand et al. 2015 [88]

Table 1. *Cont.*

Device Name	Country	Main Purpose of the Tool	Target Audience	Main Platform for Tool	Method of Data Collection and Entry	Food Composition Source	Approximate Number of Items	Time to Complete	References
DietCam	USA	Dietary intake	All ages for obesity prevention	Smartphone App	Food record from images; system calculates energy	USDA National Nutrient Database for Standard Reference	8500	Not specified	Kong and Tan, 2011 and 2012; Kong thesis, 2012; Kong et al. 2015 [16,89–91]
DIMA Dietary Intake Monitoring Application	USA	Medical management and diet	Hemodialysis patients	PDA	Food record with touch, voice, bar-code scanner	Database was created from existing nutrient database and UPC codes	Not specified	Not specified	Connelly et al. 2012; Welch et al. 2013 [92,93]
EVIDENT II app	Spain	Adherence to a Med Diet and log for step counter	Adults	Smartphone App	FFQ and Med Diet checklist	Spanish FFQ	137s	Not specified	Recio-Rodriguez et al. 2014 and 2016 [94,95]
FoodLog for dietary data collection as part of DialBetics program	Japan, Korea	Diabetes management and diet, physical activity	Adults	Smartphone App	Food record from images, text; system calculates energy, macro-nutrients	National food database: Dietary Reference Intakes, Japan (2010)	2191	Average of 35 min	Aizawa K. et al. 2014; Waki et al. 2012 and 2015 [96–98]
GoCARB	EU	Diabetes management and diet	Adults	Smartphone App	Food record based on meal images for carbohydrate intake estimates	USDA Nutrient Database for Standard Reference	5000	~1 min per image	Rhyner et al. 2016; Bally et al. 2017 [99,100]
IDQC Internet Based Diet and Lifestyle Questionnaire	China	Diet and physical activity	Adults and adolescents	Web-based	FFQ	Food Nutrition Calculator (Beijing)	135	30–40 min	Du et al. 2015 [101]
My Meal Mate (MMM)	UK	Diet, activity and body weight	Adults—Weight loss or maintenance	Smartphone App	Food record	UK Composition of Foods	40,000	Average of 22 min	Carter et al. 2013a, 2013b, 2013c [102–104]
Snap-n-Eat mobile application	USA	Dietary intake	Adults	Smartphone App	Food record from images; system calculates energy and nutrients	not reported	not reported	Not specified	Zhang et al. 2015 [15]
SuperTracker [3]	USA	Diet and physical activity	All ages	Web-based	Food records, diet recall	USDA's FNDDS	8500	Not specified	Post et al. 2012; Tsompanakis, 2015 [105,106]

[1] Now called Globo-Diet. [2] Formerly called SCRAN24, which was a PC-based platform. [3] Formerly called MyPyramid Tracker; discontinued as of July 2018. (Long et al., 2012 was for MyPyramid Tracker, the predecessor of SuperTracker). [4] FNDDS is the US Department of Agriculture's Food and Nutrient Database for Dietary Studies; FNDDS provides the nutrient values for foods and beverages reported in the dietary intake component of the National Health and Nutrition Examination Survey (NHANES). [5] Australian Food, Supplement and Nutrient Database (AUSNUT). [6] European Prospective Investigation into Cancer and Nutrition (EPIC). [7] WISP (Tinuviel Software) is nutritional analysis software for the UK and Ireland (http://www.tinuvielsoftware.co.uk/wisp4.htm). [8] Food Composition Table (FCT). [9] Based on personal communication with M. Gurinović, University of Belgrade, Serbia. [10] Based on personal communication with S. Pigat, CremeGlobal, Dublin, Ireland.

3.3. Comparison of Tools Used for Research versus Those for Consumer Use

Figure 2 compares the 25 attributes evaluated according to use in research (n = 33) vs. those intended for consumer use (n = 10). The greatest differences in summary ratings occurred in the category 'Data entry,' where half of consumer access tools made use of photos for data entry, compared to less than a third of tools used in research or surveillance. In addition, information on health characteristics and physical activity were more prevalent in tools for consumer access (60%, six tools), compared to only 36% (12 tools) and 33% (11 tools) of research or surveillance tools, respectively. The possibility to set personal goals was identified as a unique feature in tools for consumer access. In the category 'Food description' differences were observed for the automated identification of foods, in particular, with 50% (5) of consumer access tools offering this functionality, compared to only 9% (3) of research and surveillance tools. With regard to the category 'Customization,' research and surveillance tools had proportionally more options to add missing items, customize recipes, and report use of dietary supplements. Research and surveillance tools more frequently provide detailed information on dietary intake in the 'Output' category, particularly for the features 'Food groups', 'Time of intake', and 'Meal name', but fewer of the research tools contained integrated food databases, so lacked the ability to estimate energy or nutrient intakes automatically. In contrast, all consumer access tools we identified generated automatic reports, but only 39% (13 tools) of research and surveillance tools did so. In the 'Usability and validity' category, a higher proportion in tools used for research or surveillance (91%; 30 tools) have conducted validation studies, compared to 30% (n = 3) consumer access tools.

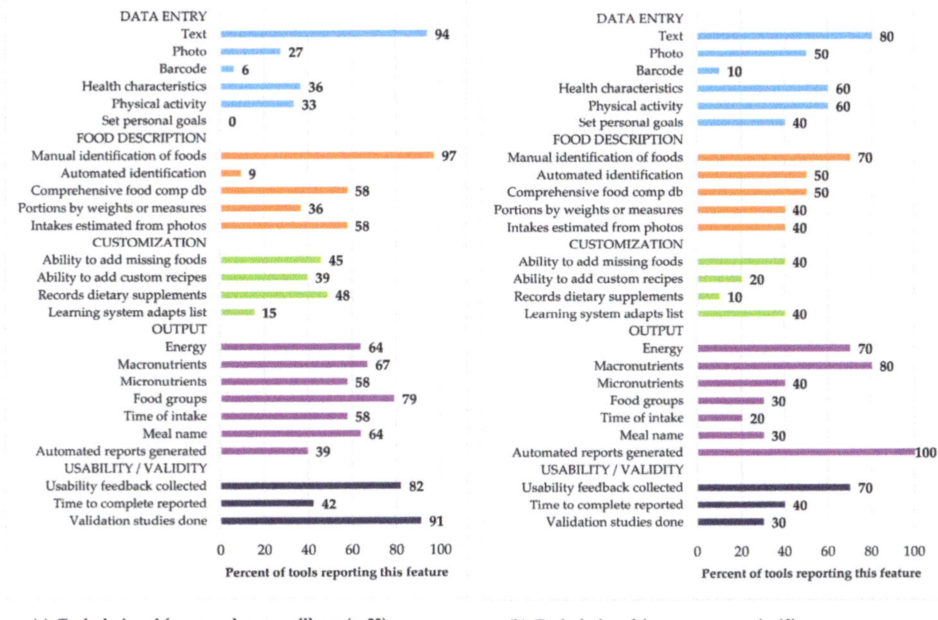

Figure 2. Summary rating of the features from the dietary assessment tools designed for research or surveillance (**A**) and for consumer use (**B**).

3.4. Validation Studies

Some type of validation study was published for 33 of the 43 new technology-based tools evaluated in this review. Seven of the tools compared energy intakes with Total Energy Expenditure (TEE) from doubly-labelled water (DLW) or accelerometers (Supplementary Table S3). In the DLW studies, energy intake estimates from the new technology tools were significantly lower than the TEE in studies using the Microsoft SenseCam [51], NuDAM [58], RFPM [64], and TADA [72] (differences ranging from 750 to 3745 kJ/day (179–895 kcal), whereas a different study with RFPM was within 636 kJ (152 kcal) [64], and two studies in children using the TECH tool were within 220–330 kJ (53–79 kcal) of TEE [74,75]. Two validation studies compared new technologies with TEE estimated from accelerometer data, showing that WebFR underestimated intakes by an average of 1840 kJ (440 kcal) in children 8–14 years [86], and FoodNow underestimated energy by 826 kJ (200 kcal) in young adults [44].

Standard methods of dietary assessment, including 24-h recalls, food records or weighed portions, were used in validation studies for 19 of the new technology tools (representing 25 individual validation comparisons), and in these studies, there was much closer agreement (Figure 3). In fact, 18 of the 25 individual comparisons were within 250 kJ (about 60 kcal) of each other when comparing the tool and the traditional method. Six of the comparisons were within 400–900 kJ (95–215 kcal), and only one had a difference greater than 1000 kJ (240 kcal) compared to the traditional method. The tools NuDAM, RFPM, and TECH were assessed using both DLW and compared with standard method of dietary assessment, e.g., 24-h recall, weighed foods, or a diary.

Macronutrient intake comparisons were available for 22 of the 25 validation comparisons (Supplemental Table S2). Protein intake estimates were the closest between traditional and new technology tools, with 18 comparisons within 5 g of the reference (average 2.1 g). Three of the protein comparisons were between 5–9 g different from the reference and only one was >10 g. Agreement was less accurate for fat with 13 comparisons within 5 g of the reference, four between 5–9 g, and three comparisons >10 g difference. Carbohydrate estimates showed the widest variation, with eight comparisons within 5 g, six between 5–9 g, and eight >10 g.

The remaining 10 tools were validated using some other method. For example, the portions estimated from the eButton were compared to actual volumes measured by seed displacement [31]. WebDASC [80] and Epic-Soft [35] were compared with biomarker data. SNAP [67], SNAPA [68], and WebCAAFE [78] compared reported foods and beverages against observations. Results from a study using DES were compared with results from a national survey in the same population [28], and DAP compared FFQs with 24-h recalls collected using the same tool [26]. VNP was evaluated by comparing the coding of 24-h recalls with DietPro 5i, a different dietary intake coding software [77]. Lastly, GoCARB was compared with self-estimates of carbohydrates and carbohydrate intakes calculated from weighed food samples [100].

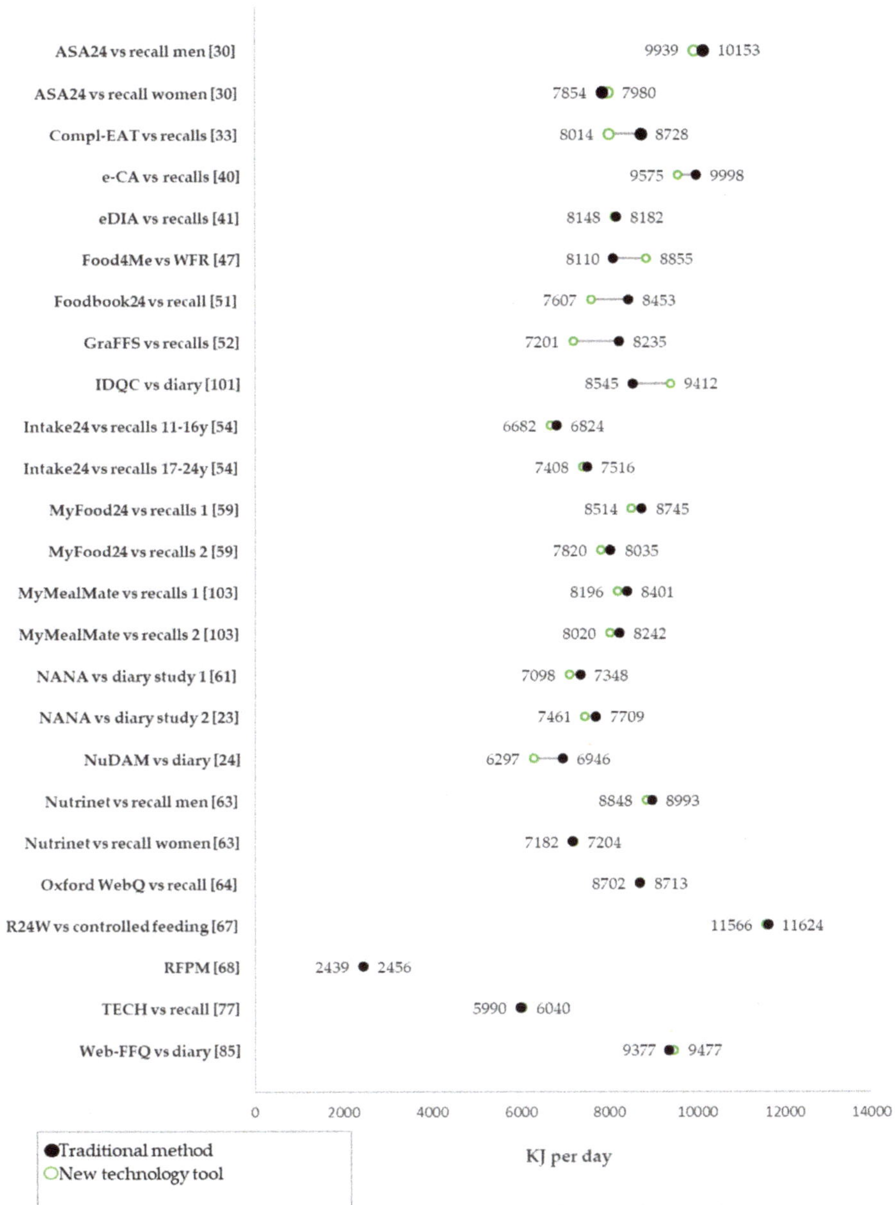

Figure 3. Energy estimations from digital tools vs. traditional methods of dietary intake assessment.

4. Discussion

The ILSI Dietary Intake and Exposure Task Force initiated this evaluation because of the rapid emergence of technologies available for dietary intake assessment coupled with concerns about a lack of quality standards for their development. Our review was anchored by a previous review and evaluation of innovative technologies for nutritional epidemiology, which assessed publications from

1995–2011 [12]. Since that review was published, personal digital assistants (PDAs) are no longer on the market, tape recorders are no longer needed for voice recording of dietary data, and cameras are integrated into smartphones, making digital image capture of foods much simpler. We focused our review on tools identified from publications in 2011–2017, and only four tools (ASA24, Nutrinet Santé, Oxford WebQ, and RFPM) were included in both this and Illner's previous assessment.

There is growing pressure in the area of dietary intake assessment to improve the accuracy and reduce costs of data collection and processing [107]. New technology tools use a variety of inputs for dietary assessment, including text, voice, digital images, and bar-code scanners. Various techniques have been implemented to enhance accuracy of portion size reporting, including automatic estimation from digital images and visualization of different sized portions on a plate, as well as the ability to report quantities by weight or common household measures. Many new technology tools, especially those designed for consumer use, provide automated feedback on the individual's nutrient intakes or dietary patterns, which may improve dietary outcomes and promote behavior change [108,109]. People are now accustomed to using technology tools, like smartphones, tablets, and computers, as part of their daily life, and usability studies indicate that many prefer technology tools for dietary intake assessment over traditional methods [20,42,71,104].

In the meantime, a number of other reviews have been published. While we deliberately chose to focus on new technologies identified from the published academic literature, other reviews have used app-store downloads as the criteria for selection [110,111]. Few of the app-store tools (4%) provided details about the sources of food composition data, and only 14% provided micronutrient estimates [111]. In contrast, half of the consumer apps in our review used a comprehensive food composition table, and 40% reported on micronutrient intakes. It is clear from the two approaches that apps with publications are more likely to include comprehensive food composition databases and, therefore, can report on a full complement of nutrients, compared to the most popular consumer apps.

Image capture can increase accuracy and ease reporting of foods and beverages consumed [14,50]. Images were used for data capture in 13 of the tools we evaluated (nine research and four consumer-based tools), either by automatically coding food intakes, passively capturing food intake throughout the day, as a method of recording intakes, or as a memory prompt. Digital images were also used to facilitate portion size estimation in over half of the tools we evaluated (53%; 19 research tools and four consumer tools). Uses ranged from automatic estimation of food volumes from digital images [14–16,30] to visualization of different portion sizes to improve portion-size reporting [20,26,40,42,45,46,52,59,81,83,84,87].

Validation studies were much more commonly reported for dietary assessment tools in the research setting than for those targeted to consumers. There was very good agreement between many of these tools and their reference method, a conclusion also drawn in another previous review [112]. We found that 30 (out of 33) of the research tools and three (out of 10) of the consumer tools conducted a validation study, although the majority of comparison methods used in validation were other self-report measures and, therefore, subject to similar errors. In 72% of the comparisons (18 of 25), the new technology was within 60 kcal of the traditional method of dietary intake assessment. The differences were somewhat wider for studies with DLW, but these differences could have been due to a variety of reasons, including estimate errors from coders manually coding from images, or because eating occasions were not reported. As pointed out previously, new technologies will not resolve all of the challenges of dietary assessment [1], but it is also reassuring that, in many cases, results are close to traditional self-reported or memory-based recalls, which have received recent criticism for their accuracy [113]. Objective biomarkers of dietary intake, such as DLW, urinary nitrogen or potassium, or plasma vitamin levels, are still lacking for most tools [1,112], and care must be taken to interpret validation by other means, such as direct data entry into two comparable tools, or comparison of results from a national survey, for example.

The technology tools we reviewed were developed for use across a wide variety of geographies, including both higher and lower-income countries. Two tools in particular were developed to facilitate interviewer-assisted data collection in lower-middle income countries [54,87], illustrating the utility of technology tools, even in countries where individuals may not have access to a smartphone, personal computer, or other technology for personal monitoring. However, technology tools will have limited use for self-monitoring in countries where smartphone or personal computers are not widely available.

Our evaluation has several notable strengths. As new tools and technologies are constantly changing, we have updated previous reviews with new tools identified from the literature and added a comprehensive evaluation of features. We have also compared features of research-based tools with those designed primarily for consumers, highlighting differences across all of our assessment topics. However, we must also acknowledge limitations in our review. The review was completed in September 2017, and it is possible that more recent publications have not been included in our review. For example, an in-depth validation of myfood24 including biomarkers was published after our assessment was completed [114], and others may have been missed as well. Results from validation studies comparing new technology tools to TEE or with daily energy estimations from conventional methods studies were presented, but further assessment of the quality of those studies was not assessed. We also focused on dietary assessment, per se, and have not included other new methods for assessing intakes, such as bite counters, tools that measure chews and swallows, or wrist-tracking devices that measure feeding [115]. It is also possible that there could be other attributes that are also important, but were not covered in this review, such as ethical issues or privacy when digital devices include other identifying features [111]. The impact of new technologies on cost will depend on the specific study design and the tools used, and this was rarely addressed in any of the publications. Finally, the search strategy may have missed some apps if key word searches did not pick up the studies, however, we used several search engines and different key word searches to minimize this risk.

The quality of tools cannot be assessed if this information is considered to be proprietary, or is omitted from scientific publications. Our assessment included 25 attributes in the areas of data entry, food description, customization, output, and usability/validity. Based on our evaluation of new technology-based tools for dietary intake assessment we have developed best practice guidelines for reporting on new technologies for dietary assessment (Figure 4), which add to existing STROBE-nut guidelines (referring to Strengthening the Reporting of Observational Studies in Epidemiology, for nutrition epidemiology) [13].

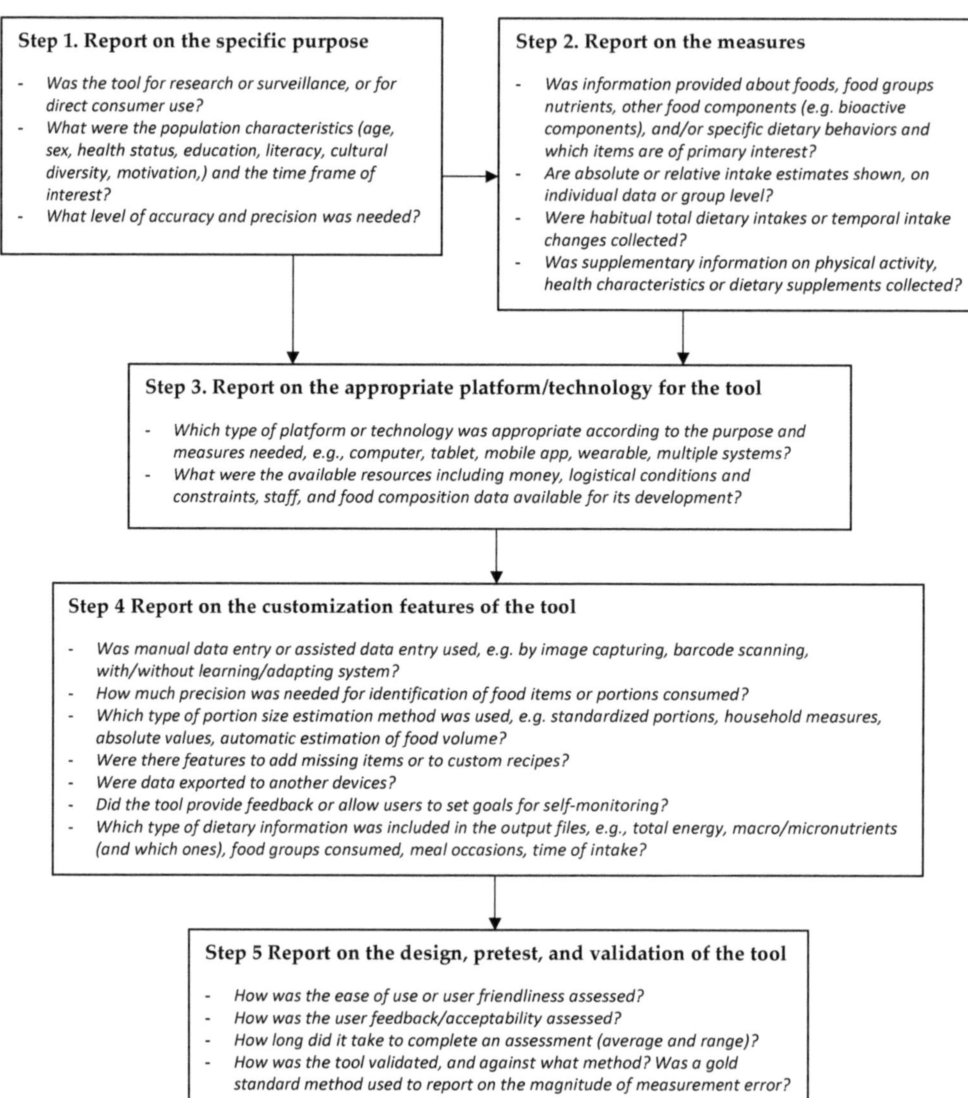

Figure 4. Best practice guidelines for reporting new technologies for dietary assessment.

4.1. Best Practice Guidance for Reporting on New Technologies for Dietary Assessment

4.1.1. Step 1: Report on the Specific Purpose

The goal of the first step is to report on the purpose of the dietary assessment tool. This depends primarily on the context in which the tool has been used. Issues related to the assessment of dietary data needed for research or surveillance purposes may differ from those needed for consumer access settings. Report what you aimed to measure, in what population, and over what period of time. In addition, the definition of the specific purpose of a tool implies the identification of the population characteristics, e.g., age, sex, health status, educational level. It is also important to inform about what level of accuracy and precision was needed. For example, if a higher level of precision was required, it may be necessary to administer repeated measurements.

4.1.2. Step 2: Report on the Measures

The goal of the second step is to inform about the main measurement features of a given tool. These relate to the information about individual foods (e.g., generic foods or branded products), food coding systems (e.g., LanguaL) and/or standardized food classification and description system (e.g., Food EX2), nutrients or other food components reported, the number of food items contained in the tool (e.g., comprehensive food lists or specific foods rich in a specific nutrient or bioactive component), and features of the response section (e.g., whether eating occasions or time is recorded, if food groups are included). We recommend reporting not only the source of the food composition data, but also to report the number of nutrients it contains, the coverage, and how the tool has been customized to best meet the population-specific needs.

We recommend defining the context for the tool and report if (1) a targeted tool provides relative or absolute intake estimates and (2) whether you are estimating daily intakes, habitual total dietary intakes, or temporal intake changes. It is also important to report if a given tool queries about supplementary information on physical activity, health characteristics, or use of dietary supplements.

4.1.3. Step 3: Report on the Appropriate Platform/Technology for the Tool

The goal of the third step is to report on the selection of the appropriate platform or technology of the tool. The choice for or against a specific technology type (e.g., tablet, computer, smartphone, wearable devices or multiple systems) depends strongly on the purpose and measures' needs. Factors affecting this step are the available resources (i.e., financial, logistical and staff conditions). The level of technology-literacy of the targeted population needs to be taken in careful consideration. Other considerations include data sharing needs (i.e., how the participant/user data are exported and to whom), data storage structure and access, statistical analysis, programming language used for scripting the tool, how the individual will access the tool, and how their privacy will be maintained.

4.1.4. Step 4: Report on the Customization Features of the Tool

The fourth step is to report on the customization of the features of the tool. These features, such as the type of data entry (e.g., text, voice, image capture, barcode scanning), list of foods and source of food composition data, type of portion size estimation (e.g., standardized portions, household measures or weights, pictures, automatic food volume estimations), need to be evaluated with respect to their adequacy to capture the purpose- and measures-specific needs of a given tool. One evaluation approach is to specifically assess the completeness and adequacy of the foods/recipes included in the tool in order to evaluate whether or how missing items could be added or recipes could be customized. Furthermore, the relevance of the dietary information in the output needs to be evaluated, as well as the need to provide feedback or to set goals for self-monitoring. Overall, details of the features that can be customized should be reported, and if there are any, an individual customization protocol should be developed and followed.

4.1.5. Step 5: Report on the Design, Pretest, and Validation of the Tool

The fifth step is to report on the design and pre-test of the tool. User interface, tool format, wording and order of questions (as appropriate) as well as browsers and battery storage are likely to affect design features of the platform and technology tool. When studying culturally diverse populations, these aspects become even more important (e.g., does the wording have the same meaning in different languages). As with any dietary assessment method, technology tools should be pre-tested, ideally on a sample of subjects similar to those who will ultimately be studied. The purpose is to report on the ease of use or user friendliness and to identify questions that are poorly understood, ambiguous, or evoke implausible or other undesirable responses. We recommend reporting on the completion time and acceptability for implementing the tool. In addition, report how the tool has been validated and against what standard.

5. Conclusions

Dietary assessment methods that utilize technology provide rapid feedback to users and offer potential cost-savings for researchers. Dietary assessment methods that utilize new technology may be more appealing and engaging than paper-based methods, particularly for children and young adults. Online methods can be deployed to large groups with minimal resources compared with methods requiring in-field researchers. In addition, many of these tools provide rapid feedback to participants that may improve compliance with diet plans or research. Connectivity enables rapid and remote interaction with the participants and nutrition professionals or researchers. Combination methods may enhance the accuracy of dietary intake reporting (such as the use of digital images to improve memory and portion size estimates).

Many of the new technology tools assessed here showed close agreement to traditional methods of dietary intake, but gaps are wider when compared to more objective measures, like TEE from doubly-labelled water, though studies using this method are limited in number. We encourage developers and researchers to publish details about their dietary assessment tools, including those designed for consumer use, and call on the research community to evaluate the validity of the tools they create and use. While we were able to extract details about many features from the tools evaluated, it often required more than one publication to find the necessary information. We recommend that descriptions of tool development and features be clearly written in publications, covering all aspects of tool development, including data entry, food description, customization features, output characteristics, sources of food composition data, and results of usability and validity studies, following the guidance provided here.

Supplementary Materials: The following are available online at http://www.mdpi.com/2072-6643/11/1/55/s1, Table S1. Search strategies used to identify technology-based tools for dietary intake assessment; Table S2. Details of data extraction and evaluation criteria used to evaluate new technology tools for dietary intake assessment; Table S3. Validation methods for total energy and macronutrients for the technology-based tools used in dietary intake assessment.

Author Contributions: All authors were involved in the design and discussions about study approach and evaluation criteria to be used. A.-K.I. completed the literature searches using PubMed, PLOS, BioMED, and Science Direct. A.L.E. completed the literature searches from OVID. All authors were involved in initial data extraction and coding. A.L.E. duplicated data extraction for all tools. All authors contributed to interpretation of the results. J.E.C., C.P., A.K.-I., and A.L.E. wrote the first manuscript draft, and all reviewed and contributed to the final manuscript.

Funding: This research was funded by the International Life Sciences Institute (ILSI) Europe's Dietary Intake and Exposure Task Force. C.P.'s time in this project is supported by the National Institute for Health Research Collaboration for Applied Health Research (CLAHRC). The views expressed are those of the author(s) and not necessarily those of the NHS, the NIHR, or the Department of Health and Social Care.

Acknowledgments: The authors, members of the ILSI Expert Group involved in this research, wish to thank Jonathon M. Taberner (Leeds University) for his assistance in the initial literature search, setting up the spreadsheet for the data extraction, and contributing to the scoring of the first 24 tools. We also wish to thank Mariah L. Tabar, international student at UniLaSalle in 2016, for her additional literature search. In addition, we thank the ILSI Europe Dietary Intake and Exposure Task Force, as well as task force manager Nevena Hristozova, for their support.

Conflicts of Interest: The funders had no role in the design of the study; in the collection, analyses, or interpretation of data; in the writing of the manuscript; or in the decision to publish the results. A.L.E. is the Chair of the ILSI Dietary Intake and Exposure Task Force and the expert group responsible for this research. C.P. and A.K.I. declare no conflicts of interest. M.J.G. leads the Food4Me Consortium responsible for the development, research pipeline, and validation of Food4Me. M.A.G. was involved as the nutritional researcher proving professional advice to IT in the creation and testing of the Diet Assess and Plan (DAP). J.H.M.V. was involved in the development and validation of Compl-Eat. J.E.C. is a director of a University of Leeds spin-out private company, Dietary Assessment Ltd., supporting the development of myfood24. She also led the project that developed MyMealMate.

References

1. Cade, J.E. Measuring diet in the 21st century: Use of new technologies. *Proc. Nutr. Soc.* **2017**, *76*, 276–282. [CrossRef] [PubMed]
2. Cade, J.E.; Warthon-Medina, M.; Albar, S.; Alwan, N.A.; Ness, A.; Roe, M.; Wark, P.A.; Greathead, K.; Burley, V.J.; Finglas, P.; et al. DIET@NET: Best practice guidelines for dietary assessment in health research. *BMC Med.* **2017**, *15*, 202. [CrossRef] [PubMed]
3. Freedman, L.S.; Potischman, N.; Kipnis, V.; Midthune, D.; Schatzkin, A.; Thompson, F.E.; Troiano, R.P.; Prentice, R.; Patterson, R.; Carroll, R.; et al. A comparison of two dietary instruments for evaluating the fat-breast cancer relationship. *Int. J. Epidemiol.* **2006**, *35*, 1011–1021. [CrossRef] [PubMed]
4. Freedman, L.S.; Schatzkin, A.; Midthune, D.; Kipnis, V. Dealing with dietary measurement error in nutritional cohort studies. *J. Natl. Cancer Inst.* **2011**, *103*, 1086–1092. [CrossRef] [PubMed]
5. Blanton, C.A.; Moshfegh, A.J.; Baer, D.J.; Kretsch, M.J. The USDA Automated Multiple-Pass Method accurately estimates group total energy and nutrient intake. *J. Nutr.* **2006**, *136*, 2594–2599. [CrossRef] [PubMed]
6. Grandjean, A.C. Dietary intake data collection: Challenges and limitations. *Nutr. Rev.* **2012**, *70* (Suppl. 2), S101–S104. [CrossRef] [PubMed]
7. Burrows, T.L.; Rollo, M.E.; Williams, R.; Wood, L.G.; Garg, M.L.; Jensen, M.; Collins, C.E. A systematic review of technology-based dietary intake assessment validation studies that include carotenoid biomarkers. *Nutrients* **2017**, *9*, 140. [CrossRef] [PubMed]
8. Karlsen, M.C.; Lichtenstein, A.H.; Economos, C.D.; Folta, S.C.; Rogers, G.; Jacques, P.F.; Livingston, K.A.; Rancano, K.M.; McKeown, N.M. Web-based recruitment and survey methodology to maximize response rates from followers of popular diets: The Adhering to Dietary Approaches for Personal Taste (ADAPT) feasibility survey. *Curr. Dev. Nutr.* **2018**, *2*, nzy012. [CrossRef] [PubMed]
9. Carter, M.C.; Hancock, N.; Albar, S.A.; Brown, H.; Greenwood, D.C.; Hardie, L.J.; Frost, G.S.; Wark, P.A.; Cade, J.E. Development of a new branded uk food composition database for an online dietary assessment tool. *Nutrients* **2016**, *8*, 480. [CrossRef]
10. Subar, A.F.; Crafts, J.; Zimmerman, T.P.; Wilson, M.; Mittl, B.; Islam, N.G.; McNutt, S.; Potischman, N.; Buday, R.; Hull, S.G.; et al. Assessment of the accuracy of portion size reports using computer-based food photographs aids in the development of an automated self-administered 24-h recall. *J. Am. Diet Assoc.* **2010**, *110*, 55–64. [CrossRef]
11. Ngo, J.; Engelen, A.; Molag, M.; Roesle, J.; Garcia-Segovia, P.; Serra-Majem, L. A review of the use of information and communication technologies for dietary assessment. *Br. J. Nutr.* **2009**, *101* (Suppl. 2), S102–S112. [CrossRef]
12. Illner, A.K.; Freisling, H.; Boeing, H.; Huybrechts, I.; Crispim, S.P.; Slimani, N. Review and evaluation of innovative technologies for measuring diet in nutritional epidemiology. *Int. J. Epidemiol.* **2012**, *41*, 1187–1203. [CrossRef] [PubMed]
13. Lachat, C.; Hawwash, D.; Ocke, M.C.; Berg, C.; Forsum, E.; Hornell, A.; Larsson, C.; Sonestedt, E.; Wirfalt, E.; Akesson, A.; et al. Strengthening the reporting of observational studies in epidemiology-nutritional epidemiology (STROBE-nut): An extension of the strobe statement. *PLoS Med.* **2016**, *13*, e1002036. [CrossRef] [PubMed]
14. Boushey, C.J.; Spoden, M.; Zhu, F.M.; Delp, E.J.; Kerr, D.A. New mobile methods for dietary assessment: Review of image-assisted and image-based dietary assessment methods. *Proc. Nutr. Soc.* **2017**, *76*, 283–294. [CrossRef] [PubMed]
15. Zhang, W.; Yu, Q.; Siddiquie, B.; Divakaran, A.; Sawhney, H. "Snap-n-eat": Food recognition and nutrition estimation on a smartphone. *J. Diabetes Sci. Technol.* **2015**, *9*, 525–533. [CrossRef] [PubMed]
16. Kong, F.; He, H.; Raynor, H.A.; Tan, J. Dietcam: Multi-view regular shape food recognition with a camera phone. *Pervasive Mob. Comput.* **2015**, *19*, 108–121. [CrossRef]
17. Baranowski, T.; Islam, N.; Baranowski, J.; Martin, S.; Beltran, A.; Dadabhoy, H.; Adame, S.H.; Watson, K.B.; Thompson, D.; Cullen, K.W.; et al. Comparison of a web-based versus traditional diet recall among children. *J. Acad. Nutr. Diet.* **2012**, *112*, 527–532. [CrossRef]

18. Baranowski, T.; Islam, N.; Douglass, D.; Dadabhoy, H.; Beltran, A.; Baranowski, J.; Thompson, D.; Cullen, K.W.; Subar, A.F. Food Intake Recording Software System, version 4 (FIRSSt4): A self-completed 24-h dietary recall for children. *J. Hum. Nutr. Diet.* **2014**, *27* (Suppl. 1), 66–71. [CrossRef]

19. Kirkpatrick, S.I.; Subar, A.F.; Douglass, D.; Zimmerman, T.P.; Thompson, F.E.; Kahle, L.L.; George, S.M.; Dodd, K.W.; Potischman, N. Performance of the automated self-administered 24-h recall relative to a measure of true intakes and to an interviewer-administered 24-h recall. *Am. J. Clin. Nutr.* **2014**, *100*, 233–240. [CrossRef]

20. Thompson, F.E.; Dixit-Joshi, S.; Potischman, N.; Dodd, K.W.; Kirkpatrick, S.I.; Kushi, L.H.; Alexander, G.L.; Coleman, L.A.; Zimmerman, T.P.; Sundaram, M.E.; et al. Comparison of interviewer-administered and automated self-administered 24-h dietary recalls in 3 diverse integrated health systems. *Am. J. Epidemiol.* **2015**, *181*, 970–978. [CrossRef]

21. Kerr, D.A.; Harray, A.J.; Pollard, C.M.; Dhaliwal, S.S.; Delp, E.J.; Howat, P.A.; Pickering, M.R.; Ahmad, Z.; Meng, X.; Pratt, I.S.; et al. The connecting health and technology study: A 6-month randomized controlled trial to improve nutrition behaviours using a mobile food record and text messaging support in young adults. *Int. J. Behav. Nutr. Phys. Act.* **2016**, *13*, 52. [CrossRef] [PubMed]

22. Kerr, D.A.; Pollard, C.M.; Howat, P.; Delp, E.J.; Pickering, M.; Kerr, K.R.; Dhaliwal, S.S.; Pratt, I.S.; Wright, J.; Boushey, C.J. Connecting Health and Technology (CHAT): Protocol of a randomized controlled trial to improve nutrition behaviours using mobile devices and tailored text messaging in young adults. *BMC Public Health* **2012**, *12*, 477. [CrossRef] [PubMed]

23. Pollard, C.M.; Howat, P.A.; Pratt, I.S.; Boushey, C.J.; Delp, E.J.; Kerr, D.A. Preferred tone of nutrition text messages for young adults: Focus group testing. *JMIR mHealth uHealth* **2016**, *4*, e1. [CrossRef] [PubMed]

24. Meijboom, S.; van Houts-Streppel, M.T.; Perenboom, C.; Siebelink, E.; van de Wiel, A.M.; Geelen, A.; Feskens, E.J.M.; de Vries, J.H.M. Evaluation of dietary intake assessed by the Dutch self-administered web-based dietary 24-h recall tool (Compl-Eat) against interviewer-administered telephone-based 24-h recalls. *J. Nutr. Sci.* **2017**, *6*, e49. [CrossRef] [PubMed]

25. Gurinovic, M.; Milesevic, J.; Kadvan, A.; Djekic-Ivankovic, M.; Debeljak-Martacic, J.; Takic, M.; Nikolic, M.; Rankovic, S.; Finglas, P.; Glibetic, M. Establishment and advances in the online Serbian food and recipe data base harmonized with EuroFIR standards. *Food Chem.* **2016**, *193*, 30–38. [CrossRef] [PubMed]

26. Gurinovic, M.; Milesevic, J.; Kadvan, A.; Nikolic, M.; Zekovic, M.; Djekic-Ivankovic, M.; Dupouy, E.; Finglas, P.; Glibetic, M. Development, features and application of Diet Assess & Plan (DAP) software in supporting public health nutrition research in central eastern european countries (CEEC). *Food Chem.* **2018**, *238*, 186–194. [PubMed]

27. Zekovic, M.; Djekic-Ivankovic, M.; Nikolic, M.; Gurinovic, M.; Krajnovic, D.; Glibetic, M. Validity of the food frequency questionnaire assessing the folate intake in women of reproductive age living in a country without food fortification: Application of the method of triads. *Nutrients* **2017**, *9*, 128. [CrossRef]

28. Jung, H.J.; Lee, S.E.; Kim, D.; Noh, H.; Song, S.; Kang, M.; Song, Y.J.; Paik, H.Y. Improvement in the technological feasibility of a web-based dietary survey system in local settings. *Asia Pac. J. Clin. Nutr.* **2015**, *24*, 308–315.

29. Sun, M.; Fernstrom, J.D.; Jia, W.; Hackworth, S.A.; Yao, N.; Li, Y.; Li, C.; Fernstrom, M.H.; Sclabassi, R.J. A wearable electronic system for objective dietary assessment. *J. Am. Diet. Assoc.* **2010**, *110*, 45–47. [CrossRef]

30. Sun, M.; Burke, L.E.; Mao, Z.H.; Chen, Y.; Chen, H.C.; Bai, Y.; Li, Y.; Li, C.; Jia, W. Ebutton: A wearable computer for health monitoring and personal assistance. *Proc. Des. Autom. Conf.* **2014**, *2014*, 1–6. [CrossRef]

31. Jia, W.; Chen, H.C.; Yue, Y.; Li, Z.; Fernstrom, J.; Bai, Y.; Li, C.; Sun, M. Accuracy of food portion size estimation from digital pictures acquired by a chest-worn camera. *Public Health Nutr.* **2014**, *17*, 1671–1681. [CrossRef] [PubMed]

32. Bucher Della Torre, S.; Carrard, I.; Farina, E.; Danuser, B.; Kruseman, M. Development and evaluation of e-ca, an electronic mobile-based food record. *Nutrients* **2017**, *9*, 76. [CrossRef] [PubMed]

33. Rangan, A.M.; O'Connor, S.; Giannelli, V.; Yap, M.L.; Tang, L.M.; Roy, R.; Louie, J.C.; Hebden, L.; Kay, J.; Allman-Farinelli, M. Electronic Dietary Intake Assessment (e-DIA): Comparison of a mobile phone digital entry app for dietary data collection with 24-h dietary recalls. *JMIR mHealth uHealth* **2015**, *3*, e98. [CrossRef] [PubMed]

34. Rangan, A.M.; Tieleman, L.; Louie, J.C.; Tang, L.M.; Hebden, L.; Roy, R.; Kay, J.; Allman-Farinelli, M. Electronic Dietary Intake Assessment (e-DIA): Relative validity of a mobile phone application to measure intake of food groups. *Br. J. Nutr.* **2016**, *115*, 2219–2226. [CrossRef] [PubMed]

35. de Boer, E.J.; Slimani, N.; van't Veer, P.; Boeing, H.; Feinberg, M.; Leclercq, C.; Trolle, E.; Amiano, P.; Andersen, L.F.; Freisling, H.; et al. Rationale and methods of the European food consumption validation (EFCOVAL) project. *Eur. J. Clin. Nutr.* **2011**, *65* (Suppl. 1), S1–S4. [CrossRef]

36. Huybrechts, I.; Geelen, A.; de Vries, J.H.; Casagrande, C.; Nicolas, G.; De Keyzer, W.; Lillegaard, I.T.; Ruprich, J.; Lafay, L.; Wilson-van den Hooven, E.C.; et al. Respondents' evaluation of the 24-h dietary recall method (EPIC-Soft) in the EFCOVAL project. *Eur. J. Clin. Nutr.* **2011**, *65* (Suppl. 1), S29–S37. [CrossRef]

37. Freisling, H.; Ocke, M.C.; Casagrande, C.; Nicolas, G.; Crispim, S.P.; Niekerk, M.; van der Laan, J.; de Boer, E.; Vandevijvere, S.; de Maeyer, M.; et al. Comparison of two food record-based dietary assessment methods for a pan-European food consumption survey among infants, toddlers, and children using data quality indicators. *Eur. J. Nutr.* **2015**, *54*, 437–445. [CrossRef]

38. Park, M.K.; Freisling, H.; Huseinovic, E.; Winkvist, A.; Huybrechts, I.; Crispim, S.P.; de Vries, J.H.; Geelen, A.; Niekerk, M.; van Rossum, C.; et al. Comparison of meal patterns across five European countries using standardized 24-h recall (GloboDiet) data from the EFCOVAL project. *Eur. J. Nutr.* **2017**, *57*, 1045–1057. [CrossRef]

39. Fallaize, R.; Forster, H.; Macready, A.L.; Walsh, M.C.; Mathers, J.C.; Brennan, L.; Gibney, E.R.; Gibney, M.J.; Lovegrove, J.A. Online dietary intake estimation: Reproducibility and validity of the Food4Me food frequency questionnaire against a 4-day weighed food record. *J Med Internet Res* **2014**, *16*, e190. [CrossRef]

40. Forster, H.; Fallaize, R.; Gallagher, C.; O'Donovan, C.B.; Woolhead, C.; Walsh, M.C.; Macready, A.L.; Lovegrove, J.A.; Mathers, J.C.; Gibney, M.J.; et al. Online dietary intake estimation: The Food4Me food frequency questionnaire. *J. Med. Internet Res.* **2014**, *16*, e150. [CrossRef]

41. Celis-Morales, C.; Livingstone, K.M.; Marsaux, C.F.; Macready, A.L.; Fallaize, R.; O'Donovan, C.B.; Woolhead, C.; Forster, H.; Walsh, M.C.; Navas-Carretero, S.; et al. Effect of personalized nutrition on health-related behaviour change: Evidence from the Food4me European randomized controlled trial. *Int. J. Epidemiol.* **2017**, *46*, 578–588. [CrossRef] [PubMed]

42. Timon, C.M.; Blain, R.J.; McNulty, B.; Kehoe, L.; Evans, K.; Walton, J.; Flynn, A.; Gibney, E.R. The development, validation, and user evaluation of Foodbook24: A web-based dietary assessment tool developed for the Irish adult population. *J. Med. Internet Res.* **2017**, *19*, e158. [CrossRef] [PubMed]

43. Timon, C.M.; Evans, K.; Kehoe, L.; Blain, R.J.; Flynn, A.; Gibney, E.R.; Walton, J. Comparison of a web-based 24-h dietary recall tool (Foodbook24) to an interviewer-led 24-h dietary recall. *Nutrients* **2017**, *9*, 425. [CrossRef] [PubMed]

44. Pendergast, F.J.; Ridgers, N.D.; Worsley, A.; McNaughton, S.A. Evaluation of a smartphone food diary application using objectively measured energy expenditure. *Int. J. Behav. Nutr. Phys. Act.* **2017**, *14*, 30. [CrossRef] [PubMed]

45. Kristal, A.R.; Kolar, A.S.; Fisher, J.L.; Plascak, J.J.; Stumbo, P.J.; Weiss, R.; Paskett, E.D. Evaluation of Web-based, self-administered, graphical food frequency questionnaire. *J. Acad. Nutr. Diet.* **2014**, *114*, 613–621. [CrossRef] [PubMed]

46. Foster, E.; Hawkins, A.; Delve, J.; Adamson, A.J. Reducing the cost of dietary assessment: Self-completed recall and analysis of nutrition for use with children (SCRAN24). *J. Hum. Nutr. Diet.* **2014**, *27* (Suppl. 1), 26–35. [CrossRef]

47. Bradley, J.; Simpson, E.; Poliakov, I.; Matthews, J.N.; Olivier, P.; Adamson, A.J.; Foster, E. Comparison of INTAKE24 (an online 24-h dietary recall tool) with interviewer-led 24-h recall in 11-24 year-old. *Nutrients* **2016**, *8*, 358. [CrossRef]

48. Simpson, E.; Bradley, J.; Poliakov, I.; Jackson, D.; Olivier, P.; Adamson, A.J.; Foster, E. Iterative development of an online dietary recall tool: INTAKE24. *Nutrients* **2017**, *9*, 118. [CrossRef]

49. O'Loughlin, G.; Cullen, S.J.; McGoldrick, A.; O'Connor, S.; Blain, R.; O'Malley, S.; Warrington, G.D. Using a wearable camera to increase the accuracy of dietary analysis. *Am. J. Prev. Med.* **2013**, *44*, 297–301. [CrossRef]

50. Gemming, L.; Doherty, A.; Kelly, P.; Utter, J.; Ni Mhurchu, C. Feasibility of a sensecam-assisted 24-h recall to reduce under-reporting of energy intake. *Eur. J. Clin. Nutr.* **2013**, *67*, 1095–1099. [CrossRef]

51. Gemming, L.; Rush, E.; Maddison, R.; Doherty, A.; Gant, N.; Utter, J.; Ni Mhurchu, C. Wearable cameras can reduce dietary under-reporting: Doubly labelled water validation of a camera-assisted 24 h recall. *Br. J. Nutr.* **2015**, *113*, 284–291. [CrossRef] [PubMed]

52. Carter, M.C.; Albar, S.A.; Morris, M.A.; Mulla, U.Z.; Hancock, N.; Evans, C.E.; Alwan, N.A.; Greenwood, D.C.; Hardie, L.J.; Frost, G.S.; et al. Development of a UK online 24-h dietary assessment tool: Myfood24. *Nutrients* **2015**, *7*, 4016–4032. [CrossRef] [PubMed]

53. Albar, S.A.; Alwan, N.A.; Evans, C.E.; Greenwood, D.C.; Cade, J.E. Agreement between an online dietary assessment tool (Myfood24) and an interviewer-administered 24-h dietary recall in british adolescents aged 11-18 years. *Br. J. Nutr.* **2016**, *115*, 1678–1686. [CrossRef] [PubMed]

54. Daniel, C.R.; Kapur, K.; McAdams, M.J.; Dixit-Joshi, S.; Devasenapathy, N.; Shetty, H.; Hariharan, S.; George, P.S.; Mathew, A.; Sinha, R. Development of a field-friendly automated dietary assessment tool and nutrient database for India. *Br. J. Nutr.* **2014**, *111*, 160–171. [CrossRef] [PubMed]

55. Astell, A.J.; Hwang, F.; Brown, L.J.; Timon, C.; Maclean, L.M.; Smith, T.; Adlam, T.; Khadra, H.; Williams, E.A. Validation of the NANA (Novel Assessment of Nutrition and Ageing) touch screen system for use at home by older adults. *Exp. Gerontol.* **2014**, *60*, 100–107. [CrossRef] [PubMed]

56. Timon, C.M.; Astell, A.J.; Hwang, F.; Adlam, T.D.; Smith, T.; Maclean, L.; Spurr, D.; Forster, S.E.; Williams, E.A. The validation of a computer-based food record for older adults: The Novel Assessment of Nutrition and Ageing (NANA) method. *Br. J. Nutr.* **2015**, *113*, 654–664. [CrossRef] [PubMed]

57. Rollo, M.E.; Ash, S.; Lyons-Wall, P.; Russell, A. Trial of a mobile phone method for recording dietary intake in adults with type 2 diabetes: Evaluation and implications for future applications. *J. Telemed. Telecare* **2011**, *17*, 318–323. [CrossRef] [PubMed]

58. Rollo, M.E.; Ash, S.; Lyons-Wall, P.; Russell, A.W. Evaluation of a mobile phone image-based dietary assessment method in adults with type 2 diabetes. *Nutrients* **2015**, *7*, 4897–4910. [CrossRef] [PubMed]

59. Touvier, M.; Kesse-Guyot, E.; Mejean, C.; Pollet, C.; Malon, A.; Castetbon, K.; Hercberg, S. Comparison between an interactive Web-based self-administered 24 h dietary record and an interview by a dietitian for large-scale epidemiological studies. *Br. J. Nutr.* **2011**, *105*, 1055–1064. [CrossRef]

60. Liu, B.; Young, H.; Crowe, F.L.; Benson, V.S.; Spencer, E.A.; Key, T.J.; Appleby, P.N.; Beral, V. Development and evaluation of the Oxford WebQ, a low-cost, web-based method for assessment of previous 24 h dietary intakes in large-scale prospective studies. *Public Health Nutr.* **2011**, *14*, 1998–2005. [CrossRef]

61. Galante, J.; Adamska, L.; Young, A.; Young, H.; Littlejohns, T.J.; Gallacher, J.; Allen, N. The acceptability of repeat internet-based hybrid diet assessment of previous 24-h dietary intake: Administration of the Oxford WebQ in UK Biobank. *Br. J. Nutr.* **2016**, *115*, 681–686. [CrossRef] [PubMed]

62. Jacques, S.; Lemieux, S.; Lamarche, B.; Laramee, C.; Corneau, L.; Lapointe, A.; Tessier-Grenier, M.; Robitaille, J. Development of a Web-based 24-h dietary recall for a French-Canadian population. *Nutrients* **2016**, *8*, 724. [CrossRef] [PubMed]

63. Lafrenière, J.; Benoît Lamarche, B.; Catherine Laramée, C.; Julie Robitaille, J.; Lemieux, S. Validation of a newly automated webbased 24-h dietary recall using fully controlled feeding studies. *BMC Nutr.* **2017**, *3*, 34. [CrossRef]

64. Martin, C.K.; Correa, J.B.; Han, H.; Allen, H.R.; Rood, J.C.; Champagne, C.M.; Gunturk, B.K.; Bray, G.A. Validity of the Remote Food Photography Method (RFPM) for estimating energy and nutrient intake in near real-time. *Obesity* **2012**, *20*, 891–899. [CrossRef] [PubMed]

65. Martin, C.K.; Nicklas, T.; Gunturk, B.; Correa, J.B.; Allen, H.R.; Champagne, C. Measuring food intake with digital photography. *J. Hum. Nutr. Diet.* **2014**, *27* (Suppl. 1), 72–81. [CrossRef] [PubMed]

66. Nicklas, T.; Saab, R.; Islam, N.G.; Wong, W.; Butte, N.; Schulin, R.; Liu, Y.; Apolzan, J.W.; Myers, C.A.; Martin, C.K. Validity of the Remote Food Photography Method against doubly labeled water among minority preschoolers. *Obesity* **2017**, *25*, 1633–1638. [CrossRef] [PubMed]

67. Moore, H.J.; Hillier, F.C.; Batterham, A.M.; Ells, L.J.; Summerbell, C.D. Technology-based dietary assessment: Development of the Synchronised Nutrition and Activity Program (SNAP). *J. Hum. Nutr. Diet.* **2014**, *27* (Suppl. 1), 36–42. [CrossRef]

68. Hillier, F.C.; Batterham, A.M.; Crooks, S.; Moore, H.J.; Summerbell, C.D. The development and evaluation of a novel internet-based computer program to assess previous-day dietary and physical activity behaviours in adults: The Synchronised Nutrition and Activity Program for adults (SNAPA). *Br. J. Nutr.* **2012**, *107*, 1221–1231. [CrossRef]

69. Daugherty, B.L.; Schap, T.E.; Ettienne-Gittens, R.; Zhu, F.M.; Bosch, M.; Delp, E.J.; Ebert, D.S.; Kerr, D.A.; Boushey, C.J. Novel technologies for assessing dietary intake: Evaluating the usability of a mobile telephone food record among adults and adolescents. *J. Med. Internet Res.* **2012**, *14*, e58. [CrossRef]

70. Ahmad, Z.; Kerr, D.A.; Bosch, M.; Boushey, C.J.; Delp, E.J.; Khanna, N.; Zhu, F. A mobile food record for integrated dietary assessment. *MADiMa16 (2016)* **2016**, *2016*, 53–62.

71. Boushey, C.J.; Harray, A.J.; Kerr, D.A.; Schap, T.E.; Paterson, S.; Aflague, T.; Bosch Ruiz, M.; Ahmad, Z.; Delp, E.J. How willing are adolescents to record their dietary intake? The mobile food record. *JMIR mHealth uHealth* **2015**, *3*, e47. [CrossRef] [PubMed]

72. Boushey, C.J.; Spoden, M.; Delp, E.J.; Zhu, F.; Bosch, M.; Ahmad, Z.; Shvetsov, Y.B.; DeLany, J.P.; Kerr, D.A. Reported energy intake accuracy compared to doubly labeled water and usability of the mobile food record among community dwelling adults. *Nutrients* **2017**, *9*, 312. [CrossRef] [PubMed]

73. Delisle, C.; Sandin, S.; Forsum, E.; Henriksson, H.; Trolle-Lagerros, Y.; Larsson, C.; Maddison, R.; Ortega, F.B.; Ruiz, J.R.; Silfvernagel, K.; et al. A Web- and mobile phone-based intervention to prevent obesity in 4-year-olds (MINISTOP): A population-based randomized controlled trial. *BMC Public Health* **2015**, *15*, 95. [CrossRef] [PubMed]

74. Henriksson, H.; Bonn, S.E.; Bergstrom, A.; Balter, K.; Balter, O.; Delisle, C.; Forsum, E.; Lof, M. A new mobile phone-based tool for assessing energy and certain food intakes in young children: A validation study. *JMIR mHealth uHealth* **2015**, *3*, e38. [CrossRef] [PubMed]

75. Delisle Nystrom, C.; Forsum, E.; Henriksson, H.; Trolle-Lagerros, Y.; Larsson, C.; Maddison, R.; Timpka, T.; Lof, M. A mobile phone based method to assess energy and food intake in young children: A validation study against the doubly labelled water method and 24 h dietary recalls. *Nutrients* **2016**, *8*, 50. [CrossRef] [PubMed]

76. da Silva, M.M.; Sala, P.C.; Torrinhas, R.S.; Waitzberg, D.L. Efficiency of the 24-h food recall instrument for assessing nutrient intake before and after Roux-en-Y gastric bypass. *Nutr. Hosp.* **2014**, *30*, 1240–1247. [PubMed]

77. da Silva, M.M.; Sala, P.C.; Cardinelli, C.S.; Torrinhas, R.S.; Waitzberg, D.L. Comparison of Virtual Nutri Plus® and Dietpro 5i® software systems for the assessment of nutrient intake before and after Roux-en-Y gastric bypass. *Clinics (Sao Paulo)* **2014**, *69*, 714–722. [CrossRef]

78. Davies, V.F.; Kupek, E.; de Assis, M.A.; Natal, S.; Di Pietro, P.F.; Baranowski, T. Validation of a web-based questionnaire to assess the dietary intake of brazilian children aged 7–10 years. *J. Hum. Nutr. Diet.* **2015**, *28* (Suppl. 1), 93–102. [CrossRef]

79. Kupek, E.; de Assis, M.A. The use of multiple imputation method for the validation of 24-h food recalls by part-time observation of dietary intake in school. *Br. J. Nutr.* **2016**, *116*, 904–912. [CrossRef]

80. Biltoft-Jensen, A.; Bysted, A.; Trolle, E.; Christensen, T.; Knuthsen, P.; Damsgaard, C.T.; Andersen, L.F.; Brockhoff, P.; Tetens, I. Evaluation of web-based dietary assessment software for children: Comparing reported fruit, juice and vegetable intakes with plasma carotenoid concentration and school lunch observations. *Br. J. Nutr.* **2013**, *110*, 186–195. [CrossRef]

81. Biltoft-Jensen, A.; Trolle, E.; Christensen, T.; Islam, N.; Andersen, L.F.; Egenfeldt-Nielsen, S.; Tetens, I. Webdasc: A Web-based dietary assessment software for 8-11-year-old danish children. *J. Hum. Nutr. Diet.* **2012**, *27* (Suppl. 1), 43–53. [CrossRef]

82. Andersen, R.; Biltoft-Jensen, A.; Christensen, T.; Andersen, E.W.; Ege, M.; Thorsen, A.V.; Knudsen, V.K.; Damsgaard, C.T.; Sorensen, L.B.; Petersen, R.A.; et al. What do Danish children eat, and does the diet meet the recommendations? Baseline data from the Opus School Meal Study. *J. Nutr. Sci.* **2015**, *4*, e29. [CrossRef] [PubMed]

83. Labonte, M.E.; Cyr, A.; Baril-Gravel, L.; Royer, M.M.; Lamarche, B. Validity and reproducibility of a Web-based, self-administered food frequency questionnaire. *Eur. J. Clin. Nutr.* **2012**, *66*, 166–173. [CrossRef] [PubMed]

84. Medin, A.C.; Astrup, H.; Kasin, B.M.; Andersen, L.F. Evaluation of a Web-based food record for children using direct unobtrusive lunch observations: A validation study. *J. Med. Internet Res.* **2015**, *17*, e273. [CrossRef] [PubMed]

85. Medin, A.C.; Carlsen, M.H.; Andersen, L.F. Associations between reported intakes of carotenoid-rich foods and concentrations of carotenoids in plasma: A validation study of a web-based food recall for children and adolescents. *Public Health Nutr.* **2016**, *19*, 3265–3275. [CrossRef] [PubMed]

86. Medin, A.C.; Hansen, B.H.; Astrup, H.; Ekelund, U.; Frost Andersen, L. Validation of energy intake from a web-based food recall for children and adolescents. *PLoS ONE* **2017**, *12*, e0178921. [CrossRef] [PubMed]

87. Caswell, B.L.; Talegawkar, S.A.; Dyer, B.; Siamusantu, W.; Klemm, R.D.; Palmer, A.C. Assessing child nutrient intakes using a tablet-based 24-h recall tool in rural Zambia. *Food Nutr. Bull.* **2015**, *36*, 467–480. [CrossRef]

88. Arsand, E.; Muzny, M.; Bradway, M.; Muzik, J.; Hartvigsen, G. Performance of the first combined smartwatch and smartphone diabetes diary application study. *J. Diabetes Sci. Technol.* **2015**, *9*, 556–563. [CrossRef]

89. Kong, F.; Tan, J. Dietcam: Regular shape food recognition with a camera phone. In Proceedings of the 2011 International Conference on Body Sensor Networks, Dallas, TX, USA, 23–25 May 2011; pp. 127–132.

90. Kong, F. Automatic Food Intake Assessment Using Camera Phones. Master's Thesis, Michigan Technological University, Houghton, MI, USA, 2012.

91. Kong, F.; Tan, J. Dietcam: Sutomatic dietary assessment with mobile camera phones. *Pervasive Mob. Comput.* **2012**, *8*, 147–163. [CrossRef]

92. Connelly, K.; Siek, K.A.; Chaudry, B.; Jones, J.; Astroth, K.; Welch, J.L. An offline mobile nutrition monitoring intervention for varying-literacy patients receiving hemodialysis: A pilot study examining usage and usability. *J. Am. Med. Inform. Assoc.* **2012**, *19*, 705–712. [CrossRef]

93. Welch, J.L.; Astroth, K.S.; Perkins, S.M.; Johnson, C.S.; Connelly, K.; Siek, K.A.; Jones, J.; Scott, L.L. Using a mobile application to self-monitor diet and fluid intake among adults receiving hemodialysis. *Res. Nurs. Health* **2013**, *36*, 284–298. [CrossRef] [PubMed]

94. Recio-Rodriguez, J.I.; Martin-Cantera, C.; Gonzalez-Viejo, N.; Gomez-Arranz, A.; Arietaleanizbeascoa, M.S.; Schmolling-Guinovart, Y.; Maderuelo-Fernandez, J.A.; Perez-Arechaederra, D.; Rodriguez-Sanchez, E.; Gomez-Marcos, M.A.; et al. Effectiveness of a smartphone application for improving healthy lifestyles, a randomized clinical trial (EVIDENT II): Study protocol. *BMC Public Health* **2014**, *14*, 254. [CrossRef] [PubMed]

95. Recio-Rodriguez, J.I.; Agudo-Conde, C.; Martin-Cantera, C.; Gonzalez-Viejo, M.N.; Fernandez-Alonso, M.D.; Arietaleanizbeaskoa, M.S.; Schmolling-Guinovart, Y.; Maderuelo-Fernandez, J.A.; Rodriguez-Sanchez, E.; Gomez-Marcos, M.A.; et al. Short-term effectiveness of a mobile phone app for increasing physical activity and adherence to the Mediterranean diet in primary care: A randomized controlled trial (EVIDENT Ii study). *J. Med. Internet Res.* **2016**, *18*, e331. [CrossRef]

96. Aizawa, K.; Maeda, K.; Ogawa, M.; Sato, Y.; Kasamatsu, M.; Waki, K.; Takimoto, H. Comparative study of the routine daily usability of FoodLog: A smartphone-based food recording tool assisted by image retrieval. *J Diabetes Sci. Technol.* **2014**, *8*, 203–208. [CrossRef] [PubMed]

97. Waki, K.; Fujita, H.; Uchimura, Y.; Aramaki, E.; Omae, K.; Kadowaki, T.; Ohe, K. Dialbetics: Smartphone-based self-management for type 2 diabetes patients. *J. Diabetes Sci. Technol.* **2012**, *6*, 983–985. [CrossRef] [PubMed]

98. Waki, K.; Aizawa, K.; Kato, S.; Fujita, H.; Lee, H.; Kobayashi, H.; Ogawa, M.; Mouri, K.; Kadowaki, T.; Ohe, K. Dialbetics with a multimedia food recording tool, FoodLog: Smartphone-based self-management for type 2 diabetes. *J. Diabetes Sci. Technol.* **2015**, *9*, 534–540. [CrossRef] [PubMed]

99. Bally, L.; Dehais, J.; Nakas, C.T.; Anthimopoulos, M.; Laimer, M.; Rhyner, D.; Rosenberg, G.; Zueger, T.; Diem, P.; Mougiakakou, S.; et al. Carbohydrate estimation supported by the GoCARB system in individuals with type 1 diabetes: A randomized prospective pilot study. *Diabetes Care* **2017**, *40*, e6–e7. [CrossRef]

100. Rhyner, D.; Loher, H.; Dehais, J.; Anthimopoulos, M.; Shevchik, S.; Botwey, R.H.; Duke, D.; Stettler, C.; Diem, P.; Mougiakakou, S. Carbohydrate estimation by a mobile phone-based system versus self-estimations of individuals with type 1 diabetes mellitus: A comparative study. *J. Med. Internet Res.* **2016**, *18*, e101. [CrossRef]

101. Du, S.S.; Jiang, Y.S.; Chen, Y.; Li, Z.; Zhang, Y.F.; Sun, C.H.; Feng, R.N. Development and applicability of an internet-based diet and lifestyle questionnaire for college students in China: A cross-sectional study. *Medicine (Baltimore)* **2015**, *94*, e2130. [CrossRef]

102. Carter, M.C.; Burley, V.J.; Cade, J.E. Development of 'My Meal Mate'—A smartphone intervention for weight loss. *Nutr. Bull.* **2013**, *38*, 80–84. [CrossRef]

103. Carter, M.C.; Burley, V.J.; Nykjaer, C.; Cade, J.E. 'My Meal Mate' (MMM): Validation of the diet measures captured on a smartphone application to facilitate weight loss. *Br. J. Nutr.* **2013**, *109*, 539–546. [CrossRef] [PubMed]

104. Carter, M.C.; Burley, V.J.; Nykjaer, C.; Cade, J.E. Adherence to a smartphone application for weight loss compared to website and paper diary: Pilot randomized controlled trial. *J. Med. Internet Res.* **2013**, *15*, e32. [CrossRef] [PubMed]

105. Post, R.C.; Herrup, M.; Chang, S.; Leone, A. Getting plates in shape using Supertracker. *J. Acad. Nutr. Diet.* **2012**, *112*, 354–358. [CrossRef] [PubMed]

106. Tsompanakis, S. Supertracker. *J. Consum. Health Internet* **2015**, *19*, 132–142. [CrossRef]

107. Thompson, F.E.; Subar, A.F.; Loria, C.M.; Reedy, J.L.; Baranowski, T. Need for technological innovation in dietary assessment. *J. Am. Diet. Assoc.* **2010**, *110*, 48–51. [CrossRef] [PubMed]

108. Forster, H.; Walsh, M.C.; Gibney, M.J.; Brennan, L.; Gibney, E.R. Personalised nutrition: The role of new dietary assessment methods. *Proc. Nutr. Soc.* **2016**, *75*, 96–105. [CrossRef] [PubMed]

109. Afshin, A.; Babalola, D.; McLean, M.; Yu, Z.; Ma, W.; Chen, C.Y.; Arabi, M.; Mozaffarian, D. Information technology and lifestyle: A systematic evaluation of internet and mobile interventions for improving diet, physical activity, obesity, tobacco, and alcohol use. *J. Am. Heart Assoc.* **2016**, *5*, e003058. [CrossRef]

110. Franco, R.Z.; Fallaize, R.; Lovegrove, J.A.; Hwang, F. Popular nutrition-related mobile apps: A feature assessment. *JMIR mHealth uHealth* **2016**, *4*, e85. [CrossRef]

111. Maringer, M.; Van't Veer, P.; Klepacz, N.; Verain, M.C.D.; Normann, A.; Ekman, S.; Timotijevic, L.; Raats, M.M.; Geelen, A. User-documented food consumption data from publicly available apps: An analysis of opportunities and challenges for nutrition research. *Nutr. J.* **2018**, *17*, 59. [CrossRef]

112. Timon, C.M.; van den Barg, R.; Blain, R.J.; Kehoe, L.; Evans, K.; Walton, J.; Flynn, A.; Gibney, E.R. A review of the design and validation of Web- and computer-based 24-h dietary recall tools. *Nutr. Res. Rev.* **2016**, *29*, 268–280. [CrossRef]

113. Subar, A.F.; Freedman, L.S.; Tooze, J.A.; Kirkpatrick, S.I.; Boushey, C.; Neuhouser, M.L.; Thompson, F.E.; Potischman, N.; Guenther, P.M.; Tarasuk, V.; et al. Addressing current criticism regarding the value of self-report dietary data. *J. Nutr.* **2015**, *145*, 2639–2645. [CrossRef] [PubMed]

114. Wark, P.A.; Hardie, L.J.; Frost, G.S.; Alwan, N.A.; Carter, M.; Elliott, P.; Ford, H.E.; Hancock, N.; Morris, M.A.; Mulla, U.Z.; et al. Validity of an online 24-h recall tool (Myfood24) for dietary assessment in population studies: Comparison with biomarkers and standard interviews. *BMC Med.* **2018**, *16*, 136. [CrossRef] [PubMed]

115. Archundia Herrera, M.C.; Chan, C.B. Narrative review of new methods for assessing food and energy intake. *Nutrients* **2018**, *10*, 1064. [CrossRef] [PubMed]

Review

Narrative Review of New Methods for Assessing Food and Energy Intake

M. Carolina Archundia Herrera [1] and **Catherine B. Chan [1,2,***

1   Department of Agriculture, Food and Nutritional Science, Alberta Diabetes Institute, University of Alberta, 6-002 Li Ka Shing Centre for Health Innovation Research, Edmonton, AB T6G 2E1, Canada; archundi@ualberta.ca
2   Department of Physiology, Alberta Diabetes Institute, University of Alberta, 6-002 Li Ka Shing Centre for Health Innovation Research, Edmonton, AB T6G 2E1, Canada
*   Correspondence: cbchan@ualberta.ca; Tel.: +1-780-492-9939

Received: 15 June 2018; Accepted: 6 August 2018; Published: 10 August 2018

Abstract: Dietary self-report instruments are essential to nutritional analysis in dietetics practice and their use in research settings has facilitated numerous important discoveries related to nutrition, health and chronic diseases. An important example is obesity, for which measuring changes in energy intake is critical for assessing efficacy of dietary interventions. However, current methods, including counting calories, estimating portion size and using food labels to estimate human energy intake have considerable constraints; consequently, research on new methodologies/technologies has been encouraged to mitigate the present weaknesses. The use of technologies has prompted innovation in dietary analysis. In this review, the strengths and limitations of new approaches have been analyzed based on ease of use, practical limitations, and statistical evaluation of reliability and validity. Their utility is discussed through the lens of the 4Ms of Obesity Assessment and Management, which has been used to evaluate root causes of obesity and help select treatment options.

Keywords: dietary assessment; energy intake; validity; reliability

---

1. Introduction

On a global scale, life expectancy has increased steadily for the past 35 years; however, in association with the global rise of obesity, the number of deaths from most non-communicable causes like diabetes mellitus rose by 32.1%, increasing the burden on health systems [1]. During the past two decades, different Intensive Lifestyle Intervention programs have consistently shown that modest but clinically significant weight loss of 5% in individuals with overweight, obesity or diabetes can yield a variety of health, disease prevention and treatment benefits [2]. Prescription of a hypocaloric diet (500–750 calories less than baseline), increased physical activity (90–175 min/week) and long term behavior change, are common techniques use in Intensive Lifestyle Intervention [3–6], which also have been described previously as individual techniques for weight control [7–9].

Assessment of dietary and/or energy intake (EI) is crucial to understand the impact of clinical trials on the management of obesity and its comorbidities [2,10]. To date, food records (FR), food frequency questionnaires (FFQ), and 24-h recalls (24HR) are the most common methods used to assess dietary and EI during treatment and follow-up [11]. These self-reported data methodologies have provided valuable information to use as a base to develop public health policy, comprehend and identify consumption of different food groups, understand relationship with diseases and determine eating patterns associated with weight loss, information that until recently could not be obtained in any other way [10].

However, a major challenge of these methods is that they rely on self-reported data. Human memory is not 100% accurate in recalling past behavior, consequently these measurements do not directly or objectively measure dietary intake or EI and do not comply with the standards of scientific methodology [12,13]. One issue is that the actual process of doing food records can lead individuals to change their food behavior patterns and therefore, misreport information resulting in an inaccurate report of foods, nutrients and energy consumed [14]. Using data from The National Health and Nutrition Examination Survey (NHANES) 2003–2012, researchers analyzed the prevalence of under and over-reporting of EI, finding that in the US adult population ($\geq$20 years) 25.1% misreported EI [15], results consistent with European countries where prevalence of under-reporting ranges from 20% to a high of 45% [16–19] with a predominance of obese populations under-reporting. Part of the limitation of behavior modification presented in food records can be overcome though the use of 24HR, since they can be unannounced so that the diet is not changed; however, estimation of the usual diet is weakened by recall bias (food omission or forgetfulness, erroneous estimation of portion size) [20]. In addition to recall bias, these methods impart a substantial researcher/individual burden and high cost of administration [21]. These methodologies used for dietary assessment have been severely criticized to the point of calling the resultant data "pseudoscientific and inadmissible in scientific research", and what "constitutes the single greatest impediment to actual scientific progress in the fields of obesity and nutrition research" [12].

Thus, the accuracy of dietary assessments or modifications in dietary or EI is full of challenges and the development of new technologies to try to overcome current limitations has been encouraged [21]. The objective of this review is to present the strengths and weaknesses of innovative new tools or methodologies that could replace, improve or complement current self-report dietary assessment instruments.

2. Materials and Methods

2.1. Search for Innovative Food and EI Assessment Tools and Methodologies

Medline, CINAHL and PsychINFO were searched for English-language articles, using the following keywords separately or in combination: diet, diet records, dietary intake, energy intake, innovate *, meals, measurement, metabolism, method, models, new, nutrition assessment, optimiz *, recent, self-report, technolog *, test reliability, test validity, trend, validation studies. The search resulted in 337 articles (Figure 1). The output was then narrowed by imposing search criteria of "2012–October 2016" and "Adults". This search resulted in 73 articles. These article titles and abstracts were screened by one author (MCAH) to determine if they fulfilled the eligibility criteria. The articles included had to describe or validate a new method, or use new technology tools that could capture food or EI. The methodologies or tools were assessed to determine their benefits and limitations as well as their reliability or validity. Studies using text messaging or mobile phone applications that required manual introduction of information were not considered because using this type of technology imposes the same limitations as the current methods, requiring a self-reported measurement with a burdensome and impractical framework for the subject.

Of the 73 articles uncovered with this search, 17 were considered potentially eligible. These articles were cross-listed in PubMed for articles related to the topic, which identified 8 other articles. Reference lists of relevant articles were also hand-searched but no other relevant articles were found. For these 25 articles, review of the full text was used to identify those meeting the criteria ($n = 11$).

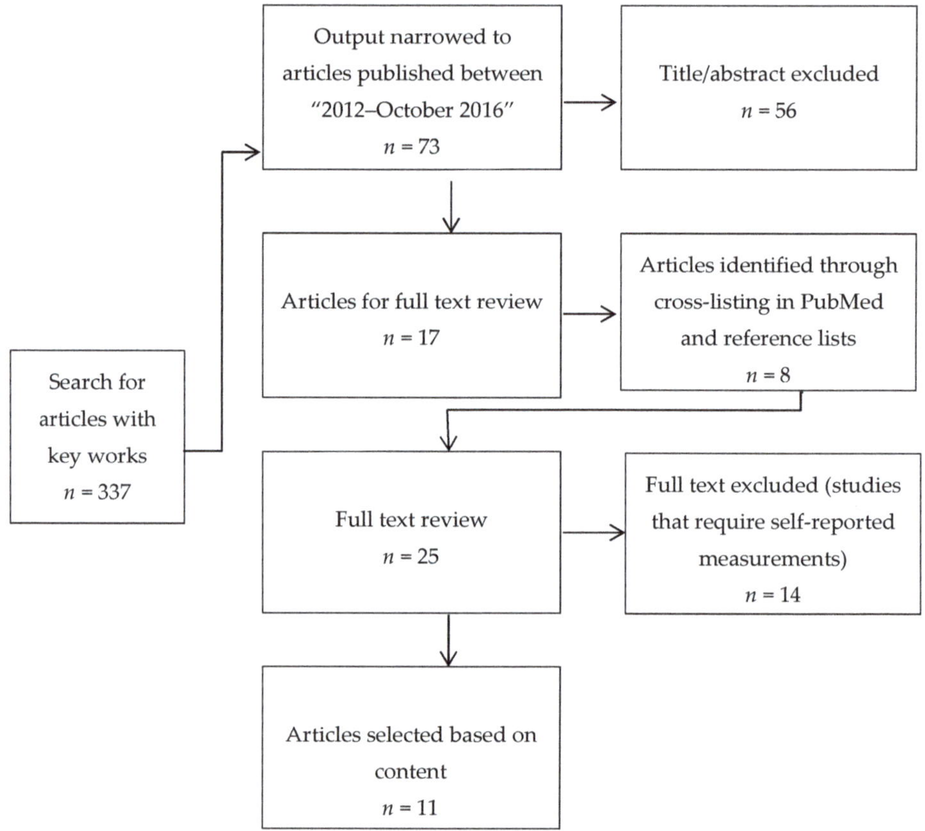

Figure 1. Flow diagram of the articles selection process and exclusion reasons.

2.2. Evaluation

A relative evaluation of the innovative technology tools and methodologies was carried out. They were assessed to ensure that the main weaknesses of present methodologies: recall biases, measurement discrepancies, lack of scientific rigor, were being acknowledged.

When developing tools to collect dietary information, specific statistical methods must be used to evaluate their reliability and validity in order to test the accuracy of the method and avoid bias [22]. Thus, utilization of these recommended statistical methods to assess the different tools and methodologies was noted when drafting this manuscript. *Reliability* refers to "the consistency of a measuring instrument" [22], in different situations; inter-rater, test-retest, inter-method and internal consistency. *Validity* refers to "how close the tool can measure the actual (true) value"; in this case, a measure of true EI when compared to the gold standard [22].

Benefits and limitations: When describing benefits and limitations of each tool/methodology, the focus was on the following criteria: Easy to administer—Referring to reducing participant burden. Current methods rely on information reported by the subjects recalling what they ate for the past week/month/year, or keeping a diary for various days. This decreases the quality of the reports, and the process itself can make subjects change their eating habits [23]. Easy to score—Observed and weighed-food records [24], doubly labelled water (DLW) [25], and FFQ [11], are some of the methods currently used to determine/estimate EI and/or eating behavior. These methods are expensive and time-consuming, making them less feasible to use and hard to score. New methodologies should minimize practitioners' or researchers' burden and expense. Capture change over more than one day—Because assessment of day-to-day variability in food intake is an important limitation of current methodologies. New methodology should overcome these limitations and be able to capture fluctuations in habitual energy and nutrient intake on free-living subjects.

3. Results

Table 1 summarizes the studies that were included in the review. Five described different types of monitors and sensors; five described camera-scan-sensor-based technologies; and one described a mathematical method. Details of the statistical methods used to assess validity and reliability are noted in Table 2.

3.1. Food/Energy Intake Monitoring Devices and Tools

Use of body sensors as a direct measurement of human eating behavior is quite recent. Body monitors and sensors have been developed with the hope of improving and facilitating measurement of daily food and EI [26,27].

3.1.1. Automated Wrist Motion Tracking

The Automated Wrist Motion Tracking, also called a "bite counter" is worn like a watch and automatically tracks wrist motion for monitoring eating in humans [28]. Reliability was tested in both controlled meal and semi-controlled settings. The sensitivity was >85% in both settings. Bites measured by the device were >80% detected compared with bites counted by direct observation. The equations used to measure sensitivity and performance are reported in Table 2 [28]. A third experiment in free-living situations was performed to examine the correlation between bites detected and EI, with $r = 0.6$. This experiment was only exploratory and was done to seek any possible relationships between these factors for further research [28].

Use of this device resulted in improved accuracy of measuring EI in free-living situations compared with 24HR and FFQ, which typically under-report EI in men by 16–20% and 31–36%, and in women by 16–20% and 34–38% respectively [29]. Participant burden was minimal because the user only needed to turn it on and off before eating, and thereafter, bites were registered automatically by the device; thus, researcher and administrative costs are ameliorated since no food weight or labour-intensive laboratory techniques are needed [28,30,31]. Forgetting to use the device, accuracy in different social settings and loss of data when both hands are used to eat are present limitations that the bite counter tool needs to address. Importantly this device's main benefit is its use as a food intake-monitoring and ingestive behavior tracking system in a real-world setting to improve users, researches and HCP understanding of food intake behaviors. Furthermore, lessen the burden of manual measurements; however, no input regarding the type or quality of the food consumed is tracked.

3.1.2. The Bite-Based Model of Kilocalorie Intake

The bite-counter described above was used for the development of a kilocalorie per bite equation (using bite counts, individual demographic and physical characteristics) that allowed EI to be estimated. The relationship obtained was estimated as kilocalories per bite = -0.128 age + 6.167 sex (female = 0) + 0.034 height + 0.035 weight − 12.012 WHR + 22.294; where WHR = waist-to-hip ratio [30]. The feasibility of using the formula was then systematically evaluated. Two trials were run using a train and test paradigm, in which the training group was used to develop the model and the test group was used to determine reliability of the regression model.

Table 1. Summary of New Methods for Assessing Food and Energy Intake.

Reference	Objective	Brief Description	Key Findings
[28]	Evaluate a new method of automated dietary intake monitoring.	The "Bite Counter" device was worn like a watch. Before eating, the user pressed a button to turn it on (and off afterwards). While operating, the device used a micro-electro-mechanical gyroscope to track wrist motion, automatically detecting when the user had taken a bite.	The method worked across a reasonably large number of subjects, and variety of foods, and there was modest correlation with EI on a per-meal level.
[30]	Evaluate accuracy of an individualized bite-based equation of kilocalorie intake compared to participant estimates of kilocalorie intake.	Subjects' real kilocalorie intake was compared to predicted kilocalories estimated by: (a) the bite-based equation of kilocalorie intake, (b) participants' kilocalorie estimate when provided with kilocalorie information of the foods eaten, (c) participants' kilocalorie estimate without kilocalorie information.	The bite-based equation measure of kilocalorie intake outperformed human estimates with and without menu kilocalorie information.
[31]	Evaluate the: Automatic Ingestion Monitor (AIM) for objective detection of food intake in free-living individuals.	The AIM integrated three sensor modalities and a pattern recognition method for subject-independent food intake recognition.	The AIM can detected food intake with an average accuracy of 89.8% suggesting that it can be used to monitor eating behavior in free-living individuals. AIM could be used as a behavioral modification tool.
[32]	Estimate EI using individualized models based on Counts of Chews and Swallows (CCS).	EI was estimated by the CCS mathematical model and compared to the weighed food records, diet diaries and photographic food records methods.	Mathematical models based on the CCS could be potentially used to estimate EI.
[33]	Present an intelligent food-intake monitoring system that can automatically detect eating activities	The multi-sensor monitor detected chewing activity via its integrated ear-microphone, consequently the camera was activated, snapshots for food detection were taken.	The high correlation rates reported (r not shown) suggested the usefulness of the proposed method to provide with an overall understanding of eating behavior characteristics (speed, type and amount of food consumed).
[34]	To compare mean EI of overweight and obese young adults assessed by a Digital Photography + Recall method (DP + R), to the mean total daily energy expenditure assessed by $TDEE_{DLW}$.	Two digital still photographs (90° and 45° angle) were taken by a digital camera approximately 30 inches above the tray. Notes were placed on the tray to identify types of beverages and standard measures were included to guide the assessment of portion size. The type and amounts of food and beverages consumed and results from recalls were entered into the Nutrition Data System for Research to quantification for EI. $TDEE_{DLW}$ was assessed in all participants to compare mean daily EI.	The mean EI estimated by DP + R and $TDEE_{DLW}$ was not significantly different (p = 0.42). On average, DP + R overestimated EI compared to $TDEE_{DLW}$ by 6.8 $\pm$ 28%.

Table 1. *Cont.*

Reference	Objective	Brief Description	Key Findings
[35]	To validate the Remote Food Photography Method (RFPM)	Developed for automating dietary assessment. Participants include a reference card placed next to the food plate as well as labels of not easily recognizable foods for the portion size estimation to take place. A barcode reader phone app and a voice message option are innovations included to facilitate identification of foods. Participant received feedback about their food intake behavior and recommendations to achieve weight goals. To maximize and promote usage of RFPM in free-living conditions, ecological momentary assessment (EMA) methods were adopted, which involves sending small reminders or prompts to the user via email or text message. EMA was tested by comparing two groups; the standard prompts (2 or 3 prompts a day send to their smartphones around meal time) versus customized prompts (3 to 4 personalized prompts, send at participants' specific meal time).	The RFPM and DLW did not differ significantly at estimating free-living EI (-152 ± 694 kcal/day, $p = 0.16$) nor did it differ when estimating energy and macronutrient intake.
[36]	To evaluate a mobile food recognition system which estimates calorie and nutritional components of food intake.	(1) User pointed the smartphone camera to the food (2) Drew bounding boxes to delimit food regions (3) Food item recognition started within the indicated bounding boxes. To recognize them more accurately each food item region is segmented by GrubCut. The recognition process results in a display of the top 5 food item candidates. The user selects the most accurate candidate and indicates the relative approximate volume of the food.	A 79.2% classification rate was achieved. The recognition processing time was only 0.065 s.
[37]	To present Snap-n-Eat, a mobile food recognition system.	The user took a photo of the plate. The system detects the salient regions corresponding to the food items. Hierarchical segmentation was performed to segment the images into regions. The system estimated the portion size of the food and uses it to determine the EI and nutritional content.	The Snap-n-Eat application achieved a 85% accuracy when detecting 15 different categories of food items. Snap-n-eat recognized foods presented on a plate and estimated their caloric EI and nutrition content automatically without any user intervention.

Table 1. *Cont.*

Reference	Objective	Brief Description	Key Findings
[38]	To assess the accuracy of the GoCARB prototype when used by individuals with type 1 diabetes and to compare it to their own performance in carbohydrate counting.	The user placed a reference card next to the dish and took two images using a mobile phone. A series of computer vision modules detected the plate and automatically segmented and recognized the different food items, while their 3D shape was reconstructed. The carbohydrate content was calculated by combining the volume of each food item with the nutritional information provided by the USDA Food and Nutrient Database.	GoCARB was more accurate at estimating carbohydrates content than individuals with type 1 diabetes. The mean absolute estimation error while using GoCARB was reduced by more than 50% than without using GoCARB.
[39]	To validate a mathematical method to measure long-term changes in free-living EI	DLW was used to assess Energy Expenditure (EE) at months 6, 12, 18, and 24. DXA and body weight measurements were taken twice at baseline, twice at month 6, and once at months 12, 18, and 24. Body weight measurements were taken at months 1, 3, 6, 9, 12, 18, and 24 in the CALERIE study. Then, they compared the ΔEI values calculated by using DLW/DXA with those obtained by using the mathematical model	The mean (95% CI) ΔEI values calculated by the model were within 40 kcal/day of the DLW/DXA method and were not significantly different throughout the 4 times segment ($p = 0.14$, $p = 0.34$, $p = 0.32$, $p = 0.11$). Most of the model-calculated ΔEI values were within 132 kcal/day of the DLW/DXA method.

Table 2. Summary of the Reliability and Validity of New Methods for Assessing Food and Energy Intake.

Reference	Name of Tool	What Is Measured	Reliability		Validity	
			Statistical Method Used	Result	Statistical Method Used	Result
[28]	Automated Wrist Motion Tracking	EI	Sensitivity (true detection rate) = (total true detection)/(total true detection + total undetected bites); Positive Predicted Value (PPV) = (total true detection)/(total true detection + total false detection); compared recorded bites with direct observation.	Control setting: Sensitivity = 94% PPV = 80% Semi-controlled setting: Sensitivity = 86% PPV = 81%	Pearson correlation of EI estimated by device vs. direct observation (r)	R = 0.6
[30]	The bite-based model of kilocalorie intake	EI	Pearson's correlation of device compared with direct observation; shrinkage value	R = 0.374 Shrinkage value (difference in R^2) = 0.014	Independent t test Paired sample t test	Mean estimation error kilocalorie information group: -185 ± 501 kcal; Mean estimation error no kilocalorie information group: -349 ± 748 kcal ($p < 0.05$); Best human-based estimation (kilocalorie information group) mean estimation error: -257 ± 790 kcal; Bite-based method (predicted formula) mean estimation error: 71 ± 562 kcal; ($p < 0.001$).
[31]	Automatic Ingestion Monitor (AIM)	EI	N/A	N/A	Accuracy = average between precision (P) and recall (R).	Accuracy of food ingestion = 89.9%, range from 75.82–97.7%.
[32]	Counts of Chews and Swallows Model	EI	A 3-fold cross validation technique, one sided Wilcoxon-Mann-Witney, Bland-Altman analysis and t-Test analysis.	Reporting error for the CCS model was lower than that of the diet diary ($p < 0.01$). The model underestimated EI. Energy intake estimation had the lowest bias.	A 3-fold cross validation technique, one-sided Wilcoxon-Mann-Witney, Bland-Altman analysis and t-Test analysis.	No statistical differences were found between the CCS model and either diet diary or photographic records.
[33]	Intelligent food-intake monitor	Food intake	Correlation: Proportion of food consumed from sound (auditory based) and image sequence (vision based) compared to the ground truth: proportion of food consumed.	Data not shown	N/A	N/A

Table 2. *Cont.*

Reference	Name of Tool	What Is Measured	Reliability		Validity	
			Statistical Method Used	Result	Statistical Method Used	Result
[34]	DP + R	EI	Inter-rater reliability coefficients	Error rate $\leq 5\%$, Recall assessments ≥ 0.95	Dependent t-test comparing device to DLW method; Bland-Altman plots; Limits of agreement	Differences between methods in the total sample was not significantly different (DP + R = 2912 ± 661 kcal/day; $TDEE_{DLW}$ = 2849 ± 748 kcal/day, $p = 0.42$); DP + R was found to overestimate EI compared to $TDEE_{DLW}$ by 63 ± 750 kcal/day ($6.8 \pm 28\%$; limits of agreement: $-1437, 1564$ kcal/day). The Bland-Altman plot indicated no proportional bias variation as a function of the level of EI in the total sample (R = -0.13, $p = 0.21$).
[35]	RFPM	EI	Bland & Altman analysis	Significant difference: $p < 0.0001$ between the RFPM and DLW in the standard prompt group. No significant difference in the customized group: $p = 0.22$. The level of bias in both groups was not influenced by the amount of EI (Adj. $R^2 = -0.03$, $p = 0.55$; Adj. $R^2 = -0.08$, $p = 0.78$)	Independent sample t-test for error between methods = EI estimated with the RFPM-EI measured with DLW	Significant smaller underestimation in the customized group (270 ± 748 kcal/day or $8.8 \pm 29.8\%$) when compared to the standard prompt group (895 ± 770 kcal/day or $34.3 \pm 28.2\%$), $t (33) = -2.35$, $p < 0.05$ with RFPM.
[36]	Real-time Food Recognition System	EI	Test-retest reliability	79.2% classification rate	N/A	N/A
[37]	Snap-n-Eat	Energy/dietary intake	Test-retest reliability	Classification accuracy (% of correctly classified images categories) = 85%	N/A	N/A

Table 2. *Cont.*

Reference	Name of Tool	What Is Measured	Reliability		Validity	
			Statistical Method Used	Result	Statistical Method Used	Result
[38]	GoCARB	Carb EI	Comparison to actual foods/database	Automatic segmentation (portion size) = 75.4% (86/114); Food item recognition = 85.1% (291/342)	Mean absolute error; Relative error	Mean absolute error = 27.89 (SD 38.20) and 12.28 (SD 9.56) grams of carbohydrates; Mean relative error = 54.8% (SD 72.3%) and 26.2% (SD 18.7%). A significant error between estimations was found ($p = 0.001$). In general, 60.5% (69/114) of the participants underestimated carbohydrate content.
[39]	Mathematical method	Change in EI	Test-retest reliability; Mean difference	40 kcal/day of mean difference between the gold standard and the mathematical model; No significant difference between the methods for any of the time segments was found (weeks 0–26: $p = 0.14$; weeks 26–52: $p = 0.34$; weeks 52–78: $p = 0.32$; weeks 78–104: $p = 0.11$).	Paired, 2-sided t test; Pearson correlation (r) Spearman's corrected (rs)	Change in EI values calculated by the mathematical method or the gold standard DLW/DXA weren't significantly different; The mathematical model had an accuracy within 132kcal/day for predicting changes in EI; The magnitude of correlation of the change in EI values between models were correlated (weeks 0–26: $r = 0.57$ (95% confidence interval 0.45, 0.68); $p = \leq 0.0001$; weeks 78–104: $r = 0.19$ (0, 0.36); $p = 0.05$).

[1] N/A = Not applicable.

When comparing the reliability of the formula-predicted EI values to the staff-observed values within the test group the Pearson correlation was $r = 0.374$. For the reliability of the model between training and test groups, the difference in r^2 (which is called the shrinkage value) was 1.4% [30]. To assess validity, researchers assessed participants' estimation error of EI compared with the equation. The bite-based equation method was more effective at estimating EI than the best human estimation [30].

The bite counter along with the bite-based method formula can provide individuals with a EI estimation that is more accurate than an individual's estimation even when EI information is available, which potentially could help improve their adherence to recommended dietary changes. The bite counter also has the benefit of being a non-invasive device, which allows tracking of free-living situations for research and also has the potential to improve the understanding of food ingestion patterns including snacking, night eating, and weekend overeating, as pointed out by Fontana 2014 [31] as one of the benefits of food intake monitoring devices and tools. However, reliability was relatively low and internal and external validity of the method needs to be further elucidated. This tool is for monitoring EI purposes only and does not provide information or feedback on diet quality.

3.1.3. The Automatic Ingestion Monitor (AIM)

The Automatic Ingestion Monitor (AIM) integrated hand gestures, jaw motions and accelerometer sensors to detect food intake in free-living individuals [31]. It was designed for objective 24-h monitoring of food intake in free-living conditions without depending on any input from the subjects. The monitor was 90% accurate in its ability to detect specific food intake epochs in free-living individuals compared with self-reported signal (push-button) indicating food intake events, and self-reported food journals [31].

When developing and validating this device, the data were obtained from monitoring free-living situations that included a wide variety of foods and activities, increasing its feasibility for everyday use and research purposes. Its use could provide insight into overall eating behavior patterns where participants burden is minimal. Nonetheless, the use of self-report as the gold-standard method, rather than direct observation, prompts caution regarding reliability. Furthermore, subject compliance with and acceptability of wearing the AIM needs to be established [31]. Insight obtain from this and previous studies [40] encourage further research to build mathematical models to obtain estimated EI using individualized models on counts of chews and swallows (CCS) [32].

3.1.4. Intelligent Food-Intake Monitor

The intelligent food-intake monitor integrates multi-sensor monitors to track chewing speed, and images of the type and amount of food consumed, giving an overall understanding of eating behavior characteristics [33]. The tool was tested for its ability to correctly detect the proportion of food consumed in real life scenarios but results were not reported [33].

The development of the device took into consideration the general process and pattern of food-intake activities to directly target their process (food ingestion, chewing and swallowing). The experiments were conducted in a real-world setting to increase the feasibility of being used in such settings. Valuable information involving eating behavior can be obtained from the use of this device because it doesn't assume that the food on the plate is consumed, thus providing a more reliable measure than capturing images of food alone to assess food consumption, thanks to the integration of chewing and swallowing detection in the process. Further research needs to be conducted to increase participants' comfort levels when using the device to ensure compliance with its use for longer periods. Even though a high level of correlation is reported between ground truth and auditory and vision predictors, no r-values were given and no strong statistical bases were presented. Participant characteristics were not supplied [33], leaving to speculation the age range that could benefit from using this monitor, and whether it would be feasibility to use in older adults, youth and children.

3.2. Camera-Scan-Sensor Based Technologies or Food/Energy Intake Assessment Tools

Sixty-four percent of the American population own a smartphone, a 35% increase since 2011 [41]. Since the use of smartphones is steadily increasing in daily life, mobile phone camera-scan-sensors are being proposed to contribute novel approaches to the measurement of food and EI.

3.2.1. DP + R

Ptomey et al. [34] developed and evaluated a pre-post meal photographic method for assessing EI in overweight and obese individuals in a cafeteria setting. Foods consumed outside this setting were assessed by recall methods.

Nutrition research staff underwent rigorous training for estimating portion size and EI from pre- and post-meal digital photographs and dietary recalls, with inter-rater reliability >95%. DP + R procedure includes taking notes and delineating standard measurements as guidelines for the portion size assessment [34]. The DP + R during ad libitum eating in a cafeteria was compared to measurement of total daily energy expenditure assessed by doubly labelled water (TDEE$_{DLW}$) method [34] with no significant differences found; thus, the method was considered valid.

DP + R method is a reliable and validated method for estimating EI in overweight or obese participants in a cafeteria setting. The main advantage over a food record/recall alone is verification of the written record by the photograph. This method was judged to provide an acceptable level of burden for both participant and research team when compared to previous procedures but a considerable burden is still present for the researcher because of the need to enter nutritional information into a database to quantitate EI [34]. The capacity of DP + R to capture change over time is limited since the procedures are done in cafeteria settings, therefore when the subjects stop attending, the change will not be captured. However, the authors point out the possibility of modifying the DP + R method to use in conjunction with smartphone photos to make the method portable [34].

3.2.2. Remote Food Photography Method (RFPM)

Participants send images taken on their smartphone wirelessly [35] to a Food Photography Application© [42], which is linked to the Food and Nutrient Database for Dietary Studies 3.0 [43]. Trained raters use the application to oversee the semi-automated process of food and nutrient intake estimation [35]. In this trial, some participants received prompts to use the Application customized to their specific meal times, or generic prompts in the morning, at noon and in the late afternoon.

Analyses were run to evaluate any significant differences between the RFPM and DLW estimation of EI, and if they were influenced by the EI consumed; no significant differences were found when participants received customized reminder messages but device reliability was decreased when participants received generic prompts [35]. The RFPM and DLW were used to measure EI in free-living individuals during a 6-day period. The error between methods (EI estimated with the RFPM minus EI measured with DLW) was calculated and was smaller in the participants receiving customized prompts [35].

The underestimation of EI by RFPM improves drastically compared to self-report methods, particularly when accompanied by customized prompts, allowing monitoring of habitual EI in free-living individuals. The method also offers the opportunity to detect missing data (due to technical problems or no compliance) promptly, and take pertinent action (contact participant) to improve data quality and compliance, thereby reducing recall bias [44]. The ability of RFPM to provide users feedback about their behavior is another benefit worth mentioning. In general, the user burden is keep to a minimal and 82% of users rated overall satisfaction 5 or higher (based on a six-point scale) [35]. However, since the method is only semi-automated, it remains expensive to analyze. The RFPM has also been used to estimate EI in children in both research and free-living settings [35,42,44].

3.2.3. Real-Time Food Recognition System

The user points the smartphone camera at the food plate for the food recognition process. After selection of the food from a database and indication of its approximate volume, the calorie and nutrition values are displayed [36]. The real-time recognition of foods was approximately 80% correct.

This system utilizes a real-time image recognition system, and the processing time only takes 0.065 s once the user enters the input. A fully automated interface with a food database completes its system. Evaluation of its usability was carried out, where adjustment of the bounding boxes on the different food items wasn't as positively rated (2.4 out of 5) as for the item recognition itself, which was done automatically without additional user input, obtaining an average score of 4.2 out of 5 [36]. However, this tool hasn't been validated and has a limited number of food categories and it does not specify the database used for the nutrition information.

3.2.4. "Snap-n-Eat"

A "snap-shot" (photograph) of participants' plate is captured. The analytical system is based on predefined EI and nutritional density for each food category. Depth images are used to estimate the portion size of the food and the EI and nutritional content are displayed on the user's screen in ~4 s [37]. A classification accuracy (the percentage of the test images of each category correctly classified) of 85% was obtained for 15 different food categories.

Snap-n-Eat presents a food recognition system for which users only need to take a snapshot of their food in order for the system to estimate its EI and nutritional content allowing participants to track their daily food intakes helping to understand their eating habits in a cost and time effective manner. However, in order to be a feasible tool, a scale-up to hundreds of food items and a validation process is needed [37].

3.2.5. GoCARB

The user photographs their food from at least two angles. The food items are segmented and recognised and their carbohydrate content is estimated based on the nutritional information of the USDA Nutrient Database for Standard Reference [38]. GoCARB's portion sizing and individual food item recognition accuracy ratings were 75% and 85%, respectively [38]. To validate the device, adult participants with type 1 diabetes were asked to calculate the carbohydrate content of the meals by themselves and subsequently with the help of the GoCARB. The error using GoCARB error was approximately half of that without any aid [38].

The application is overall better than participants at estimating carbohydrate content of meals. In the GoCARB app the carbohydrate content estimation is done automatically so the burden on researchers and participants is minimal; thus, 90% (17/19) qualify the tool as easy to use and would like to use the application on a regular basis. These measurements were done in a clinical setting that may not represent real-life situations where the meals may have more complex composition than the test meals. The overall nutrient content of the meal is not analyzed, and for individuals with diabetes it is important to consider the influence of the overall meal in determining their postprandial glycemia [38].

3.3. Mathematical Algorithm

A totally different, novel approach to assess EI is through mathematical algorithms.

Mathematical Method

Sanghvi's group [39] validated a mathematical formula originating from the National Institute of Diabetes and Digestive and Kidney Diseases (NIDDK) [45], to measure long-term changes in free-living EI of humans by using repeated DLW/DXA measurements collected over 2 years in 140 free-living subjects from the Comprehensive Assessment of Long-term Effects of Reducing Intake of Energy (CALERIE) study [46]. The formula inputs were baseline demographic (age, sex, height) and repeated body weight data, which was used to obtain the change in body weight over time and the rolling average.

Measured body weight and EI changes for the participants were documented over 2 years at 4 different time intervals. During the course of the study, the test-retest reliability was obtained by comparing the gold standard with the mathematical model, differing by only 40 kcal/day [39]. The change in EI values calculated by the mathematical method was compared (paired, 2-sided *t*-test) to the gold standards DLW/DXA and found to be similar [39].

In order for the formula to measure long-term changes in free-living EI, easily acquired initial information regarding age, sex, height and physical activity are required. However, baseline DLW measurements are also needed to establish energy requirements if one wishes to know absolute EI as well as changes in EI over time, limiting its use to researchers with the ability to obtain this parameter [39]. If all the mathematical parameters are available, the formula is an easy-to-score tool that captures changes over more than one day; however, the model might require adjustments for use in children or older adults. In addition, specific nutrient/food intake information is not known, therefore without co-administration of a diet record or FFQ it would not be possible to obtain this information [39]. Important limitations must be considered. The study was conducted on normal weight individuals and validation in individuals with obesity was not demonstrated even though the authors were confident the model could be used on this population because the model was built to measure changes in metabolism and body composition.

4. Discussion

From a research perspective, the first and foremost goal of evaluating food and EI is to be able to increase our understanding of diet-disease associations. Validated and reliable measures of food and EI are crucial to understand their relationship with health, especially with the overwhelming increase in obesity prevalence [1]. Individuals with obesity present different problems ranging from the physiological to the psychological aspects, which represent barriers to their treatment. The 4Ms of Obesity Assessment and Management (Mental, Mechanical, Metabolic and Monetary) has been proposed as a framework to help identify the root cause and help obesity treatment [47].

The methodologies/tools presented in this review have the potential to aid in the understanding and treatment of obesity within this framework. This review identified 3 main new modalities for estimating food and EI. These include devices that monitor intake through sensors that detect movement of the arm and/or jaw, counts of chews and swallows, smartphone-based photographic methods linked to food databases and a mathematical formula.

In order to come to a consensus of which methodology/technology would be the most highly recommended it is important not to lose sight of why EI is being assessed or monitored. The overall objective should guide opting for one or the other.

In the context of the 4Ms of Obesity Assessment and Management, if the individual being treated is believed to have psychological (Mental) issues influencing their eating behavior, then the main objective is to understand their eating behaviors or food intake patterns in order to detect and/or modify eating habits. Food intake-monitoring devices and tools (Bite Counter, AIM, Intelligent food-intake monitor) would be recommended in this context because they could provide useful insight regarding food intake behaviors (e.g., timing and size of meals). In general, food intake-monitoring devices and tools can count the number of bites an individual takes, track the approximate EI and monitor episodes of food intake. Several benefits to the understanding of food intake behaviors may accrue from these methods. These devices could fill a gap in providing timely monitoring and feedback to individuals wishing to change eating habits, similar to the way the use of pedometers and/or accelerometers has been validated to promote and assess physical activity [48], by establishing and monitoring personal goals achievement, a behavior that according to Social Cognitive Theory, is an effective behavior change strategy [49] aiding behavioral change. Therefore, these tools could be used for monitoring, controlling and correcting eating behaviors and portion size in obese or overweight individuals as well as for chronic disease management. However, their effectiveness in eliciting behavior change has yet to be documented. Future work includes the possible addition of a vibrotactile alarm, similar to the technology used on intelligent watches or pedometers so that subjects can self-adjust their eating behavior based on the estimated EI per bite [28]. Moreover, the commercial cost of these devices has not been established since they are still on the development phase and have not gone further to establish a market cost. Further, wearing some devices may be more acceptable to participants than others.

On the other hand, if the aspect of obesity treatment is within the Metabolic category of the 4Ms, as in the case of individuals with T2D or hypertension, then the intent would shift the focus to understanding specific macro/micronutrient intakes (sugars, salt, fats). Similarly, within the Mechanical category (such as osteoarthritis), weight loss could be desired to reduce pain. For both approaches, camera-scan-sensor (Snap-n-Eat, GoCARB) could be useful. RFPM could be applied in a hospital setting where monitoring individuals' nutrition intake is essential but difficult to do on a routine basis. Registered dietitians and nurses could use this tool to oversee adequate food intake essential for hospitalized individuals' wellbeing. If the overall objective is a focus on measures of long-term changes in EI in free-living individuals undergoing a research or lifestyle intervention, the mathematical method would highly be recommended since its accuracy lies within 40 kcal/day of mean difference with the gold standard, as long as the initial DLW measurement is possible to obtain, which could be a potential limitation.

Overall, the studies included in this review presented new devices designed to improve how EI is measured, analyzed and registered. However, the devices and methods have usually undergone pilot testing in small numbers of participants and various limitations elicit caution. Food intake-monitoring tools have limited ability to assess day-to-day variability in food intake [28,30,31]. They do not take into account the type of food consumed, its EI density nor its consistency; therefore, no information about the macro/micronutrient is obtained, resulting in a inability to capture change in type of food or nutritional intake over time. As mentioned previously, a current limitation with present methodologies used to assess food intake or EI is individual reactivity causing changes in food behavior patterns, thereby resulting in inaccurate reporting. None of the present studies addressed these issues, therefore the question arises: could bias play a role in the use of these devices? That is, would peoples' consumption of food intake be modified by simply wearing these tools? And if so, what would be the differences compared with current methodologies? Certainly, more accurate data of consumption patterns seems possible, but to date none of the devices has gone beyond pilot testing nor addressing potential bias. To our knowledge, the application of these methodologies to clinical settings or outside of the original developers' laboratories has not been reported.

Regarding smartphone-based apps, additional limitations applying to one or more include participants forgetting to take the photographs, or not having the smartphone with them [35]. In general, two major limitations need to be addressed with camera-scan-sensor methods. First, they cannot quantify all food ingredients or beverages. These tools only work with the food items in the database of each individual tool, and their validity is also dependent on the food nutrient value on which the databases are built. However, with current food record/recall databases, there are acknowledged differences between what a person consumes and what the database contains [38]; even the Canadian Nutrient File or the USDA database cannot keep up with constantly evolving food possibilities. The use of these technologies is not advanced enough to correctly and accurately estimate 100% of food intake since the best achieved accuracy was 85% based on a small number of foods [37]. Second, they cannot judge quality since a photograph doesn't convey information about ingredients that are hidden or blended [35]. Nevertheless, these methods show improvement in estimating, on average, the nutrient content of meals more accurately, easier and faster than individuals' self-report measures (24HR, FFQ, etc.) but caution must be taken when using and analyzing these methods. Bearing in mind the strong link between food intake and health, continuing to document the improved validity and reliability of the food item recognition and nutritional information provided by these tools would undoubtedly lead to better outcome measurement in the fields of obesity and nutrition. However, the feasibility creating comprehensive databases for food recognition is problematic in an environment of incessantly increasing food possibilities. On the other hand, the ubiquity of smartphone ownership means that affordability and acceptability are of less concern with the main investment being the data processing.

5. Conclusions

In conclusion, these innovative dietary assessment tools are able to record food/energy intake more accurately than participants' estimates and are an improvement on important weaknesses of conventional methods (paper-based records/recalls), particularly regarding the burden of recording by participants and collecting/administering and evaluating/scoring the information by researchers. However, caution is needed when using them since they are still being refined. Future work should look at combining body monitor sensors and camera-scan-sensors to work together in order to counter their strengths and weaknesses. This work should eventually progress outside of research settings and promote the collaboration of dietitians with engineers to co-develop the design, development, evaluation and implementation of these new tools, since this would likely increase their effectiveness, acceptability and validity. Lastly, this research field should take into consideration changing formats of national nutrition recommendations, such as the 2014 Brazilian dietary guidelines, 2015 Dietary Guidelines for Americans and the American Heart Association, which are shifting the focus from single nutrients or kilocalorie counting into healthy eating patterns [50–52]. Therefore, future development should aim at being able to detect overall eating patterns.

Author Contributions: Conceptualization, M.C.A.H. and C.B.C.; Methodology, M.C.A.H.; Data Curation, M.C.A.H.; Writing-Original Draft Preparation, M.C.A.H.; Writing-Review & Editing, C.B.C.

Funding: This research received no external funding.

Acknowledgments: M.C.A.H. was supported by the National Council of Science and Technology of Mexico.

Conflicts of Interest: The authors declare no conflict of interest.

References

1. Wang, H.; Naghavi, M.; Allen, C.; Barber, R.M.; Bhutta, Z.A.; Carter, A.; Casey, D.C.; Charlson, F.J.; Chen, A.Z.; Coates, M.M.; et al. Global, regional, and national life expectancy, all-cause mortality, and cause-specific mortality for 249 causes of death, 1980–2015: A systematic analysis for the global burden of disease study 2015. *Lancet* **2016**, *388*, 1459–1544. [CrossRef]
2. Williamson, D.A. Fifty years of behavioral/lifestyle interventions for overweight and obesity: Where have we been and where are we going? *Obesity* **2017**, *25*, 1867–1875. [CrossRef] [PubMed]
3. Das, S.K.; Roberts, S.B.; Bhapkar, M.V.; Villareal, D.T.; Fontana, L.; Martin, C.K.; Racette, S.B.; Fuss, P.J.; Kraus, W.E.; Wong, W.W.; et al. Body-composition changes in the comprehensive assessment of long-term effects of reducing intake of energy (CALERIE)-2 study: A 2-y randomized controlled trial of calorie restriction in nonobese humans. *Am. J. Clin. Nutr.* **2017**, *105*, 913–927. [CrossRef] [PubMed]
4. The Diabetes Prevention Program (DPP) Research Group. The Diabetes Prevention Program (DPP): Description of lifestyle intervention. *Diabetes Care* **2002**, *25*, 2165–2171.
5. The Look AHEAD Research Group. Eight-year weight losses with an intensive lifestyle intervention: The Look AHEAD study. *Obesity* **2014**, *22*, 5–13.
6. Ma, W.; Huang, T.; Zheng, Y.; Wang, M.; Bray, G.A.; Sacks, F.M.; Qi, L. Weight-loss diets, adiponectin, and changes in cardiometabolic risk in the 2-year pounds lost trial. *Clin. Endocrinol. Metab.* **2016**, *101*, 2415–2422. [CrossRef] [PubMed]
7. Sharma, B.K.; Patil, M.; Satyanarayana, A. Negative regulators of brown adipose tissue (BAT)-mediated thermogenesis. *J. Cell Physiol.* **2014**, *229*, 1901–1907. [CrossRef] [PubMed]
8. Ryan, D.H.; Jensen, M.D. What the new obesity guidelines will tell us. *Curr. Opin. Endocrinol. Diabetes Obes.* **2013**, *20*, 429–433. [CrossRef] [PubMed]

9. Blair, S.N. Physical inactivity: The biggest public health problem of the 21st century. *Br. J. Sports Med.* **2009**, *43*, 1–2. [PubMed]

10. Subar, A.F.; Freedman, L.S.; Tooze, J.A.; Kirkpatrick, S.I.; Boushey, C.; Neuhouser, M.L.; Thompson, F.E.; Potischman, N.; Guenther, P.M.; Tarasuk, V.; et al. Addressing current criticism regarding the value of self-report dietary data. *J. Nutr.* **2015**, *145*, 2639–2645. [CrossRef] [PubMed]

11. Johnson, R.K. Dietary intake—How do we measure what people are really eating? *Obes. Res.* **2002**, *10*, 63S–68S. [CrossRef] [PubMed]

12. Archer, E.; Pavela, G.; Lavie, C.J. The inadmissibility of what we eat in America and NHANES dietary data in nutrition and obesity research and the scientific formulation of national dietary guidelines. *Mayo Clin. Proc.* **2015**, *90*, 911–926. [CrossRef] [PubMed]

13. Dhurandhar, N.V.; Schoeller, D.; Brown, A.W.; Heymsfield, S.B.; Thomas, D.; Sorensen, T.I.; Speakman, J.R.; Jeansonne, M.; Allison, D.B. Energy Balance Measurement Working Group. Energy balance measurement: When something is not better than nothing. *Int. J. Obes.* **2015**, *39*, 1109–1113. [CrossRef] [PubMed]

14. Ioannidis, J.P. Implausible results in human nutrition research. *BMJ* **2013**, *347*, f6698. [CrossRef] [PubMed]

15. Murakami, K.; Livingstone, M.B. Prevalence and characteristics of misreporting of energy intake in US adults: NHANES 2003–2012. *J. Nutr.* **2015**, *114*, 1294–1303. [CrossRef] [PubMed]

16. Berta Vanrullen, I.; Volatier, J.L.; Bertaut, A.; Dufour, A.; Dallongeville, J. Characteristics of energy intake under-reporting in French adults. *J. Nutr.* **2014**, *111*, 1292–1302. [CrossRef] [PubMed]

17. Gemming, L.; Jiang, Y.; Swinburn, B.; Utter, J.; Mhurchu, C.N. Under-reporting remains a key limitation of self-reported dietary intake: An analysis of the 2008/09 New Zealand adult nutrition survey. *J. Clin. Nutr.* **2014**, *68*, 259–264. [CrossRef] [PubMed]

18. Murakami, K.; McCaffrey, T.A.; Livingstone, M.B. Associations of dietary glycaemic index and glycaemic load with food and nutrient intake and general and central obesity in British adults. *J. Nutr.* **2013**, *110*, 2047–2057. [CrossRef] [PubMed]

19. Johansson, L.; Solvoll, K.; Bjorneboe, G.E.; Drevon, C.A. Under- and overreporting of energy intake related to weight status and lifestyle in a nationwide sample. *Am. J. Clin. Nutr.* **1998**, *68*, 266–274. [CrossRef] [PubMed]

20. Basiotis, P.P.; Welsh, S.O.; Cronin, F.J.; Kelsay, J.L.; Mertz, W. Number of days of food intake records required to estimate individual and group nutrient intakes with defined confidence. *J. Nutr.* **1987**, *117*, 1638–1641. [CrossRef] [PubMed]

21. Thompson, F.E.; Subar, A.F.; Loria, C.M.; Reedy, J.L.; Baranowski, T. Need for technological innovation in dietary assessment. *J. Am. Diet. Assoc.* **2010**, *110*, 48–51. [CrossRef] [PubMed]

22. Bountziouka, V.; Panagiotakos, D.B. Statistical methods used for the evaluation of reliability and validity of nutrition assessment tools used in medical research. *Curr. Pharm. Des.* **2010**, *16*, 3770–3775. [CrossRef] [PubMed]

23. Rebro, S.M.; Patterson, R.E.; Kristal, A.R.; Cheney, C.L. The effect of keeping food records on eating patterns. *J. Am. Diet. Assoc.* **1998**, *98*, 1163–1165. [CrossRef]

24. Hise, M.E.; Sullivan, D.K.; Jacobsen, D.J.; Johnson, S.L.; Donnelly, J.E. Validation of energy intake measurements determined from observer-recorded food records and recall methods compared with the doubly labeled water method in overweight and obese individuals. *Am. J. Clin. Nutr.* **2002**, *75*, 263–267. [CrossRef] [PubMed]

25. Schoeller, D.A. Measurement of energy expenditure in free-living humans by using doubly labeled water. *J. Nutr.* **1988**, *118*, 1278–1289. [CrossRef] [PubMed]

26. Amft, O.; Troster, G. Recognition of dietary activity events using on-body sensors. *Artif. Intell. Med.* **2008**, *42*, 121–136. [CrossRef] [PubMed]

27. Sazonov, E.; Schuckers, S.; Lopez-Meyer, P.; Makeyev, O.; Sazonova, N.; Melanson, E.L.; Neuman, M. Non-invasive monitoring of chewing and swallowing for objective quantification of ingestive behavior. *Physiol. Meas.* **2008**, *29*, 525–541. [CrossRef] [PubMed]

28. Dong, Y.; Hoover, A.; Scisco, J.; Muth, E. A new method for measuring meal intake in humans via automated wrist motion tracking. *Appl. Psychophysiol. Biofeedback* **2012**, *37*, 205–215. [CrossRef] [PubMed]

29. Subar, A.F.; Kipnis, V.; Troiano, R.P.; Midthune, D.; Schoeller, D.A.; Bingham, S.; Sharbaugh, C.O.; Trabulsi, J.; Runswick, S.; Ballard-Barbash, R.; et al. Using intake biomarkers to evaluate the extent of dietary misreporting in a large sample of adults: The open study. *Am. J. Epidemiol.* **2003**, *158*, 1–13. [CrossRef] [PubMed]

30. Salley, J.N.; Hoover, A.W.; Wilson, M.L.; Muth, E.R. Comparison between human and bite-based methods of estimating caloric intake. *J. Acad. Nutr. Diet.* **2016**, *116*, 1568–1577. [CrossRef] [PubMed]

31. Fontana, J.M.; Farooq, M.; Sazonov, E. Automatic ingestion monitor: A novel wearable device for monitoring of ingestive behavior. *IEEE Trans. Biomed. Eng.* **2014**, *61*, 1772–1779. [CrossRef] [PubMed]

32. Fontana, J.M.; Higgins, J.A.; Schuckers, S.C.; Bellisle, F.; Pan, Z.; Melanson, E.L.; Neuman, M.R.; Sazonov, E. Energy intake estimation from counts of chews and swallows. *Appetite* **2015**, *85*, 14–21. [CrossRef] [PubMed]

33. Liu, J.; Johns, E.; Atallah, L.; Pettitt, C.; Lo, B.; Frost, G.; Yang, G.-Z. An intelligent food-intake monitoring system using wearable sensors. In *2012 Ninth International Conference on Wearable and Implantable Body Sensor Networks*; IEEE: London, UK, 2012; pp. 154–160.

34. Ptomey, L.T.; Willis, E.A.; Honas, J.J.; Mayo, M.S.; Washburn, R.A.; Herrmann, S.D.; Sullivan, D.K.; Donnelly, J.E. Validity of energy intake estimated by digital photography plus recall in overweight and obese young adults. *J. Acad. Nutr. Diet.* **2015**, *115*, 1392–1399. [CrossRef] [PubMed]

35. Martin, C.K.; Correa, J.B.; Han, H.; Allen, H.R.; Rood, J.C.; Champagne, C.M.; Gunturk, B.K.; Bray, G.A. Validity of the remote food photography method (RFPM) for estimating energy and nutrient intake in near real-time. *Obesity* **2012**, *20*, 891–899. [CrossRef] [PubMed]

36. Kawano, Y.; Yanai, K. Foodcam: A real-time food recognition system on a smartphone. *Multimed. Tools Appl.* **2015**, *74*, 5263–5287. [CrossRef]

37. Zhang, W.; Yu, Q.; Siddiquie, B.; Divakaran, A.; Sawhney, H. "Snap-n-eat": Food recognition and nutrition estimation on a smartphone. *J. Diabetes Sci. Technol.* **2015**, *9*, 525–533. [CrossRef] [PubMed]

38. Rhyner, D.; Loher, H.; Dehais, J.; Anthimopoulos, M.; Shevchik, S.; Botwey, R.H.; Duke, D.; Stettler, C.; Diem, P.; Mougiakakou, S. Carbohydrate estimation by a mobile phone-based system versus self-estimations of individuals with type 1 diabetes mellitus: A comparative study. *J. Med. Int. Res.* **2016**, *18*, e101. [CrossRef] [PubMed]

39. Sanghvi, A.; Redman, L.M.; Martin, C.K.; Ravussin, E.; Hall, K.D. Validation of an inexpensive and accurate mathematical method to measure long-term changes in free-living energy intake. *Am. J. Clin. Nutr.* **2015**, *102*, 353–358. [CrossRef] [PubMed]

40. Sazonov, E.S.; Schuckers, S.A.C.; Lopez-Meyer, P.; Makeyev, O.; Melanson, E.L.; Neuman, M.R.; Hill, J.O. Toward Objective Monitoring of Ingestive Behavior in Free-living Population. *Obesity* **2009**, *17*, 1971–1975. [CrossRef] [PubMed]

41. Pew Research Center. The Smartphone Difference. Available online: http://www.pewinternet.org/2015/04/_1/us-smartphone-use-in-2015/ (accessed on 1 April 2015).

42. Martin, C.K.; Nicklas, T.; Gunturk, B.; Correa, J.B.; Allen, H.R.; Champagne, C. Measuring food intake with digital photography. *J. Hum. Nutr. Diet.* **2014**, *27*, 72–81. [CrossRef] [PubMed]

43. USDA Food and Nutrient Database for Dietary Studies, version 3.0 [internet], Beltsville (MD): USDA, Agricultural Research Service, Food Surveys Research Group. 2008. Available online: http://www.ars.usda.gov/Services/docs.htm?docid=17031 (accessed on 1 April 2015).

44. Martin, C.K.; Han, H.; Coulon, S.M.; Allen, H.R.; Champagne, C.M.; Anton, S.D. A novel method to remotely measure food intake of free-living individuals in real time: The remote food photography method. *J. Nutr.* **2009**, *101*, 446–456. [CrossRef] [PubMed]

45. Hall, K.D.; Chow, C.C. Estimating changes in free-living energy intake and its confidence interval. *Am. J. Clin. Nutr.* **2011**, *94*, 66–74. [CrossRef] [PubMed]

46. Rochon, J.; Bales, C.W.; Ravussin, E.; Redman, L.; Holloszy, J.O.; Racette, S.B.; Roberts, S.B.; Das, S.K.; Romashkan, S.; Galan, K.M.; et al. Design and conduct of the CALERIE Study: Comprehensive assessment of the long-term effects of reducing intake of energy. *J. Gerontol. Ser. A* **2010**, *66A*, 97–108. [CrossRef] [PubMed]

47. Sharma, A.M.M. A mnemonic for assessing obesity. Obesity reviews. *Off. J. Int. Assoc. Study Obes.* **2010**, *11*, 808–809. [CrossRef] [PubMed]

48. Van Remoortel, H.; Giavedoni, S.; Raste, Y.; Burtin, C.; Louvaris, Z.; Gimeno-Santos, E.; Langer, D.; Glendenning, A.; Hopkinson, N.S.; Vogiatzis, I.; et al. Validity of activity monitors in health and chronic disease: A systematic review. *Int. J. Behav. Nutr. Phys. Act.* **2012**, *9*, 84. [CrossRef] [PubMed]

49. Bandura, A. *Social Foundations of Thought and Action: A Social Cognitive Theory*; Prentice-Hall, Inc.: Englewood Cliffs, NJ, USA, 1986; p. 617.

50. Monteiro, C.A.; Cannon, G.; Moubarac, J.C.; Martins, A.P.; Martins, C.A.; Garzillo, J.; Canella, D.S.; Baraldi, L.G.; Barciotte, M.; Louzada, M.L.; et al. Dietary guidelines to nourish humanity and the planet in the twenty-first century. A blueprint from brazil. *Public Health Nutr.* **2015**, *18*, 2311–2322. [CrossRef] [PubMed]
51. Millen, B.E.; Abrams, S.; Adams-Campbell, L.; Anderson, C.A.M.; Brenna, J.T.; Campbell, W.W.; Clinton, S.; Hu, F.; Nelson, M.; Neuhouser, M.L.; et al. The 2015 dietary guidelines advisory committee scientific report: Development and major conclusions. *Adv. Nutr.* **2016**, *7*, 438–444. [CrossRef] [PubMed]
52. Lloyd-Jones, D.M.; Hong, Y.; Labarthe, D.; Mozaffarian, D.; Appel, L.J.; Van Horn, L.; Greenlund, K.; Daniels, S.; Nichol, G.; Tomaselli, G.F.; et al. Defining and setting national goals for cardiovascular health promotion and disease reduction: The american heart association's strategic impact goal through 2020 and beyond. *Circulation* **2010**, *121*, 586–613. [CrossRef] [PubMed]

Brief Report

Dietary Intake Reporting Accuracy of the Bridge2U Mobile Application Food Log Compared to Control Meal and Dietary Recall Methods

Jennifer L. Lemacks *[ID]**, Kristen Adams** and **Ashley Lovetere**

The University of Southern Mississippi, 118 College Drive #5142, Hattiesburg, MS 39406-0001, USA; anna.k.lee@usm.edu (K.A.); ashley.lovetere@usm.edu (A.L.)
* Correspondence: jennifer.lemacks@usm.edu; Tel.: +1-601-266-6825

Received: 17 December 2018; Accepted: 17 January 2019; Published: 19 January 2019

Abstract: Mobile technology introduces opportunity for new methods of dietary assessment. The purpose of this study was to compare the reporting accuracy of a mobile food log application and 24 h recall method to a controlled meal among a convenience sample of adults (18 years of age or older). Participants were recruited from a community/university convenience sample. Participants consumed a pre-portioned control meal, completed mobile food log entry (mfood log), and participated in a dietary recall administered by a registered dietitian (24R). Height, weight, and application use survey data were collected. Sign test, Pearson's correlation, and descriptive analyses were conducted to examine differences in total and macronutrient energy intake and describe survey responses. Bland Altman plots were examined for agreement between energy intake from control and 24R and mfood log. The 14 included in the analyses were 78.6% female, 85.7% overweight/obese, and 64.3% African American. Mean total energy, protein, and fat intakes reported via the mfood log were significantly ($p < 0.05$) lower compared to the control, by 268.31kcals, 20.37 g, and 19.51 g, respectively. Only 24R mean fat intake was significantly ($p < 0.01$) lower than the control, by 6.43 g. Significant associations (r = 0.57–0.60, $p < 0.05$) were observed between control and mfood log mean energy, carbohydrate, and protein intakes, as well as between control and 24R mean energy (r = 0.64, $p = 0.01$) and carbohydrate (r = 0.81, $p < 0.001$) intakes. Bland Altman plots showed wide limits of agreement, which were not statistically significant but may have practical limitations for individual dietary assessment. Responses indicated the ease of and likelihood of daily mfood log use. This study demonstrates that the Bridge2U mfood log is valid for the assessment of group level data, but data may vary too widely for individual assessment. Further investigation is warranted for nutrition intervention research.

Keywords: diet; assessment; food log; recall

1. Introduction

Dietary assessment is a critical component used to identify relationships between nutrients and chronic disease, such as obesity, cancer, cardiovascular disease, and diabetes. Self-report dietary assessment instruments are widely used in research to estimate energy, macronutrient, and micronutrient intakes, and to relate nutrient intake to various psychosocial and clinical factors to determine disease risk and association [1,2]. Compared to gold standard methods that introduce the least bias, such as the doubly labeled water to determine energy expenditure [3], self-report dietary assessment instruments are much less invasive and costly and are easily conducted in real world settings. The accuracy of dietary assessment methods is particularly important for identifying associations between diet and disease and determining changes in dietary behavior as a result of intervention.

Self-report dietary assessment methods include 24-h recalls (24R), food frequency questionnaires, a food record/diary/log, and food screeners. The 24R method is considered the best self-report dietary assessment instrument due to the least bias reported of self-report instruments [1], especially when recalls are interviewer-administered and conducted using the multi-pass method [4]. However, the method is not without limitations, which are largely due to an individual's inability to recall what they ate the previous day or accurately estimate portion sizes [5]. Additionally, numerous factors have been linked to dietary intake misreporting from self-reported instruments. For example, it is well-known that overweight or obese individuals commonly underreport energy intake compared to normal weight counterparts from self-administered food frequency questionnaires [6–8].

Mobile technology assisted dietary assessment has emerged as a method for collecting dietary data and allows for real-time recording of food intake [9]. A review of mobile dietary assessment methods examined for feasibility and validity determined three main assessment methods: self-reported dietary intake entry, food photograph analyses by trained research dietitians, and auto-analyzed food images [10]. While electronic food logs may exhibit similar limitations to pen and paper methods (i.e., participant's inability to estimate portion sizes or recall foods, resulting in omission of foods), mobile food logs offer vast opportunities, including reduced researcher burdens and costs due to automatic, real-time data entry and allow for the real-time detection of procedural issues/non-compliance and communication with participants to alleviate issues [10]. Despite the potential, there is limited research on the validity of mobile self-report food logs. Thus, the purpose of this study was to examine the reporting accuracy of a mobile dietary recall food log application (a self-reported dietary intake entry method) and 24R method for a controlled meal among a convenience sample of adults (18 years of age or older).

2. Materials and Methods

2.1. Setting and Technology Development

The study presented in this manuscript was a secondary study of the Church Bridge Project, which was a weight management intervention delivered in a church-based setting and targeted young to middle aged (18 to 50 years of age) adult African Americans [1]. The intervention included the development of a mobile and web application (Bridge2U) to facilitate survey, anthropometric, and dietary data collection. The platform also allowed interventionists to deliver dietary feedback to participants received via the mobile application and participants to monitor their weight loss progress. A mobile food log (mfood log, depicted in Figure 1) allowed participants to enter dietary data in real time or prospectively and was used for this secondary study. The mfood log utilized the USDA Food Composition Database API [2] to support the search location of foods in the application and includes 7793 standard reference foods and 229,064 branded food products. According to the National Institutes of Health definition, this original study was considered a clinical trial and registered at ClinicalTrials.gov (identifier: NCT02773069).

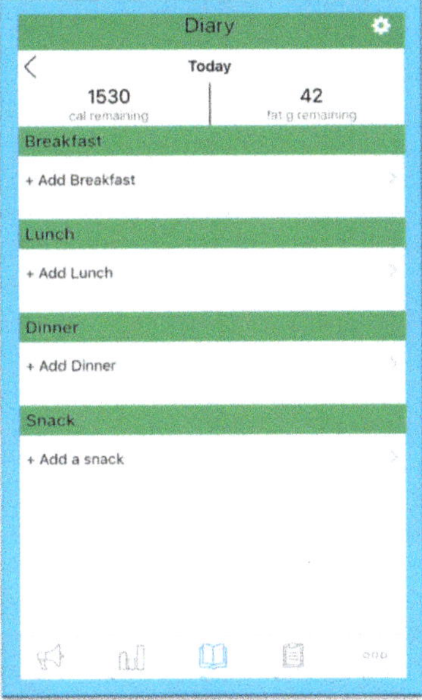

Figure 1. Screen capture of the Bridge2U mobile food log.

2.2. Recruitment

Two groups were recruited for the study from the Church Bridge Project research sample [1] and university population to result in a community and university convenience sample, respectively. Neither group was currently enrolled in an active dietary intervention. General inclusion criteria were adults 18 years of age or older and either Church Bridge Project participants or university students, faculty, or staff. Participants were excluded from the university sample if they were nutrition or computer science students, faculty, or staff, and individuals that the researchers knew personally. Flyers were utilized to recruit participants in the study, as well as word of mouth through church and university leaders. Interested participants completed an online enrollment form to collect basic contact information, food allergy information, inclusion criteria (age, department), and how they heard about the study. All instruments and protocol were approved by The University of Southern Mississippi Institutional Review Board.

2.3. Control Meal Preparation

Researchers and registered dietitians prepared a control meal to include spaghetti and meat sauce; the option for a parmesan cheese topping; steamed broccoli; a roll; a dessert option of a sugar cookie or chocolate pudding; and a beverage choice of Coke, Diet Coke, Sprite, or water. All items were pre-portioned and a nutrient analysis of the pre-portioned items (including the optional cheese) was conducted using the USDA Food Composition Database [11].

2.4. Data Collection

Upon arrival to the study site, participants received an oral consent form, a test mobile phone, and random participant identifier to log into the Bridge2U mobile application. After all participants were checked in, a researcher delivered an oral consent presentation and consent was obtained from participants. While there were no hidden video or audio recordings, the researcher provided minimal information regarding the project research questions (i.e., what was being monitored and measured) in an attempt to maintain the integrity of the study and preserve natural behaviors in a controlled setting. Participants were given an overview of the study to include the major study requirements: meal participation, mfood log entry, and 24R administration the following day. Participants were not informed of the purpose for mfood log entry or given specifics as to why the dietitian would be calling them the next day, other than "to conduct a follow up interview".

2.5. Meal Participation, mFood Log Entry and 24R Administration

Prior to meal consumption, participants were instructed to login to the mobile application using a random identifier. Participants were then instructed to enter food items and portions consumed whenever they deemed appropriate during the study meal using the mfood log. The pre-portioned control meal was provided to each participant with the option to request additional pre-portioned servings of any food or beverage offered. Research dietitians recorded consumption as a percentage of each meal component consumed once participants stated they were done consuming their meal and both prior to and after any additional servings were provided. For example, if the participant consumed all of an entrée, the researcher would record 100% of the entrée as consumed. If the participant consumed half of the roll and none of the broccoli, the researcher would record 50% and 0% consumed for those meal components, respectively.

After completion of the meal and mfood log, participants were reminded that they would be contacted by a registered dietitian the following day to complete a 24R of the study meal. The day after the meal consumption study, a registered dietitian contacted each participant to complete the food recall interview of the study meal consumed following standard, 24R methodology [12]. Data were recorded and analyzed using the USDA Food Composition Database [11], in alignment with the meal analysis and mfood log database used.

2.6. Anthropometric and Survey Data

Height and weight data were collected using a portable SECA 217 stadiometer and SECA 869 digital weight scale, respectively, and measured to the nearest tenth. Body mass index (BMI) was calculated as weight in kilograms divided by height in meters, squared. After using the mfood log, participants completed a self-administered survey with Likert response items to determine the ease of use of the application (with a scale of 1 being most difficult and 10 being easiest), likelihood of using the application on a daily basis (with a scale of 1 being least likely to 10 being most likely), and whether they had previously used a dietary intake mobile application (yes or no); responses "yes" to previous use of a dietary intake application were followed by a request to name the application previously used.

2.7. Statistical Analysis

Data from the control, mfood log, and 24R methods were analyzed using the USDA Food Composition database [2] to maintain data consistency and determine total energy intake in kilocalories, as well as macronutrient (carbohydrate, fat, and protein) grams consumed. Scatterplots and boxplot analyses were examined for potential outliers. A sign test was conducted to examine mean differences in energy intake between the control meal, 24R, and mfood log data. Pearson's correlation estimates were examined to note any significant linear associations between total and macronutrient energy intake computed between control, mfood log, and 24R methods and ease and likelihood of use and energy reporting differences between control, mfood log, and 24R; the method was used as an indicator of group level agreement between methods. A Bland Altman plot was used to examine individual level agreement between control and mfood log, which is useful to identify a relationship between differences and magnitude of measurements systematic bias. Descriptive data were reported to describe responses to survey items. All analyses were conducted using IBM SPSS 18.0 (Armonk, NY, USA). Statistical significance was determined based on an alpha level less than or equal to 0.05.

3. Results

Six and 21 participants were reached for the community and university samples, respectively, and invited to participate in the controlled study. Six participants enrolled from the Church Bridge Project and one had to withdraw due to a family emergency. Twenty-one participants enrolled from the university setting; however, three were excluded due to scheduling conflicts, two withdrew for personal reasons, and six did not attend data collection. Therefore, there were five and 10 participants from the church and university settings, respectively. There were 15 participants in total included in the study from both settings. Examination of scatterplots revealed one potential outlier. Boxplot analysis revealed the value was an extreme value (greater than quartile 1 multiplied by 1.5 and added to the interquartile range value). The participant was removed from the sample, resulting in a final 14 participants included in the study.

3.1. Participant Characteristics

Participant ages ranged from 19 to 45 years, with a mean of 26.2 years. Participants were 78.6% ($n = 11$) female, 85.7% ($n = 12$) overweight/obese, and 64.3% ($n = 9$) African American; race, gender, and body mass index class counts and percentages are reported in Table 1.

Table 1. Race, gender, and body mass index class of participants, $n = 15$.

Characteristic	n	%
Race		
Caucasian	3	21.4
African American	9	64.3
Other Race	2	14.1
Gender		
Female	11	78.6
Male	3	21.4
Body Mass Index Class		
Normal (18.0–24.9 kg/m^2)	2	14.3
Overweight (25.0–29.9 kg/m^2)	2	14.3
Obese Class 1 (30.0–34.9 kg/m^2)	4	28.6
Obese Class 2 (35.0–39.9 kg/m^2)	4	28.6
Obese Class 3 (>/=40.0 kg/m^2)	2	14.3

3.2. Inferential Analyses

Reported mean energy intake was lower than the control meal for both the 24R and mfood log. Sign test results showed that there was no significant difference in energy intake reported between 24R and mfood log methods ($p = 0.18$). Mean energy intake reported using the mfood log was statistically significantly ($p = 0.002$) lower than the control meal; 24R was not significantly ($p = 0.09$) lower than the control meal. The mean differences between the mfood log and control method were also significantly lower for protein ($p < 0.001$) and fat ($p = 0.001$) intakes. mFood log mean protein ($p = 0.01$) and fat intakes ($p = 0.002$) were also significantly lower than the 24R method. Table 2 displays all means and differences. Pearson's correlation analyses revealed significant, medium, positive associations between control and mfood log total energy ($p = 0.03$), carbohydrate ($p = 0.02$), and protein ($p = 0.04$) intakes. A significant, medium, positive association was also noted between the control and 24R total energy intake ($p = 0.01$); additionally, a strong, positive association between the control and 24R carbohydrate intake was observed ($p < 0.001$, Table 3). Significant, medium, positive associations were also observed between 24R and mfood log mean total energy (r = 0.59, $p = 0.03$) and protein (r = 0.62, $p = 0.02$) intakes.

Table 2. Mean energy and macronutrient intake of each method and difference from control meal, $n = 14$.

Reporting Method	Energy Intake (Kcal) Mean (±SD)	Energy Intake (Kcal) Mean Difference from Control	Carbohydrate Intake (g) Mean (±SD)	Carbohydrate Intake (g) Mean Difference from Control	Protein Intake (g) Mean (±SD)	Protein Intake (g) Mean Difference from Control	Fat Intake (g) Mean (±SD)	Fat Intake (g) Mean Difference from Control
Control Meal	838.88 (239.47)		108.54 (34.99)		36.16 (10.34)		28.57 (8.62)	
mFood Log	570.57 (252.22) [1]	268.31 (227.53)	100.50 (46.60)	8.04 (38.02)	15.79 (7.87) [1,2]	20.37 (8.91)	9.07 (7.94) [1,2]	19.51 (10.76)
24R	739.21 (183.32)	99.67 (187.17)	104.84 (27.11)	3.70 (20.38)	29.23 (13.45) [2]	6.93 (12.53)	22.15 (7.94) [1,2]	6.43 (8.92)

[1] Significant mean difference between mfood log or 24R and control methods ($p < 0.05$); [2] Significant mean difference between mfood log and 24R methods ($p < 0.01$); Abbreviations: g, grams; Kcal, kilocalories; SD, standard deviation.

Table 3. Pearson's correlation analyses (with reported r- and p-values) of total energy and macronutrient intake between control, mfood log, and 24R methods.

	r-Value (p-Value)			
	Control Meal			
	Total Energy (Kilocalories)	Carbohydrate Intake (Grams)	Protein Intake (Grams)	Fat Intake (Grams)
mFood Log				
Total Energy (Kilocalories)	0.57 (0.03) [1]			
Carbohydrate Intake (grams)		0.60 (0.02) [1]		
Protein Intake (grams)			0.55 (0.04) [1]	
Fat Intake (grams)				0.16 (0.59)
24R Method				
Total Energy (Kilocalories)	0.64 (0.01) [1]			
Carbohydrate Intake (grams)		0.81 (<0.001) [1]		
Protein Intake (grams)			0.47 (0.09)	
Fat Intake (grams)				0.42 (0.13)

[1] Significant Pearson's correlation (r-value) coefficients.

3.3. Level of Agreement between Variables

Statistically, both the mfood log and 24R methods are considered to be in agreement with the control as the values are within the upper and lower confidence intervals of the mean $+/-$ standard deviation, respectively (See Figure 2). The regression line of differences was insignificant for 24R ($\beta = 0.32$, $p = 0.25$) and mfood log ($\beta = -0.07$, $p = 0.83$). Practically, the limits are wide considering the mean differences were 177.7 to 267.2 kcals below the mean and 466.5 to 714.2 kcals above the mean for 24R and mfood log, respectively.

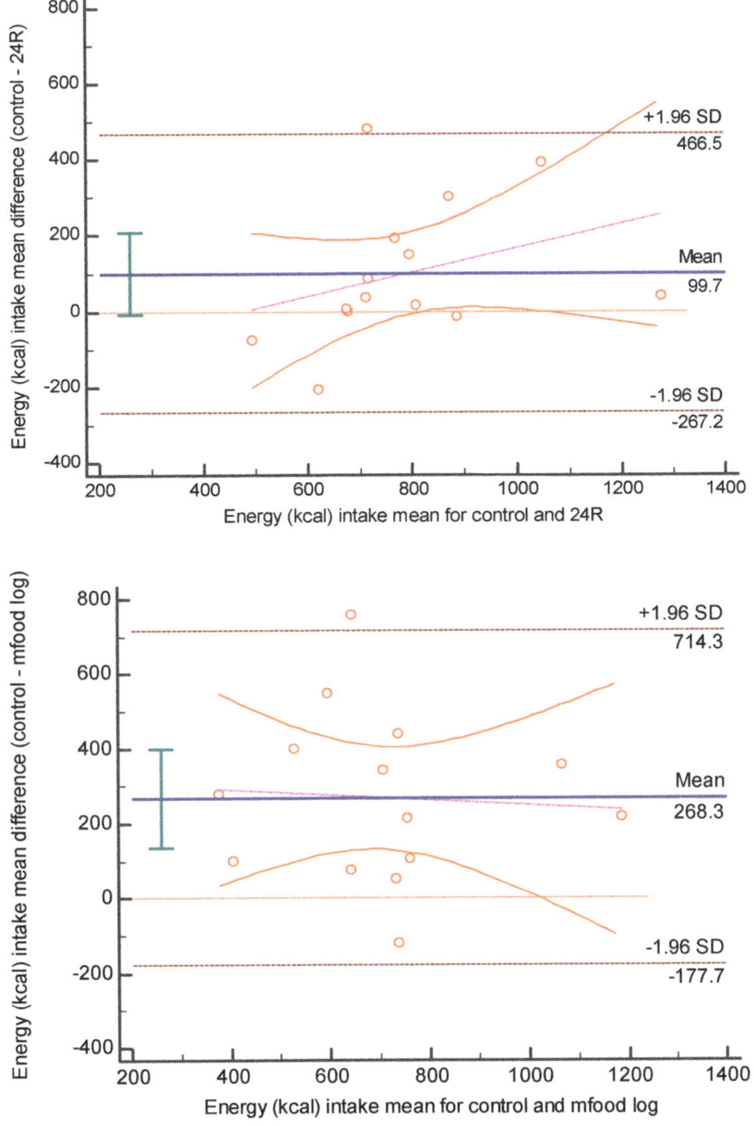

Figure 2. Bland Altman plot with regression lines for mfood log (control—mfood log) and 24R (control—24R).

3.4. Survey Descriptives

While responses to the likelihood of using the mfood log on a daily basis had a wide range, with a lower mean than ease of use, the mfood log was generally reported as easy to use (see Table 4). Only three participants (of the 15) reported having used another dietary intake mobile application and all three reported having previously used MyFitnessPal. Pearson's correlation analyses revealed a significant inverse association (r = −0.60, p = 0.03) between ease of use and mean energy intake differences between control and mfood log; no association (r = −0.45, p = 0.11) was identified between the mean difference and likelihood of use.

Table 4. Survey results for ease and likelihood of use for the Bridge2U mfood log, n = 15.

Survey Item	Response Range [1]	Mean
Ease of use		
	3–10	7.5
Likelihood of Daily Use		
	1–9	5.9

[1] Higher numerical value indicates a positive response (for example, "most likely").

4. Discussion

The purpose of this study was to examine the reporting accuracy of an mfood log and in comparison with a control meal and gold standard food recall method (24R). The study results showed that participants significantly underreported energy intake when data were entered into the mfood log compared to what was observed at the control meal. The 24R method also underreported, but was not significantly different from the control meal; energy intake was also not significantly different between the 24R and mfood log methods. Limits of agreement were similar between both control and mfood log and control and 24R, with no statistical importance, but may have practical limitations.

While there is limited research examining the dietary intake reporting accuracy of mobile applications, a similar study has found strong correlations (r = 0.69–0.86, p < 0.001) between energy macronutrients captured via a food diary mobile application and 24 h recalls over two days [13]. Another study examined the accuracy of a mobile application to measure the intake of food groups and found strong correlations (mean r = 0.79, range: 0.69–0.88) between a 3-day, 24-h recall and the mobile application [14]. Similarly, our study did find a significant linear relationship between the control meal and both mfood log and 24R methods. Our results also align with another study that reported small mean differences and medium correlations between the 24R and mfood log [15].

Reported results showed promise for estimation of group means and thus, group dietary assessment with no proportional bias; however, further research is warranted for individual dietary assessment. Very few studies have examined the level of agreement using the Bland Altman plot; however, one study did examine Bland Altman plots and from a practical perspective, reported wide agreement between a mobile food log and 24R (their method of comparison) [15]. While our study indicated that statistically, the methods could be used interchangeably, from a practical standpoint, the caloric range would have definite implications for inaccuracy at the individual level. It is difficult to directly compare the two studies since our data represents one meal, whereas Carter et al. [15] examined dietary data for an entire day and thus, multiple meals.

Our sample was largely female, two-thirds African American, and mostly overweight/obese. Mean values for both the mfood log and 24R reflected under-reporting of energy intake compared to the control method. In the United States, under-reporting of energy intake has been associated with female sex, non-Hispanic blacks, and overweight and obesity [16]. While it is difficult to determine these impacts on reporting in this study due to the relative homogeneity of the sample, this study provides key preliminary data toward the improvement of intake reporting using mfood logs among these populations.

As for the acceptance of mobile applications for dietary assessment, our results indicated a general acceptance for ease and likelihood of using the mfood log. The findings correspond with research reporting participant preference for using a mobile application to log food intake instead of paper/pen methods [17]. A 2014 review of dietary assessment found that user satisfaction was high for six studies using mobile phones for dietary assessment, with one study reporting a low user satisfaction [10]. Two newer studies also corroborate with the general acceptance of mobile food records among adults, including in a community setting [18,19].

Sharp and Allman-Farinelli [10] found three predominant methods for dietary intake assessment, including a mobile phone electronic food diary (similar to our method), food photograph recall aids, analysis of food photographs by trained dietitians, and automated food photograph or video analysis. Compared to conventional methods (paper/pencil), reliability and validity were similar, but not inferior, to mobile methods; however, participant satisfaction and preference for mobile methods were higher [10]. While it seems that participant satisfaction with mobile dietary assessment tools echoes throughout the literature, this may not directly translate to implications for participant burden. Among adolescents reporting general satisfaction with a mobile dietary assessment tool, the majority (70%) of participants reported that the use of the application was burdensome or it was difficult to remember to record food intake [19]. Our own results suggest that those who found the application easier to use also had a smaller mean energy difference between the control method and mfood log.

As it pertains to dietary self-monitoring, research has shown that the use of mobile applications versus conventional (typically paper records) techniques may result in better self-monitoring adherence and improvements in dietary intake and anthropometrics [20]. Clearly, more research is needed in larger, more diverse samples, and longer duration studies to determine the feasibility and validity of using mobile food diaries or logs to assess dietary intake at the group and individual levels. Additionally, mobile technologies offer other advantages, such as standardized and automated, real-time data entry [21]. Research should focus on innovative, mobile-supported solutions to reduce participant underreporting, improve food log adherence and ease of use, and mitigate portion size estimation error. There are several limitations of the reported study that must be noted. Sample size was small, with only 14 participants and a convenience sample; therefore, results may not be generalizable to a larger population, which is a typical limitation [10]. Due to the controlled component of the study, our data only captures data entry at one meal and not multiple meals or days. As a result, our methods also do not account for a possible learning effect that may have improved reporting for both the 24R and mfood log methods. Additionally, the controlled nature of the study lends itself to smaller sample sizes, which is a strength and limitation similar in the limited related research [15,22]. Lastly, similar to another study [15], we cannot determine how available nutrition information, that is viewable while using the mfood log, impacts dietary intake reporting.

Our study is one of the few studies that examines Bland Altman plot and Pearson correlation analyses for the validity of dietary intake reporting using a mobile application dietary assessment food log tool. The results warrant further exploration to determine the ability of mobile technology to serve as a valid and reliable method for dietary data collection to be used for intervention and epidemiological purposes. Future studies should consider larger, more generalizable sample sizes, as well as the incorporation of gold standard assessment or controlled methods for comparison and longitudinal designs.

Author Contributions: Conceptualization, J.L.L., K.A., and A.L.; methodology, J.L.L.; formal analysis, J.L.L.; investigation, J.L.L., K.A., and A.L.; resources, J.L.L.; writing—original draft preparation, J.L.L.; writing—review and editing, J.L.L., K.A., and A.L.; supervision, J.L.L.; project administration, J.L.L.; funding acquisition, J.L.L.

Funding: Research reported in this publication was supported by the National Institute On Minority Health and Health Disparities of the National Institutes of Health under Award Number R15MD010213. The content is solely the responsibility of the authors and does not necessarily represent the official views of the National Institutes of Health.

Acknowledgments: We would like to acknowledge other grant staff, intervention participants and community partners for facilitating this research.

Conflicts of Interest: The authors declare no conflict of interest. The funders had no role in the design of the study; in the collection, analyses, or interpretation of data; in the writing of the manuscript, or in the decision to publish the results.

References

1. Thompson, F.E.; Kirkpatrick, S.I.; Subar, A.F.; Reedy, J.; Schap, T.E.; Wilson, M.M.; Krebs-Smith, S.M. The national cancer institute's dietary assessment primer: A resource for diet research. *J. Acad. Nutr. Diet.* **2015**, *115*, 1986–1995. [CrossRef] [PubMed]
2. Thompson, F.E.; Subar, A.F. Dietary assessment methodology. In *Nutrition in the Prevention and Treatment of Disease*; Coulston, A.M., Boushey, C.J., Ferruzzi, M.G., Eds.; Academic Press: Cambridge, MA, USA, 2017; pp. 5–45.
3. Institute of Medicine (US) Committee on Military Nutrition Research. *Emerging Technologies for Nutrition Research*; National Academies Press: Washington, DC, USA, 1997.
4. Moshfegh, A.J.; Rhodes, D.G.; Baer, D.J.; Murayi, T.; Clemens, J.C.; Rumpler, W.V.; Paul, D.R.; Sebastian, R.S.; Kuczynski, K.J.; Ingwersen, L.A.; et al. The US department of agriculture automated multiple-pass method reduces bias in the collection of energy intakes. *Am. J. Clin. Nutr.* **2008**, *88*, 324–332. [CrossRef] [PubMed]
5. Subar, A.F.; Freedman, L.S.; Tooze, J.A.; Kirkpatrick, S.I.; Boushey, C.; Neuhouser, M.L.; Thompson, F.E.; Potischman, N.; Guenther, P.M.; Tarasuk, V.; et al. Addressing current criticism regarding the value of self-report dietary data. *J. Nutr.* **2015**, *145*, 2639–2645. [CrossRef] [PubMed]
6. Samuel-Hodge, C.D.; Fernandez, L.M.; Henríquez-Roldán, C.F.; Johnston, L.F.; Keyserling, T.C. A comparison of self-reported energy intake with total energy expenditure estimated by accelerometer and basal metabolic rate in African-American women with type 2 diabetes. *Diabetes Care* **2004**, *27*, 663–669. [CrossRef] [PubMed]
7. Voss, S.; Kroke, A.; Klipstein-Grobusch, K.; Boeing, H. Obesity as a major determinant of underreporting in a self-administered food frequency questionnaire: Results from the EPIC-Potsdam Study. *Z. Ernahrungswiss* **1997**, *36*, 229–236. [CrossRef] [PubMed]
8. Bedard, D.; Shatenstein, B.; Nadon, S. Underreporting of energy intake from a self-administered food-frequency questionnaire completed by adults in Montreal. *Public Health Nutr.* **2004**, *7*, 675–681. [CrossRef] [PubMed]
9. Stumbo, P.J. New technology in dietary assessment: A review of digital methods in improving food record accuracy. *Proc. Nutr. Soc.* **2013**, *72*, 70–76. [CrossRef] [PubMed]
10. Sharp, D.B.; Allman-Farinelli, M. Feasibility and validity of mobile phones to assess dietary intake. *Nutrition* **2014**, *30*, 1257–1266. [CrossRef] [PubMed]
11. National Agricultural Library USDA Food Composition Databases. Available online: https://ndb.nal.usda.gov/ndb/search/list?home=true (accessed on 22 June 2018).
12. 24-hour Dietary Recall (24HR) At a Glance. Available online: https://dietassessmentprimer.cancer.gov/profiles/recall/ (accessed on 22 June 2018).
13. Carter, M.C.; Burley, V.J.; Cade, J.E. Development of 'My Meal Mate'—A smartphone intervention for weight loss. *Nutr. Bull.* **2013**, *38*, 80–84. [CrossRef]
14. Rangan, A.M.; Tieleman, L.; Louie, J.C.Y.; Tang, L.M.; Hebden, L.; Roy, R.; Kay, J.; Allman-Farinelli, M. Electronic Dietary Intake Assessment (e-DIA): Relative validity of a mobile phone application to measure intake of food groups. *Br. J. Nutr.* **2016**, *115*, 2219–2226. [CrossRef] [PubMed]
15. Carter, M.C.; Burley, V.J.; Nykjaer, C.; Cade, J.E. 'My Meal Mate' (MMM): Validation of the diet measures captured on a smartphone application to facilitate weight loss. *Br. J. Nutr.* **2013**, *109*, 539–546. [CrossRef] [PubMed]
16. Murakami, K.; Livingstone, M.B.E. Prevalence and characteristics of misreporting of energy intake in US adults: NHANES 2003–2012. *Br. J. Nutr.* **2015**, *114*, 1294–1303. [CrossRef] [PubMed]
17. Carter, M.C.; Burley, V.J.; Nykjaer, C.; Cade, J.E. Adherence to a smartphone application for weight loss compared to website and paper diary: Pilot randomized controlled trial. *J. Med. Internet Res.* **2013**, *15*, e32. [CrossRef] [PubMed]
18. Fowler, L.A.; Yingling, L.R.; Brooks, A.T.; Wallen, G.R.; Peters-Lawrence, M.; McClurkin, M.; Wiley, K.L., Jr.; Mitchell, V.M.; Johnson, T.D.; Curry, K.E.; et al. Digital food records in community-based interventions: Mixed-Methods pilot study. *JMIR mHealth uHealth* **2018**, *6*, e160. [CrossRef] [PubMed]

19. Lee, J.-E.; Song, S.; Ahn, J.; Kim, Y.; Lee, J. Use of a mobile application for self-monitoring dietary intake: Feasibility test and an intervention study. *Nutrients* **2017**, *9*, 748. [CrossRef] [PubMed]
20. Lieffers, J.R.L.; Hanning, R.M. Dietary Assessment and Self-monitoring: With Nutrition Applications for Mobile Devices. *Can. J. Diet. Pract. Res.* **2012**, *73*, e253–e260. [CrossRef] [PubMed]
21. Forster, H.; Walsh, M.C.; Gibney, M.J.; Brennan, L.; Gibney, E.R. Personalised nutrition: The role of new dietary assessment methods. *Proc. Nutr. Soc.* **2016**, *75*, 96–105. [CrossRef] [PubMed]
22. Arab, L.; Estrin, D.; Kim, D.H.; Burke, J.; Goldman, J. Feasibility testing of an automated image-capture method to aid dietary recall. *Eur. J. Clin. Nutr.* **2011**, *65*, 1156–1162. [CrossRef] [PubMed]

MDPI

St. Alban-Anlage 66

4052 Basel

Switzerland

Tel. +41 61 683 77 34

Fax +41 61 302 89 18

www.mdpi.com

Nutrients Editorial Office

E-mail: nutrients@mdpi.com

www.mdpi.com/journal/nutrients